Cost-Justifying Usability
An Update for an Internet Age

Cost-Justifying Usability

An Update for an Internet Age

Edited by

Randolph G. Bias and Deborah Mayhew

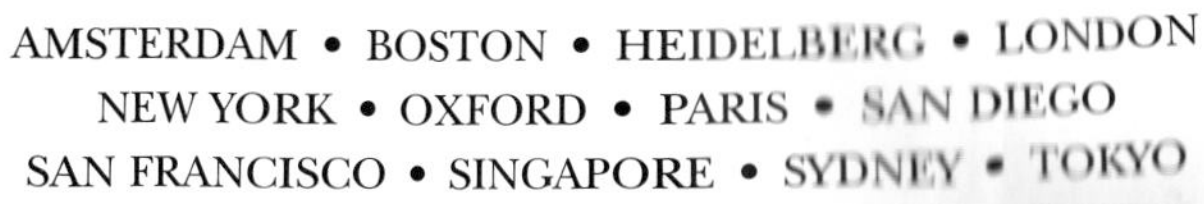

Publisher Diane Cer[illegible]
Editorial Coordinator [illegible]
Editorial Assistant [illegible]on Crump
Publishing Services [illegible]eiro
Project Manager J[illegible]esign
Cover Design Yvo[illegible]pesetter Ltd., Hong Kong
Composition SNP[illegible] Inc.
Copyeditor Grap[illegible], Inc.
Proofreader Gra[illegible]nc.
Indexer Graph[illegible]ess
Interior printer [illegible]Color
Cover printer [illegible]

Morgan Kauf[illegible]ublishers is an imprint of Elsevier.
500 Sansome[illegible], Suite 400, San Francisco, CA 94111

This book i[illegible]ted on acid-free paper.

Library of Congress Cataloging-in-Publication Data

Application submitted

ISBN: 0-12-095811-2

For information on all Morgan Kaufmann publications,
visit our Web site at www.mkp.com or www.books.elsevier.com

Printed in the United States of America
05 06 07 08 09 5 4 3 2 1

Contents

Preface

GREETINGS, AND BEST WISHES

Thank you, gentle reader, for buying, checking out, or borrowing this book. Your interest means that there is some impossible-to-measure increase in the odds that some user, when confronting some future Web site, Web-based application, or traditional software user interface (UI) will actually be able to carry out his or her task without error, without help, without frustration.

Though that increase in odds is not measurable, the benefits of the application of usability engineering to software UIs definitely are measurable. In our 1994 first edition, we argued for a cost-benefit analysis approach to usability engineering, and attempted to arm usability professionals and their managers with the tools to compete for development resources. Early reviews of our book, and scores of personal anecdotes, would suggest we were at least somewhat successful. The state of usability of human-computer interfaces, however, would suggest we were not anywhere near totally successful.

Since our first book several things have happened, none more influential to the computing world than the broad availability of the Internet. This, plus another decade of case studies of usability engineers attempting a cost-justification approach and the advent of new usability engineering methods, motivate this second edition.

THE BOOK

When we started addressing a second edition, we assumed we would reprise about half of the 14 chapters from the first edition and supplement those with some new chapters. In fact, there are only four chapters here from the first edition, and those (Mayhew and Tremaine, Karat, Rohn, and Mauro) bear little resemblance to their earlier incarnations.

We have divided the 22 chapters of this book into five sections. Section 1, the Introduction, consists of two chapters. In Chapter 1, Clare-Marie Karat joins me (Bias) to place the new book in context. We review the history of the topic of cost-justifying usability and present arguments explaining why a cost-justification approach to usability is even more important in the Internet age. In Chapter 2 we asked Aaron Marcus to reprise and update an article he had written for his Aaron Marcus and Associates Web site, as we felt his summary of cost-justification statistics and examples served as an excellent overview of the first edition and therefore a nice launchpad for the new book.

Section 2 is the Framework Section. In Chapter 3, Marilyn Tremaine joins me (Mayhew) to offer a basic framework chapter. First, we summarize a generic sample usability-engineering project plan, based on an overall approach to usability engineering laid out in my 1999 book *The Usability Engineering Lifecycle* (1999, Morgan Kaufmann Publishers). Then we go on to offer four detailed, hypothetical cost-benefit analysis examples, thereby illustrating how to cost-justify usability work on an internal application, a vendor application, an e-commerce site, and a product information site. Karat follows this up, in Chapter 4, with an update of her 1994 "business case approach" chapter from our first edition.

Like other service providers, usability professionals sometimes must market themselves. But *unlike* most other service providers, we must spend time convincing would-be clients that they have a problem. In Chapter 5, Richard Henneman offers a chapter, "Marketing Usability," with no analog in our earlier edition. We round out the "Framework" section with another totally new chapter, wherein David Crow addresses the special case of "Valuing Usability for Start-Ups."

Janice Rohn kicks off the third section "Organizational and Design Context," with Chapter 7, a much updated and expanded version of her "Cost-Justifying Usability in Vendor Companies" from the first edition. In Chapter 8, Chauncey Wilson and Stephanie Rosenbaum team up to offer "Practical ROI Issues for UCD Teams," in which they distinguish between internal returns, external returns, and social return on investment (ROI), and teach us what a "GUI Roll" is. Charles Mauro updates his chapter from the first edition with a consultant's perspective in Chapter 9, "Usability Science: Tactical and Strategic Cost Justifications in Large Corporate Applications." In this chapter he addresses, among other things, the cost of litigation and usability as a litigation deterrent. In Chapter 10, Karat and Arnie Lund address special cost-justification concerns for Web-based applications. Next are two more totally new chapters addressing internationalization. In Chapter 11, David Siegel and Susan Dray offer "The Business Case for International User-Centered Design." Then in Chapter 12, I

(Mayhew) address what is unique about adding usability engineering to an international development project and present how to adapt the general cost justification technique in this case. In Chapter 13, "The ROI of Accessibility," Tom Brinck wraps up this section with an extension into the field of usability for people with disabilities.

Whereas the chapters in the "Organizational and Design Context" section generally cover contexts or domains that require some tailoring of the cost-justification approach to usability, the fourth section, "Methods and Approaches," offers examples that are more generally applicable. In Chapter 14, Anne Kirah and some Microsoft colleagues address ethnography, presenting some specific examples of projects. In Chapter 15, "Outside the Box: Approaches to Good Initial Interface," Doug Gillan and Merrill Sapp fly in the face of all of our preaching about "iterative design" and argue that regardless of how many times you're going to iterate, it still will be cost-effective to take care to apply the principles of perception and cognition in the early design work. I (Mayhew) complete my contributions, in Chapter 16, with "Keystroke Level Modeling as a Cost-Justification Tool," in which I address the value of cognitive modeling even before a prototype is built. Michael Medlock and Dennis Wixon team up for another chapter from Microsoft, this one on their "RITE Method" of software development, where they talk of foraging for formative usability data. Jurek Kirakowski considers how to intelligently decide on sample size, and thereby be most efficient in your usability testing, in Chapter 18. And Scott Weiss completes the "Methods and Approaches" section with his consideration of online surveys and suggests that the question is not "if" to survey, but "when."

Then we offer an "End-Game" section comprised of three chapters. We envision that Nigel Bevan's Chapter 20, "Cost Benefits Framework and Case Studies," serves as a bit of a book-end chapter when paired with Chapter 2. That is, Nigel summarizes the types of benefits addressed throughout *this* edition of the book and illustrates them with some examples. As there are often a couple of books leaning on the bookcase, outside of the bookends, so there are two final chapters. Clyde Heppner and his colleagues have provided us with a particularly detailed case study of how usability was embraced at their company in Chapter 21, "At Sprint, Understanding the Language of Business Gives Usability a Positive Net Present Value." Finally, I (Bias) interviewed four software development executives to get their input on what makes a good usability argument in Chapter 22, "Cost-Justifying Usability: The View From the Other Side of the Table."

Whenever I (Bias) tell someone who is *not* in the computer industry that I teach usability, it tends to fetch blank stares. I elaborate that it is my job to help people make computer screens "user-friendly" so that people can carry out their tasks on the computer with ease. At this point the person with whom I'm talking usually offers some version of, "That sounds great—when do you start?" Despite our first edition; despite half a century of the Human Factors and Ergonomics Society, two decades of ACM SIGCHI, and over one decade of UPA; despite the best efforts of scores of usability educators, hundreds of usability professionals, and thousands of people in the software industry who value usability engineering; when users turn on the computer, or open their Web browser, they tend to get frustrated. Let's get started, again.

THE AUDIENCE

We have gathered these contributors (and the contributors have written their chapters) with three different audiences in mind. The first and most obvious audience would be usability professionals and their managers, whom we wish to help be successful in securing (where appropriate!) funding for their usability engineering efforts. A second audience would be any software development managers, directors, or executives who are trying to better understand how to decide at what level to fund usability efforts. Finally, usability and software engineering educators will consider, we hope, all or parts of this book as they train the next generation of usability professionals and software engineers.

ACKNOWLEDGEMENTS

We would both like to thank Morgan Kaufmann's Diane Cerra, Asma Stephan, and Mona Buehler for their patience, talent, and diligence. Did we mention "patience"?

Thanks, too, to reviewers Ben Shneiderman and Chauncey Wilson, for their time and shared insights. Ben's comments were valuable, as expected. And Chauncey's were valuable and unbelievably voluminous, and we cannot imagine the time he put into this task.

Thanks, too, to scores of individual usability professionals who told us that they found our first edition valuable to them.

I (Bias) would like to thank my fellow empty-nester, Cheryl Bias, for her patience and ambient beauty.

I (Mayhew) would like to thank Katie Ann, for remaining in my nest a while longer and providing general inspiration.

1 CHAPTER Justifying Cost-Justifying Usability

Randolph G. Bias School of Information, The University of Texas at Austin
Clare-Marie Karat IBM TJ Watson Research Center

1.1 INTRODUCTION

Sir Walter Scott wrote, in *Marmion*,

> "Oh! what a tangled web we weave
>
> When first we practise to deceive!" (Scott, 1805)

Sometimes it is quoted as ". . . what a *wicked* Web we weave."

This year is the bicentennial anniversary of the publication of Scott's book. Did he anticipate the World Wide Web, 200 years ago? Scott's words are relevant to today's networked world and the development activities that support it, including designing for ease of use.

Usability is *not* the end-all-be-all; rather, it must be considered alongside (and in equal measure with!) functionality and schedule. The problem is *not* that Web site developers are *wicked*; rather, they are too often in a hurry and are not operating in a reward structure that motivates attention to usability.

So the Web is not wicked. But *tangled*? Oh yeah.

Most of us are wowed by the Internet. It is amazing what we can accomplish while sitting at home in our pajamas. But when we say "us" or "we," we are referring to that small subset of the population who have some expectation of what will happen if we "right click," or those of us who have heard of, for example, Doug Engelbart. When one of "us" sits down with one of "them" (the great unwashed masses), the madcap hilarity begins. "They" are so stupid! Can you believe it?, some of them don't even know that an underline, on a Web page, means the text is a link to another page (except when it isn't). "They" don't even know that they can't "break" the Internet (except that, when they double-click on several links quickly, they might crash their own application, basically breaking the Internet, as far as they're concerned).

In his understandably popular *Design of Everyday Things*, Donald Norman (1990) characterized usability as ". . . the next competitive frontier." His point was that hardware and software were becoming commodities, and that the differentiating characteristic for computer systems (among other things) would be their usability.

In 1994, the first edition of this book (Bias and Mayhew, 1994a) was published, and, by many accounts, the ideas and methods covered therein enabled many usability professionals to bring to bear quantitative cost and benefit data to justify expenditures on usability support, using a metric understood by all subgroups of the product design, development, deployment, marketing, sales, and support teams—that is, dollars.

In the mid- to late-1990s the Internet came into being (or at least, became available to the masses), and the number and breadth of computer users grew exponentially. Expanding commensurately were the user interface (UI) design challenges, as Web site and Web-based application designers had to consider a user population about which they had less clarity and which had unprecedented levels of diversity in terms of previous computing experience, domain expertise, native language, cultural membership, level of disability, network connection speed, task motivation, and other variables not yet considered.

We are happy to report that, in response to these design challenges, usability engineering has become a routine, expected component of the software development process, as *de facto* a member of the product development team as quality assurance or marketing.

Well, not so much.

The truth of the matter is, in our should-be-humbler, but we hope not-unduly-pessimistic opinion, here in the first few years of the new millenium usability still struggles to find, as we characterized it in our first edition, "a seat at the table" (Bias and Mayhew, 1994b). Many companies (including Sprint, as you'll read about in Chapter 21) have gotten the usability fever, and have integrated usability engineering into their product or site development process. Many companies routinely (if not blindly) adhere to the tenets of usability engineering and user-centered design (UCD) in the development of their Web sites and other software user interfaces (UIs). But it does not take long, when you are trying to carry out tasks on the Internet, to realize that good usability is not standard for most Web sites and applications. In this chapter we offer a short history of cost-justifying usability, and present a short list of why the Internet age presents us with an even stronger need to employ such an approach.

1.2 LESSONS FROM THE PAST: SEMINAL RESEARCH ON RETURN ON INVESTMENT IN HUMAN FACTORS IN THE 1980s AND 1990s

Marilyn Mantei (now Marilyn Tremaine, coauthor of Chapter 3) and Toby Teorey formally introduced the topic of the cost benefit of usability to the human-computer interface (HCI) field in a 1988 paper, in which they discussed the costs of incorporating usability engineering into a product's development cycle (Mantei and Teorey, 1988). Their analysis suggested that an investment of $250,000 might be necessary to cover a wide range of usability work on a typical product. The second author of this chapter (Clare-Marie Karat) was struck by these findings and thought that HCI practitioners might be able to have more financial impact with a lower investment in usability. As a practitioner in an industrial software development organization, Karat elected to employ low-fidelity prototyping of a UI with representative end users, and iterated for two iterations on the design of the UI until the usability objectives were met (e.g., 95% success rate for users within 3 minutes of initial use of the application) (Karat, 1989). Karat collected cost-benefit data and determined that for the first three uses of the system by the target user group, there was a $2 return for every dollar invested in usability on the project. A more complete analysis of the return on investment (ROI) showed a $10 return for every dollar invested in usability.

At about the same time, the first author of this chapter (Bias) invited a group of usability engineers (Deborah J. Mayhew, Susan Dray, Page O'Neal, and Clare-Marie Karat) to participate in a symposium at the Human Factors Society conference to discuss and debate further developments in the area of usability cost-benefit analysis (Bias, 1990). This symposium brought together a group of people who were interested in this topic, and the synergy was evident. (And by the size of the audience, we inferred the topic had also struck a chord with usability practitioners.) At Interact '90 in Cambridge, England, Karat further discussed the research with Marilyn Mantei and a number of European researchers who were now interested in the topic (Karat, 1990).

Karat taught tutorials on the cost-benefit methodology she developed, at the ACM SIGCHI Conferences on Human Factors in Computing Systems and at the Human Factors and Ergonomics Society conferences, for a number of years (e.g., Karat, 1991). The methodology covered a business case approach to analyzing the cost benefit of usability, identified a framework of costs and benefits for the analysis, and provided the approach for calculating the ROI in usability using both simple and complex financial methods. Karat became a focal point for collecting de-identified usability engineering case studies for the HCI community, and these case studies illustrated the various concepts in the methodology and documented the financial value of usability.

Bias and Mayhew built a proposal for a book of collected works on the topic, using the 1990 HFS symposium presentations as a core. That collection became a seminal book on the topic of cost-justifying usability (Bias and Mayhew, 1994a), and was characterized by one reviewer as "the bible for human-computer interface professionals." The chapters in that first edition provided the usability field with the accumulated research in the area. Karat's methodology and case studies served as the basis for her cornerstone chapter in the first edition of this book. Practitioners and researchers around the world have used the content of this book for many years to facilitate justification of usability work and to advance knowledge on the topic in the HCI field.

Also in the early 1990s, Robert Pressman published a book on software development that documented that 80% of the software life cycle costs are spent in the postrelease maintenance phase (Pressman, 1992). He also explained that the relative cost of a change escalates during development from 1.5 units of project resource in the concept phase, to 6 units during the development phase, to 100 units of resource during the postrelease maintenance phase. Dennis Wixon and Sandra Jones disseminated a case study of the positive financial ROI in usability on a Digital Equipment Corporation (DEC) product (Wixon and Jones, 1992). Revenues were 80% higher for the second release of a product done with usability engineering as a primary focus as compared to the first release without usability engineering. Interviews with customers showed that buying decisions were made primarily based on usability.

Products undergoing usability engineering can come to market faster than those without it, or they can be completed on time with higher quality. There is financial value in being early to market. Conklin documented that speeding up market introduction can result in 10% higher revenues because of increased volume or increased profit margins (Conklin, 1991). Bringing a product to market 6 months late may cost companies 33% of after-tax profits (House and Price, 1991). Completing a product on time with higher quality was demonstrated to provide a significant ROI in usability (Karat, 1989; Wixon and Jones, 1992).

Members of the HCI community began to focus on the cost benefit of particular HCI methods and tools during the early 1990s as well. Jakob Nielsen held a workshop at the ACM SIGCHI CHI'92 conference on usability inspection methods. The workshop became the basis of a book that included research by many authors on the comparative effectiveness and cost of different discount usability methods (Nielsen and Mack, 1994). Around this time, Robert Virzi published a number of papers on the effectiveness of smaller sample sizes and low-fidelity prototyping methods that complemented the growing cost benefit of usability knowledge base (Virzi, 1990). Later in the decade, Arnold Lund published a paper (Lund, 1997) on an alternative method of justifying the cost of usability. This paper focused on the value of new ideas and the improvement in the usability of systems and products.

Recent publications illustrate the cost benefit of specific applications of usability to Web applications and services (e.g., Karat *et al.*, 2002, 2003). In the application of HCI methods to the Web and the resulting analysis of the cost benefit of that investment, a couple of trends are emerging. There are basic HCI cost-benefit frameworks that apply to the Web as well as they do to traditional software development projects; however, there are some unique aspects of cost-justifying usability for the Web as well. Examples of the ROI of usability engineering for Web applications and discussion of these factors are covered in the Wilson and Rosenbaum chapter (Chapter 8), the Karat and Lund chapter (Chapter 10), the Heppner *et al.* chapter (Chapter 21), and others.

1.3 WWW.SOWHAT

It has been a decade since we first attended to cost-justifying usability. Of course, in the realm of HCI design, the most significant event was the advent of (or, perhaps more accurately, the wide availability of) the Internet. In the mid- to late-1990s there was a heated debate over whether the Web would, or should, affect the conduct of usability or user-centered design. Some said that the need to collect user data to inform, early on, and later validate, the quality of UI designs was not influenced by the particular form or medium of the UI. Others argued that the Web, requiring as it did breaks with graphic user interface (GUI) interaction standards and including a wide variety of users, would demand a fresh look at usability. As we have already stated, we believe that many of the HCI-field frameworks and methods apply to usability engineering for the Web. Given the unique context of Web applications, though, there are some new HCI models, and enhancements to HCI methods and practice that are necessary for the Web.

We believe that the presence of the Web has made a systematic, professional approach to usability engineering even more important. Thus, the ability to cost justify usability is an even more important skill. Let us offer seven reasons why we believe cost-justifying usability in 2005 is even more important than it was in 1994. (These are an expansion of a list Bias offered in the Foreword for Vredenburg *et al.*, 2002.)

1.3.1 Who *Are* All Those People?

The most salient new aspect introduced by the Internet is the breadth of potential user audience. When you "put up" a Web site, you immediately make your design available to every person in the world who can find his or her way to a public library or, perhaps even more amazingly, a Starbucks. Imagine determining the user profile for *that* user test! How does the usability engineer respond to the nay-sayer stakeholder who asks, "But did you test left-handed Brazilian Capricorns?" Well, if such right- and Southern-hemisphered Capricorns comprise an important segment of the site's intended audience, the usability engineer would do well to be able to quantify the cost of *not* testing any.

The breadth of, and indeed, lack of knowledge about, the characteristics of new or anonymous site visitors (a site may start collecting information about visitors once they begin to use a site, but visitors may turn cookies off in order to retain privacy), or Web-based application users, argues for new best practices in usability studies (e.g., remote testing methods), new accepted levels of statistical confidence, and new vigilance as to the user profiles of potential visitors, actual visitors, and buyers. But whatever those practices, confidence levels, and profiles are, this breadth makes it all the more important to know the costs and benefits of employing usability engineering for the site or application, and the costs of *not* employing usability engineering.

1.3.2 If I Had a Hammer, I Wouldn't Necessarily Be a Carpenter

But I would be a Web designer and developer, even if I couldn't even spell "HTML." Relatively easy-to-use Web development tools make it the case that the person with the merest minimum of computing skills and, more to the point, absolutely no proven design skills can be a Web site designer or developer.

Imagine a company owner (let's say it's a woman) who chooses to invest in a Web presence for her company. Where should she go to find the person to design the new company Web site? Should she go to her most valued product designers? No, that won't happen, because those people are busy enough doing their current jobs. (And there is no guarantee that just because they are good at designing the company's products, they will be good at designing a Web site.) Should she go to a professional association to find a list of expert Web designers who have the well-acknowledged certification? Well, no. There is no such accepted certification. Suffice it to say, there is a wide range in the usability of Web sites within and across the domains of e-commerce, and informational, nonprofit, and government sites. In many cases, new hires with little experience and skill are given responsibility for a Web site. In contrast, some companies such as Amazon put a high value on usability of their Web sites, and their Web site team is rewarded monetarily for UI improvements that increase the throughput at the site and motivate users to click on suggested recommendations (Karat, 2004).

The point is that the presence of the Internet and the easy-to-use Internet site development tools means that more and more less-qualified people are serving as designers, thus making usability engineering even more necessary.

1.3.3 The Tug of "Internet Time"

In 1994, when we published our first edition, software development teams were decrying the ever-shortening development cycles as market pressures were shortening them, for some simpler applications or follow-up releases, to mere months. Since the advent of the Internet, some new sites go up in, say, an hour. What percentage of the UI design cycle should be devoted to usability engineering when the total development cycle is 60 minutes?

We will leave it to others to debate the wisdom of putting up a site in 1 hour. Instead, we would like to observe simply that "Internet time" has demanded the further diminution of development cycles, and increased the concomitant pressure to ship, or "go live," without even the most discount course of usability engineering. As stated in Bias's Foreword to *User-Centered Design,* "When you are in a hurry, it is even *more* important to follow a (perhaps constrained, but intelligently selected) course of User-Centered Design . . ." (Bias, 2002). And, we might add, it is even more important to be able to cost-justify any proposed usability efforts. How much will it cost, in lost sales or increased customer support, if the hurried design is hard to use? These situations make it all the more important that Web site owners collect data on the usability of their sites on an ongoing basis so that they have knowledge and context (over time) regarding the usability of the site as well as other system and business issues and activity related to it. Amazon.com is continually collecting and analyzing user data on more than 100 variables from visitors to the site to determine needed changes, which are phased in at regular intervals. If you have done your homework and are using a disciplined approach, you can move more quickly to make changes. That homework requires an investment in quality measurement and monitoring activities that is akin to building high-quality management processes into the design of a site. A popular business adage that pertains here as well is Peter Drucker's "you can't manage what you can't measure."

1.3.4 The Old "Get Something Out There" Approach Doesn't Work Anymore

Web sites and Web applications are software. Before there was the Internet, software developers developed applications (and application infrastructure). At some level, for most of the various sorts of applications, here's the way that development worked.

1. Someone had a good idea for an application.
2. A subset of the good functionality was chosen to go into Release 1.0.
3. The funding source (venture capitalists, angel investor, or the guys in the garage) agreed to fund a certain effort by a certain date.
4. The team scrambled to design, develop, test, and deploy the first release on time.
5. Software development being the inexact science that it is, deadlines started to slip, and the team made trade-offs to get a release out, with less functionality, with less testing, at a later date, or all three.

But all this was justifiable because, for the most part, the consumers understood how it worked. No one except the very brave bought Release 1. By Release 2 the application had more functionality and underwent more thorough testing so that then, if there was no more mature competition, those with a strong need for the functionality may have stepped up to the plate. According to their knowledge of the process, most customers expected Release 3 to be a product that was stable and usable. (Of course, for products now, including Web applications, there is a tendency to put out Version 2, after a "limited" Version 1.)

With the advent of the Internet, many of the people who had been responsible for the development of client-server or desktop or mainframe products were now responsible for the development of Web sites or Web-based applications. Naturally enough, they brought with them the best practices from their earlier software development world. Alas, whereas *some* of those practices served them well—user requirements gathering, software development processes, quality assurance testing—the practice of "getting something out there" and then improving it back at the shop while some customers serve as unofficial advanced beta testers does *not* work in the Internet world. For example, if a visitor to an e-commerce Web site (let's say it's a man) shows up to buy a book and the site crashes repeatedly, or if the visitor puts something in his shopping cart only to be unable to navigate the check-out, or if for any reason the experience does not meet his expectations, that visitor is gone (unless there is no other way for him to buy his book). He will *not* be back in three months, thinking "well, let's see if they've got it right now." He will find a competitor who has the same functionality, the same inventory, and a *better* user experience.

Following is a short, true story to illustrate the danger of "just getting something up." One Christmas I (Bias) wanted to buy a particular music CD for one of my sons. About that time I received a $10-off electronic coupon for a Web site that specialized in CDs, so I decided to give it a try. Upon first arriving at the company's Web site, I was asked for my demographic data. Well, I hate that, because if they don't have the product I want, then I'll have wasted that time completing the form. But I figured "10 bucks," so I completed the form. I searched for my CD, found it, and went to check out. At this point I was given the same blank demographic data form to complete. I gritted my teeth, feeling vaguely whorish, now, wondering what else I would do for $10, and completed the form again. At some subsequent point in the check out process, I received the blank form a third time. Now, I believe I'm not stupid. But even if I *am* stupid, I was apparently a representative of their target audience, as I had received the coupon. So, I should have been able to figure out how to buy a CD without receiving the same blank form three times. Upon the third appearance of the blank form I went immediately to the address line, entered the URL of a competitor site, and in 1 minute my desired CD was working its way to my house. I have *never* been back to that Web site, and never will; they lost that sale, every potential sale to me in perpetuity, and, it is my hope, every potential sale to anyone who has ever heard me tell this story in a presentation (in which I mention the site name). What makes this story even more tragic (from a company profit standpoint) is that this was a usability problem that would have been identified in the first minute of usability testing. Forrester research data document that 65% of people give up on a site if they cannot quickly find what they want (Souza, 2001). Customers will give an e-commerce site two to three clicks (data from a de-identified 2003 case study provided to Karat). Customers do not return if they do not have a good initial experience. These sites and organizations can go under quickly. You don't get a second chance to make a first impression on the Web.

1.3.5 The Dangers of Amateur Usability Engineering

A bad software developer is discovered at least by system test time. A bad usability engineer may not be discovered until the product has shipped or the site has gone live, and the customer-support phones start ringing off the hook. Poor usability means (for traditional software UIs) the customers' training costs go up, productivity goes down, and the total cost of ownership (TCO) goes up. For Web sites it means visitors leave in frustration, never to return, customer dissatisfaction goes up, and the trade press, online forums, and email groups help to spread the bad news.

If a product development manager receives poor usability support, he or she does not think, "Oh, I received poor usability support." Rather, he or she thinks, "Usability isn't worth the investment." And then it's *years* until "we" get back in there. If we may, let us join the chorus of those who have debunked three related myths.

Myth 1: Usability Is Just Common Sense

If usability were "just common sense," then how would you explain the rampant poor usability of applications and Web sites?

Myth 2: Good Intentions and an Awareness of the Importance of the User Is All You Need

An awareness of users, and a proclivity to test them, is a good first step, but a good usability engineer must also know a few other things. A partial list of methods, skills, and knowledge necessary to use them includes:

Task analysis

Contextual inquiry

Interview research

Usability objectives specification

Cost-feature tradeoff analysis

User profiling

Architectural design guidelines

User interface design guidelines

Domain expertise in the field the project addresses

Knowledge of previous research

Heuristic evaluation

Usability walkthrough

Codiscovery

Paper-and-pencil testing

Laboratory testing

Group facilitation

Survey generation

Naturalistic observation

Field study

Remote-usability testing

Automated-usability testing

Knowledge of which method to use when

Human perception, cognition, learning, memory

Decision making

Motivation

Mental modeling

Anthropometry

Descriptive and inferential statistics

Content analysis of qualitative data and affinity diagramming

Cost-justification methods

Research ethics

Advocacy

Written and oral communications

Prioritization of issues

Software-development process

Organizational decision-making process

Organizational and strategic business goals and how the project supports them

Science

Correlation does not equal Causation

Response bias

Inadvisability of accepting the null hypothesis

A student graduating with an MSIS, with a concentration in usability/HCI, from, for instance, the University of Texas at Austin School of Information, will have:

Taken at least one research methods and statistics course.

Taken two courses in usability engineering, both of which involved carrying out usability evaluations.

Visited *several* local usability laboratories and learned from these practitioners.

Taken as many as four courses in information architecture and digital media design.

Completed a thesis or capstone entailing an industrial-strength design/evaluation project.

Myth 3: Anyone Who Says Such Things About Amateur Usability Engineering Is Trying to Drive Toward Certification

This might be the case, but certification is not a "silver bullet." There's no widely acknowledged certification for usability professionals. Yes, the Board Certification of Professional Ergonomists offers two certifications (a Certified Public Ergonomist and a Certified Human Factors Practitioner), and there are certifications in Europe, but none of these is widely accepted. Ultimately, we are worried about competence, not certification. And usability, while perhaps not rocket science, is a profession, requiring education, mentored apprenticeship, and serious attention to empirically-based best practices.

1.3.6 More Capabilities Means More Novice Users

Because the Web enables people to do more things online (e.g., pay taxes, renew drivers' licenses, buy a book, find a spouse) there are necessarily more times that the user will be a novice. Novice users need to have interfaces that are "walk up and use" for many tasks and that provide graceful recovery when errors are made. It is always a good business decision to build the usability into the site initially (a sentiment repeated by a software company CEO in Chapter 22). It will cost the organization less to design its site for the novice user from the beginning than to redesign the site for novice users after going live.

1.3.7 The World Is More Complex, so It Is Harder to Know What All the Possibilities Are

An unusable Web site can cause an organization to fail. Even if your organization is employing skilled usability engineers to design your site and the team has sufficient time and resources to complete the work, there are still other factors that can cause a Web site to fail that may not be within the team's control. For example, in using a National Science Foundation (NSF) site recently to purchase an NSF-sponsored book, Karat was unable to download a purchased item because of a local configuration setting on the browser on her laptop. After calling the support number, and finding out a few hours later that all the help team could say was that the site was working properly, she experimented with changing a couple of local configuration parameters on her browser. It turned out that one option needed to be changed. That resolved the issue and she was able to successfully download the document within the time window allocated by the site. She called back to inform the support people of the resolution to the issue and they were pleased and thankful for the feedback. Other than that one step, the site had seemed straightforward and usable. This example highlights the critical importance of making the parts of the user experience that a Web site has control over as usable as possible and then being prepared to troubleshoot sources of failure outside of the Web site's control that nonetheless have a large impact on the user's overall experience and satisfaction with a Web site. This type of thoughtful and rigorous usability engineering requires that the organization have HCI experts leading the work effort.

1.4 ISN'T IT OBVIOUS?

Why did we put the effort into generating this book? Would *anyone* doubt the wisdom of such an approach? In the concluding chapter of the first edition of this book, we wrote, "Ten—or maybe only five—years from now, this book will hardly be necessary" (Bias and Mayhew, 1994b, p. 321). For a "landmark book" (Rosenberg, 2004, p. 24), we were weak on prophesy.

We assert that the obviously poor state of usability of Web and other software UIs is evidence that either the ROI approach we championed in the first book was wrong, or it wasn't followed broadly or effectively. Rosenberg (2004) seems to argue not that the ROI approach is a bad idea, but that it tends to be poorly, and thinly, applied, and he calls for attention to TCO. The presence of this second edition implies that we believe strongly that the general approach, to use the universally accepted metric of dollars, is the way for usability to gain additional and appropriate influence in the software development process. And the 21 chapters that follow are intended to better arm the usability professional, or manager, with the tools to effectively ply this trade.

1.5 NOT "IF" BUT "WHICH"

Truly, we say with confidence, we must accept that usability is of potential value in all software development. The question has changed from "should there be *any* usability engineering" to "which methods should we employ on this project?" In usability engineering for the Web, practitioners still need to focus on identifying and understanding the target users, user goals, and context of use as they did and still do in traditional development projects. Then they must take into account the context of use on the Web to determine what HCI methods to employ in the design of a usable Web site. The context variable is similar to tradeoff decisions made in the past regarding traditional development projects. For example, HCI practitioners have needed to understand the types of risks involved when negotiating for allocation of resources for HCI work on different projects or when deciding themselves about where to put their effort and what types of methods and tools to use within a project. The Web seems to bring more complexity to the topic of context, making the decision-making process more difficult. We can certainly build on what the field has learned to date to help make the case for resources for usability engineering in Web site design and to help practitioners make tradeoffs within a project.

1.6 THE SOLUTION: APPROACH, TOOLS, AND COMMUNICATION

To give project teams the best chance of success in designing usable Web sites, there needs to be a clear understanding of the organization's goals for the site, and then the user's goals, background, and context of use. HCI experts can make the necessary tradeoffs in determining the methods and tools to use in completing the usability engineering work with end users of a site, given time and resource constraints. Usability engineers can communicate the user data from the site to senior management and the organization can make decisions to improve the site as needed as user and organizational goals change. There are both minor and major updates to sites and these must be managed with organizational goals. We hope that this overview of the topic gives you the perspective necessary to gain a richer understanding and incorporation of the information in the chapters ahead.

Usability is usually "worth it." For your products or site, do you think so, or do you know so?

REFERENCES

Bias, R. G. (1990). Chair, Cost-justifying human factors support. In *Proceedings of the Human Factors Society 34th Annual Meeting*, Human Factors Society: Santa Monica, CA, 832–833.

Bias, R. G. (2002). Foreword. In K. Vredenburg, S. Isensee, and C. Righi (Eds.), *User-Centered Design: An Integrated Approach*. Englewood Cliffs, NJ: Prentice-Hall.

Bias, R. G. (2003). The dangers of amateur usability engineering. In "Usability in Practice: Avoiding Pitfalls and Seizing Opportunities," S. Hirsh (Chair). Presented at ASIST 2003 Annual Meeting, Long Beach, California, October 20–23, 2003.

Bias, R. G., and Mayhew, D. J. (Eds.) (1994a). *Cost Justifying Usability*. Academic Press: Boston.

Bias, R. G., and Mayhew, D. J. (Eds.) (1994b). Summary: A place at the table. In R. G. Bias and D. J. Mayhew (Eds.), *Cost Justifying Usability*. Academic Press: Boston.

Conklin, P. (1991). Bringing usability effectively into product development. *Human-Computer Interface Design: Success Cases, Emerging Methods, and Real-World Context*. Boulder, Co., July, 24–36.

House, C. H., and Price, R. L. (1991). The return map: Tracking product teams. *Harvard Business Review*, 69 (1), 92–100.

Karat, C. (1989). Iterative testing of a security application. In *Proceedings of the Human Factors Society*, Denver, CO, 273–277.

Karat, C. (1990). Cost-benefit analysis of iterative usability testing. In D. Diaper (Ed.), *Human Computer Interaction—Interact 90*. Elsevier: Amsterdam, 351–356.

Karat, C. (1991). Cost-benefit and business case analysis of usability engineering. Tutorial presented at the ACM SIGCHI Conference on Human Factors in Computing Systems. New Orleans, LA, April 28–May 2.

Karat, C., Brodie, C., Karat, J., Vergo, J., and Alpert, S. (2003). Personalizing the user experience on ibm.com. In Vredenburg, K. (Ed.), *IBM Systems Journal*, 42 (4), 686–701.

Karat, C., Karat, J., Vergo, J., Pinhanez, C., Riecken, D., and Cofino, T. (2002). That's entertainment! Designing streaming, multimedia web experiences. In *International Journal of Human-Computer Interaction*, 369–385.

Karat, J. (2004). Personal communication with Amazon usability engineers.

Lund, A. M. (1997). "Another Approach to Justifying the Cost of Usability." *Interactions*, 4 (3), 49–56.

Mantei, M. M., and Teorey, T. J. (1988). Cost/benefit analysis for incorporating human factors in the software lifecycle. *Communications of the ACM*, 31 (4), 428–439.

Nielsen, J., and Mack, R. (Eds.) (1994). *Usability Inspection Methods*. New York: John Wiley and Sons.

Norman, D. A. (1990). *The Design of Everyday Things*. New York: Doubleday.

Pressman, R. S. (1992). *Software Engineering: A Practitioner's Approach.* New York: McGraw Hill.

Rosenberg, D. (2004). The myths of usability ROI. *Interactions.* New York: Association of Computing Machinery. September, pp. 22–29.

Souza, R. (2001). Get ROI from Design. *Forrester Report*: Cambridge, MA.

Virzi, R. (1990). Streamlining the design process: Running fewer subjects. In *Proceedings of the Human Factors Society*, Orlando, FL, 291–294.

Vredenburg, K., Isensee, S., and Righi, C. (2002). *User-Centered Design: An Integrated Approach.* Upper Saddle River, NJ: Prentice Hall.

Wixon, D., and Jones, S. (1992). *Usability for Fun and Profit: A Case Study of the Design of DEC RALLY Version 2.* Internal Report, Digital Equipment Corporation.

2 CHAPTER User Interface Design's Return on Investment: Examples and Statistics

Aaron Marcus Aaron Marcus and Associates, Inc.

2.1 INTRODUCTION: WHAT DO WE MEAN BY THE RETURN ON INVESTMENT OF USABILITY?

Making computer-based products (and services) more usable is smart business. Usability increases customer satisfaction and productivity, leads to customer trust and loyalty, and contributes to tangible cost savings and profitability. User interface (UI) development is part of a product's development cost anyway, and it pays to do it right.

Most software and Web site development managers view usability costs as added effort and expense, but more commonly the reverse is true. Because the first 10% of the design process, when key system-design decisions are made, can determine 90% of a product's cost and performance, usability techniques help keep the product aligned with company goals (Smith and Reinersten, 1991). Usability returns many benefits (return on investment [ROI]) to products developed for internal use or sale (Lund, 1997; Mayhew and Mantei, 1994; Wilson and Rosenbaum, Chapter 8).

The following are some of the benefits that drive internal ROI:

- Increased user productivity
- Decreased user errors
- Decreased training costs
- Increased savings from making changes earlier in design lifecycle
- Decreased user support

The following are some of the benefits that drive external ROI:

- Increased sales
- Decreased customer support costs
- Increased savings from making changes earlier in the design lifecycle
- "Reduced cost of providing training (if training is offered through the vendor company)" (Mayhew and Mantei, 1994, p. 126)
- Increased perception of value of company by stakeholders

Usability also plays a role in the public's perception of a company, affecting brand value and market share. About 15% (Nielsen, 1993) of the space in reviews published in trade magazines, journals, and national newspapers is devoted to user friendliness or usability. Media giants such as the *New York Times*, the *Financial Times*, and the *Wall Street Journal* publish weekly columns that evaluate software (Mayhew and Mantei, 1994, p. 25). *Info World* devotes between 18% and 30% of its software review articles to ease of learning, ease of use, and quality of documentation (Nielsen, 1993). In recent years, ease of use is an increasingly frequent theme of consumer, Web, and mobile technology reviews. The *New York Times*' negative review of BMW's iDrive vehicle UI is characteristic of such usability-oriented reviews (Smith, 2002, p. 50).

Usability can also affect a company's financial health and public perception in less obvious ways. Many companies do not understand the issues users have with their products. Because of problems caused by these oversights, manufacturers have been found liable for defective designs. To their regret, the courtroom evaluation of a product's usability was often the first time such manufacturers were exposed to human factors engineering (Mauro, 1994, p. 127). More recently, the medical industry has noted that the use of human factors has reduced device-related medical errors (www.injuryboard.com/view.cfm/Article=810, as of September 1, 2004).

Let's now look at the evidence of ROI for usability in UI design.

2.2 HOW CAN WE "PROVE" THE RETURN ON INVESTMENT? SOME EXAMPLES AND STATISTICS

With an understanding of the basic benefits of usability, let's examine the evidence for a positive ROI. In the following sections, we list key usability benefits and then define appropriate value propositions. For each of these value propositions, we present examples from the literature that help interpret the cost of usability challenges and/or we cite statistics. Although a cost-benefit analysis for every circumstance does not exist, these "proofs" of applying usability in UIs predict likely quantifiable benefits or ROI. For ease of reference, these findings are summarized in Table 2.1.

2.2.1 Overall Value of Implementing User Interface Practices

Because of the number of well-documented examples of cost savings with usability engineering, sound statistics can be applied generally to UI development. These statistics serve as benchmarks.

Value Proposition: High Return on Savings and Product Usability

Some Statistics

"The rule of thumb in many usability-aware organizations is that the cost-benefit ratio for usability is $1 : $10–$100. Once a system is in development, correcting a problem costs 10 times as much as fixing the same problem in design. If the system has been released, it costs 100 times as much relative to fixing in design" (Gilb, 1988).

"The average user interface has some 40 flaws [note: this figure is presumably based primarily on client-server software applications, as opposed to Web sites; typical Web sites might have even more flaws considering the large number of sites constructed by developers with little usability training]. Correcting the easiest 20 of these yields an average improvement in usability of 50%. The big win, however, occurs when usability is factored in from the beginning. This can yield efficiency improvements of over 700%" (Landauer, 1995).

2.2.2 Development: Reduce Costs

Usability engineering is most effective at the beginning of the product development cycle, especially if it is part of quality functional deployment (QFD), a process used for structuring the development process through a primary focus on customer (i.e., user, not purchaser) requirements. With use of QFD, reducing development time by one third to one half is possible (Scerbo, 1991).

Although full-blown versions of QFD may be too complex for a particular development project and QFD often requires a skilled facilitator to manage the process, "lighter" versions can be helpful in fast-paced and/or smaller

Table 2.1 Fast Stats

Overall Value of Implementing UI Practices	
High return on savings and product usability	"Once a system is in development, correcting a problem costs 10 times as much as fixing the same problem in design. If the system has been released, it costs 100 times as much relative to fixing in design" (Gilb, 1988).
Development: Reduce Costs	
Save development costs	"Approximately 63% of large software projects are over budget and the top four reasons rated as having the highest responsibility were related to usability engineering" (Nielsen, 1993).
Save development time	"Speeding up development is a key goal for integrating usability effectively into product development; one-quarter delay in bringing a product to market may result in the loss of 50% of the product's profit" (Conklin, 1991).
Reduce maintenance costs	"It has been found that 80% of software life cycle costs occur during the maintenance phase and were associated with 'unmet or unforeseen' user requirements and other usability problems" (Nielsen, 1993).
Save redesign costs	"Sun Microsystems has shown how spending about $20,000 could yield a savings of $152 million. Each and every dollar invested could return $7,500 in savings" (Rhodes, 2000).
Sales: Increase Revenue	
Increase transactions/ purchases	"You can increase sales on your site as much as 225% by providing sufficient product information to your customers at the right time" (User Interface Engineering, 2001).
Increase product sales	"It is common for usability efforts to result in a hundred percent or more increase in traffic or sales" (Nielsen, 1999a).
Increase traffic, size of audience	"The company said in the month after the February 1999 re-launch that traffic to the Shop IBM online store

Table 2.1 Fast Stats—*Continued*

	increased 120 percent, and sales went up 400 percent" (Battey, 1999). "The change increased the traffic [at HomePortfolio.com] up 129% the week we put it up" (Interaction Design, Inc., 2001).
Retain customers	"More than 83% of Internet users are likely to leave a Web site if they feel they have to make too many clicks to find what they're looking for" (Arthur Andersen, 2001).
Attract more customers (appeal)	"When respondents were asked to list the five most important reasons to shop on the Web, 83% stated 'Easy to place an order' as the top reason" (Nielsen, 1999a).
Increase market share (competitive edge)	"The importance of having a competitive edge in usability may be even more pronounced for e-commerce sites, which commonly drive away nearly half of repeat business by making it difficult for visitors to find the information they need" (Manning, 1999).
Use: Improve Effectiveness	
Increase success rate, reduce user error	"In Jared Spool's study of 15 large commercial sites users could only find information 42% of the time even though were taken to the correct home page before they were given the test tasks" (Nielsen, 1998b).
Increase efficiency/ productivity (reduce time to complete task)	"Inadequate use of usability engineering methods in software development projects have been estimated to cost the US economy about $30 billion per year in lost productivity" (Landauer, 1995).
Increase user satisfaction	"In a Gartner Group study, usability methods raised user satisfaction ratings for a system by 40%; when systems match user needs, satisfaction often improves dramatically" (Harrison *et al.*, 1994, p. 215).
Increase job satisfaction/ decrease job turnover	"Surveys showed that video display terminal workers had twice as many complaints of neck and shoulder discomfort, eye strain was reported three times as

Continues

Table 2.1 Fast Stats—*Continued*

	often, and there were higher rates of absenteeism, less job satisfaction, and increased (30%) turnover" (Schneider, 1985).
Increase ease of use	"Incorporating ease of use into your products actually saves money. Reports have show it is far more economical to consider user needs in the early stages of design, than it is to solve them later" (IBM, 2001).
Increase ease of learning	"A study by *Computer + Software News* (1986) found that users rated ease of use second at 6.8 out of 10, while ease of learning was rated fourth at 6.4 on a scale of important purchase factors" (Harrison *et al.*, 1994, p. 211).
Increase trust in systems	"[EuroClix user trials] . . . study clearly shows that consumers' trust concerns can significantly be alleviated by providing relevant information when and where users need it" (Egger and de Groot, 2000).
Decrease support costs	"In the next release, support calls 'dropped dramatically'; Microsoft recognized 'significant cost savings' " (Ehrlich and Rohn, 1994, p. 96). "Over 50,000 users called support for assistance, at a cost to the company of nearly $500,000 a month. To correct the situation, the manufacturer . . . ended up spending $900,000 on the problem. No user testing . . . was conducted before its release . . ." (Mauro, 1994, p. 129).
Reduce training/ documentation cost	"A study by *Computer + Software News* (1986) found that information systems managers rated ease of training seventh (out of 10) on a scale of important purchase factors" (Harrison *et al.*, 1994, p. 211).
Other	
Litigation deterrence and safety	"Usability is a principal factor for determining manufacturers' liability based on expert hard evidence on how a design should have used usability" (Mauro, 1994, p. 127).

development environments. For example, new products do not require investment in any particular design—and numerous possibilities can be explored at relatively little cost (Bias and Mayhew, 1994) (Fig. 2.1).

Usability analysis can help establish key, substantive design issues, even for new versions of existing products or to counter stakeholder belief in adhering to flawed existing products. Applying human factors in the initial design can greatly reduce extensive redesign, maintenance, and customer support, which can substantially eat away profits.

Reducing costly repairs to flaws can accumulate short-term benefits during product development, and reducing customer interactions can accumulate long-term benefits during product release. Customers directly benefit from usability improvements by increases in ease of use, ease of learning, user satisfaction, and user productivity. At the same time, decreases occur in the number of "user" errors, costs for training and support, and maintenance. Taking proactive measures in usability and quality during the initial production stages can produce a cost-saving "ripple effect."

Value Proposition: Save Development Costs

Examples

"Savings from earlier vs. later changes: Changes cost less when made earlier in the development life cycle. Twenty changes in a project, at 32 hours per change and [a minimal] hourly rate of $35, would cost $22,400. Reduc-

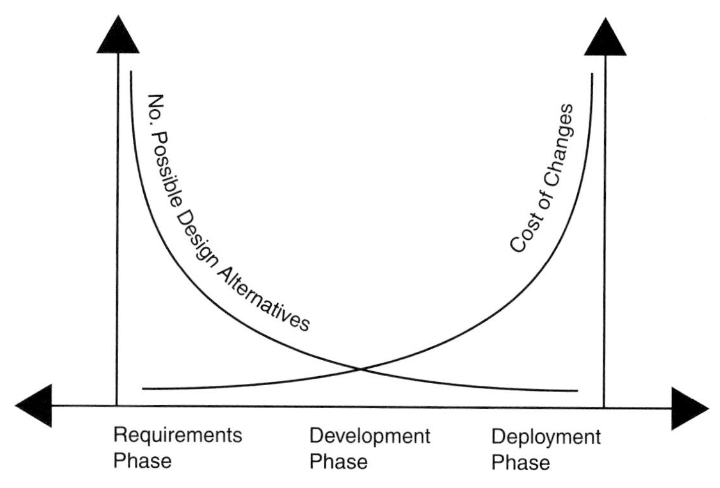

FIGURE 2.1 The number of possible designs decreases as the cost to make changes increases (Erhlich and Rohn, 1994, p. 80).

ing this to 8 hours per change would reduce the cost to $5,600. Savings = $16,800" (Human Factors International, 2001a).

"A financial services company had to scrap an application it had developed, when, shortly before implementation, developers doing a User Acceptance test found a fatal flaw in their assumptions about how data would be entered. By this time, it was too late to change the underlying structure, and the application [was] never implemented" (Dray, 1995).

Statistics

"When managers were polled regarding the reasons for the inaccurate cost estimates, the top four reasons were issues that could have been addressed by following best practices in usability engineering. These include frequent requests for changes by users, overlooked tasks, users' lack of understanding of their own requirements, and insufficient communication and understanding between users and analysts" (Barker, 2000).

"A study of software engineering cost estimates showed that 63% of large software projects significantly overran their estimates. . . . When asked to explain their inaccurate cost estimates, software managers cited 24 different reasons and, interestingly, the four reasons rated as having the highest responsibility were related to usability engineering. Proper usability engineering methodology will prevent most such problems and thus substantially reduce cost overruns in software projects" (Nielsen, 1993, citing from Lederer and Prasad, 1992).

Value Proposition: Save Development Time

Examples

Usability techniques allowed a high-tech company to reduce the time spent on one tedious development task by 40% (Ehrlich and Rohn, 1994). At another company, usability techniques helped cut development time by 33% to 50% (Bosert, 1991).

Statistics

"Conklin (1991) states that speeding up development is a key goal for integrating usability effectively into product development and that a one-quarter delay in bringing a product to market may result in the loss of 50% of the product's profit" (Karat, 1994).

"Increased revenues accrue due to the increased marketability of a product with demonstrated usability, increased end-user productivity, and lower training costs. Conklin (1991) states that another usability goal is speeding up market introduction and acceptance by using usability data to improve marketing literature, reach market influencers and early adopters, and demonstrate the product's usability and reduced training cost" (Karat, 1994). Although much of Conklin's work was based on long development cycles (lasting years in some cases), and even with the emergence of extreme programming, agile programming, and other means to achieve programming results quicker, even letting users do quality assurance by responding to the current state of Web sites or Web applications, these techniques still have value.

Value Proposition: Reduce Maintenance Costs

Example

"[Usability engineering techniques] are quite effective at detecting usability problems early in the development cycle, when they are easiest and least costly to fix. By correcting usability problems in the design phase, American Airlines reduced the cost of those fixes by 60–90%" (Harrison *et al.*, 1994, p. 217).

Statistics

"One [well-known] study found that 80 percent of software life-cycle costs occur during the maintenance phase. Most maintenance costs are associated with 'unmet or unforeseen' user requirements and other usability problems" (Pressman, 1992).

"Martin and McClure found that $20–30 billion was spent worldwide on maintenance. Studying backlogs of maintenance work shows that an 'invisible' backlog is 167% the size of the declared backlog. Anonymous case study data show that internal development organizations are spending the majority of their resources on maintenance activities and thus cannot initiate development of strategic new systems" (Martin and McClure, 1983).

Value Proposition: Save Redesign Costs

Example

"Sun Microsystems has shown how spending about $20,000 could yield a savings of $152 million. Each and every dollar invested could return $7,500 in savings" (Rhodes, 2000). Whether a login improvement saves thousands of people a few seconds every day for a year or a dialogue box eliminates the need for a limited number of people to spend an hour of time, the net savings can be dramatic.

2.2.3 Sales: Increase Revenue

Usable products often lead to substantial cost savings and sales. Unusable products most often prevent a customer from accomplishing a task or retrieving information necessary to make an e-commerce purchase. Online shoppers spend most of their time and money at Web sites with the best usability (Nielsen, 1998a). Good navigation and Web site design make it easier for users to find what they're looking for and to make a purchase once they've found it (Donahue, 2001). Because so many poorly designed Web sites exist, when customers find one that "works," they tend to return for repeat business and gain trust in the organization.

Usable products also lead to good product reviews. Publications devote space just to this one factor, and good reviews lead to increased sales.

Value Proposition: Increase Transactions/Purchases

Statistics

"You can increase sales on your site as much as 225% by providing sufficient product information to your customers at the right time. In our recent research, we found that the design of product lists directly affected sales. On sites that did not require shoppers to bounce back-and-forth between the list and individual product pages, visitors added more products to their shopping cart and had a more positive opinion of the site. By understanding your customer expectations and needs, and designing your product lists accordingly, you can significantly increase your sales" (User Interface Engineering, 2001).

"One study estimated that improving the customer experience increases the number of buyers by 40% and increases order size by 10%" (Creative Good, 2000).

Value Proposition: Increase Product Sales

Examples

Wixon and Jones did a case study of a usability-engineered software product that increased revenue by more than 80% over the first release of the product (built without usability work). "The revenues of the usability-enhanced system were 60% higher than projected. Many customers cited usability as a key factor in buying the new system." Usability activities included field studies, tracking of usability bugs, and heuristic evaluations (Wixon and Jones, 1995).

"After move.com completed the redesign of the home 'search' and 'contact an agent' features based on a UI consulting firm's recommendations, users' ability to find a home increased from 62% to 98%, sales lead generation to real estate agents increased over 150%, and [move.com's] ability to sell advertising space . . . improved significantly" (Vividence Corp., 2001).

Statistics

"The magnitude of usability improvements is usually large. This is not a matter of increasing use by a few percent. It is common for usability efforts to result in a hundred percent or more increase in traffic or sales" (Nielsen, 1999b).

"Convoluted e-commerce sites can lose up to half of their potential sales if customers can't find merchandise, according to Forrester Research, Inc." (Kalin, 1999).

Value Proposition: Increase Traffic (Size of Audience)

Examples

"IBM's Web presence has traditionally been made up of a difficult-to-navigate labyrinth of disparate subsites, but a redesign made it more cohesive and user-friendly. According to IBM, the massive redesign effort quickly paid dividends. The company said in the month after the February 1999 re-launch that traffic to the Shop IBM online store increased 120 percent, and sales went up 400 percent" (Battey, 1999).

"At HomePortfolio.com we monitored site traffic, observed consumers in usability studies and worked with internal business groups. This helped us make changes that made the site's purpose clearer and increased transaction rates measurably. The change increased the traffic up 129% the week we put it up" (Interaction Design, Inc., 2001).

Value Proposition: Retain Customers (Frequency of Use)

Statistics

"More than 83 percent of Internet users are likely to leave a Web site if they feel they have to make too many clicks to find what they're looking for, according to Andersen's latest Internet survey" (Arthur Andersen, 2001).

"A bad design can cost a Web site 40 percent of repeat traffic. A good design can keep them coming back. A few tests can make the difference" (Kalin, 1999).

Value Proposition: Attract More Customers (Increase Appeal)

Example

"Staples.com determined that the key to online success and increased market share was to make its e-commerce site as usable as possible. Staples.com spent hundreds of hours evaluating users' work environments, decision-support needs, and tendencies when browsing and buying office products and small business services through the Web. Methods included data gathering, heuristic evaluations, and usability testing." [They achieved these results]:

- 67% more repeat customers
- 31–45% reduced drop-off rates
- 10% better shopping experience
- 80% increased traffic
- Increased revenue (Human Factors International, 2001b).

A Statistic

"In a 1999 study of Web users, respondents were asked to list the five most important reasons to shop on the Web. Even though low prices definitely do attract customers, pricing was only the third-most important issue for respondents. Most of the answers were related to making it easy, pleasant, and efficient to buy. The top reason was 'Easy to place an order' by 83% of the respondents" (Nielsen, 1999a).

Value Proposition: Increase Market Share (Competitive Edge)

Example

" 'Usability is one of our secret weapons,' says Mark Thompson, vice-president of customer experience at Charles Schwab & Co., Inc. The secret weapon appears to be working. Schwab's main Web site for U.S. investors, www.schwab.com, handles more than $7 billion in securities transactions a week, with more than 2 million active customer accounts holding $174 billion in assets. With those numbers, you might wonder why Schwab would need to make any changes to its Web site at all. But Schwab knows it cannot afford to coast; as more and more newcomers get online and the competition for their dollars increases, more e-commerce sites are making ease of use a differentiator. 'A year ago, it was a rush to put up applications and functionality. . . . It's now a rush to be useful' " (Kalin, 1999).

Statistics

"The importance of having a competitive edge in usability may be even more pronounced for e-commerce sites. Such sites commonly drive away nearly half of repeat business by not making it easy for visitors to find the information they need" (Manning, 1999).

"The repeat customers are most valuable: new users at one e-commerce site studied spent an average of $127 per purchase, while repeat users spent almost twice as much, with an average of $251" (Nielsen, 1997a).

2.2.4 Use: Improve Effectiveness

User-centered design benefits users, the users' company, and the vendor's company. Increased usability increases productivity and job satisfaction while decreasing customer support needs and documentation requirements. All these benefits, with additional possible reduced employee absenteeism and turnover, align with fulfilling successful business goals.

Value Proposition: Increase Success Rate and Reduce User Error

Examples

"One study at NCR showed a 25% increase in throughput with an additional 25% decrease in errors resulting from redesign of screens to follow basic principles of good design" (Gallaway, 1981).

"On Disney.com, for example, when UIE asked users to find the hotel closest to the monorail at Disney World, about 20 percent became lost in Disneyland and didn't even know it. 'If one in five people who came to the theme parks got lost,' [Jared] Spool says, 'Disney would fix it.' Disney Online's Senior Vice President and General Manager Ken Goldstein notes that Disney Online is already committed to developing an easy-to-use Internet design. While Disney Online did not have anything to do with Spool's tests, Goldstein is interested in his findings. 'As the next generation of Disney.com evolves,' Goldstein says, 'we will continue to respond to customer input through our own usability testing'" (Kalin, 1999).

Statistics

"A study from Zona Research found that 62% of Web shoppers have given up looking for the item they wanted to buy online (and 20% had given up more than three times during a two-month period)" (Nielsen, 1998b).

"In Jared Spool's study of 15 large commercial [Web]sites, users could only find information [that they were seeking] 42% of the time even though they were taken to the correct home page before they were given the test tasks" (Nielsen, 1998b).

Value Proposition: Increase Efficiency and Productivity (Reduce Time to Complete Tasks)

Examples

"With its origins in human factors, usability engineering has had considerable success improving productivity in IT organizations. For instance, a major computer company spent $20,700 on usability work to improve the sign-on procedure in a system used by several thousand people. The resulting productivity improvement saved the company $41,700 the first day the system was used. On a system used by over 100,000 people, for a usability outlay of $68,000, the same company recognized a benefit of $6,800,000 within the first year of the system's implementation. This is a cost-benefit ratio of $1:$100" (Karat, 1994, pp. 57–58).

"To build a model intranet, Bay Networks spent $3 million and two years studying the different ways people think about the same thing. The result: all think alike about the $10 million saved each year" (Fabris, 1999).

Statistics

"Inadequate use of usability engineering methods in software development projects have been estimated to cost the U.S. economy about $30 billion per year in lost productivity (see Tom Landauer's excellent book *The Trouble with Computers*). By my estimates, bad intranet Web design will cost $50–100 billion per year in lost employee productivity in 2001 ($50B is the conservative estimate; $100B is the median estimate; you don't want to hear the worst-case estimate!). Bad design on the open Internet will cost a few billion more, though much of this loss may not show up in gross national products, since it will happen during users' time away from the office" (Nielsen, 1997b).

"On a corporate intranet, poor usability means poor employee productivity; usability guru Jakob Nielsen estimates that any investment in making an intranet easier to use can pay off by a factor of 10 or more, especially at large companies" (Kalin, 1999).

Value Proposition: Increase User Satisfaction

Example

"One airline's IFE (In-flight Entertainment System) was so frustrating for the flight attendants to use that many of them were bidding to fly shorter, local routes to avoid having to learn and use the difficult systems. The time-honored airline route-bidding process is based on seniority. Those same long-distance routes have always been considered the most desirable. For flight attendants to bid for flights from Denver to Dallas just to avoid the IFE indicated a serious morale problem" (Cooper, 1999).

Statistic

"When systems match user needs, satisfaction often improves dramatically. In a 1992 Gartner Group study, usability methods raised user satisfaction ratings for a system by 40%" (Harrison *et al.*, 1994, p. 219).

Value Proposition: Increase Job Satisfaction and Decrease Job Turnover

Example

"Humantech, Inc., studied ergonomic office environments and productivity for 4000 managerial, technical, and clerical workers in a broad cross-section of North American industries. Surveys showed that video display terminal workers had twice as many complaints of neck and shoulder discomfort, eye strain was reported three times as often, and there were higher rates of absenteeism less job satisfaction, and increased (30%) turnover" (Schneider, 1985).

Value Proposition: Increase Ease of Use

Statistic

"Incorporating ease of use into your products actually saves money. Reports have shown it is far more economical to consider user needs in the early stages of design, than it is to solve them later. For example, in *Software Engineering: A Practitioner's Approach,* author Robert Pressman shows that for every dollar spent to resolve a problem during product design, $10 would be spent on the same problem during development, and multiply to $100 or more if the problem had to be solved after the product's release" (IBM, 2001).

Value Proposition: Increase Ease of Learning

Statistic

"A study by *Computer + Software News* (1986) found that users rated ease of use second at 6.8 out of 10, while ease of learning was rated fourth at 6.4 on a scale of important purchase factors" (Harrison *et al.*, 1994, p. 211).

Value Proposition: Increase Trust in Systems

Example

"User trials were used to redesign the EuroClix Web site before its launch. In its first six months, it convinced more than 30,000 users to sign up. This study clearly shows that consumers' trust concerns can significantly be alleviated by providing relevant information when and where users need it" (Egger and de Groot, 2000).

Value Proposition: Decrease Support Costs

Examples

"At Microsoft several years ago, Word for Windows's print merge feature was generating a lot of lengthy (average = 45 minutes) support calls. As a result of usability testing and other techniques, the user interface for the feature was adjusted. In the next release, support calls 'dropped dramatically'; Microsoft recognized 'significant cost savings' " (Ehrlich and Rohn, 1994, p. 96).

"A certain printer manufacturer released a printer driver that many users had difficulty installing. Over 50,000 users called support for assistance, at a cost to the company of nearly $500,000 a month. To correct the situation, the manufacturer sent out letters of apology and patch diskettes (at a cost of $3 each) to users; they ended up spending $900,000 on the problem. No user testing of the driver was conducted before its release. The problem could have been identified and corrected at a fraction of the cost if the product had been subjected to even the simplest of usability testing" (Mauro, 1994, p. 129).

Value Proposition: Reduce Training and Documentation Cost

Examples

"In another company, business representatives did a cost-benefit analysis for a new system and estimated that a well-designed GUI front end had an Internal Rate of Return of 32%. This was realized through a 35% reduction in training, a 30% reduction in supervisory time, and improved productivity, among other things" (Dray and Karat, 1994).

"At one company, end-user training for a usability-engineered internal system was one hour compared to a full week of training for a similar system that had no usability work. Usability engineering allowed another company to eliminate training and save $140,000. As a result of usability improvements at AT&T, the company saved $2,500,000 in training expenses" (Harrison *et al.*, 1994, p. 215).

Statistic

"A study by *Computer + Software News* (1986) found that information systems managers rated ease of training seventh (out of 10) on a scale of important purchase factors" (Harrison *et al.*, 1994, p. 215).

2.2.5 Other Return on Investment Factors

Since the early 1960s, product safety–related issues have led to pro-plaintiff legal precedents in U.S. courts. A manufacturer that has not included usability factors into its product is usually found liable. If a manufacturer has assimilated human factors engineering into its development process, claims on the grounds of usability may be greatly reduced.

Value Proposition: Litigation Deterrence and Safety

Examples

"Although software makers don't seem liable to the same sorts of litigation as, for example, a manufacturer of medical equipment, poor usability may be an element in lawsuits. For example, the Standish Group reported that American Airlines sued Budget Rent-A-Car, Marriott Corporation, and Hilton Hotels after the failure of a $165 million car rental and hotel reservation system project. Among the major causes of the project's disintegration were 'an incomplete statement of requirements, lack of user involvement, and constant changing of requirements and specifications,' all issues directly within usability's purview" (Standish Group, 1995).

"Poor usability is a potential element in lawsuits and other litigation. The U.S. government's recent case against Microsoft hinged on a usability question: Are users well-served when the browser and operating system are closely integrated?" (Donahue, 2001).

Statistic

Chapanis (1991) cites two independent studies that showed a 54% reduction in rear-end accidents with the use of human factors improvement: the centered high-mount brake light on autos.

2.3 CONCLUSION

The examples and statistics listed here are limited by space to brief citations. The original documents provide more complete contexts and additional details of which specific functions led to savings of money, time, or effort or to the improvements in performance or preference. The return can be a modest but important savings in user's time or a dramatic increase in safety that is recognized as a profound industry-changing paradigm.

One other aspect not touched upon by this chapter is the concern that improved usability might mean that companies can hire users with less training and background or that increasingly usable software might put usability professionals out of a job. First, for decades inevitable business pressures have led companies to find less expensive laborers to complete business tasks. This has little to do with the goal of achieving usability. Some professionals hide behind obscure, arcane practices to preserve their privileged positions, anyway. Whoever is doing the work should have access to a UI as usable as professional designers and analysts can achieve. Everyone benefits in the end.

Another aspect of increased usability is that some of the professional work, for better or for worse, is transferred to the general public. Consider desktop publishing, which transferred specific skills of typographers, graphic designers, graphic artists, compositors, printers, and other specialists to the general public. One can argue for both a democratic spread of capabilities and a patrician concern for loss of quality and expertise. Although it merits consideration, discussing this argument would require another chapter in this book.

One might lament the state of public and professional education and expertise. They are related. More usable products and services mean greater efficiency, effectiveness, and satisfaction at whatever level one attains, plus the opportunity to go further if so desired. Lack of usability inhibits such achievements.

In short, usability remains a viable, positive goal for UI design, even if other concerns, such as branding, user experience, and preference, seem to have a greater share in the current professional spotlight of attention.

As for usability professionals being put out of a job by increasingly usable software . . . if only that were so. One might as well be concerned about teachers not being needed anymore if everyone becomes well educated. This is not likely to happen in our lifetime—or anyone else's. New technology, new organizations of society, new media, and new content always generate new usability challenges. The only thing for certain is that things will always change. This has been apparent for several decades now, as new technology inevitably brings new players into the product/service development mix. The mistakes made are always recycling, but with different twists. There will be no lack of usability work to be done. Which country it will be done in remains to be determined (Marcus, 2004).

Finally, let us not forget that usability is not the only determinant of ROI. It is one of many factors that contribute to product/service success. Very useful products/services might be badly marketed, overpriced, not delivered in time, too technologically far ahead of their time (i.e., too demanding of it), too far ahead of their time in terms of customer understanding and desire, or badly timed in terms of the national or international economies. Users of the Apple Newton personal digital assistant (PDA), for example, continue to praise its design features, but it appeared too far ahead of the marketplace that now exists for such handheld products.

Returning to the argument for basic usability in products/services: The benefits of usability engineering can be achieved throughout the lifecycle of development. By applying usability techniques to the production process, developers can make them more efficient, which, in turn, can uniquely benefit the lifecycle. Efficient development methods can result in a faster release date, allowing manufacturers to unveil their products or services to the market prior to a competitor's. A user-centered product or service can garner positive media reviews, leading to increased sales. An effective, user-friendly UI can increase customer ease of learning, ease of use, job satisfaction, and trust in the product. As noted, it might even lead to less employee absenteeism and turnover.

Each product/service will require individual usability tasks, which also may differ from country to country or culture to culture (Marcus and Baumgartner, 2004). Developers should determine appropriate techniques for UI development before beginning a project to obtain the optimal results to facilitate cost-analysis projections. Competing groups are constantly seeking budget resources, so it is crucial to identify the cost justifications of usability engineering. Usability advocates must present a solid business case to business managers who will be looking at the company's bottom line.

Customers are constantly becoming more reliant on technical tools. As these tools are upgraded, users must learn increased information, functionality, and complexity, and usability becomes ever more critical. Because most software and Web users are not technical experts, it is imperative to make accomplishing goals simpler and easier. Regardless of the activity, whether performance tasks or vendor purchases, the user must be the center of the design process.

Cost-benefit analyses consistently show healthy returns on the dollars invested in usability. As more companies understand the significant benefits of usability and do careful cost justification, usability techniques will become standard.

Planners, analysts, marketers, engineers (implementers), designers, and trainers face many challenges ahead. For example, it is worthwhile to know the best techniques for communicating ROI benefits in differing contexts and marketing/sales situations, as well as the best techniques to achieve specific kinds of benefits. Some of this information is discussed in the literature, but it is not always easily available to those who need to know.

The goal of this chapter has been to make available an initial useful compendium of information about the ROI of usability, especially for UI design.

ACKNOWLEDGMENTS

The author acknowledges the conceptual direction in this analysis provided by Eugene Chen, Director of Design, AM+A, and the research and analysis of Designer/Analysts Kathleen Donahue and Junghwa Lee, AM+A. In addition, the author acknowledges the assistance of Angela Gross, Manager of Marketing and Business Development; Karen Brown, Senior Analyst/Designer; Pia Reunala, Marketing Representative; Paul Cofrancesco, Senior Designer/Analyst; and Kimberley Chambers, Designer/Analyst, AM+A. Finally, the author acknowledges the intellectual debt to the work of Bias and Mayhew in the work cited (Bias and Mayhew, 1994). This chapter has previously appeared in *User Experience* (Marcus, 2002) and as an AM+A white paper.

REFERENCES

Arthur Andersen. (2001). *Web Site Design Survey*. Retrieved October 15, 2001, from www.arthurandersen.com/website.nsf/content/MarketOfferingseBusinessResources OnlineUserPanelWebsiteDesign.

Barker, D. T. (2000). *Cost Benefits of Usability Engineering*. Retrieved October 9, 2001, from www.interfacearchitecture.net/articles/benefits.htm.

Battey, J. (1999). *IBM's Redesign Results in a Kinder, Simpler Web Site*. Retrieved October 10, 2001, from www.infoworld.com/cgi-bin/displayStat.pl?/pageone/opinions/hotsites/hotextr990419.htm.

Bias, R. G., and Mayhew, D. J. (Eds.) (1994). *Cost-Justifying Usability.* Boston: Academic Press.

Bosert, J. L. (1991). Quality functional deployment: A practitioner's approach. In *ASQC.* New York: Quality Press.

Chapanis, A. (1991). The business case for human factors in informatics. In B. Shackel and S. Richardson (Eds.), *Human Factors for Informatics Usability* (pp. 39–71). Cambridge: Cambridge University Press.

Conklin, P. (1991). Bringing usability effectively into product development. Paper presented July 24–26, 1991, at the *Human-Computer Interface Design: Success Cases, Emerging Methods, and Real-World Context.* Boulder, CO.

Cooper, A. (1999). *The Inmates Are Running the Asylum: Why High-Tech Products Drive Us Crazy and How to Restore the Sanity.* Indianapolis: SAMS.

Creative Good. (2000). *The Dotcom Survival Guide.* Retrieved October 10, 2001, from www.creativegood.com.

Donahue, G. M. (2001). *Usability and the Bottom Line.* Retrieved October 16, 2001, from www.ieee.org.

Dray, S. M. (1995). The importance of designing usable systems. *interactions,* 2 (1), 17–20.

Dray, S. M., and Karat, C. (1994). Human factors cost justification for an internal development project. In R. G. Bias and D. J. Mayhew (Eds.). *Cost-Justifying Usability* (pp. 111–122). Boston: Academic Press.

Egger, F. N., and de Groot, B. (2000). Developing a model of trust for electronic commerce: An application to a permissive marketing Web site. Paper presented May 15–19, 2000, at the *Poster Proceedings of the Ninth International World-Wide Web Conference.* Amsterdam.

Ehrlich, K., and Rohn, J. A. (1994). Cost justification of usability engineering: A vendor's perspective. In R. G. Bias and D. J. Mayhew (Eds.), *Cost-Justifying Usability* (pp. 73–110). Boston: Academic Press.

Fabris, P. (1999). *You Think Tomaytoes, I Think Tomahtoes.* Retrieved October 10, 2001, from www.cio.com/archive/webbusiness/040199_nort_content.html.

Gallaway, G. (1981). Response times to user activities in interactive man/machine computer systems. In S. M. Dray, *The Importance of Designing Usable Systems.* Retrieved October 10, 2001, from www.dray.com/articles/usablesystems.html.

Gilb, T. (1988). Principles of software engineering management. In *Usability Is Good Business.* Retrieved October 15, 2001, from www.compuware.com.

Harrison, M. C., Henneman, R. L., and Blatt, L. A. (1994). Design of a human factors cost-justification tool. In R. G. Bias and D. J. Mayhew (Eds.), *Cost-Justifying Usability* (pp. 203–242). Boston: Academic Press.

Human Factors International. (2001a). *Some Client Experiences.* Retrieved October 10, 2001, from www.humanfactors.com/library/casestudies.asp.

Human Factors International. (2001b). *We Make Financial Software Usable.* Retrieved October 9, 2001, from www.humanfactors.com/home/finance.asp.

IBM. (2001). *Cost Justifying Ease of Use: Complex Solutions Are Problems.* Retrieved October 9, 2001, from www-3.ibm.com/ibm/easy/eou_ext.nsf/Publish/23.

Interaction Design, Inc. (2001). *Design Does Provide Return on Investment.* Retrieved October 10, 2001, from www.user.com/transaction-and-design.htm.

Kalin, S. (1999). *Mazed and Confused.* Retrieved October 10, 2001, from www.cio.com/archive/webbusiness/040199_use.html.

Karat, C. (1994). A business case approach to usability cost justification. In R. G. Bias and D. J. Mayhew (Eds.), *Cost-Justifying Usability* (pp. 45–70). Boston: Academic Press.

Landauer, T. K. (1995). *The Trouble with Computers: Usefulness, Usability, and Productivity.* Cambridge, MA: MIT Press.

Lederer, A. L., and Prasad, J. P. (1992). Nine management guidelines for better cost estimating. *Communications of the ACM,* 35 (2), 51–59.

Lund, A. M. (1997). Another approach to justifying the cost of usability. *Interactions,* 4 (3), 49–56.

Manning, H. (1999). The right way to test ease-of-use. In G. M. Donahue, S. Weinschenk, and J. Nowicki, *Usability Is Good Business.* Retrieved October 15, 2001, from www.compuware.com.

Marcus, A. (2002). Return on Investment for Usable UI Design *User Experience, Usability Professional Association's Magazine,* 1:3, Winter 2002, 25–31.

Marcus, A. (2004). The ins and outs of outsourcing. *User Experience,* 3 (7), 2.

Marcus, A., and Baumgartner, V. (2004). Mapping user-interface design components vs. culture dimensions in corporate websites. *Visible Language Journal,* MIT Press, 38 (1), pp. 1–65.

Martin, J., and McClure, C. (1983). *Software Maintenance: The Problem and Its Solution.* Upper Saddle River, NJ: Prentice Hall.

Mauro, C. (1994). Cost-justifying usability in a contractor company. In R. G. Bias and D. J. Mayhew (Eds.), *Cost-Justifying Usability* (pp. 123–142). Boston: Academic Press.

Mayhew, D., and Mantei, M. (1994). A basis framework for cost-justifying engineering. In R. G. Bias and D. J. Mayhew (Eds.), *Cost-Justifying Usability* (pp. 9–44). Boston: Academic Press.

Nielsen, J. (1993). *Usability Engineering.* San Francisco: Morgan Kaufmann.

Nielsen, J. (1997a). *Loyalty on the Web.* Retrieved October 10, 2001, from http://useit.com/alertbox/9708a.html.

Nielsen, J. (1997b). *Discount Usability for the Web.* Retrieved October 10, 2001, from www.useit.com.

Nielsen, J. (1998a). *The Web Usage Paradox: Why Do People Use Something This Bad?* Retrieved October 12, 2001, from www.useit.com.

Nielsen, J. (1998b). *Failure of Corporate Websites.* Retrieved October 10, 2001, from www.useit.com/alertbox/981018.html.

Nielsen, J. (1999a). *Why People Shop on the Web.* Retrieved October 29, 2001, from www.useit.com.

Nielsen, J. (1999b). *Web Research: Believe the Data.* Retrieved October 12, 2001, from www.useit.com/alertbox/990711.html.

Pressman, R. S. (1992). *Software Engineering: A Practitioner's Approach.* New York: McGraw-Hill.

Rhodes, J. S. (2000). *Usability Can Save Your Company.* Retrieved on October 10, 2001, from www.webword.com/moving/savecompany.html.

Scerbo, M. W. (1991). Usability engineering approach to software quality. *Annual Quality Congress Transactions,* 45, 726–733.

Schneider, M. F. (1985). Why ergonomics can no longer be ignored. *Office Administration and Automation,* 46 (7), 26–29.

Smith, E. (2002). The way we live now: Driven to distraction. *New York Times,* December 1, 2002, Section 6, p. 50.

Smith, P. G., and Reinersten, D. G. (1991). *Developing Products in Half the Time.* New York: Van Nostrand Reinhold.

Standish Group (1995). Chaos research report. In G. M. Donahue, S. Weinschenk, and J. Nowicki, *Usability Is Good Business.* Retrieved October 15, 2001 from www.compuware.com.

Use of Human Factors Has Reduced Device-Related Medical Errors, Retrieved September 1, 2004 from www.injuryboard.com/view.cfm/Article=810.

User Interface Engineering. (2001). *Are the Product Lists on Your Site Losing Sales?* Retrieved October 10, 2001, from http://world.std.com/~uieweb/whitepaper.htm.

Vividence Corp. (2001). *Moving On Up: Move.com Improves Customer Experience.* Retrieved October 15, 2001, from www.vividence.com/public/solutions/our+clients/success+stories/movecom.htm.

Wilson, C. E., and Rosenbaum, S. (2005). Internal, social, and external ROI: Practical ROI issues for UCD teams. In R. G. Bias and D. J. Mayhew (Eds.), *Cost-Justifying Usability* (2nd ed., in press).

Wixon, D., and Jones, S. (1995). Usability for fun and profit: A case study of the design of DEC Rally Version 2. In M. Rudisill *et al., Human-Computer Interface Design: Success Stories, Emerging Methods and Real-World Context* (pp. 3–35). San Francisco: Morgan Kaufmann.

3 A Basic Framework

CHAPTER

Deborah J. Mayhew Deborah J. Mayhew & Associates
Marilyn M. Tremaine New Jersey Institute of Technology

3.1 INTRODUCTION

This chapter presents a framework for cost-justifying usability engineering efforts on software development projects by describing how to *calculate the costs* and *estimate the benefits* of each of the Usability Engineering Lifecycle tasks that can potentially be applied (Mayhew, 1999). We present a general framework that is relevant to developing software for any application from commodity trading to a Web commerce site.

A usability engineering cost-benefit analysis is conducted in the software development process for two major reasons:

1. To demonstrate that usability engineering is a viable and significant cost saving approach
2. To plan the usability engineering program for a particular development project

General cost-benefit analyses of hypothetical usability engineering efforts can be prepared as a strategy to win general support for trying out usability engineering tasks and techniques in a software development organization. When an organization has no experience with usability engineering, cost-benefit analyses may be an effective way to free up resources for usability improvement efforts.

In organizations that are more mature with respect to usability engineering, cost-benefit analyses can be used to plan an optimal usability engineering program for a particular software development project. To help settle on a final usability engineering plan for a specific development project, costs are calculated for the most aggressive program that can be implemented, including the most reliable and thorough *techniques* for all Lifecycle *tasks* (see later discussion). Then benefits are predicted, using *conservative* predictions. If benefits still outweigh costs dramatically, as they invariably will when critical parameters (e.g., number of users, volume of transactions) are favorable, then even the most aggressive usability engineering program can be argued to be viable. This is because only the most conservative claims concerning potential benefits have been made. In fact, the benefits predictions can be redone using more aggressive but still realistic benefit assumptions. The new calculations will show that, in all likelihood, an even more dramatic benefit can be obtained, even from a significant investment in usability engineering.

If, however, benefits and costs in the initial analysis match up fairly closely, then the initial, aggressive usability engineering plan needs to be scaled back, possibly to even a bare-bones plan. It is still possible to achieve the very conservative benefits with even shortcut usability techniques, and thus predict with confidence a healthy return on investment (ROI) from a minimal approach to usability engineering. It is wiser, in the long run, to engage in a conservative benefit assumption and spend a small amount of money, than to barely achieve predicted benefits with a large expenditure. Large expenditures with little to show for them rapidly destroy a manager's or consultant's credibility.

Thus, cost-benefit analysis can be used to develop a cost-effective usability engineering effort for a software development project that is likely to pay off as predicted.

When an organization is first experimenting with usability engineering techniques and is still skeptical about their value, it is wise to make extremely conservative cost-benefit arguments, based on a relatively low-cost usability engineering effort and very modest predictions of benefits, and then to show, after the fact, that much larger benefits were realized. Once an organization has had several positive experiences with investing in usability engineering, it will be more receptive to more aggressive proposals for usability engineering programs, and also to more optimistic benefits predictions.

3.2 THE USABILITY ENGINEERING LIFECYCLE

The first step in cost-justifying usability engineering on a particular software development project is to lay out a usability engineering plan for that project.

The Usability Engineering Lifecycle (Mayhew, 1999) documents a structured and systematic approach to addressing usability within the product development process. It consists of a set of usability engineering tasks applied in a particular order at specified points in an overall software development lifecycle. Readers familiar with the Usability Engineering Lifecycle might wish to skip this section, which provides an overview of the Lifecycle. Readers interested in more detail than that provided in this overview are referred to Mayhew (1999).

Several types of tasks are included in the Usability Engineering Lifecycle, as follows:

- Structured usability requirements analysis tasks
- An explicit usability goal setting task, *driven directly from requirements analysis data*
- Tasks supporting a structured, top-down approach to user interface design that is *driven directly from usability goals and requirements data*
- Objective usability evaluation tasks for iterating design towards usability goals

The chart in Figure 3.1 represents, in summary, visual form, the Usability Engineering Lifecycle. The overall Lifecycle is cast in three phases: Requirements Analysis, Design/Testing/Development, and Installation. Specific usability engineering tasks within each phase are presented in boxes, and arrows show the basic order in which tasks should be carried out. Much of the sequencing of tasks is iterative, and the specific places where iterations would most typically occur are illustrated by arrows returning to earlier points in the Lifecycle.

In addition, free-floating text not enclosed in boxes makes general reference to tasks in an underlying software engineering methodology, with which the Usability Engineering Lifecycle tasks—which are the main focus in this chart—would be conducted in parallel and with which they would be integrated. For example, Lifecycle requirements analysis tasks would be conducted in parallel with function and/or data modeling tasks in many software engineering methodologies, and in particular with the development of a "Requirements Model" in the Object Oriented Software Engineering (OOSE) methodology (Jacobson *et al.,* 1992). These notes provide a general idea of how the Usability Engineering Lifecycle must be integrated with an underlying software engineering methodology.

In considering how to adapt the Usability Engineering Lifecycle to a Web development project—or any type of project for that matter—the distinction between *tasks* and *techniques* is an important one.

A usability engineering *task* can be defined as an activity that produces a concrete work product that is a prerequisite for subsequent usability engineering

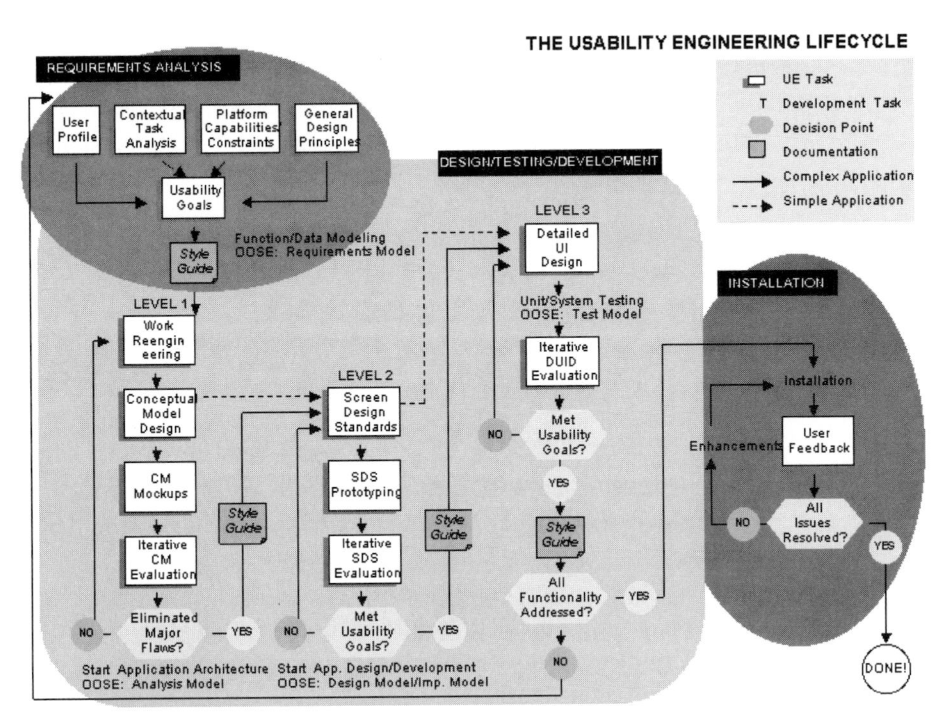

FIGURE 3.1 The Usability Engineering Lifecycle (taken from Mayhew, 1999; used with permission).

tasks. Each task has some conceptual goal that defines it. For example, the goal of the User Profile task is to gain a clear understanding of those characteristics of the intended user population that will have a direct bearing on which design alternatives will be most usable to them.

A *technique*, on the other hand, is a particular process or method for carrying out a task and for achieving a task goal. Usually there are a number of alternative techniques available for any given task. For example, for the User Profile task, alternative techniques include distributing user questionnaires and conducting user or user manager interviews. Generally, techniques vary in how costly and time consuming it is to execute them, in the quality and accuracy of the work products they generate, in how difficult they are for nonspecialists to learn and use, and in the sophistication of the technology required to carry them out.

The key to the general applicability and flexibility of the Usability Engineering Lifecycle lies in the choice of which *techniques* to apply to each task, *not* in the choice of which *tasks* to carry out. *All* the tasks identified in the Lifecycle should be carried out for every development project involving interactive software in order to achieve required levels of usability. However, the approach to any given project can be adapted by a careful selection of *techniques* based on project constraints.

Each usability task in the overall Usability Engineering Lifecycle is briefly described in the following sections, with notes on adapting the tasks to Web development projects in particular.

3.2.1 Phase One: Requirements Analysis

User Profile. A description of the specific user characteristics relevant to user interface design (e.g., computer literacy, expected frequency of use, level of job experience) is obtained for the intended user population. This will drive tailored user interface design decisions, and also identify major user categories for study in the Contextual Task Analysis task discussed later.

The problems of doing a User Profile for a Web site or Web-enabled application are similar to those of doing one for a vendor company: Users are not readily accessible, and may not be known at all. However, developing a User Profile is still possible. The marketing department is often able to identify and get access to potential users. A shortcut for the User Profile task is interviewing marketing, sales, and sales support personnel or others who may have contact with actual current and potential users. And User Profile information can be solicited through the site itself after the Web site or Web-enabled application is implemented. This information can be used to update and improve the new versions of the Web site and to build new related Web sites and applications. A link can be embedded in the Web site that leads users to a User Profile questionnaire where incentives (e.g., discounts or raffle entries) are offered to the user for filling in the online survey (see Weiss, Chapter 19).

Contextual Task Analysis. A study of users' current tasks, workflow patterns, work environments and conceptual frameworks is made, resulting in a description of current tasks and workflow and an understanding and specification of underlying user goals in their identified environments. These will be used to set usability goals and drive Work Reengineering and user interface design.

The problems of doing a Contextual Task Analysis for a Web site or Web-enabled application are, again, very like those of doing one for a vendor company: the users are not easily accessible and may not be known at all, and the "work" may not currently be being performed by intended users. However, conducting a Contextual Task Analysis is still possible.

In a Contextual Task Analysis for a Web site or Web-enabled application, the focus might be more on what people *want and/or need*, rather than on how they currently *do* tasks. You can often get help from Marketing to identify and get access to potential users. You *can* do a Contextual Task Analysis of average people doing personal tasks at home, such as catalog ordering, planning travel, or buying a new car (Vaananen-Vainio-Mattila and Ruuska, 2000; Dray and Mrazek, 1996). You can also solicit task-related information from the Web site itself after the fact. You can have a feedback page and use feedback to update and maintain a site and to build new related Web sites and Web-enabled applications. You can also conduct some task analysis techniques, such as card sorting, via a Web site (see, for example, Weiss, Chapter 19).

Usability Goal Setting. Specific, *qualitative* goals reflecting usability requirements extracted from the User Profile and Contextual Task Analysis, and *quantitative* goals defining minimal acceptable user performance and satisfaction criteria based on a subset of high-priority qualitative goals, are developed. These usability goals focus later design efforts and form the basis for later iterative usability evaluation.

In most cases, at least when designing public Web sites or applications, *ease-of-learning* and *ease-of-remembering* goals will be more important than *ease-of-use* goals, because of the infrequency of use of the Web site. Most users do not visit a given Web site daily, and many often visit a site only once.

Ease of navigation and *maintaining context* will usually be very important *qualitative* goals for Web sites and applications.

Web designers need to be aware when formulating *quantitative performance* goals that system response time will limit and affect user performance, and that system response time will vary enormously depending on the users' platforms.

In many cases of Web site or Web-enabled application design, *relative quantitative* goals may be appropriate (e.g., "It must take no longer to make travel reservations on the Web site than it does with a travel agent by phone," or "It must take less time to make travel reservations on this site than on main competitors' sites").

Platform Capabilities/Constraints. The user interface capabilities and constraints (e.g., windowing, pull down menus, frames, animation) inherent in the technology platform chosen for the product (e.g., Microsoft Windows, Web browsers, or product-unique platforms) are determined and documented. These will define the scope of possibilities for user interface design.

Unlike some of the other Usability Engineering Lifecycle tasks, the Platform Capabilities and Constraints task will often be *more* complicated when designing a Web site or Web-enabled application than when designing traditional software applications. This is because usually (with the exception of the case of some *intranet* applications) designers may have to assume a very large number and wide variety of hardware and software platforms.

Internet users' platforms will vary, possibly widely, in at least the following ways:

- Screen size and resolution
- Data transmission speed
- Activation of pop-up blockers, and security and spyware detection software
- Browser capabilities (varies by vendor and by version):
  - Controls available through the browser (vs. must be provided within the site or application)
  - Browser interpreters (e.g., version of HTML, Java)
  - Installed "helper applications" or "plug-ins" (e.g., multimedia players)

Web user interface designers need to design for the expected *range* of platform capabilities and constraints. For example, one common technique is to have a control at the entry point to a Web site or Web-enabled application that allows users to choose between a "graphics mode" and a "text mode." Thus, users with slow modems can "turn off" any graphics that would seriously degrade download time, and see an alternative text-only version of the site or application.

Similarly, if a Web site or Web-enabled application requires specific "helper applications" or "plug-ins," many are now designed to allow immediate downloading and installation of the required helper application or plug-in. The user interface to downloading and installing helper applications or plug-ins is still often not very user friendly, but at least providing the capability is a step in the right direction.

In general, while designers of intranets may be able to assume certain high-end platform parameters, designers of public Web sites need to be aware that if they take full advantage of all the latest Web capabilities, many users will find their Web site or Web-enabled application unusable. Care needs to be taken to provide alternative interfaces for users with lower-end platforms.

General Design Guidelines. Relevant general user interface design guidelines available in the usability engineering literature are gathered and reviewed. They will be applied during the design process to come, along with all other project-specific information gathered in the previous tasks.

Most general software user interface design principles and guidelines will be directly applicable to Web site and application design.

Things to bear in mind that do make designing for the Web a little different than designing traditional software include the following:

- Response times are slower and less predictable on the Web, limiting which design techniques are practical
- There are little or no existing comprehensive and widely accepted user interface *standards* for the Web (although various *guidelines* are available)
- Browsers and users, rather than designers and developers, may control much of the *appearance* of Web content
- Web users may be mainly discretionary and infrequent, increasing the need for "walk up and use" interfaces
- The World Wide Web is a huge and fluid space with fuzzy boundaries between sites. There is thus an increased need for navigational support and a "sense of place"

These differences are not quantitative, however. Rather, they are a matter of degree. Web platforms simply place more constraints on designers than do traditional platforms. Designing for the Web is somewhat like designing for traditional software 20 years ago—although the capabilities on the Web are now catching up fast.

3.2.2 Phase Two: Design/Testing/Development

Level 1 Design

Work Reengineering. Based on all requirements analysis data and the usability goals extracted from them, user tasks are redesigned at the level of organization and workflow to streamline work and exploit the capabilities of automation. No *visual* user interface design is involved in this task—just abstract organization of functionality and workflow design. This task is sometimes referred to as Information Architecture.

Sometimes you are actually simply engineering—rather than reengineering—work, because your Web site or Web-enabled application supports work unlike anything most of the intended users currently do (e.g., deciding on the structure for an information space users did not previously have access to). Nevertheless, you can still do a Contextual Task Analysis to discover users' needs and desires, and can base your initial work organization on this analysis.

In most cases, even when users do not currently do a particular job, they do already do something highly related to that job, and this can be the focus of a Contextual Task Analysis. In addition, once an initial release of a Web site or Web-enabled application is in production, you can perform another Contextual

Task Analysis to discover how it is being used and where it breaks down and use these insights to reengineer the underlying work models for later releases. And, just as when designing traditional software, you can still *validate* your reengineered work models empirically with evaluation techniques.

Conceptual Model Design. Based on all of the previous tasks, initial high-level design rules are generated for presenting and interacting with the application structure and navigational pathways (i.e., the information architecture). Screen design detail is *not* addressed at this design level.

The Conceptual Model Design is equally important in Web-site and Web-application design as in traditional software design. A Conceptual Model Design for a Web site might typically include rules that would cover the consistent presentation of:

- Site title and logo, including location
- Use of frames (e.g., for highest level links, context information, and page content)
- Links to different levels in the site map
- "You are Here" indicators on links
- Links versus other actions (e.g., "Submit" or "Search")
- Links versus non-links (e.g., illustrations)
- Inter versus intrasite links
- Inter versus intrapage links

On very simple Web site or Web-enabled application projects, it may not be necessary to formally document the Conceptual Model Design. Nevertheless, the Conceptual Model should be explicitly designed and validated.

Conceptual Model Mockups. Paper-and-pencil or prototype mockups (Snyder, 2003) of high-level design ideas generated in the previous task are prepared, representing ideas about how to present high-level functional organization and navigation. Detailed screen design and complete functional design are *not* in focus here.

Instead of paper foils or throw-away prototypes, in the case of Web sites and applications, the "mockups" could be partially coded products, for example, pages, frames, and navigational links with minimal page content detail.

Iterative Conceptual Model Evaluation. The mockups are evaluated and modified through iterative evaluation techniques such as formal usability testing, in which real, representative users attempt to perform real, representative tasks with minimal training and intervention, imagining that the mockups are a real product user interface. This and the previous two tasks are conducted in iterative cycles until identified major usability "bugs" are engineered out of the Level 1 (i.e., Conceptual Model) design. Once a Conceptual Model is relatively stable, system architecture design can commence.

Remote usability testing is particularly well suited to testing Web sites and Web applications. It can replace (or complement) traditional usability testing in which the tester and user are side by side in the same location, and is more practical when users are widely dispersed geographically. The basic technique of remote usability testing involves giving a tester access to what is happening (or did happen) on the computer of a test user in another location. There are several ways to do this, including attended, real-time evaluations similar to traditional laboratory testing but conducted in real time over the Internet, instrumented methods that are unattended but otherwise similar to traditional testing techniques, and automated methods that unobtrusively collect usage data while a user is using a site (Perkins, 2002).

Level 2 Design

Screen Design Standards. A set of application- or site-specific standards and conventions for all aspects of detailed screen or page design is developed, based on any industry and/or corporate standards that have been mandated (e.g., Microsoft Windows or Apple Macintosh), the data generated in the Requirements Analysis phase, and the application- or site-unique Conceptual Model Design arrived at during Level 1 Design. Screen Design Standards will ensure coherence and consistency—the foundations of usability—across the user interface.

Screen Design Standards are just as important and useful in Web design as in traditional software design. Besides the usual advantages of standards, in a Web site they will help users maintain a *sense of place within a site,* because your site standards will probably be different from those on other sites.

On very simple Web site or Web-enabled application projects, it may not be necessary to formally document the Screen Design Standards. Nevertheless, the Screen Design Standards must be explicitly designed and validated.

Web design techniques (both good and bad) tend to be copied—perhaps other Web designers will copy your Screen Design Standards! Perhaps someday we will even have a set of universal Web Screen Design Standards supported by Web development tools, not unlike Microsoft Windows and Apple Macintosh standards. This would contribute greatly to the usability of the Web, just as the latter standards have done for traditional software.

Screen Design Standards Prototyping. The Screen Design Standards (as well as the Conceptual Model Design) are applied to design the detailed user interface to selected subsets of product functionality. This design is implemented as a running prototype.

Instead of paper foils or throw-away prototypes, in the case of Web sites and applications, the prototypes can simply be partially coded products, for example, *selected* pages, frames and navigational links, now with complete page content detail.

Iterative Screen Design Standards Evaluation. An evaluation technique such as formal usability testing is carried out on the Screen Design Standards prototype, and then redesign and re-evaluate iterations are performed to refine and validate a robust set of Screen Design Standards. Iterations are continued until identified major usability bugs are eliminated and usability goals seem within reach.

Again, remote usability testing (Perkins, 2002) can be particularly useful when testing Web sites and applications.

Style Guide Development. At the end of the design and evaluate iterations in Design Levels 1 and 2, you have a validated and stabilized Conceptual Model Design and a validated and stabilized set of standards and conventions for all aspects of detailed screen design. These are captured in the document called the product Style Guide, which already documents the results of Requirements Analysis tasks. During Detailed User Interface Design, following the Conceptual Model Design and Screen Design Standards in the product Style Guide will ensure quality, coherence, and consistency, the foundations of usability.

For simple Web sites and applications, as long as good design processes and principles have been followed, documentation can be minimal and informal: a simple running list.

For complex Web sites and Web-enabled applications with many designers, developers and/or maintainers of a constantly evolving site or application, documenting Requirements Analysis work products and design standards is very important, just as it is on large, traditional software projects.

Level 3 Design

Detailed User Interface Design. Detailed design of the complete product user interface is carried out based on the refined and validated Conceptual Model Design and Screen Design Standards documented in the product Style Guide. This design then drives application or site development.

For simple Web sites or applications, designers might bypass documenting user interface design at the Conceptual Model Design and Screen Design Standards levels, and simply prepare Detailed User Interface Design specifications directly from standards they have informally established at these earlier design levels. Developers can then code directly from these specifications.

For more complex Web sites or applications, the Conceptual Model Design and Screen Design Standards should usually be documented before this point. Then developers can code directly from an application or site Style Guide, or from Detailed User Interface Design specifications prepared based on a product Style Guide by the user interface designer.

Iterative Detailed User Interface Design Evaluation. A technique such as formal usability testing is continued during application or site development to expand evaluation to not-yet-assessed subsets of functionality and categories of users, and also to continue to refine the user interface and validate it against usability goals.

On projects developing relatively *simple* Web sites and applications, it might be more practical to combine the three levels of the design process into a single level, in which Conceptual Model Design, Screen Design Standards, and Detailed User Interface Design are *all* sketched out in sequence *before* any evaluation proceeds. Then a single process of design and evaluation iterations can be carried out.

In this case, Iterative Detailed User Interface Design Evaluation will be the first usability evaluation task conducted. Thus, evaluation must address all levels of design simultaneously. This is practical only if the whole site or application is fairly simple, which information-only (i.e., nontransactional) Web sites often are. It is important to remember that even if Detailed User Interface Design is drafted before any evaluation commences, it is still crucial to consider all the same design issues that arise in the Conceptual Model Design and Screen Design Standards tasks when conducting design in a three-level process.

In Web sites and applications of *intermediate complexity*, the first *two* design levels (Conceptual Model Design and Screen Design Standards) may be combined into a single process of design and evaluation iterations to validate them simultaneously, and then an additional process of design and evaluation can be carried out during the Detailed User Interface Design level. Alternatively, Level 1 can be carried out with iterative evaluation, and then Levels 2 and 3 can be collapsed into one iterative cycle. In either case, this will be the *second* usability evaluation cycle conducted, and can indeed focus mainly on Screen Design Standards and Detailed User Interface Design, because Conceptual Model Design will have been focused on during an earlier evaluation task.

Also, in the case of Web site or Web-enabled application design, one alternative is that mockups, prototypes, *and* application code can *all* simply be final code at different points of completion rather than paper foils or throw-away prototypes.

As in previous design levels, remote usability testing (Perkins, 2002) can be particularly useful when testing Web sites and applications.

3.2.3 Phase Three: Installation

User Feedback. After the product has been installed and in production for some time, feedback is gathered to feed into enhancement design, the design of new releases and/or the design of new but related products.

User feedback can be solicited directly from a Web site or Web-enabled application. This can be done by providing a link on the site taking users to a structured feedback page, or by offering direct e-mail from the site and asking users to provide free-form feedback. You can even have survey questions pop up, triggered by specific usage events. An advantage of the latter is that it collects feedback while the user's experience is fresh in his or her mind. The disadvantage, of course, is that users may find it irritating to be interrupted by a solicitation of this sort.

You might need to provide some incentive for users to take the time to provide feedback (see Weiss, Chapter 19), especially if you provide a lengthy, structured form (shorter forms probably work best). Possible incentives include entry in a raffle or discounts on products or services.

The user feedback techniques that lend themselves most easily to Web sites and Web-enabled applications include questionnaires (see Weiss, Chapter 19) and usage studies. Other techniques (e.g., interviews, focus groups and usability testing) are more difficult to employ since they require the identification and recruitment of users to meet in person with project team members, which may not be difficult on *Intranet* sites, but may be difficult on *Internet* sites.

To young Web designers or developers who launched their careers in the Internet age and have worked primarily on Web-development projects, the Usability Engineering Lifecycle may, at first glance, seem much too complex and time consuming to be practical in the fast-paced world of Web development. If you consider only the traditional and most reliable and thorough techniques for Lifecycle tasks, and typical timeframes for the development of very simple read-only Web sites, this is a fair assessment. For example, whereas I (Mayhew) have often conducted task analyses techniques that took several months to complete, and formal usability tests that took a month or more, I have also worked on Web development projects that from start to launch took a total of 8 to 12 weeks. Clearly you cannot spend several months conducting task analyses—just one of the first steps in the Usability Engineering Lifecycle—when the whole project must be completed in 2 to 3 months!

Two points must be made however. First, the Usability Engineering Lifecycle is highly flexible and adaptable through the selection of *techniques* applied to each task and the collapsing of Design Levels, as previously described, and can accommodate even projects with very limited timeframes. I (Mayhew) have in fact successfully adapted it to even 8-week Web-development projects.

Second, Web site functions were initially very simple compared to most traditional software applications, and so the fact that they typically took 8 to 12 weeks to develop, as compared to months or even years for traditional software applications, made some sense. Now, however, Web sites and applications have become more and more complex, and in many cases are much like traditional applications that happen to be implemented on a browser platform. The industry needs to adapt its notion of reasonable, feasible, and effective timeframes (and budgets) for developing complex Web-based applications, which simply are not the same as simple content-only Web sites. This includes adapting its notion of what kind of usability engineering techniques it should invest in.

In a report by Forrester Research, Inc. (Sonderegger, 2000), called "Scenario Design" (their term for usability engineering), it is pointed out that:

> **Executives Must Buy Into Realistic Development Time Lines and Budgets**
> The mad Internet rush of the late 1990s produced the slipshod experiences that we see today. As firms move forward, they must shed their misplaced fascination with first-mover advantage in favor of lasting strategies that lean on quality of experience.
>
> ✦ Even single-channel initiatives will take eight to 12 months. The time required to conduct field research, interpret the gathered information, and formulate implementation specs for a new Web-based application will take four to six months. To prototype, build, and launch the effort will take another four to six months. This period will lengthen as the number of scenarios involved rises.
>
> ✦ These projects will cost at least $1.5 million in outside help. Firms will turn to eCommerce integrators and user experience specialists for the hard-to-find-experts, technical expertise, and collaborative methodologies required to conduct Scenario Design. Hiring these outside resources can be costly, with run rates from $150 K to $200 K per month. This expenditure is in addition to the cost of internal resources, such as project owners responsible for the effort's overall success and IT resources handling integrations with legacy systems (p. 12).

We agree that 8 to 12 *months* is a more realistic timeframe (than 8 to 12 *weeks*) to develop a usable Web site or Web-enabled application that will provide a decent ROI. And, if this is the overall project timeframe, there is enough time to use traditional usability engineering techniques to more reliably ensure Web site usability, a major contributor to ROI. In my experience (Mayhew), depending on the complexity of a Web site or Web-enabled application, somewhere between $100,000 and $250,000 should pay for a reliable and thorough usability engineering program. This is a small fraction of the $1.5 million estimated by Forrester for all the outside help a Web site sponsor will need. And, as the rest of this chapter and other chapters in this volume illustrate, significant time and money invested in Web site or Web-enabled application usability will usually pay off.

3.3 GENERAL APPROACH

To cost-justify a usability engineering plan, you simply adapt a very generic and widely used cost-benefit analysis technique. Having laid out a detailed usability project plan based on Lifecycle tasks (see previous sections, and Mayhew, 1999), it is a fairly straightforward matter to calculate the costs of that plan. Then you need to calculate the benefits. This is a little trickier, and it is where the adaptation of the generic analysis comes into play (see later discussion). Then you simply compare costs to benefits to find out if and to what extent the benefits outweigh the costs. If they do to a satisfactory extent, then you have cost-justified the planned effort.

More specifically, first a usability engineering plan is laid out. The plan specifies particular techniques to employ for each Usability Engineering Lifecycle task, breaks the techniques down into steps, and specifies the personnel hours and equipment costs for each step. The cost of each task is then calculated by multiplying the total number of hours for each type of personnel by their effective hourly wage (fully loaded, i.e., including salary, benefits, office space, equipment, utilities, and other facilities), and adding up personnel costs across types. (Sometimes it is hard to get data on fully loaded wages for an organization. In this case, I (Mayhew) use a rule of thumb I have heard informally and simply double the before-tax annual salary, then divide by the typical number of hours a full-time worker is paid for in a year, usually about 2000. Even if my audience is unwilling to give me actual figures for fully loaded wages, they can contest—or not—my ballpark figure based on this rule of thumb.) Any equipment and other costs can be added in. Then the costs from all tasks are added to arrive at a total cost for the plan.

Next, the overall benefits of the specific usability engineering plan are predicted by selecting relevant benefit categories, calculating expected benefits by plugging project-specific parameters and assumptions into benefit formulas, and adding benefits across categories.

The list of possible benefits to consider is long, because usability engineering can lead to tangible benefits to all concerned, regardless of the type of organization or type of application. The development team realizes savings because problems are identified early, when they are cheap to fix. In vendor companies, the customer support team realizes a reduced customer support burden (although it may take time to show up, because a new interface may cause an initial surge in calls before it settles down). More usable e-commerce Web sites will have higher buy-to-look ratios, a lower rate of abandoned shopping carts, and increased return visits. Internal user productivity will be increased, and there will be lower user training costs.

The potential benefit categories selected in a particular cost-benefit analysis will depend on the type of organization taking on the development effort, including the usability engineering costs. In the case of a *development organization serving internal users*, benefits to the company as a whole might include:

- Increased user productivity
- Decreased user errors (and faster recovery when errors are made)
- Decreased training costs
- Savings gained from making changes earlier in the development lifecycle
- Decreased customer service calls from users

Benefits of usability engineering efforts to a *vendor company* might include:

- Increased sales
- Decreased customer service calls from users
- Savings gained from making changes earlier in the development lifecycle
- Reduced cost of providing training (if training is offered through the vendor company)

Note that although the primary benefit relevant to the development organization with internal users might be increased user productivity, this is not usually of direct concern to a vendor company (even though it should be). The vendor company is more concerned with selling more products and decreasing their customer support costs. Thus, in a cost-benefit analysis, you should focus attention on the potential benefits that are of most interest to the audience for the analysis.

Note that these benefits represent just a sample of those that might be relevant in these two types of organizations. Others might be included as appropriate, given the business goals of the organization and the primary concerns of the audience, and could be calculated in a similar fashion as shown in a later section.

In the case of Web sites and Web-enabled applications, the potential benefit categories relevant to a particular cost-benefit analysis will depend on the basic business model for the site. Benefit categories potentially relevant to different types of sites are summarized in Table 3.1.

Note that the relevant benefit categories for different types of Web sites and Web-enabled applications vary somewhat. In a cost-benefit analysis, you should focus attention on the potential benefits that are *of most relevance to the bottom line business goals for the site*, whether they are short term, long term, or both.

Again, note that these benefits represent just a sample of those that might be relevant in these types of sites, and do not address other possible benefits of usability in other types of sites. Others might be included as appropriate, given the business goals of the site sponsors and the primary concerns of the audience, and could be calculated in a similar fashion as those shown later in the chapter.

Finally, overall benefits are compared to overall costs to see if, and to what extent, the overall usability engineering plan is justified.

When usability practitioners are invited to participate in projects already in progress, which is often the case for external consultants, they have less chance of including all Usability Engineering Lifecycle tasks and of influencing overall schedules and budgets. They are more likely to have to work within already-committed-to schedules, platforms, and system architectures, to use shortcut techniques for Usability Engineering Lifecycle tasks, and to minimally impact budgets. Nevertheless, it is almost always possible to create a usability engineering plan that will make a significant contribution to a software development project, even when you come into the project relatively late. You can use the cost-benefit analysis technique to prepare and support even usability engineering plans that involve only parts of the overall Usability Engineering Lifecycle, and only shortcut techniques for tasks within it.

3.4 SAMPLE COST-BENEFIT ANALYSES

Let us consider a hypothetical usability engineering plan and see how its costs can be estimated. Then we will incorporate this plan into scenarios involving

Table 3.1 Potential Benefit Categories for Different Types of Web Sites

	Site Type				
Benefits	E-Commerce	Funded by Advertising	Product Information	Customer Service	Intranets
Increased buy-to-look ratios	✓				
Decreased abandoned shopping carts	✓				
Increased number of visits		✓			
Increased return visits	✓	✓			
Increased length of visits		✓			
Decreased failed searches	✓	✓			
Decreased costs of other sales channels	✓				
Decreased use of "Call Back" button (i.e., live customer service)	✓				✓
Savings resulting from making changes in earlier development lifecycle	✓	✓	✓	✓	✓
Increased "click through" on ads		✓			

Table 3.1 *Continued*

	Site Type				
Benefits	E-Commerce	Funded by Advertising	Product Information	Customer Service	Intranets
Increased sales leads			✓		
Decreased costs of traditional customer service channels				✓	
Decreased training costs					✓
Increased user productivity					✓
Decreased user errors					✓

development in four different types of projects, and see how you would conduct cost-benefit analyses of that plan for each project. The four scenarios involve the development of:

- An application for internal users
- A commercial application by a vendor company
- An e-commerce Web site
- A product information Web site

3.4.1 An Application for Internal Users

Imagine that a development organization is planning to develop an application for use by an *internal* user organization (e.g., within a bank or an insurance company). The project is of moderate complexity and cost, and will result in an application that will be used by 250 users. Once developed and installed, the application is expected to be in production for approximately 5 years before any major revisions are made.

First, the final results of a cost-benefit analysis are presented. Then, in the steps that follow, the derivation of the final results are shown. Table 3.2 shows the overall calculation of the *cost* of a usability engineering plan proposed by the project usability engineer. The first column identifies the overall project phase. The second column identifies which Usability Engineering Lifecycle tasks (see Fig. 3.1) and techniques are planned in each phase. The third, fourth, fifth and sixth columns identify the number of work hours required by usability engineers, developers, managers, and users to complete each task. The last column summarizes the total cost of each task, based on the fully loaded hourly rates for each type of personnel. A total cost for the whole plan is given at the bottom of the table.

It is important to note that the usability engineering plan laid out in Table 3.2 is one specific to a particular project. Different plans could be devised, involving different phases, tasks and techniques, and the costs of these plans would vary accordingly. Also, for simplicity's sake, we have not included the costs of materials and equipment in this example. These could easily be estimated and added to the total cost of the usability engineering plan. Finally, note that the role "developers" is also a simplification. It might include not just engineers, but also graphic designers, business analysts, quality assurance (QA) staff, and so on, in fact any non–usability engineer and nonmanager who is expected to participate in the usability engineering tasks laid out in the plan could fall into this category.

In this hypothetical project, the project usability engineer has calculated the *predicted benefits* of carrying out this usability engineering plan *in the first year* of application installation, as shown in Table 3.3.

The predicted benefits *over the expected product lifetime* of the application (5 years) are also shown in Table 3.3. Comparing these benefits and costs, the project usability engineer argues that, as shown in Table 3.4, in the first year alone, a net benefit of $168,144.44 is predicted, and over the expected 5-year lifetime of the product, a net benefit of $1,158,422.22 is predicted. The project usability engineer expects the plan to be approved and funded based on this cost-benefit analysis.

Note that the simple analyses offered here do not consider the time value of money—that is, the money for the costs is spent at one point in time, and the benefits come later. Also, if the money was *not* spent on the costs, but instead was invested in some other way, it would likely increase in value. In our experience, the predicted benefits of usability engineering are usually so dramatic that these more sophisticated financial considerations aren't necessary to convince

Table 3.2 Cost of a Usability Engineering Plan

Phase	*Task (Technique)*	*Usability Engineer Hours at $175*	*Developer Hours at $175*	*Manager Hours at $200*	*User Hours at $25*	*Total Cost*
Requirements Analysis	User Profile (Questionnaire)	62	0	4	33	$12,475
	Contextual Task Analysis	138	8	8	60	$28,650
	Platform Capabilities and Constraints	16	6	0	0	$3,850
	Usability Goals	20	0	4	2	$4,350
Design/ Testing/ Development	Work Reengineering (Information Architecture)	80	0	0	16	$14,400
	Conceptual Model Design	80	8	0	8	$15,600
	Conceptual Model Mockups (Paper Prototype)	36	0	0	0	$6,300
	Iterative Conceptual Model Evaluation (Usability Test)	142	0	0	22	$25,400
	Screen Design Standards	80	8	0	8	$15,600
	Screen Design Standards Prototyping (Live Prototype)	28	80	0	0	$18,900
	Iterative Screen Design Standards Evaluation (Usability Test)	142	40	0	22	$32,400
	Detailed User Interface Design	80	8	0	8	$15,600
	Iterative Detailed User Interface Design Evaluation (Usability Test)	142	40	0	22	$32,400
	Totals	**1046**	**198**	**16**	**201**	**$225,925**

Table 3.3 Expected First Year and Lifetime Benefits for an Application for Internal Users

Benefit Category	*Benefit Value* First Year
Increased productivity	$199,652.78
Decreased errors	$47,916.67
Decreased training	$62,500.00
Decreased late design changes	$84,000.00
Total benefit	**$394,069.44**

Benefit Category	*Benefit Value* Lifetime (5 yrs)
Increased productivity 5 yrs	$998,263.89
Decreased errors × 5 yrs	$239,583.33
Decreased training × 1 yr	$62,500.00
Decreased late design changes × 1 yr	$84,000.00
Total benefit	**$1,384,347.22**

Table 3.4 Net Benefit Calculations for an Application for Internal Users

	Benefit	*Cost*	*Net Benefit*
First Year	$394,069.44	$225,925.00	$168,144.44
Lifetime (5 yrs)	$1,384,347.22	$225,925.00	$1,158,422.22

the audience of the analysis. However, if needed, these calculations based on the time value of money are presented in Karat (Chapter 4), and also in Bias *et al.* (2002).

In the following sections, we lay out step-by-step how the project usability engineer arrived at the final results stated previously.

1. Start with the Usability Engineering Plan

If it has not already been done, this is the first step in conducting a cost-benefit analysis. The usability engineering plan identifies which Usability Engineering Lifecycle tasks and techniques (see previous discussion and Mayhew, 1999) will be employed and breaks them down into required staff and hours. Costs can then be computed for these tasks in the next two steps.

The usability engineering plan for this sample analysis is shown in Table 3.2. It is important to note that there is not one correct usability engineering plan. This too is something that will vary across projects. The choice of technique for carrying out each task in the Usability Engineering Lifecycle will depend on project budgets, schedules, and complexity. Thus, the cost of the sample plan in the examples presented here should not be assumed—a project-unique plan must be designed around the parameters of a specific project, and then costs worked out as the example given here illustrates.

2. Establish Analysis Parameters

Most of the calculations for both planned costs and predicted benefits are based on project-specific parameters. These should be established and documented before proceeding with the analysis. Sample analysis parameters for our hypothetical project are given in Table 3.5.

It should be emphasized that when using the general cost-benefit analysis technique illustrated here, these particular parameter *values* should not be assumed. The particular parameter values of *your* project and organization should be substituted for those in Table 3.5. They will almost certainly be different from the parameters used in this example. For example, your application

Table 3.5 Analysis parameters for an application for internal users

Analysis Parameters	*Values*
Number of end users	250
User work days per year	230
User fully loaded hourly wage	$25
Developer fully loaded hourly wage	$175
Usability engineer fully loaded hourly wage	$175
Manager fully loaded hourly wage	$200
Ratio of early-to-late design changes	0.25
Expected system lifetime (yrs)	5
Current transactions per day	100
Current recovery time per error (2 min expressed as hrs)	0.033333333
Time per early design change (hrs)	8
Ratio of late-to-early design changes	4
Usability lab	In place

may be intended for many more (or fewer) than 250 users, and the fully loaded hourly wage (the costs of salary plus benefits, office space, equipment, utilities, and other facilities) of your personnel may be significantly lower or higher than those assumed in these sample analyses.

Note that, in general, certain parameters in a cost-benefit analysis have a major impact on the magnitude of potential benefits. For example, when considering *user productivity*—of primary interest to internal development organizations—the critical parameters are the *number of users*, the *volume of transactions*, and, to some extent also the *users' fully loaded hourly wage*. When there is a large number of users and/or a high volume of transactions, even very small performance advantages (and low hourly wages) in an optimized interface will add up quickly to significant overall benefits. On the other hand, where there is a small number of potential users, and/or a low volume of transactions, benefits may not add up to much even when the potential per-transaction performance advantage seems significant and the user hourly wage is higher.

For example, consider the following two scenarios. First, imagine a case in which there are 5000 users and 120 transactions per day per user. Even a half second advantage per transaction in this case adds up.

$$5000 \text{ users} \times 120 \text{ transactions} \times 230 \text{ days} \times 1/2 \text{ second} = 19{,}167 \text{ hours}$$

If the users' hourly rate is \$25, the annual savings are:

$$19{,}167 \text{ hours} \times \$25 = \$479{,}175$$

This is a pretty dramatic benefit for a tiny improvement on a per transaction basis! On the other hand, if there were only 25 users, and they were infrequent users, with only 12 transactions per day, even if a per-transaction benefit of 1 minute could be realized, the overall benefit would be minor.

$$25 \text{ users} \times 12 \text{ transactions} \times 230 \text{ days} \times 1 \text{ minute} = 1{,}150 \text{ hours}$$

At \$25 per hour, the overall annual productivity benefit will only be:

$$1{,}150 \text{ hours} \times \$25 = \$28{,}750$$

Thus, in the case of productivity benefits, costs associated with optimizing the user interface are more likely to pay off when there are many users and many transactions.

In the case of the *sales* benefit for a *vendor*, as in a later example, the critical parameter is usually *profit margin.* If the profit margin per product is low, then a very large number of additional sales would have to be achieved from usability alone for the usability costs to pay off. On the other hand, if the profit margin per product is high, then only a small number of increased sales from usability would be necessary to pay for the usability program. Similarly, in the case of an e-commerce or product information Web site, as in other examples given later, the critical parameters will be volume of visitors and profit margin per completed sales transaction. Thus, critical analysis parameters will directly determine how much can be invested in usability and still pay off.

3. Calculate the Cost of Each Usability Engineering Lifecycle Task in the Usability Engineering Plan

The cost of each individual task/technique listed in Table 3.2 was estimated by breaking the task/technique down into small steps, estimating the number of hours required for each step by different types of personnel, and multiplying these hours by the known fully loaded hourly wage of each type of personnel (if outside consultants or contractors are used, their simple hourly rate plus travel expenses would apply, and if external users are recruited to participate, they will be paid at some simple hourly rate or flat fee).

Fully loaded hourly wages are calculated by adding together the cost of salary, benefits, office space, equipment, and any other relevant overhead for a type of personnel, and dividing this by the number of hours paid for each year for that personnel type. The hourly rate used here for usability engineering staff is based on an informal average of typical current salaries of senior-level internal usability engineering staff and external consultants in my recent experience. (See www.upassoc.org/upa_publications/upa_voice/survey/2000_survey.html for a fairly recent salary survey of usability practitioners.) The hourly rate of developers was similarly estimated. (See, for example, www.payscale.com/salary-survey/vid-18644/fid-6886.) However, the fully loaded hourly rate figures used to generate this and the other sample cost-benefit analyses below are just examples, and you would have to substitute the actual hourly rates of personnel in your own organization in an actual analysis. Additional costs, such as equipment and supplies, could also be estimated and added into the total cost of each task/technique, although that was not done here for simplicity's sake.

Cost estimates for each usability engineering task/technique included in the usability engineering plan presented in Table 3.2 were calculated as presented in Tables 3.6 through 3.18. In these calculations, as shown in Table 3.6, usability engineers and developers are estimated to cost $175 per hour, managers are

Text continues on p. 72

Table 3.6 Cost of User Profile (Questionnaire)

Step	*Usability Engineer Hours*	*Developer Hours*	*Manager Hours*	*User Hours*	
Conduct needs finding	4		2	2	
Draft questionnaire	6				
Management feedback	2		2	2	
Revise questionnaire	4				
Pilot questionnaire	4			4	
Revise questionnaire	2				
Select user sample	4				
Distribute questionnaire/ get responses	8			25	
Data analysis	8				
Data interpretation/ presentation	20				
Total hours	62	0	4	33	
Times hourly rate	× $175	× $175	× $200	× $25	
Equals	$10,850	+ $0	+ $800	+ $825	= **$12,475**

Table 3.7 Cost of Contextual Task Analysis

Step	*Usability Engineer Hours*	*Developer Hours*	*Manager Hours*	*User Hours*	
Review requirements specs	12				
Interview project team/ user reps	16	8	6	2	
Identify key actors/ use cases	6		2	2	
In-context observations	40			40	
Card sorting	32			16	
Task Analysis documentation	32				
Total hours	138	8	8	60	
Times hourly rate	× $175	× $175	× $200	× $25	
Equals	$24,150	+ $1,400	+ $1,600	+ $1,500	= **$28,650**

Table 3.8 Cost of Platform Constraints and Capabilities

Step		*Usability Engineer Hours*		*Developer Hours*		*Manager Hours*		*User Hours*		
Review documentation		4								
Interview developers		6		6						
Document constraints/ capabilities		6								
Total hours		16		6		0		0		
Times hourly rate	×	$175	×	$175	×	$200	×	$25		
Equals		$2,800	+	$1,050	+	$0	+	$0	=	**$3,850**

Table 3.9 Cost of Usability Goals

Step		*Usability Engineer Hours*		*Developer Hours*		*Manager Hours*		*User Hours*		
Draw from User Profile		4								
Draw from Contextual Task Analysis		4								
Research business goals		4				2		2		
Formulate/prioritize/ document goals		8				2				
Total hours		20		0		4		2		
Times hourly rate	×	$175	×	$175	×	$200	×	$25		
Equals		$3,500	+	$0	+	$800	+	$50	=	**$4,350**

Table 3.10 Cost of Work Reengineering (Information Architecture)

Step	*Usability Engineer Hours*	*Developer Hours*	*Manager Hours*	*User Hours*	
Review all Requirements Data and Usability Goals	8				
Design Draft Information Architecture	24				
Validate Draft Information Architecture (Reverse Card Sorting)	24			16	
Document Draft Information Architecture	24				
Total hours	80	0	0	16	
Times hourly rate	× $175	× $175	× $200	× $25	
Equals	$14,000	+ $0	+ $0	+ $400	= **$14,400**

Table 3.11 Cost of Conceptual Model Design

Step	*Usability Engineer Hours*	*Developer Hours*	*Manager Hours*	*User Hours*	
Review all Requirements Data, Usability Goals and Information Architecture	8				
Design Draft Conceptual Model	32	8		8	
Document Draft Conceptual Model	40				
Total hours	80	8	0	8	
Times hourly rate	× $175	× $175	× $200	× $25	
Equals	$14,000	+ $1,400	+ $0	+ $200	= **$15,600**

Table 3.12 Cost of Conceptual Model Mockup (Paper Prototype)

Step		*Usability Engineer's Hours*		*Developer's Hours*		*Manager's Hours*		*User's Hours*		
Select functionality		4								
Create paper prototype foils		32								
Total Hours		36		0		0		0		
Times Hourly Rate	×	$175	×	$175	×	$200	×	$25		
Equals		$6,300	+	$0	+	$0	+	$0	=	**$6,300**

Table 3.13 Cost of Iterative Conceptual Model Evaluation (Usability Test)

Step		*Usability Engineer Hours*		*Developer Hours*		*Manager Hours*		*User Hours*		
Design/develop test materials		32								
Design/assemble test environment		4								
Pilot test/revise materials		10						6		
Run test/collect data (2 usability engineers)		32						16		
Collate data		16								
Analyze/interpret data, formulate Redesign		24								
Document/present conclusions		24								
Total hours		142		0		0		22		
Times hourly rate	×	$175	×	$175	×	$200	×	$25		
Equals		$24,850	+	$0	+	$0	+	$550	=	**$25,400**

Table 3.14 Cost of Screen Design Standards

Step	*Usability Engineer Hours*	*Developer Hours*	*Manager Hours*	*User Hours*	
Review all requirements Data, Usability Goals, Information Architecture and Conceptual Model Design	8				
Design Draft Screen Design Standards	32		8	8	
Document Draft Screen Design Standards	40				
Total hours	80	8	0	8	
Times hourly rate	× $175	× $175	× $200	× $25	
Equals	$14,000	+ $1,400	+ $0	+ $200	= **$15,600**

Table 3.15 Cost of Screen Design Standards Prototyping (Live Prototype)

Step	*Usability Engineer Hours*	*Developer Hours*	*Manager Hours*	*User Hours*	
Select functionality	4				
Prepare design specification	24				
Build live prototype		80			
Total hours	28	80	0	0	
Times hourly rate	× $175	× $175	× $200	× $25	
Equals	$4,900	+ $14,000	+ $0	+ $0	= **$18,900**

Table 3.16 Cost of Iterative Screen Design Standards Evaluation (Usability Test)

Step	*Usability Engineer Hours*	*Developer Hours*	*Manager Hours*	*User Hours*	
Design/develop test materials	32				
Design/assemble test environment	4	32			
Pilot test/revise materials	10	8		6	
Run test/collect data (2 usability engineers)	32			16	
Collate data	16				
Analyze/interpret data, formulate Redesign	24				
Document/present conclusions	24				
Total hours	142	40	0	22	
Times hourly rate	× $175	× $175	× $200	× $25	
Equals	$24,850	+ $7,000	+ $0	+ $550	= **$32,400**

Table 3.17 Cost of Detailed User Interface Design

Step	*Usability Engineer Hours*	*Developer Hours*	*Manager Hours*	*User Hours*	
Review all Requirements Data, Usability Goals, Information Architecture, Conceptual Mode Design and Screen Design Standards	8				
Design Draft Detailed User Interface Design	32	8		8	
Document Draft Detailed User Interface Design	40				
Total hours	80	8	0	8	
Times hourly rate	× $175	× $175	× $200	× $25	
Equals	$14,000	+ $1,400	+ $0	+ $200	= **$15,600**

Table 3.18 Cost of Iterative Detailed User Interface Design Evaluation (Usability Test)

Step		*Usability Engineer Hours*		*Developer Hours*		*Manager Hours*		*User Hours*		
Design/Develop Test Materials		32								
Design/Assemble Test Environment		4		32						
Pilot test/revise materials		10		8				6		
Run test/collect data (2 usability engineers)		32						16		
Collate data		16								
Analyze/interpret data, formulate redesign		24								
Document/present conclusions		24								
Total hours		142		40		0		22		
Times hourly rate	×	$175	×	$175	×	$200	×	$25		
Equals		$24,850	+	$7,000	+	$0	+	$550	=	**$32,400**

estimated to cost $200 per hour and users to cost $25 per hour. The total cost of each task/technique shown in Tables 3.6 through 3.18 are used in Table 3.2 to calculate the total cost of the whole usability engineering plan.

4. Select Relevant Benefit Categories

As shown in Table 3.3, the project usability engineer selected four benefit categories relevant to this application for internal users to include in the cost-benefit analysis:

1. Increased productivity
2. Decreased errors
3. Decreased training
4. Decreased late design changes

These seemed to be of most relevance to building an application for an internal user population. Other benefits might have been included, for example, decreased cost of user support time, but just these four were selected to keep the analysis simple and conservative. As already discussed, the best benefit categories to include in a cost-benefit analysis will depend on the type of project and the intended audience for the analysis.

In this case the project usability engineer expects to achieve increased productivity by focusing on streamlining across-screen navigation within tasks, by minimizing typing and mouse clicks on individual screens, and by designing to facilitate scanning and interpreting displays. He or she expects to decrease errors both by following well-established design principles during design and by detecting and eliminating common errors through usability testing. He or she expects to decrease training time by designing a consistent, rule-based user interface architecture which matches users' knowledge and expectations and in which the smallest number of rules accounts for the widest scope of functionality. Finally, late design changes will be minimized and replaced by less expensive early design changes by following an iterative design process which incorporates usability inspection and testing.

Table 3.3 summarizes the predicted magnitude of each of these benefits and adds them to predict a total benefit. What follows is an explanation of how benefit predictions in each category were derived.

5. Predict Benefits

Benefits are predicted in each selected benefit category by doing some simple arithmetic based on project-specific analysis parameters and some simple assumptions. The project parameters in this case are laid out in Table 3.5. The benefits assumptions are given in Table 3.19.

In the case of productivity, the relevant *parameters* are (from Table 3.5):

- The total number of users
- The number of days each user works per year

Table 3.19 Benefits Assumptions for an Application for Internal Users

Benefit Assumptions			
Increased Productivity	Decreased Errors	Decreased Training	Decreased Late Design Changes
Decr. time/ transaction (5 sec = 0.001389 hrs)	1 error eliminated/ day	10 hrs saved off current 1 wk training	20 changes made early

- The number of transactions each user currently performs each working day
- The users' fully loaded hourly wage

The *assumption* made regarding increased productivity (from Table 3.19) is that:

- Each transaction will take 5 seconds (0.001389 hours) less on a user interface developed with the usability engineering plan than on a user interface developed without the usability engineering plan

This single assumption is the crux of the whole cost-benefit analysis. While *costs* can be calculated with a high degree of confidence based on past experience, and all the *parameters* fed into the analysis are known facts, the *assumptions* made are just that—assumptions—rather than known facts or guaranteed outcomes. The audience for the analysis is asked to accept that these assumptions are reasonable, and they must to be convinced by the overall analysis.

Note that *any* cost-benefit analysis for *any* purpose must ultimately include some assumptions that are really only predictions of the likely outcome of investments of various sorts. The whole point of a cost-benefit analysis is to try to evaluate in advance, in a situation in which there is some element of uncertainty, the likelihood that an investment will pay off. The trick is basing the predictions of uncertainties on a firm foundation of known facts. In the case of a cost-benefit analysis of usability engineering, there are several foundations upon which to formulate sound assumptions regarding benefits.

First, there is 25 years of published research that shows measurable and significant performance advantages of specific user interface design alternatives under certain circumstances. Examples of design alternatives for which performance data exist include the following:

- Use of color
- Choice of input devices
- Use of windowing
- Use of direct manipulation
- Screen design details
- Menu structure

Benefit assumptions can thus be defended by referring to studies of such design alternatives. Available studies that explore the relative benefits of different design alternatives typically vary one narrow aspect of design, such as fill-in form design, use of windows, use of color, or system response time, keeping all other design variables constant, and measure human performance on some simple, well-defined tasks.

From these studies, we can extrapolate to make some reasonable predictions about the order of magnitude of differences we might expect to see in user interfaces that have been optimized through the execution of a usability engineering plan. The research does not provide simple, generic answers to design questions. However, what the research does provide are *general ideas* of the *magnitude of performance differences that can occur between optimal and suboptimal interface design alternatives.* The basic benefit assumptions made in any cost-benefit analysis can thus be generated and defended in part by simply referring to the wide body of published research data that exists.

Besides citing relevant general research literature, there are other ways to arrive at and defend one's benefit assumptions in a cost-benefit analysis. Actual case histories of the benefits achieved as a result of applying usability engineering techniques are very useful in helping to defend the benefits assumptions of a particular cost-benefit analysis. A few published case histories exist (e.g., Karat, 1989); Wixon and Wilson (1997) and Whiteside *et al.* (1988) reported that across their experience with many projects over many years, they found that they averaged an overall performance improvement of about 30% when at least 70 to 80% of the problems they identified during testing were addressed by designers. Landauer (1996, pp. 221–223) also cites an impact of usability engineering between zero and several hundred percent (with an average of 50 percent) on a set of 18 projects reported in the literature, depending on the complexity and type of product. Also, in Chapter 16, I (Mayhew) describe a case study based on a real project, in which task times were improved on average by 21% on a redesigned user interface as compared to an original user interface. Across eight tasks, which took an average of 2 minutes on the original interface, this translated into an average time savings of 26 seconds per task. This is just another example of a documented case in which an alternative interface to a specific task significantly increased productivity.

But even anecdotes are useful. For example, a colleague working at a vendor company once told me (Mayhew) that she had compared customer support calls on a product for which they had recently developed and introduced a new, usability-engineered release. Calls to customer support *after* the new release were decreased by 30%. This savings greatly outweighed the cost of the usability engineering effort.

Nielsen (1993, p. 84) informally reports a case involving the interface to the installation process for an upgrade of a spreadsheet package. When the upgrade was shipped, customers needed an *average of two 20 minute calls each* to customer support to correctly install the upgrade. Support calls to the vendor cost them an average of $20 per 5 minutes to service—thus, the support cost *per customer* for this product was about $160. Unfortunately, the profit margin on the upgrade product was only $70 per customer—thus, not only was the ROI on the product eroded but the upgrade product actually *cost* the vendor nearly $100 per customer! The cost of the support costs was all a result of usability problems that probably could have been detected and fixed fairly cheaply prior to releasing the upgrade product.

In fact, some e-commerce Web sites have failed and been shut down in large part because of poor usability. (Souza, 2000, cites boo.com and levi.com as examples.) All these anecdotes can serve to strengthen specific cost-benefit analyses that make conservative assumptions regarding benefits.

In addition, experienced usability engineers can draw upon their own general experience evaluating and testing software user interfaces and their specific experiences with a particular development organization to defend benefit assumptions that they have incorporated into cost-benefit analyses. Familiarity with typical interface designs from a development organization allows the usability engineer to decide how much improvement to expect from applying usability engineering techniques in that organization. If the designers are generally untrained and inexperienced in interface design and typically design poor interfaces, the usability engineer would feel comfortable and justified defending more aggressive benefits assumptions. On the other hand, if the usability engineer knows the development organization to be quite experienced and effective in interface design, then more conservative predictions of benefits would be appropriate, on the assumption that usability engineering techniques will result in fine tuning of the interface but not radical improvements. The usability engineer can assess typical interfaces from a given development organization against well-known and accepted design principles, against past usability test results, and against the research literature to help defend specific assumptions made when estimating benefits.

In general, it is usually wise to make *very conservative* benefit assumptions for several reasons. First, any cost-benefit analysis has an intended audience, who must be convinced that benefits will most likely outweigh costs. Assumptions that are very conservative are less likely to be challenged by the relevant audience, thus increasing the likelihood of acceptance of the analysis conclusions. In addition, conservative benefits assumptions help to manage expectations. It is always better to achieve a greater benefit than was predicted in the cost-benefit analysis, than to achieve less benefit, even if the benefits still outweigh the costs. Having underestimated benefits will likely make future cost-benefit analyses more credible and more readily accepted. Also, it is important to realize that some validly predicted benefits may be canceled out by other non–usability-related changes, such as decreases in user morale and motivation, decreased system reliability, an economic downturn, new competition in the marketplace and so on (see Wilson and Rosenbaum, Chapter 8). Having made conservative benefits predictions decreases the possibility that other factors will completely wipe out any benefits that result from improved usability.

Returning to the explanation of the derivation of benefit predictions, we look at the analysis parameters used in Table 3.5, and the benefit assumptions for each benefit category given in Table 3.19. The project usability engineer selected these assumptions believing they are very conservative. In presenting the analysis, he or she cites some of the literature mentioned previously in this chapter that shows up to 20 to 30% savings in task time on one interface relative to another, and points out that this analysis assumes only a modest 4% increase in productivity. He or she also points out that most current internal applications take 1 week to train because there is a lack of consistency in the user interfaces of those applications, and users must memorize many cryptic codes and unclear error messages. Knowing this allows the usability engineer to make the case that an interface in which a small number of rules explains a wide scope of functionality, and in which the user needs to memorize less will be teachable in a significantly shorter period of time. The assumption that the new interface will eliminate one error per user per day is extremely conservative, and the usability engineer can cite internal statistics showing high typical user error rates on current internal applications. Finally, to defend the assumption about the relative cost of early versus late design changes, the project usability engineer cites a classic paper in the literature (Mantei and Teorey, 1988).

Table 3.20 shows the calculation of the total predicted benefit in each benefit category, based on parameters and assumptions. In the case of increased productivity, multiplying the number of users by days per user, transactions per day, hours saved per transaction, and hourly rate results in the total benefit given in Table 3.3 for this benefit category: $199,652.78. Benefit assumptions for the other three benefit categories can be seen in Table 3.19, calculations of total benefit predictions in each of these three other categories can be seen in Table 3.20, and these total benefits per category are added together in Table 3.3 to arrive at a total benefit in the first year alone, and a lifetime benefit over an assumed 5-year application lifetime.

6. Compare Costs to Benefits

Having calculated the costs of a particular usability engineering plan and the total benefits predicted to result from executing that plan as compared to not executing it, the next step is simply to subtract the total costs from the total

Table 3.20 Benefits Calculations for an Application for Internal Users

Increased Productivity										
No. Users		No. Days		No. Transactions		Hrs Saved/ Transaction		Hourly Rate		Total
250	×	230	×	100	×	0.001389	×	$25	=	**$199,652.78**

Decreased Errors										
No. Users		No. Days		No. Eliminated Errors		Hrs Saved per Error		Hourly Rate		Total
250	×	230	×	1.0	×	0.033333	×	$25	=	**$47,916.67**

Decreased Training						
No. Users		Hours Saved per User		Hourly Rate		Total
250	×	10	×	$25	=	**$62,500.00**

Decreased Late Design Changes Cost of Early Changes						
No. Changes		Hours per Change		Hourly Rate		Total
20	×	8	×	$175	=	**$28,000.00**

Cost of Late Changes				
Cost of Early Changes		Ratio of Late to Early Changes		Total
$28,000.00	×	4	=	**$112,000.00**

Savings of Early Changes Relative to Late Changes				
Cost of Late Changes		Cost of Early Changes		Total
$112,000.00	–	$28,000.00	=	**$84,000.00**

benefits to arrive at a net benefit. In this example, this calculation is shown in Table 3.4. The analysis predicts a clear net benefit ($168,144.44) in the first year alone, and a dramatic net benefit ($1,158,422.22) over the expected application lifetime.

Our project usability engineer's initial usability engineering plan appears to be well justified. It is a fairly aggressive plan in that it includes all Lifecycle tasks, and the most reliable and thorough techniques for each task. Given the very clear net benefit, the usability engineer would be wise to stick with this aggressive plan and submit it to project management for approval and funding.

If the net benefit had been marginal, or if there had been a net cost, then the usability engineer would be well-advised to go back and rethink the plan, scaling back to shortcut techniques for some tasks. Perhaps, for example, the usability engineer should plan to do only a shortcut User Profile by interviewing user management, a shortcut Task Analysis consisting of just a few rounds of contextual observations and interviews with users, and then do just one iterative cycle of usability testing on a complete, detailed design, to catch major flaws and be sure the predicted benefits have been achieved. Of course, this would make the predictions more risky, and suggest an even more conservative analysis.

As explained earlier, to plan the budget for a usability engineering program, it makes sense to start out by calculating the costs of the most aggressive usability engineering program that you would like to implement, including the more reliable and thorough techniques for most, if not all, Lifecycle tasks. If predicted benefits outweigh costs dramatically, as they usually will when critical parameters are favorable, then you can easily make a good argument for even the most aggressive usability engineering program, because only the most conservative claims concerning potential benefits have been made, and therefore can be defended easily.

If, however, benefits and costs in the initial calculation seem to match up fairly closely, then you might want to consider scaling back the planned usability engineering program, maybe even to just a bare-bones plan, with more shortcut techniques applied for each Lifecycle task.

To illustrate this planning strategy, consider the following two scenarios. First, revisit the example above, which involved building a system for 250 internal users. Fairly conservative assumptions were made concerning benefits: task time reduced by 5 seconds, training time reduced by 10 hours, one error eliminated per day per user at 2 minutes saved per error. Even with these conservative assumptions, the aggressive usability engineering plan was predicted to payoff in the first year, with net benefits continuing to accrue dramatically after that.

In fact, if you had made the more aggressive, yet still realistic, benefits assumptions of training time reduced by 20 hours (rather than by 10), two errors eliminated per user per day (rather than just 1), and task time reduced by 15 seconds (rather than just by 5 seconds), the benefits would have added up to $903,839.58 in the first year alone, outweighing the costs of $225,925 by $677,914.58, and to $3,683,197.92 over 5 years, outweighing the costs by $3,457,272.90. Thus, you could argue, even the most conservative assumptions predict a fairly dramatic payoff of a comprehensive usability engineering program, but the likelihood is that the payoff will be higher still.

In contrast, suppose you again started out by estimating a comprehensive usability engineering program to cost $225,925. In this case, however, suppose that there are only 50 intended users (instead of 250) performing 50 transactions per user per day (instead of 100). In this case, calculations using the original, more conservative, benefits assumptions would show a loss until well into the second year, and a 5-year lifetime net benefit of only $18,318.06.

Even though the benefits assumptions were conservative, and although a first year loss is not necessarily a bad thing, it still seems risky to make an aggressive investment that, based on conservative assumptions, really doesn't show a significant payoff even over the course of 5 years. In this case, you would want to scale back the planned usability engineering program and its associated costs. Because the benefits assumptions made were so conservative, it is likely that they will be achieved even with a minimal usability effort. Thus, the cost-benefit analysis technique can be used to "what if" in order to plan a level of usability engineering effort that is most likely to pay for itself.

3.4.2 A Commercial Application by a Vendor Company

In this section, we turn to another hypothetical example—this time, cost justification of a usability engineering plan in the case of a vendor company selling a software package. The analysis process will be very similar—it is primarily the benefit categories that will be different. In the case of a software vendor company, the primary benefit of interest is not user productivity or user errors, but increased sales and decreased customer support costs.

We will assume exactly the same usability engineering plan laid out in the previous example and presented in Table 3.2. Table 3.21 provides the summary of predicted benefits, and Table 3.22 provides the net benefit calculations. Here it can be seen that a net benefit of $361,408.33 is expected in the first year, and a net benefit of $728,075.00 is expected over a system lifetime of 3 years.

In the following sections we lay out step by step how the project usability engineer arrived at these conclusions.

Table 3.21 Expected First Year and Lifetime Benefits for a Commercial Application by a Vendor Company

Benefit Category	*Benefit Value* First Year
Increased sales	$100,000.00
Decreased customer service calls from users	$183,333.33
Decreased training	$220,000.00
Decreased late design changes	$84,000.00
Total benefit	**$587,333.33**

Benefit Category	*Benefit Value* Lifetime (3 Yrs)
Increased Sales × 1	$100,000.00
Decreased customer service calls from users × 3 yrs	$550,000.00
Decreased Training × 1 yr	$220,000.00
Decreased Late Design Changes × 1 yr	$84,000.00
Total benefit	**$954,000.00**

Table 3.22 Net Benefit for a Commercial Application by a Vendor Company

	Benefit	*Cost*	*Net Benefit*
First year	$587,333.33	$225,925.00	$361,408.33
Lifetime (3 yrs)	$954,000.00	$225,925.00	$728,075.00

1. Start with the Usability Engineering Plan

We will assume the same usability engineering plan as in the previous example, shown in Table 3.2.

2. Establish Analysis Parameters

Analysis parameters for this sample analysis are given in Table 3.23. Here it can be seen that while some parameters are the same as in the previous example

Table 3.23 Analysis Parameters for a Commercial Application by a Vendor Company

Analysis Parameters	*Values*
Current typical sales (in units)	10,000
User fully loaded hourly wage:	$25
Developer fully loaded hourly wage:	$175
Usability engineer fully loaded hourly wage:	$175
Manager fully loaded hourly wage:	$200
Customer support fully loaded hourly wage:	$50
Trainer fully loaded hourly wage:	$50
Max no. users per training class	20
Expected system lifetime (yrs):	3
Profit margin per unit	$100
Average length of customer support call (10 min expressed as hrs)	0.166667
Time per early design change (hrs):	8
Ratio of late to early design changes:	4
Usability lab:	In place

(e.g., personnel hourly wages), some are different (e.g., current typical annual sales and profit margin per unit), in part because of the different benefit categories that will be included in this analysis.

3. Calculate the Cost of Each Usability Engineering Lifecycle Task in the Usability Engineering Plan

Because all analysis parameters relative to computing costs are assumed to be the same as in the previous example, the cost calculations for each task in the plan, and thus the total cost, are the same, as shown in Table 3.2.

Note that this time the user rate of $25 per hour is not based on a typical user's fully loaded hourly wage, which it was in the previous analysis of traditional software development for internal users. Instead, the user rate is based on the assumption that test users in the case of a commercial software package will have to be recruited from the general public or from customer organizations to participate in usability engineering tasks/techniques, and that they will be paid at a rate of $25 an hour for their time.

4. Select Relevant Benefit Categories

In this example, the benefit categories differ somewhat from those in the previous example. They include the following:

- Increased sales
- Decreased customer service calls from users
- Decreased training costs
- Decreased late design changes

A wide variety of trade magazines and product review companies purchase and/or receive for free the new software releases that are sent to market. Approximately 15% (Nielsen, 1993) of reviews in microcomputer trade journals were devoted to analyzing the *user friendliness* or *usability* of new software products a decade ago—undoubtedly more are now. Many newspapers, such as the *New York Times,* the *Financial Times,* and the *Wall Street Journal* have weekly columns that evaluate software, and even general discussion of functionality in such columns often make comments on ease-of-use. Potential customers read these columns and make purchase decisions based on the evaluations, including the user interface portion. *Consumer Reports* has been including evaluations of the user interfaces to consumer products, as well as investigating their reliability and safety. Trade journals for products in the industry perform similar services for the customer.

In addition to publications, software products often undergo reviews from user groups and companies whose sole purpose is to review major expensive software products. The trend has been to include a review of the user interface as well as the functionality of the software. Later adopters call companies who have installed the product earlier for reviews.

While it is difficult to predict exactly what impact usability engineering efforts will have on product sales, it should be clear that usability is now an aspect of competitive edge. Making a conservative prediction of increased sales in your cost-benefit analysis is likely to be accepted by the relevant audience, especially if they are reminded of the facts just described.

To predict increased sales benefits, we can consider relevant market forces, such as current market share, trends of the market, and strengths and weaknesses of the competition, and then choose a conservative and realistic assumption concerning the number of new sales that could potentially be attributed to increased usability alone. Then we simply multiply this number by the known profit margin of the product.

Another significant benefit category for vendor companies is decreased customer support costs. Today's customer service operation is an extensive business. One major software vendor has so many callers on its customer hotline that the company has hired its own disk jockey to play music and give software advertisements while customers are on hold. An interface that is understandable and easy to learn will generate fewer customer requests for help and thus lower the number of customer support staff needed to handle the customer hotline. This reduction in personnel costs can be estimated as a potential benefit of applying usability engineering techniques during development.

In addition, we will assume in this example that the vendor provides training. Whether the vendor bundles the cost of training in with the application or sells the training as a separate product, it is crucial to keep the cost down to compete in the marketplace. Finally, vendors are just as interested in keeping the cost of late design changes down as any other kind of development organization, so we will include that benefit in our analysis as well.

The project usability engineer expects to increase sales by accomplishing a user interface design that has usability features that are easily demonstrable during the sales process and that will get good reviews for usability in relevant trade journals. He or she expects to decrease customer support calls by designing an interface that is more rule-based, consistent, and predictable, and that also builds off users' current knowledge and skills more effectively. Arguments regarding how he or she can decrease the costs of training and late design changes would be the same as those given in the previous example of an application for internal use.

5. *Predict Benefits*

Table 3.24 shows the basic assumptions made regarding the magnitude of the benefit in each chosen benefit category. Table 3.25 shows the benefit calculations

Table 3.24 Benefits Assumptions for a Commercial Application by a Vendor Company

Benefit Assumptions			
Increased Sales	Decreased Customer Service Calls	Decreased Training	Decreased Late Design Changes
Incr. sales by 10%	Eliminated 2 calls per customer per yr	8 hrs saved off current 2 day training	20 changes made early

Table 3.25 Benefits Calculations for a Commercial Application by a Vendor Company

Increased Sales						
Current Sales		Rate of Increase		Profit margin per unit		Total
10,000	×	0.10	×	$100.00	=	**$100,000.00**

Decreased customer service calls from users								
No. Customers		No. Calls eliminated		No. Hours saved per call		Customer support hourly rate		Total
11,000	×	2	×	0.166667	×	$50	=	**$183,333.33**

Decreased Training								
No. Customers		Max No. Customers/Class		No. Hrs Saved/Class		Trainer Hrly Rate		Total
11,000	÷	20	×	8	×	$50	=	**$220,000.00**

Decreased Late Design Changes Cost of Early Changes:						
No. Changes		Hrs/Change		Developer Hrly rate		Total
20	×	8	×	$175	=	**$28,000.00**

Cost of Late Changes:				
Cost of Early Changes		Ratio of Late to Early Changes		Total
$28,000.00	×	4	=	**$112,000.00**

Savings of Early Changes Relative to Late Changes:				
Cost of Late Changes		Cost of Early Changes		Total
$112,000.00	–	$28,000.00	=	**$84,000.00**

for each benefit category, arrived at by incorporating both analysis parameters and benefit assumptions.

The project usability engineer based the conservative assumption regarding increased sales on statistics from the marketing department regarding the current rate of lost sales opportunities attributed to usability issues. He or she based the very conservative assumption regarding decreased customer support costs on statistics from the customer support organization showing a very high rate of calls on existing products attributable to usability issues. He or she knows the typical training class on existing products is currently two days, and argues a more rule-based interface with less required rote memorization can be taught in half the time.

6. Compare Costs to Benefits

Again, Table 3.22 shows the net benefit calculations based on comparing the total benefits calculated in Table 3.21 with the total costs calculated in Table 3.2. The fairly aggressive usability engineering plan seems more than justified, even with relatively conservative benefits assumptions. An intangible benefit not explicitly factored into this analysis is the additional customer loyalty that will undoubtedly be built by providing a more user-friendly product. The usability engineer can point this out in the presentation of the analysis to the stakeholder audience.

3.4.3 An E-Commerce Site

The previous two examples illustrate a cost-benefit analysis conducted for a usability engineering plan in the context of software development projects based on traditional platforms such as Microsoft Windows. The basic analysis process is really no different for any kind of application on any sort of platform—the main difference is in the choice of benefit categories. We now present two examples of justifying a usability engineering effort on Web development projects.

In this example, imagine a Web development organization is planning to redesign an existing e-commerce site that is not producing the ROI hoped for. Traffic statistics are available from the existing site. As before, we first present the final results of a cost-benefit analysis of including a usability engineering effort in the redesign project. Then, in the steps that follow, the derivation of the final results are shown.

Table 3.26 Expected Monthly Benefits for an E-Commerce Web Site

Benefit Category	*Benefit Value per Month*
Increased buy-to-look ratio	$12,500.00
Decreased abandoned shopping carts	$12,500.00
Decreased usage of "Call Back" button	$3,125.00
Total monthly benefit	**$28,125.00**

Table 3.27 Net Benefit Calculations for an E-Commerce Web Site

Benefits/Month	*Total Cost*	*Payoff Period in Months (Cost ÷ Benefit)*	*Net Benefit* (First Year)
$28,125.00	$225,925.00	8.03	$111,575.00

Again, for simplicity's sake, we will assume the same usability engineering plan with its associated cost as shown in Table 3.2. The project usability engineer estimated that for this project the usability engineering plan will produce a new site design with the *predicted benefits every month* summarized in Table 3.26.

Comparing these benefits and costs, the project usability engineer argued that the proposed usability engineering plan will pay for itself in the first 8 months after launch, as shown in Table 3.27, with a net benefit of $111,575.00 in the first year, and monthly benefits of $28,125.00 thereafter.

Note that an alternative net benefit calculation is used in this example. Instead of estimating benefits on an annual basis and then computing a net benefit for both the first year and for the expected application lifetime, in this case we predict benefits on a monthly basis and then divide the total cost of the usability engineering plan by the total predicted monthly benefits to determine a "payoff period," that is, the point at which the predicted benefits equal the cost. This is simply an alternative way to express the net benefit. After that point, net benefits are predicted to accrue on a monthly basis in the amount of the total predicted monthly benefit.

Below is a step-by-step description of how the project usability engineer arrived at the final results.

1. Start with the Usability Engineering Plan

Again, we follow the usability engineering plan laid out in Table 3.2.

2. Establish Analysis Parameters

In the case of an *e-commerce site*, the critical parameters are usually *volume of visitors* and *profit margin per purchase*. If the profit margin per online purchase is low, then a very large number of additional purchases would have to result from usability alone for the usability engineering costs to payoff. On the other hand, if the profit margin per online purchase is high, then only a small number of increased purchases must result from improved usability to pay for the usability engineering plan. Thus, these critical parameters will directly determine how much can profitably be invested in usability.

The analysis parameters for this sample project are summarized in Table 3.28. Again, some are the same or similar to those for the previous sample analyses, while some are different because of the use of different benefit categories (e.g., current average visitors per month).

Table 3.28 Analysis Parameters for an E-Commerce Web Site

Analysis Parameters	*Values*
Current average visitors per month:	125,000
Current buy-to-look ratio:	2%
Current rate of usage of the "Call Back" button:	2%
Profit margin per unit:	$10
Average length of servicing each use of "Call Back" button (3 minutes expressed as hours)	0.050000
User fully loaded hourly wage:	$25
Developer fully loaded hourly wage:	$175
Usability engineer fully loaded hourly wage:	$175
Manager fully loaded hourly wage:	$200
Customer support fully loaded hourly wage:	$50
Usability lab:	In place

3. Calculate the Cost of Each Usability Engineering Lifecycle Task in the Usability Engineering Plan

We are again simply using the usability engineering plan and associated costs given in Table 3.2 in this sample analysis.

The hourly rate for users used in this cost estimate is $25. Note that in this sample analysis, this is not based on a typical user's fully loaded hourly wage at their job, which it was in the case of a cost justification of traditional software development for internal users, or would be in the case of an analysis for intranet development for internal users. Instead it is based on the assumption that test users in the case of an e-commerce Web site will have to be recruited from the general public to participate in usability engineering tasks or techniques, and that they will be paid at a rate of $25 an hour for their time.

4. Select Relevant Benefit Categories

In this example, the project usability engineer decided to include the following benefits:

- Increased buy-to-look ratio
- Decreased abandoned shopping carts
- Decreased use of "Call Back" button

These benefit categories were selected because of their relevance to the audience for the analysis: the business sponsors of the site. There would undoubtedly be other very real potential benefits of the usability engineering plan in this case, but these were chosen for simplicity and to make a conservative prediction of benefits (as shown later). In particular, it should be noted that the benefit of decreased late design changes—included in the previous two examples—has been omitted in this example. This is simply because its benefit cannot easily be computed on a monthly basis, which all other benefits can in this case. If the net benefit calculations had been computed as a site lifetime benefit, as in the previous examples, rather than as a payoff period and first year benefit, then it could easily have been included, increasing the overall benefit. The usability engineer can point out this and other additional but omitted potential benefit categories to the relevant audience to argue that the real net benefit is actually likely to be even larger than the one presented, which is based on very conservative assumptions.

Comparing the new site design to the existing site design, the usability engineer anticipated that in the course of redesign, the usability engineering effort would decrease abandoned shopping carts by insuring that the checkout process is clear, efficient, provides all the right information at the right time, and does not bother users with tedious entry of information they do not want or need to provide. He or she expected to improve the buy-to-look ratio by insuring that the right product information is contained on the site, and that navigation to find products is efficient and always successful. He or she also expected to decrease the use of the "Call Back" button by insuring that the information architecture matched users' expectations and by designing and validating a clear conceptual model, so that navigation of and interactions with the site are intuitively obvious. Accomplishing all these things depends on conducting the requirements analysis and testing activities in the proposed plan, as well as on applying general user interface design expertise.

5. Predict Benefits

Next the project usability engineer predicted the magnitude of each benefit that would be realized *if* the usability engineering plan (with its associated costs) is implemented. For example, he or she predicted how much *higher* the buy-to-look ratio would be on the site if it were re-engineered to be more usable than the existing site.

Benefit assumptions made in this analysis are given in Table 3.29. Benefit calculations based on the analysis parameters and these assumptions are given in Table 3.30.

The usability engineer based the benefit assumptions in this analysis on statistics available in the literature. In particular, he or she began with the often quoted average e-commerce Web site buy-to-look ratio of 2 to 3% (Sonderegger, 1998; Souza, 2000), then based the assumption that this ratio could be improved by a minimum of 2% (1% from improving the product search process, and 1% from improving the checkout process) through usability engineering tech-

Table 3.29 Benefit Assumptions for an E-Commerce Web Site[a]

Increased Buy-to-Look Ratio	*Decreased Abandoned Shopping Carts*	*Decreased Use of "Call Back" Button*
1% incr. in visitors who decide to buy (and checkout successfully)	1% incr. in visitors who have already decided to buy but who also now checkout successfully	1% decr. in visitors who use the "Call Back" button

[a] All percentages are relative to total monthly visitors.

Table 3.30 Benefit Calculations for an E-Commerce Web Site

Increased Buy-to-Look Ratio						
Current monthly visitors		Rate of increase in buyers		Profit margin per unit		Total
125,000	×	1%	×	$10	=	**$12,500.00**

Decreased Abandoned Shopping Carts						
Current monthly visitors		Rate of increase in buyers		Profit margin per unit		Total
125,000	×	1%	×	$10	=	**$12,500.00**

Decreased Usage of "“all Back" button								
Current monthly visitors		Rate of decrease in use of "Call Back" button		# Hours saved per call eliminated		Customer support hourly rate		Total
125,000	×	1%	×	0.050000	×	$50	=	**$3,125.00**

niques, based on a variety of statistics available in the literature. For example, Souza (2001) suggests that it is typical for as many as 5% of online shoppers to fail to find the product and offer they are looking for, and cites one study in which 65% of shopping attempts at a set of prominent e-commerce sites ended in failure. Sonderegger (1998) suggests that sales underperform on e-commerce sites by as much as 50% or more because of poor site usability. GVU (1999) survey data suggests that almost 50% of Web site users cannot find the information they are looking for, and that over 80% of Web shoppers have left one site for another when they had dissatisfying experiences with site usability. Souza (2001) also noted that (at the time he interviewed for his report) companies spent between $100,000 and $1,000,000 to redesign specific sites, but few had any sense of which specific design changes might have paid off.

The project usability engineer based the assumption of reduced usage of the "Call Back" button by 1% on statistics suggesting that as many as 20% of e-commerce site users typically call in to get more information (Souza, 2001.)

Most of us have experienced all these problems—difficulty finding products, difficulty checking out, and need to use a "Call Back" button to complete transactions—and would have little argument with the idea that they are typical. Given all the dramatic statistics cited, the modest assumptions made in this analysis seem very conservative indeed. And given the fact that companies typically spend a great deal of money on redesign with no process in place that can ensure improvements in usability, the notion of a highly structured and goal-oriented usability engineering process starts to make a lot of sense.

6. *Compare Costs to Benefits*

The net benefit calculations for this analysis were given in Table 3.27. In this example, it can be seen that the expected payoff period is 8 months, and that after that, benefits are predicted to continue to accrue at a rate of $28,125 per month. Because the new site is expected to have a lifetime much longer than 8 months, the project usability engineer expected the usability engineering plan to be approved based on this cost justification.

Given the very clear net benefit, the project usability engineer would be wise to stick with this fairly aggressive plan and submit it to project management for approval.

If the estimated payoff period had been long, or if there were no reasonable payoff period, then the team would be well advised to go back and rethink the plan, scaling back to shortcut techniques for certain tasks, and perhaps collapsing the design process from three to two or even one design level, to reduce the costs. However, even with very conservative benefit assumptions, a short payoff period is predicted. Benefits could very likely be even more robust than those predicted, further shortening the payoff period and increasing the ongoing accrual of benefits after the payoff period.

3.4.4 A Product Information Site

This example is based on a hypothetical scenario given in a Forrester report from June 2001 called "Get ROI from Design" (Souza, 2001). It involves an automobile manufacturing company that has put up a Web site that allows customers to get information about the features of the different models of cars they offer and options available on those cars. It allows users to configure a base model with options of their choice and get sticker price information. Users cannot purchase a car online through this Web site—it is meant to generate leads, and points users to dealerships and salespeople in their area.

Again, for simplicity's sake, we will assume the same usability engineering plan with its associated cost as is shown in Table 3.2. The project usability engineer estimated that for this project the usability engineering plan will produce a new site design with the *expected benefits every month* summarized in Table 3.31.

Table 3.31 Expected Monthly Benefits for a Product Information Web Site

Benefit Category	*Benefit Value* per Month
Increased lead generation	$37,500.00
Total monthly benefit	**$37,500.00**

Table 3.32 Net Benefit Calculations for a Product Information Web Site

Benefits/Month	*Total Cost*	*Payoff Period in Months (Cost ÷ Benefit)*	*Net Benefit (First Year)*
$37,500.00	$225,925.00	6.02	$224,075.00

Comparing these benefits and costs, the project usability engineer argued that the proposed usability engineering plan would pay for itself in the first 6 months after launch, as shown in Table 3.32, and that after that period benefits would continue to accrue at a rate of $37,500 per month. Because the new site is expected to have a lifetime of more than 6 months, the project usability engineer expected the usability engineering plan to be approved based on this cost justification.

Note that again, instead of estimating benefits on an annual basis and then computing a net benefit for both the first year and for the expected application lifetime, in this case we predict benefits on a monthly basis, and then divide the total cost of the usability engineering plan by the total predicted monthly benefits to determine a "payoff period," that is, the point at which the benefits accrue to equal the cost. After that point, net benefits are predicted to accrue on a monthly basis in the amount of the total monthly benefit each month.

1. Start with the Usability Engineering Plan

In this example, we again start with the same assumed plan as in all previous examples, presented in Table 3.2.

2. Establish Analysis Parameters

Analysis parameters for this example are presented in Table 3.33. Again, we are assuming there is an existing site with known traffic statistics, and that the project involves a redesign.

Table 3.33 Analysis Parameters for a Product Information Web Site

Analysis Parameters	*Values*
Current average visitors per month:	250,000
Current percent of visitors that result in a concrete sales lead:	1%
Current percent of leads generating a sale	10%
Profit margin per sale:	$300
User fully loaded hourly wage:	$25
Developer fully loaded hourly wage:	$175
Usability engineer fully loaded hourly wage:	$175
Manager fully loaded hourly wage:	$200
Customer support fully loaded hourly wage:	$50
Usability lab:	In place

3. Calculate the Cost of Each Usability Engineering Lifecycle Task in the Usability Engineering Plan

We will use the same cost calculations as in the previous examples, shown in Table 3.2.

4. Select Relevant Benefit Categories

Since this is a *product information site,* only certain benefit categories are of relevance to the business goals of this redesign project. The project usability engineer decides to include just one benefit category in the analysis: increased lead generation.

The project usability engineer selected this benefit category because he or she knows it will be of most relevance to the audience for the analysis: the business sponsors of the site. There may be other potential benefits of the usability engineering plan (such as decreased late design changes), but the usability engineer chose just this one for simplicity and to make a conservative prediction of benefits.

As compared to the existing site design, the usability engineer anticipated that in the course of redesign, the usability engineering effort would increase leads by ensuring that visitors can find basic information and successfully configure models with options. Achieving this will depend on conducting the requirements analysis and testing activities in the proposed plan, as well as on applying general user interface design expertise.

5. *Predict Benefits*

Next the project usability engineer predicted the magnitude of the benefit that would be realized *if* the usability engineering plan (with its associated costs) were implemented. In this case, he or she predicted how much *higher* the lead generation rate would be on the site if it were re-engineered for usability. Table 3.34 presents the benefit assumption made in the analysis, and Table 3.35 presents the benefit calculations, which incorporate both the benefit assumption and analysis parameters.

The project usability engineer makes the very conservative assumption that a better interface will result in 1% more leads each month. Traffic statistics from the current site show that considerably more than 1% of configuration attempts are currently abandoned, representing lost opportunities for leads. To support the benefits estimate, the usability engineer points out existing design flaws that might account for abandoned configuration attempts and that could be rectified in the redesign process.

6. *Compare Costs to Benefits*

Next the usability engineer compared benefits and costs to determine the payoff period. This was shown in Table 3.32.

Again, the project usability engineer's initial usability engineering plan appears to be well justified. It was an aggressive plan, in that it included all Lifecycle tasks and used the more reliable and thorough techniques for each task.

Table 3.34 Benefit Assumptions for a Product Information Web Site

Increased Lead Generation
1% incr. in visitors who generate a lead

Table 3.35 Benefit Calculations for a Product Information Web Site

Increased Lead Generation								
Current Monthly Visitors		Increase in Visitors Who Generate a Lead		Percent of Leads that Generate a Sale		Profit Margin/ Sale		Total
250,000	×	1%	×	10%	×	$300	=	**$37,500.00**

In addition, the benefit assumptions were conservative. Given the short estimated payoff period, he or she would be wise to stick with this aggressive plan and submit it to project management for approval and funding.

3.5 SUMMARY

The particularly critical value of usability to the ROI of Web sites and Web-enabled applications is illustrated by the following case study.

A contract development team was building a Web site for a client organization. The Web site was to include up-to-date drug information and was intended to be used by physicians as a substitute for the standard desk references they currently use to look up drug information such as side effects, interactions, appropriate uses, and data from clinical trials. The business model for the site was an advertising model. Physicians were expected to visit the site regularly because it offered more current and more easily found information than the published desk references (such as The Physician's Desk Reference, or PDR). The marketing plan was to have pharmaceutical companies buy advertising for their drug products on the site because the visitors to the site (physicians) represented their target market. Regular and increasing traffic from repeat visitors, and new visitors joining based on word-of-mouth amongst physicians, would drive up the value of advertising, generating a profit—and ROI—for the client.

The development team generated a prototype design that the client would use to pursue venture capital to support the full-blown development and initial launch and maintenance of the site. The client paid for this prototype development.

Once the prototype was ready, a usability engineer was brought in to design and conduct a usability test. Eight physicians were paid to fly to the development center and participate in usability testing. Several basic search tasks were designed for the physicians to perform. They were pointed to the prototype's homepage, and then left on their own to try to successfully find the drug information that was requested in the first task.

Within 45 seconds of starting their first search task, *seven out of the eight physicians gave up*, and announced, unsolicited, that the site was unusable and that if it were a real site, they would abandon it at that point and never return.

Clearly, if the site had launched as it was designed prior to this test, not only would an optimal ROI not have been realized, but in fact, the site would have failed altogether and a complete loss of the clients' investment would have resulted. If seven eighths of all visitors never returned, the Web site would not have generated enough traffic to have motivated companies to buy advertising. The entire investment would have been lost.

Instead, the test users were asked to continue with the entire test protocol, and the data generated revealed insights into the problem that was a show stopper on the first task, as well as other problems uncovered in other test tasks. The site was redesigned to eliminate the identified problems. Clearly, the usability test, which had an associated cost, was worth the investment in this case.

This true anecdote illustrates something that distinguishes Web sites from commercial software products. In a commercial software product, the buyers discover the usability problems only after they have paid for the product. Often they cannot return it once they have opened the shrink-wrapped package and installed the product. Even if it has a money-back guarantee, returning it takes some effort and they are not likely to do so, especially if there are not many alternative products on the market with noticeably greater usability.

On a Web site, on the other hand, it costs the visitor nothing to make an initial visit. On a Web site based on an advertising model, such as the one just described, the site sponsor makes nothing at all unless there is sufficient *ongoing traffic* to attract and keep advertisers. On an e-commerce site, the sponsor makes nothing at all unless the visitors actually find and successfully *purchase products*, and unless usage of the Web channel increases sales or reduces the costs of other sales channels.

A Web site is not a product and the user does not have to buy it to use it. The Web site is just a channel, like a TV show, magazine, or catalog, and if users do not find and repeatedly and successfully use the channel, the investor gets no ROI for having developed the channel. Thus, usability can make or break the ROI for a Web site even more so than for traditional software products.

It is also true that success competing in the marketplace is even more dependent on relative usability on Web sites than is the case with traditional software products or sales channels. Someone wishing to buy a book may be inclined to buy from a particular brick-and-mortar bookstore that is easy to get to, even if it is not the best bookstore around. On the other hand, if customers cannot easily find the desired book through, for example, the barnesandnoble.com Web site, they need not even get out of their chairs to shop at a competitor's site instead—for example, amazon.com. It is not enough to simply have a Web site that supports direct sales; your site must be more usable than the competition's site (as well as have equivalent or superior content), or business will be lost based on the relative usability of the selling channel alone. For example, 60% of a sample of consumers shopping for travel online stated that if they cannot find what they are looking for quickly and easily on one travel site, they will simply leave and try a competitor's site (Harteveldt, 2000).

In addition, if you are a catalog order company such as L.L. Bean or Land's End, and your product is good but your Web site is bad, customers will not use your Web site and will resort to traditional sales channels (fax, phone) instead. This will result in a poor ROI for the Web site that was intended to be justified by relatively low operating costs compared with more traditional catalog sales methods.

Site usability is the equivalent of good—or *great*—customer service. Consider a company's current level of investment in traditional customer service channels. They need to make an equivalent investment in the usability of a site meant to replace or augment traditional channels of customer service. Site usability, like good traditional customer service, ensures that customers can find what they want to buy—an obvious prerequisite to sales. Just as salespeople in stores help you find the book you want, a good user interface on a bookseller's Web site must make browsing and searching easy and successful. Usability, like good customer service, also reduces errors in business transactions and the corresponding costs to fix those errors. For example, just as a good catalog offers accurate pictures of clothing and size charts to minimize the costs of processing returns, a clothing Web site must also ensure that customers don't order clothing in the wrong size or color to minimize returns. This might entail providing multiple views of the garment and sizing charts. In addition, usability, like good customer service, motivates customers to choose to use a Web site over traditional methods of doing business, ensuring an ROI in the Web site itself. Usability helps ensure that customers will return repeatedly to a Web site, just as good salespeople help ensure that customers will return to brick-and-mortar stores—again contributing to ROI. And usability done right in initial development is always cheaper than fixing usability problems identified after launch—or going out of business.

By contrast, lack of usability on an e-commerce site is the equivalent of poor customer service. Imagine having the following telephone conversation with a human customer service representative (CSR):

CSR:	Hello. Thank you for calling XYZ Shopping; how can I help you?
Shopper:	Hello, I would like to place an order.
CSR:	Do you know the name and extension number of the order taker?
Shopper:	Of course not! I just want to order something! Can't you take my order?
CSR:	No, you need to know who to ask for.

Much later, after finally finding out how to reach the order taker, then being put on hold repeatedly, and finally completing giving the required order information, imagine that this customer has the following conversation with customer service:

Shopper: Oh, I just realized I need to make a change—can you do that?

CSR: If you don't know the name and extension of the order changer, you will just have to wait until your order arrives, then send it back and reorder . . .

Would *you* order by phone with this company again after such an experience? Unfortunately, this hypothetical interaction is analogous to many online shopping experiences. E-commerce Web sites too often make it very difficult to figure out how to place an order and make changes in midstream, and there are often long periods of being "on hold," waiting for graphics-heavy pages to download.

Thus, there are real risks associated with *not* investing in Web usability. When customers are unsatisfied with the quality of customer service on Web sites, customer loyalty erodes. Even companies with well-established brand loyalty may find this loyalty beginning to decline with the launch of an unusable Web site meant to replace traditional customer service channels. Customer dissatisfaction may result from an unacceptable learning curve to accomplish desired goals (as in the example of the physician's drug reference Web site cited earlier), from unacceptable task times necessary to accomplish desired goals (e.g., too many clicks or download times that are too long), or from an unacceptable rate of errors and confusion during task completion (e.g., abandoned shopping carts). Potential sales may be lost because customers can't find what they want to buy or get the information they need to make buy decisions. Customers may make errors in business transactions that cost time and money to rectify, and create customer dissatisfaction. For example, a vendor recently agreed to pay return shipping costs because the Web site misled one of the authors (Mayhew) into ordering an item twice. The second author (Tremaine) was charged California sales tax on a product shipped to New Jersey because the "out of state" button was placed in a nonobvious place on the screen. Both authors will think twice about using these Web sites again because of the time lost in repairing the needless errors. Customers may remain loyal but refuse to use a Web site and return to traditional methods of doing business, reducing the ROI on the Web site investment. Customers may also defect to competitor companies whose Web sites are more usable. Costly rework of a site to fix problems discovered after initial launch will also eat into the ROI.

The four cost-benefit analysis examples offered in this chapter are based on simple subsets of all actual costs and potential benefits and very simple and basic assumptions regarding the value of money over time. More complex and sophisticated analyses can be calculated (see Karat, Chapter 4). However, usually a simple and straightforward analysis of the type offered in the examples above is sufficient for the purpose of winning funding for usability engineering investments in general, or planning appropriate usability engineering programs for specific development projects.

The sample cost-justification analyses offered here suggest that it is usually fairly easy to justify a significant investment of time and money in usability engineering during the development of software applications. The framework and examples presented in this chapter and elsewhere in this volume should help you demonstrate that this is the case for your software development project, Web-based or otherwise.

(Portions of this chapter are excerpted or adapted from Mayhew and Tremaine [1994], Mayhew [1999], and Mayhew and Bias [2003]. Used with permission.)

REFERENCES

Bias, R. G., and Mayhew, D. J. (1994). *Cost Justifying Usability.* Chestnut Hill, MA: Academic Press.

Bias, R. G., Mayhew, D. J., and Upmanyu, D. (2003). Cost Justification. In J. Jacko, and A. Sears (Eds.), *The Handbook of Human-Computer Interaction.* NJ: Lawrence Erlbaum Associates.

Dray, S., and Mrazek, D. (1996). A Day in the Life of a Family: An International Ethnographic Study. In D. R. Wixon, and J. Ramey (Eds.), *Field Methods Casebook for Software Design.* New York City: John Wiley & Sons, Inc.

GVU. (1999). www.gvu.gatech.edu/user_surveys. Atlanta, GA: Georgia Institute of Technology.

Harteveldt, H. H. (2000). Travel Data Overview. Cambridge, MA: Forrester Research.

Jacobson, I., Christerson, M., Jonsson, P., and Overgaard, G. (1992). *Object-Oriented Software Engineering.* Harlow, England: Addison-Wesley.

Karat, C. M. (1989). Iterative Usability Testing of a Security Application. *Proceedings of the Human Factors Society 33rd Annual Meeting* (pp. 273–277). HFES.

Landauer, T. K. (1996). *The trouble with computers: Usefulness, usability, and productivity.* Cambridge: MIT Press.

Mantei, M. M., and Teorey, T. T. J. (1988). Cost/benefit for incorporating human factors in the software lifecycle. *ACM Communications,* 31 (4), 428–439.

Mayhew, D. J. (1999). *The Usability Engineering Lifecycle.* San Francisco: Morgan Kaufmann Publishers.

Mayhew, D. J., and Tremaine, M. M. (1994). A Basic Framework for Cost-Justifying Usability Engineering. In R. G. Bias, and D. J. Mayhew (Eds.), *Cost justifying Usability.* Boston: Academic Press.

Mayhew, D. J., and Bias, R. G. (2003). Cost-Justifying Web Usability. In J. Ratner (Ed.), *Human Factors and Web Development* (2nd ed.). NJ: Lawrence Erlbaum Associates.

Nielsen, J. (1993). *Usability Engineering.* Boston: Academic Press.

Perkins, R. (2002). Remote Usability Evaluations Using the Internet: Real Time vs. Instrumented and Automated Methods. *Proceedings of the 1st European UPA Conference* (pp. 79–86), September.

Snyder, C. (2003). *Paper Prototyping.* San Francisco: Morgan Kaufmann Publishers.

Sonderegger, P. (1998). The Age of Net Pragmatism. Cambridge: Forrester Research.

Sonderegger, P. (2000). Scenario Design. Cambridge: Forrester Research.

Souza, R. K. (2000). The Best of Retail Site Design. Cambridge: Forrester Research.

Souza, R. K. (2001). Get ROI From Design. Cambridge: Forrester Research.

Vaananen-Vainio-Mattila, K., and Ruuska, S. (2000). Designing mobile phones and communicators for consumers' needs at Nokia. In E. Bergman (Ed.), *Information Appliances and Beyond.* San Francisco: Morgan Kaufmann Publishers.

Whiteside, J., Bennett, J., and Holtzblatt, K. (1988). Usability Engineering: Our Experience and Evolution. In M. Helander (Ed.), *Handbook of Human-Computer Interaction.* Amsterdam: North-Holland.

Wixon, D., and Wilson, C. (1997). The Usability Engineering Framework for Product Design and Evaluation. In M. Helander, T. K. Landauer, and P. Prabhu (Ed.), *Handbook of Human-Computer Interaction* (2nd ed.). Amsterdam: North-Holland.

4 CHAPTER A Business Case Approach to Usability Cost Justification for the Web

Clare-Marie Karat IBM TJ Watson Research Center

4.1 INTRODUCTION

This chapter focuses on the merits of cost-benefit analysis of usability engineering with a new emphasis on Web applications and services. It covers the financial benefits resulting from usability engineering, cost-benefit analysis and its relationship to business cases, the different published cost-benefit methodologies, and anonymous case study data. Professionals working in usability engineering have recognized for some time the need to communicate the cost benefit and value of this work more effectively. More than a decade ago, Chapanis (1991) discussed the human factors profession's need to improve its communication of the content and value of human factors and usability engineering. Curtis (1992) called on usability engineers to make better business cases to executives about the effect of a product's user interface on the market and to search actively for human factors opportunities that will have visible financial impacts.

Toward achieving successful integration of human factors considerations in development projects, I have found it important to address project-management concerns regarding time, personnel, and financial resources required for human factors work, as well as the resulting economic benefits from its inclusion in projects (Karat, 1990, 1993b, 1994). In 1994, the first edition of *Cost-Justifying Usability* became a seminal book on the topic. The collection of information in the book guided human-computer interaction (HCI) professionals, and the data and methods in the book have been used and quoted around the world.

After publication, I received several e-mails and phone calls each month regarding requests for further information about the cost-justification methods.

I continued to receive correspondence each year about case study data that I was given permission to use in communication with others as long as it was de-identified to protect confidentiality. Also, I was in contact with a number of graduate students who were focusing on the cost-benefit analysis of usability engineering in their Master's and dissertation research (e.g., Johnston, 2004; Riechenauer, 2004). It was surprising to me that the frequency of requests has not diminished over the years. I thought that the financial case for usability would have become accepted in the IT field by now, but this has not yet occurred. The change in focus on Web applications, new types of technology and growing uses of technology in everyday life, the bursting of the dot.com Bubble, and the economic conditions of the past few years have required that the HCI profession continue to educate decision makers in their organizations about the value of usability. Although significant progress has been made, incorporating usability in development and understanding how to derive the best return on investment (ROI) with it is by no means standard practice in companies around the world.

In addressing cost justification of usability for the Web, many factors remain the same, and there are some new factors to consider. For example, understanding the benefits of HCI work for e-commerce sites involves calculating the value of improvements in basic measures, such as completion rates on key task flows on a Web site and increases in site traffic, and understanding the cross-channel value to the customer, the trust the customer has in the Web site and brand, and the value proposition in purchasing on the Web site for a repeat customer (Chatham, 2004; Herman and Raju, 2004; Karat *et al.*, 2003, 2004).

The costs of HCI activities related to Web applications are generally the same as those that practitioners use to calculate for more traditional development work. They include the cost of the HCI activities related to understanding customer tasks, goals, and context of use, and the costs associated with designing, developing, and testing user experiences (Karat, 1994). The goal of this chapter is to prepare HCI professionals to build business cases for investment in usability for Web applications and services and to be able to communicate the value of usability in these contexts to the internal and external customers they serve.

Cost and revenue accountability are high priorities for organizations in both the public and private sectors. The value of human factors activities in improving the usability of systems and products must be quantified in financial terms for these organizations. This analysis is relevant for products developed by an organization for internal use and for products developed for external customers. I urge human factors professionals involved with development projects to include their work's cost-benefit analysis as a standard component of communication activities. I propose that these activities become proactive; these analyses can be a part of the initial business cases for proposed projects and a part of the reviews of projects that are underway or completed.

It is also important to have the "big picture" of what it will take to influence one or more decision makers within an organization to invest in usability. Beyond the quality of the data on which the ROI calculation for usability is based, I have found that a few factors are critical to determine the effectiveness of ROI arguments. It depends first and foremost on the people involved. Even in today's world, the heart of business is the relationships among people. The person who hears the argument must have strategic business vision or must be a pragmatist who will experiment with new methods. If the usability professional is able to make a credible presentation of the ROI argument for usability and build trust with the decision maker, then the stage is set for the next critical factor. How well do the ROI arguments tie to the hierarchically ordered business goals and strategies of the organization? If the usability goals help to achieve organizational goals, or even take them a step further, then this approach demonstrates to the organization a new type of partnership that they find very valuable.

A final critical factor involves world events that are affecting the business world. One is the global marketplace and the movement of jobs and work to areas of the world where people can complete the work at lower cost. Other significant events are the movement of the "Baby Boom" generation toward retirement, the impending skills shortage, and the increasing life expectancy rates in many parts of the world. If an ROI argument can be heard by a visionary or pragmatic decision maker, will help to achieve business goals or make strategies successful, and can help the organization in its approach to a global trend, then the usability professional has hit a home run!

Let's begin with the core building block, the solid business case that the HCI professional can use to communicate the value of the usability work on a project. Please see Wilson and Rosenbaum's Chapter 8 on Internal, Social, and External ROI; Henneman's Chapter 5 on Marketing Usability; Chapter 21 on Respecting the Shareholder; and other chapters of this book related to successful ROI arguments for investment in usability.

4.2 WHY MEASURE THE COST BENEFIT OF HUMAN FACTORS?

Whether usability engineers are working on Web-centric or traditional applications, they can increase their value to organizations by providing financial data regarding impact of human factors work on organizational areas of concern.

These areas of concern may include time and cost reductions for product development, post-release revenue increases and cost reductions, and customer satisfaction and trust. Providing human factors cost-benefit information to project managers gives them a more complete and accurate basis for decisions, helps human factors professionals successfully compete for resources, and helps them become and remain a part of the critical development path. Informing management about human factors work's contribution to fiscal, organizational, and strategic goals demonstrates an understanding of those goals and illustrates the supporting role that human factors play in achieving them.

Providing human factors information about products to other groups in an organization such as education, system maintenance, and marketing is an efficient and effective way to support them in achieving their goals and helps these groups reduce the time and monetary resource necessary to complete their product-related work.

Cost-benefit data on human factors can be used by practitioners to improve the efficiency of human factors methods used in particular situations. Human factors engineers can use these data to understand the trade-offs involved in using different usability engineering techniques within project resource constraints and development schedules (Karat, 1990, 1993a; Karat *et al.*, 1992). For example, HCI practitioners can collect information about the relative value of group design walkthrough sessions with a prototype compared to individual hands-on interactive sessions with a similar prototype. Our team has found through research that it is more cost effective to use group walkthrough sessions than to use individual sessions during the early and middle portions of the development cycle. We have found that the data from the two methods have a high positive correlation and we obtain the design information at the appropriate level of detail (Karat *et al.*, 2002, 2003, 2004). We employ individual user evaluation sessions with a high-fidelity prototype or beta system when we are further in the development cycle and need more in-depth design information.

Human factors cost-benefit data (preproject estimates and postproject feedback) can guide management decisions regarding the prioritization of human factors resource allocation among various projects and decisions about the number of human factors laboratories and jobs appropriate for an organization. This feedback is also important to document unsuccessful use of human factors resource because of factors such as practitioners' impractical decisions or lack of product team commitment to usability. Organizations can use both success and failure data to improve the efficient and effective use of human factors resource.

Almost no case studies show failures in HCI development work. It is clear that in the current worldwide economic climate, it is risky to publicize any failure.

However, if an alternative approach is devised that can help HCI investments be more valuable in the future and save the organization money in the project lifecycle, then that is a powerful and productive way to bring this type of information forward. The communication, therefore, is about the greater impact that HCI work can have using one method or process over another in the described context and the resulting positive impact on organizational goals. There is no reason to throw good money after bad, and senior management recognizes and rewards improvements that help the organization reach financial and strategic goals.

4.3 USABILITY ENGINEERING IN WEB SITE AND APPLICATION LIFECYCLES

The scope of opportunity for usability engineering's contribution to the financial status of organizations and development projects can be demonstrated by discussing the application lifecycle; an analogous case may be made for human factors work in hardware development. Past research documented that the user interface is a very significant or major portion of the application code, development effort, and budget, with estimates ranging from 40 to 60% (MacIntyre *et al.*, 1990; Rosenberg, 1989; Wixon and Jones, 1992). Prior research has also documented that a majority of application lifecycle costs occur in the maintenance phase, and that a majority of maintenance activities (80%) focus on unmet or unforeseen user requirements (Martin and McClure, 1983; Pressman, 1992).

Although these data apply to traditional applications, a similar case can be made for Web applications. While it is expected that Web sites will be updated frequently, many of these updates involve straightforward changes to the content (text and graphics). Addressing usability problems that require changes to the underlying architectural and task flows on the Web site are similar in cost and time to maintenance-stage changes to other types of application software. Working on the design of a personalized user experience for a major e-commerce site, our team learned firsthand about the infrastructure, architecture, and task flow changes necessary to implement the new capability and user experience (Karat *et al.*, 2004). Several authors who contributed to this book also address these architectural issues. In summary, because a substantial amount of project lifecycle effort and resources is devoted to the project's user interface, usability engineering has the potential to make significant contributions to Web application quality and the financial success of most projects.

4.3.1 What Contributions Can Usability Engineering Demonstrate?

Organizations can reap financial benefits by including human factors work when they develop systems for organizational use and products for external customers. Inclusion of human factors work in software application and hardware development can reap both short-term and long-term benefits (Karat, 1994; Karat *et al.*, 2002, 2003, 2004 Souza, 2001a). Short-term benefits are defined as those that accrue during product development; long-term benefits are those observed in the maintenance phase after release. A crucial short-term benefit is the reduction in development cost and time. This key benefit can have a rippling effect throughout the application's lifecycle. Pressman (1992) estimates the increasing cost of a change during development as one unit during project definition, 1.5 to 6 units during project development, and 60 to 100 units during maintenance after project release. Defining user requirements, testing usability prototypes, and performing usability walkthroughs early in development can significantly reduce the cost of identifying and resolving usability problems and can save time in application development.

An application's usability and quality are interdependent. Just as the application code has defect-free objectives (for most companies the goal is the elimination of level 1 and 2 bugs, which are critical problems), product usability merits explicit objectives that can be quantified and tested. Product usability objectives can be created and measured for a range of dimensions. You can't manage what you can't measure!

Design reviews, in which requirements and the rationale for them are reviewed with key members of the development team, enable development teams to identify issues and gain consensus on goals; they are cost effective and save development time (Gilb, 1988; Pressman, 1992). Similarly, measurable usability objectives, usability design walkthroughs, and usability testing early in development also reduce development time and cost. Usability engineering can help improve the development process as well as the product itself. Streamlining the development process can save time, money and human resource by correctly identifying product requirements, improving communication and documentation of requirements, and improving the rationale for design decisions.

Scerbo (1991) and Bosert (1991) make the case that usability engineering is a part of quality functional deployment (QFD), a process used for structuring development process through a primary focus on customer requirements. Through QFD, reducing development time by one third to one half is possible. As with other methods at the disposal of the HCI practitioner, it is possible to design and use "light" versions of QFD that are adapted and extended to the context of use.

Together, the short- and long-term financial benefits of usability-engineered internal and external applications may include, for example:

- Decreased development costs and time to market
- Increases in sales, market share, and revenues
- Faster learning and improved information retention (learning before productivity)
- Increases in customer or employee productivity
- Increases in customer or employee satisfaction
- Improved decision making and effectiveness
- Space savings, equipment and consumable cost reductions
- Decreased customer or employee service and support costs
- Decreased maintenance costs
- Increases in customer or employee loyalty, trust, and retention

These variables can be illustrated through examples involving Web applications. A usable and trustworthy interface to a Web application that provides a satisfying experience to users can increase sales, market share, and revenue for an organization. Souza (2001a) has reported that 65% of online shopping attempts end in failure because users cannot find what they are looking for. The majority of these users will not attempt to purchase from the site after a failure experience. Chatham (2004) documents that more than a third of consumers say they would make more purchases online if the Web sites provided information and user control during the shopping experience that enabled them to feel the site was trustworthy. When usability engineers are involved in the design of a user interface, they can design the interaction methods and flow so that the users will be successful in completing tasks. It is critical to build trust with the users of an e-commerce site. A brand developed through brick-and-mortar channels can provide an initial level of trust for an online shopper; however, that trust must be confirmed in the online environment through the user experience. If the user has a good experience online, that experience can translate into trust and a growing relationship, increased sales, repeat business, and higher profitability for the organization (Karat *et al.*, 2002, 2003; Kobsa and Teltzrow, 2004; Teltzrow and Kobsa, 2004). The user experience can be influenced by many factors. Kobsa and Teltzrow found that a group of e-commerce customers provided with usable privacy statements on Web sites have increased trust in the site and make purchases 33% more often than a group of shoppers presented with a standard privacy policy that is difficult to understand or use. The first group was also 20% more likely to share personal information with the site.

When an organization's only channel for customer transactions is through the Web site, the usability and sense of trust that the customer has in the site is critical. The Web site must deliver a good user experience and purchasing value to attract and retain customers. The Web site's user interface represents the company's brand online. There is no safety net in the form of another channel, such as a brick-and-mortar retail store. Herman and Raju (2004) have provided examples of how user experience improvements are prioritized and launched based on business case analysis of the cost-benefit of these improvements for customers of eBay. These user interface practitioners have had significant success in working within the company's product planning process using business case justification to get approval for HCI activities. They then measure the cost benefit of the HCI improvements after launch with the same metrics that they based the business case on to determine the actual ROI of the HCI projects.

Recent research (Karat *et al.*, 2003, 2004) showed that trust in the brand of a large multinational IT company was a critical variable that would lead target customers to initially explore their online site. When target customers determined that their trust in the brand was valid on a new usability-engineered version of the site as well, the customers stated that this would lead to increased visits to the site and additional purchases. These users stated that they greatly appreciated the control they were given over their information on the user interface, and that this contributed to their trust in the brand running the site. The users said that the fact that they could be comfortable that the company would not sell their information or use it without their consent strengthened their trust in the site and would directly translate into additional purchases from that brand.

The target customers, whose jobs required them to perform tasks on the Web site every day, declared that the site did some of the steps in their jobs for them, thereby saving them time. This was of real value to them, and they stated that the site had won their business on this basis. Souza (2001a) and other researchers have identified the business value of repeat customers on a site; repeat customers spend twice as much as new users. Unpublished market research across various industries has also shown that repeat customers provide a significantly higher level of profitability to an organization because they spend more money and require a lower proportion of resource for sales and fulfilment activities.

For example, in the design of the user experience on a multimedia cultural site for entertainment, researchers determined that a usable and entertaining interface that addresses user goals would attract customers for initial visits, and these customers, experiencing the value of the site, would visit the site repeatedly (Karat *et al.*, 2002). Furthermore, this site acted as an additional channel for visits and sales in the brick-and-mortar organizations represented there. The site's affect on cross-channel sales had been a key concern of the organizations participating in it.

The finding that the site would help to generate additional sales across the channels supported the notion of an "expanding" pie rather than a "zero sum" pie of opportunity (in which the channels would cannibalize each other, as some had feared). In a zero sum game there is a fixed amount of resource to be distributed. Many brick-and-mortar retailers initially feared that consumers viewed the money they spent with a store in a zero sum manner. For example, if a customer was shopping online, he or she might spend $150 and then not shop in the brick-and-mortar store in the area. Research that others and I have completed in this area has found that this assumption is not correct. There is an expanding opportunity with customers, at least within certain ranges. Competitors may lose customer business at some point, as customers consolidate their purchasing with a smaller set of valued and reliable providers, but for any one retailer, the opportunity with a customer to expand the retailer's value and market share is open and flexible. This finding of the positive effect of a well-designed Web site on cross-channel sales and visits was supported in usability research conducted for a major American retailer known for their customer service in brick-and-mortar stores (de-identified research study). Sales revenue rose 15% overall after a usable Web site was designed and implemented for them.

The benefits of offering customers and users quality products that are highly usable and satisfying may cover other aspects, because particular benefits vary from product to product. Reducing the development time required to bring a usability-engineered product to market can result in substantial financial returns to the organization beyond the initial savings attributable to the time reduction. Increased sales, revenues, and profit margins are long-term benefits of usability-engineered products. House and Price (1991) and Conklin (1991) document the loss of 33 to 50% of a product's profit when it was brought to market 3 to 6 months late. Wixon and Jones (1992) documented a case study of a usability-engineered product with a revenue increase of 80% over the first release of the product, which was built without human factors professionals. The second release achieved revenues 60% over project projections, and customers cited the usability of the product as part of their buying decisions.

For e-commerce sites, designing usable and satisfying sites for customers through the use of HCI methods can translate in the ability to get to market ("go live") earlier and build market share, revenue, and profitability. It is prohibitively expensive to attempt to attract customers back to a Web site when they have had an unsatisfying experience on an earlier visit to the site. In contrast, if the organization invests in attracting customers to a site initially, and these customers have usable and satisfying Web site experiences and find what they want, the positive first impression can promote repeat visits by these customers.

Furthermore, Karat *et al.* (2003) and other de-identified case studies that have been provided to me or that I have been directly involved with in the last few years have shown that a multichannel organization can provide additional value to their customers through their Web site with incentives such as time savings and flexibility as compared to going to the organization's brick-and-mortar store or calling customer support for catalogue ordering. For example, on a usable and personalized Web site, a customer may elect to set up periodic replenishment orders of certain products or may set his or her profile to receive notification when new shipments of specified brands in a specific clothing size or cosmetic or jewelry type are available. In these situations, the organization can increase the customer's use of self service on the site and thereby lower operational costs while providing additional value to their customers.

It is critical in Web applications that the organization employ HCI methods to listen to the voice of the customer to understand when it is necessary to refresh the site to meet customers' changing needs and expectations and deliver value. Some e-commerce sites have brief pop-up questionnaires that aim to capture this information, some give customers the ability to provide feedback from many pages. Still others recruit feedback from different customer segments that receive different forms of compensation for providing information. Others work hard to capture information from customer-service interactions related to their Web sites. Tracking and analysing click-stream data is useful as well, although it generally needs to be augmented with customer interview or survey data to understand user goals and motivation.

The discussion of increased revenues resulting from usability-engineered products has focused thus far on applications developed for external customers. Most internal development organizations (including human factors team members) are now charging other departments for their services and products, reaping benefits from their "sales" of usability-engineered products to internal customers similar to those described for external products.

In some instances, the financial benefits to the organization variables are manifested differently depending on whether the usability-engineered product is developed for internal use or is marketed to external customers and end users.

For example, for internal development projects (for which someone in the organization is the customer for the system developed), increased user productivity and decreased user errors are direct benefits to the organization and may help the organization lower or stabilize personnel costs. This increased productivity may be the result of faster learning and better retention of the information presented through a usability-engineered application. To an organization developing applications and products for external sale, increased user productivity and decreased errors are indirect benefits in that the productivity and organizational information can be used to market the product's value to customers, thus generating higher revenues (please see Wilson and Rosenbaum's Chapter 8 for further information). The customers realize the direct benefits of these improvements. Because a rising standard of living depends on improvements in productivity (Roach, 1991), this potential benefit of usability engineering merits attention from internal and external customers.

Investment in usability engineering for the development of external applications for Web sites may provide value to both the organization hosting the site and the customers of the site. The usability-engineered application may actually help users make better decisions and in a more timely manner, as in the case of personalized applications, which simplify and streamline the identification of desired products and handle purchase and fulfilment aspects (Karat *et al.*, 2004). These types of increases in human performance through supporting, highly usable technology can result in exponential improvements in organizational effectiveness instead of incremental improvements. For example, in one de-identified case study from 2004, the inclusion of human factors in the design of an e-commerce site led to a 90% reduction in customer errors on the site because customers were able to simply and effectively locate the specific products they needed and purchase them. This improvement in productivity benefited both the hosting organization and the external customer organizations. Customer satisfaction on the site improved dramatically. There was a ripple effect on many related variables, including additional purchases from repeat visitors, lower operating costs for the customer organizations, lowered operating costs for the hosting organization, and increased profitability for both customer and hosting organizations. The speed and accuracy of the ordering process allowed customer organizations to increase their own cost effectiveness and productivity, and the same accuracy reduced support costs for the host company by reducing the number of returned products.

Usability-engineered internal development projects can result in decreased costs for training, support, service, product documentation, and personnel, and increased user satisfaction. In an anonymous case study, the investment in usability engineering allowed an organization to eliminate the development of a 10-hour education course to support the application, resulting in a savings of $140,000. The $140,000 savings did not include savings in employee time for those people who would have taken the course (Karat, 1993b).

The financial benefits of some of the variables mentioned may result from a combination of factors. For example, the improved usability of a dedicated system and increased user satisfaction may result in lower employee turnover, which represents a large financial savings to the organization. Schlesinger and Heskett (1991) cite data illustrating that the total costs of employee turnover are 1.5 times an employee's annual salary. In a published case study, Schlesinger and Heskitt (1991) state that a 10% reduction in employee turnover in two divisions of a hotel chain was worth more than the combined profits made by the two divisions.

For organizations that develop and market products for external customers, the financial benefits in the service support areas mentioned are realized in more indirect ways. For example, organizations may be able to set more competitive service rates for customers, while realizing acceptable rates of ROI. These rates may help the organization gain new customers and retain current ones.

Usability-engineered products may result in financial benefits for development organizations (direct for internal products; indirect for external products) related to space savings, equipment and consumable cost reductions, lowered risk, and improved work-process control. Space savings may result from improved task flows for usability-engineered applications and their related business processes. For example, streamlined business processes may eliminate steps and their related office space, desks, equipment and consumables. Similarly, improved just-in-time manufacturing processes in factories drastically reduce the need for warehouse space. Improved security, audit trails, and safety (lowered risk of human error) may result in decreased costs as well. (See Casey, 1998, for famous case studies of design, technology, and human error.)

Finally, decreased maintenance costs can be a large, direct financial benefit to an organization that produces usability-engineered internal and external products. Usability engineering may never be able to identify and resolve all user requirement issues during product development; however, if half of the major issues currently handled during maintenance were resolved during development, it would represent a very large cost reduction. Moreover, it would provide the organization's software maintenance employees with an opportunity to complete more rewarding and productive activities. Martin and McClure (1983) found that $20 to 30 billion was spent worldwide on maintenance. Studying backlogs of maintenance work shows that an "invisible" backlog is 167% the size of the declared backlog. Anonymous case study data from the late 1990s show that internal development organizations were spending the majority of their resources on maintenance activities and thus could not initiate development of strategic new systems. As mentioned earlier regarding Web applications, changes to the architectural flow of the application require efforts similar to maintenance changes to other kinds of application software. For example, in a de-identified 2003 case study, it was determined that the changes necessary to a large e-commerce retail site would require the majority of the IT budget, and the development of valued new capability could not be addressed until these changes were made. The needed architectural changes were estimated to require 18 months to implement. A plan was proposed to roll out the highly valued capability in 2 years. The business owner was very concerned about the changes his major competitors would make while his hands were tied for 2 years, an eternity on the Web.

This overview of usability engineering's financial benefits defines the scope of opportunity available to organizations investing in this resource. Based on the published and anonymous case study data and my 20 years of experience in software design and usability work, I believe that an organization either pays relatively little to invest in usability engineering during product development (and reaps the associated financial benefits), or pays more before, and much more after, product release to fix usability problems and associated problems left unaddressed in the product. Investment in usability engineering is generally a good business decision. I state "generally" a good business decision, because a capable HCI practitioner must have the skills to identify the key user requirements, select appropriate HCI methods to complete the work, translate the user information into good design, and work well within the organization. Another key reason I urge HCI practitioners to collect cost-benefit data on their usability work is to learn about the relative value of different methods in different situations. Simply putting HCI funding in the budget doesn't guarantee success.

I recommend that human factors cost-benefit data be included as part of a proposed project's business case and as part of the regular management review of a project underway. Investment in the usability of Web sites and applications is critical and must be analysed in the context of the organization's strategy for managing multi-channel access for customers. Web sites provide additional opportunities and complexities given multi-channel access as compared to other types of application development that must be addressed in the design to ensure the best chances of success.

The remainder of this chapter focuses on methodology for capturing and analyzing the data for a business case approach to cost-justifying usability engineering, and examples of published and anonymous case study data that use this methodology. Please note that the de-identified case study data have been provided to me over the last few years and need to remain anonymous to protect confidential data.

4.4 WHAT IS COST-BENEFIT ANALYSIS?

Cost-benefit analysis is a method of analyzing projects for investment purposes, and proceeds as follows (Burrill and Ellsworth, 1980):

1. Identify the financial value of expected project cost and benefit variables
2. Analyze the relationship between expected costs and benefits using simple or sophisticated selection techniques
3. Make the investment decision

4.5 HOW DO COST-BENEFIT ANALYSES RELATE TO BUSINESS CASES?

Business cases are a mechanism for proposing projects, tracking projects, and communicating with the organization at large. A business case includes:

1. A statement of the project purpose and tie to strategic business goals
2. A project description
3. A market analysis regarding the proposed project
4. An analysis of expected benefits and costs for "business as usual" as compared with "business as proposed"
5. Staffing and equipment requirements for the proposed project
6. A project timetable
7. An analysis of project dependencies
8. A risk analysis for the project

Organizations use business cases as a means of making investment decisions. A company or group generally allocates resources to projects that will accomplish organizational goals. These goals may include, for example, financial, social, or legal milestones. Groups are forced to make decisions about the distribution of limited resources among their many organizational goals and related benefits. Business cases provide an objective and explicit basis for these investment decisions. Competing projects are judged against each other, based on standards for ROI of resources. Other political factors may influence the decision, but the business case provides a documented statement of the case for investment.

In the case of investment in usability engineering, cost-benefit judgments have been made largely in the absence of formal cost-benefit analyses and business cases. In these instances, the data on which the decision is made and the standard being used to judge usability engineering's value are unclear at best. At the project management level, usability engineering is competing for resources against other groups who *do* have objective cost-benefit data available for management review. It is critical that usability engineers assume responsibility for collecting and providing data that will benefit our profession as well as the project teams and larger organizations that we support.

I have been asked to provide some guidance about how to obtain the necessary information to build the usability business case. In many organizations, the organization's business goals are made available to all employees. Where they are not, I encourage you to ask your management chain for them in order to provide the best possible support to the organization. Your manager may be able to summarize them at an appropriate level of abstraction if they are highly confidential. In terms of the cost estimates for business case purposes, if these data are not readily available, ask human resources personnel for general estimates, again at a level of abstraction that will be acceptable for release to you and for your purposes. These rates will need to be fully loaded with benefit and overhead costs. For other cost estimates, contact the financial officer responsible for budgets and forecasts.

I ask for references from potential customers or from someone in my management chain in order to gain access to the right individual to make the business case estimates. It is often possible to create a partnership with and provide value to these individuals by explaining your purpose and providing the data you have to date. I have been told many times by financial officers, senior management, and customers that the data I have collected is the best and most up-to-date information the organization has at that point—an additional source of value on the job. You may be asked to sign an additional non-disclosure agreement with your customer regarding these data, if your contract does not cover this area sufficiently. For internal projects and if all else fails, have a conversation with your manager about the fact that it is difficult to provide the best value to the organization without the information you need to do your job. Ask your manager to attend a meeting for you, ask your questions for you, and then provide you with acceptable summary information. It is likely that you will be allowed to contact the people to follow up with a couple of questions once the business purpose is clear, and then you have the opening to develop your own relationship with the individuals.

4.6 ASSESSING COSTS AND BENEFITS OF USABILITY ENGINEERING

In human factors cost-benefit analysis, one general approach is to compare the costs and benefits of a proposed usability-engineered application or product with those for a product developed without explicit human factors work by the organization. Alternatively, if there are several competing usability engineering proposals for a product's development, they may be contrasted with each other. Sometimes you and your team may be able to propose two different approaches, perhaps a scaled-down version and a more complete version, or there may be two very different technical approaches to review and consider. Some organizations solicit proposals for a project from competing groups. Having choices is helpful to management in that it highlights the particular value in the different approaches and the trade-offs to be considered. I have seen instances where 2 or 3 proposals are funded for the first stage, then one is selected for continued funding.

The goal of the analysis is to determine the cash value of the positive difference in a project that would result from human factors work. The first task is to identify and financially quantify each of the expected costs (e.g., human factors and programming personnel, user, and HCI activities such as interviews, usability design walkthroughs, and usability evaluations) and the benefits (e.g., lower training costs, higher sales, increased productivity) of usability engineering. Estimates of all significant human factors cost and benefit variables in the project's lifecycle are required to analyze the relationship between the two factors. In identifying and quantifying human factors costs and benefits, consider the following guidelines:

1. The variables that are relevant for the analysis will vary depending on the focus of the project and the context of its use.
2. There will be differences in the key benefit and cost variables for Web sites, applications, and other products for internal and external use (e.g., the pre-

viously described difference in the benefit of increased productivity for internal and external products).

3. Initial analyses by the HCI practitioner often overlook benefits and costs, so distribute a draft analysis of the benefit and cost variables to project team members and people with different backgrounds in your organization to gain the wisdom of their different perspectives and expertise.

After the variables have been identified, the second step is to separate the tangible and intangible benefits and costs. Tangible variables are those that can be quantified financially; intangibles such as organizational image or brand are not easily measured (see Henneman's Chapter 5, Wilson and Rosenbaum's Chapter 8, and others in this book, and Due, 1989). I recommend that the list of intangible variables be kept and referred to periodically, because methods for quantifying the variables and their relationships to financial measures may become applicable later.

The third step is to determine the value of the tangible benefits and costs. The goal is a unit or per-hour cash value of benefits and costs and estimates of the total numbers involved for use in the summary analysis. To define the financial value of the variables, establish contacts in development, maintenance, marketing, personnel, financial analysis, business planning, education, and service groups in your organization. As mentioned earlier, you may need to work through your management chain to gain access to the people and the information. Also contact users and customers for available data regarding cost-benefit calculations. For example, customers can provide information under confidential agreements about their transaction rates, customer satisfaction, successful customer search rates, abandoned shopping cart rates at particular points in the transaction process, error rates, and costs of error rates in terms of services and support. Use past data, estimates from projections, or group decision data (Galegher *et al.*, 1990) and document your sources. The social psychology method of group decisions has been found to produce reasonable estimates. The method relies on individuals with varied skills and perspectives making anonymous estimates and sharing their rational for the estimates with others. After a few iterations in which group members read and think about the other people's scores and rationales, a group of individuals will reach consensus on the estimate.

Regarding personnel data, ensure that your estimates are based on fully loaded rates (i.e., all benefits and overhead included), and account for raise projections and productivity ratios (i.e., the value used by your human resources personnel or business group to reflect that people are not productively engaged 100% of the time at work). Identify savings and costs as initial or ongoing by year. Finally, review and update your cost-benefit data periodically and as significant changes in the project occur.

4.7 BENEFIT CALCULATIONS EXAMPLES

The following are five brief examples that illustrate how to calculate the benefit of usability engineering. The examples are a mixture of actual case study data and composite cases, and come from Web and traditional application projects. There is value in understanding the logic, the fundamental aspects, and business value of both new and older examples. Those who do not understand history are doomed to repeat it.

4.7.1 Increased Sales or Revenues Resulting from Increased Completion Rates on a Web site

In a de-identified case study, a company employed usability engineering to improve the ability of its customers to find what they were looking for on the Web site. When the improved user interface was launched, the completion rate increased 15%. The projected value of the increase was:

- Most recent yearly revenue on the site = $300,000,000 per year
- Previous completion rate = 75%
- Value of increase in one percentage point in completion rate: $4,000,000 per year
- New completion rate = 90%
- Proportion of those who purchase at completion point remains steady at 10%
- Financial benefit of improved completion rate =

$$15 \times 4{,}000{,}000 = \$60{,}000{,}000 \text{ for the first year}$$

Note: This example could have included variables for year-to-year growth in visits to the site and changes in site revenue.

The figure just presented was the projected additional revenue. For this de-identified case study, the true value of the additional revenue was $55 million in the first year after the updated site went live.

4.7.2 Increased Sales and Revenue from Repeat Visits to a Web Site

In a de-identified case study, a company with a Web site invested in HCI activities to improve the user experience on the site. The HCI practitioners collected business case data from target customers on the value of the user-experience improvements to the site. The business value of the improvements in the user experience on the site were calculated as follows:

- Average amount of current transaction on the site = $150
- Proportion of visits that result in a sale (throughput) = 10%
- Number of additional purchases target customers stated they would make if user experience launched = 1 to 3 per year
- Field test data on proportion of target customers who actually make additional purchases = 0.33
- Number of target customers for user experience = 500,000
- Value of the improvement in user experience ranges from = 150 × 1 × 0.33 × 500,000 = $2,500,000 on 1 additional purchase to 150 × 3 × 0.33 × 500,000 = $7,500,000 on 3 additional purchases in a year.

Note: Again, there may be several other relevant factors in play. For example, the improvement in the user experience may attract more visitors to the site and also increase the throughput percentage. This calculation is simplified for illustrative purposes.

4.7.3 Increased Sales or Revenues

Case study data on usability engineering involvement in the copier business in the 1980s demonstrate impressive revenue increases for these products (Brown, 1991; Wasserman, 1991). This is a seminal case study with valuable business and usability points that usability professionals should know about and understand.

The initial situation was as follows:

- A large, international copier company was experiencing declining revenues and unnecessary service calls because of customer perceptions of the unreliability of their copy machines.

- Usability data that identified significant problems with the copier machines was initially ignored, rejected, or suppressed because of management's unwillingness to recognize or address the problem.
- Economic forces in the form of a severe downturn in revenues and profit and strong new competition from Japanese companies winning market share dictated the need for change, and a new focus on the social context of copier use and iterative prototyping led to product redesign.
- Revenue rose immediately when redesigned machines reached the market; a $1 billion increase over 3 years.
- The cost-benefit of usability engineering was very positive; if human factors costs are estimated at $2 million, then the cost-benefit ratio = 1 : 500
- The company was awarded the Malcolm Baldrige Award for Quality in 1989 and the lessons learned resulted in a strong commitment to human factors in this company for many years.

4.7.4 Increased User Productivity

Three iterative usability prototype tests were conducted on a system (Application 2) that is employed by users to complete 1,308,000 tasks per year (Karat, 1990). The usability work resulted in an average reduction of 9.6 minutes per task. What were the first-year benefits in increased user productivity resulting from human factors?

Reduction in time on task for Application 2 from initial design to final user interface design:

- Average savings = 9.6 minutes
- Projected first-year benefits resulting from increased user productivity:
- Tasks per year = 1,308,000
- 1,308,000 × 9.6 minutes = 209,280 hours
- 209,280 hours × Productivity ratio × Personnel costs = $6,800,000

Note: A productivity ratio is a number between 0 and 1 that documents the proportion of time that people are working productively while on the job. Human resources departments and other departments within an organization use the productivity ratio for a variety of purposes, including workload management and compensation calculations. These ratios are confidential; however, a usability engineer may be able to learn an appropriate range to use for this value in a particular organization.

4.7.5 Decreased Personnel Costs

As stated earlier, Schlesinger and Heskett's (1991) data show that total costs of employee turnover are 1.5 times the employee's annual salary. A business has 500 employees using a dedicated software system in the performance of their jobs. There is currently a 25% employee turnover rate per year. Average annual employee salary is $20,000. The user interface of the system is improved, based on user requirements, and employee turnover is reduced by 20%. What is the value of the resulting 20% reduction in employee turnover the first year?

- Previous yearly employee turnover = 125
- New yearly employee turnover = 100
- Reduction in employee turnover = 25
- Average annual employee salary = $20,000
- Value of reduction in employee turnover = 25 × 20,000 × 1.5 = $750,000

Note: This is a conservative estimate, because another benefit of the reduced turnover rate may be a reduction in the need for training as well as increased productivity from the skills and experience of the employees. A more complete assessment might include these values. This purpose of this example was to demonstrate the financial benefit of the one area of decreased personnel costs.

4.8 USABILITY ENGINEERING COSTS

The costs of usability engineering include costs for one or more of the following activities (Karat, 1990, 1991, 1993a; Karat et al., 2004; Mantei and Teorey, 1988; Mauro, 2002):

- End user requirements definition
- User profile definition
- Focus groups
- Task analysis

- Interviews and surveys of end user requirements
- Benchmark studies
- Usability objectives specification
- Field studies of end user work context
- Style guide development
- Initial design development
- Design walkthroughs with small groups of users
- Paper-and-pencil simulation testing
- Thinking-aloud studies
- Heuristic evaluations and other types of inspections
- Prototype development (low, medium, or high fidelity)
- Usability tests (laboratory or field) with individual users
- Prototype redesign
- Online tools that enable remote data collection and usability evaluation and are used in combination with individual and small group HCI work in the laboratory or field settings
- Surveys and questionnaires of users on live sites or for deployed applications

This list provides some perspective on the range of activities that may be employed in developing a usability-engineered product. In my experience as a usability engineer, I have not observed more than six of these activities completed for any one project. The usability work on a project is tailored to a project's requirements, time frame, and resources.

When calculating the costs of usability engineering activities, consider the following additional cost calculation guidelines:

1. Regarding personnel costs, include costs for all development team support, other support, or contract services. For participant costs in particular, there can be hidden costs. If you are lucky enough to be able to hire a recruiter to identify participants for you, budget at least the same amount to pay the recruiter for each participant as you do to pay the participants for their time. Participant incentives vary depending on the uniqueness of the user profile and difficulty in identifying and recruiting them. It is not unreasonable to pay a recruiter $250 per participant or to pay participants between $100 and $250 each for 2 to 3 hours of their time. If you are doing the recruiting your-

self, estimate with great care. I have found that I spend 3 days across several weeks recruiting each participant when my target user group is a select and difficult-to-find subpopulation.

2. If a permanent usability laboratory is built, costs may be prorated based on the number of usability tests to be conducted for a given period of time.

The guideline regarding usability laboratory costs applies to the purchase of new equipment as well. As an alternative, the costs of equipment and the laboratory may be included in fully loaded personnel costs as overhead expense.

4.9 COST AND COST-BENEFIT CALCULATION EXAMPLES

4.9.1 Usability Engineering Costs for Repeat Visits to a Web Site

Usability work for the de-identified study previously described included interviews with target customers and stakeholders, development of user scenarios, development of a low- and mid-fidelity prototypes, three iterations of usability sessions with target customers (laboratory design walkthroughs, field design walkthroughs, and individual usability testing of the prototype system with target customers). The work was completed across 7 months. Usability costs were as follows:

- Usability resource (fully loaded salary rate including all travel and development work for three people for 7 months) = $450,000
- Cost-benefit ratio for usability work:
  - Projected increased revenue for first year = $60,000,000
  - Cost-benefit ratio = 1 : 133

Or, stated another way,

- Return factor (cost/benefit) = 133

Note: See House and Price, 1991 for a discussion of the Return Factor.

4.10 USABILITY ENGINEERING COSTS FOR APPLICATION 2

Usability work for the internal product called Application 2 (Karat, 1990) included a benchmark test (field test), development of a high-technology prototype, three iterations of usability prototype testing (all laboratory), and redesign. The work was completed across 21 months. Usability costs were as follows:

- Usability resource = $23,300
- Participant travel = $9,750
- Test-related development work = $34,950
- Total cost for 2 years = $68,000
- Cost-benefit ratio of usability work:
- Projected first-year benefits in increased productivity = $6,800,000
- Costs of usability activities = $68,000
- Cost-benefit ratio = 1 : 100
- Return factor = 100

As it turned out, the estimates of the benefits of the HCI work were underestimated. The financial benefits in increased productivity were 10% higher than anticipated in the first year. In the second year, additional streamlining of business processes occurred that incorporated the use of the tool and lead to greater financial return on investment. The design of the tool was updated and further usability improvements were made. As part of the streamlining effort, certain job roles were eliminated and these employees were shifted into more productive work for the organization. Regular employees became responsible for handling certain work tasks for themselves and thus the usability of the tool was all the more critical to the success of operational goals for the organization.

4.11 SIMPLE COST-BENEFIT ANALYSIS TECHNIQUES

The analysis technique used to compare the quantified costs and benefits for a project may be either simple or sophisticated. One simple analysis technique called the cost-benefit ratio has been demonstrated in the preceding cost-calculation examples. To calculate a cost-benefit ratio, the benefit amount is divided by the cost amount to determine the ratio of costs to $1 of benefits. In the cost-calculation example for Application 1, $40,700 was divided by $20,700 to achieve a cost-benefit ratio of 1 : 2. For every $1 spent on usability engineering, $2 were saved. In this example, the cash benefits were for the first year of system use. Return factor is another way of reporting the same analysis of costs and benefits.

4.11.1 Payback Period

Payback period is an investment evaluation method based on determining the amount of time (e.g., year units) that it will take to generate net cash flows (cash benefits) to recover the initial investment in the project (Burrill and Ellsworth, 1980). The procedure for using the payback period analysis technique is as follows:

- Payback period is the smallest value of k so that

$$R_1 + R_2 + \cdots + R_k \geq C,$$

  where R = cash benefits in a year minus the costs in that year and C = initial development cost.

- Projects are selected for investment if the payback period is less than the organization's standard (e.g., 2–4 years depending on type of Web or traditional application).
- Competing projects are ranked on the basis of increasing payback periods; those with shorter payback periods are judged as better investments, other things being equal.

4.11.2 Payback Period Example

A project requires an initial outlay of $75,000, and net cash inflows for the project in the first 6 years are $10,000, $25,000, $20,000, $20,000, $20,000, and $10,000, respectively. The project will be withdrawn or replaced after 6 years. The organization's standard payback period for projects is 4 years. Should this project be selected for investment purposes?

Answer: According to the payback method, the answer is yes, because the sum of the net income for the first 4 years is greater than the initial investment outlay.

$$\$10{,}000 + \$25{,}000 + \$20{,}000 + \$20{,}000 \geq \$75{,}000$$

Although the method is widely used, easy to compute, and provides some control over risk exposure, it is not a measure of project profitability (e.g., cash inflows from years 5 and 6 are not included in the analysis in the previous example), and it does not adjust for the timing of cash inflows. The time value of money is the focus of sophisticated analysis techniques.

4.12 SOPHISTICATED SELECTION TECHNIQUES

4.12.1 Background

Cost-benefit techniques that adjust for the timing of cash inflows are interest-based or sophisticated selection techniques. Time-adjusted cash flow selection techniques are based on the idea that money has a time value and that, by calculating the present value of future cash inflows from a project, including an acceptable rate of return, and specifying the time period in which the cash returns are received, better investment decisions can be made (Gordon and Pinches, 1984). The concept focuses on the analysis of present and future costs and benefits for a project in terms of the present-day value of all the monies. The formula for calculating the present value of one future cash flow is as follows:

$$P = F_n(1/(1+i))^n$$

where

- P = present value
- F = future cash inflow in time period n
- i = discount rate, the minimum acceptable rate of return for investments
- n = number of time periods in years

To simplify the calculation, the expression $(1/(1+i))^n$ is called the present value interest factor (PVIF). A chart of the computed values of PVIF is found in Table 4.1. To find a value of PVIF, select the value for a particular discount rate

Table 4.1. PVIF Chart (Present Value of $1 in Period *x*)

Period	*12%*	*13%*	*14%*	*15%*	*16%*	*17%*	*18%*	*19%*
1	.8929	.8850	.8772	.8696	.8521	.8547	.8475	.8403
2	.7972	.7831	.7695	.7561	.7432	.7305	.7182	.7062
3	.7118	.6931	.6750	.6575	.6407	.6244	.6086	.5934
4	.6355	.6133	.5921	.5718	.5523	.5337	.5158	.4987
5	.5674	.5428	.5194	.4972	.4761	.4561	.4371	.4187
6	.5066	.4803	.4556	.4323	.4104	.3898	.3704	.3521
7	.4523	.4251	.3996	.3759	.3538	.3332	.3139	.2959
8	.4039	.3762	.3506	.3269	.3050	.2848	.2660	.2487
9	.3606	.3329	.3075	.2643	.2630	.2434	.2255	.2090
10	.3202	.2946	.2697	.2472	.2267	.2080	.1911	.1756

Period	*20%*	*21%*	*22%*	*23%*	*24%*	*25%*	*26%*	*27%*
1	.8333	.8270	.8197	.8130	.8065	.8000	.7937	.7874
2	.6944	.6830	.6719	.6610	.6504	.6400	.6299	.6200
3	.5787	.5645	.5507	.5374	.5245	.5120	.4999	.4882
4	.4823	.4665	.4514	.4369	.4230	.4096	.3968	.3844
5	.4019	.3860	.3700	.3552	.3411	.3277	.3149	.3027
6	.3349	.3186	.3033	.2888	.2751	.2621	.2499	.2383
7	.2791	.2633	.2486	.2348	.2218	.2097	.1983	.1877
8	.2326	.2176	.2038	.1909	.1789	.1678	.1574	.1478
9	.1938	.1799	.1670	.1552	.1443	.1342	.1249	.1164
10	.1615	.1486	.1369	.1262	.1164	.1074	.0992	.0916

Period	*28%*	*29%*	*30%*	*32%*	*34%*	*35%*
1	.7813	.7752	.7692	.7576	.7463	.7407
2	.6104	.6009	.5917	.5739	.5569	.5487
3	.4768	.4658	.4552	.4348	.4156	.4064
4	.3725	.3611	.3501	.3294	.3102	.3011
5	.2910	.2799	.2693	.2495	.2315	.2230
6	.2274	.2170	.2072	.1890	.1727	.1652
7	.1776	.1682	.1594	.1432	.1289	.1224
8	.1388	.1304	.1226	.1085	.0962	.0906
9	.1084	.1011	.0943	.0822	.0718	.0671
10	.0847	.0784	.0725	.0623	.0536	.0497

(between 12% and 35%) and a period of time (between 1 and 10 years) on the chart and read the resulting PVIF value.

4.12.2 Present Value of a Future Cash Flow Example

If a project will result in one future cash inflow of $50,000 in the second year and the discount rate is 15%, what is the present value of the future yield?

- $F = 50{,}000$
- PVIF = 0.7561
- P = 50,000 (0.7561) = $37,805.

The next step is to calculate the present value of several cash inflows across time. The present value of project benefits (cash inflows) across a number of years is calculated by adding the separate cash flows. The formula for the calculation is as follows:

$$P = F_1\ (1/(1+i))^1 + F_2\ (1/(1+i))^2 + \cdots F_n\ (1/(1+i))^n$$

where

P = present value

F = future cash flow in project years $1 - n$

i = discount rate

n = number of project years

4.12.3 Present Value of Cash Inflows Example

If the projected cash flows from a project are $50,000, $100,000, and $200,000 for years 1, 2, and 3, respectively, and the discount rate is 17%, what is the present value of the projected benefits?

- F_1 = $50,000
- F_2 = $100,000
- F_3 = $200,000

- $i = 17\%$
- $n = 3$
- $P = 50{,}000\ (0.8547) + 100{,}000\ (0.7305) + 200{,}000\ (.6244)$
- Total = 42,735 + 73,050 + 124,880 = $240,665

4.12.4 Net Present Value

Net present value (NPV) is a sophisticated selection technique. NPV is the present value of the benefits (inflows) from a project minus the present value of the project cost (outflows). The formula for NPV is as follows:

$$NPV = F_1\ (1/(1+i))^1 + F_2\ (1/(1+i))^2 + \cdots F_n\ (1/(1+i))^n - C$$

where

- F = future cash inflows in years $1 - n$
- i = discount rate
- n = number of years the project runs
- C = present value of project cost (outflows)

The procedure for using NPV to make an organizational investment decision is that a project is selected for investment if and only if the NPV is positive. Competing projects can be ranked in order of decreasing NPV if investment amounts are relatively equal. Otherwise, order can be ranked according to the profitability index (see following section).

NPV Investment Decision Example

In the previous example, the present value of the project's projected inflows for the first 3 years was $240,665. If the present value of the project's cost (outflows) is $200,000, should the project be accepted for investment? *Answer:* Yes, because the NPV is positive.

$$\begin{aligned} NPV &= 42{,}735 + 73{,}050 + 124{,}880 - 200{,}000 \\ &= 240{,}665 - 200{,}000 = \$40{,}665 \end{aligned}$$

When competing projects are of unequal sizes, compare them by means of a *profitability index*. The profitability index (PI) is simply another way of stating the relationship between the present value of project inflows and outflows. PI is calculated as follows:

$$\text{PI} = \text{Present value of project inflows} \div \text{Present value of project outflows}$$

Using data from the previous example,

$$\text{PI} = 240{,}665 \div 200{,}000 = 1.20$$

4.12.5 Internal Rate of Return

The most popular sophisticated analysis technique is the internal rate of return (IRR). IRR is closely related to NPV. IRR is the actual rate of return that an investment in a project will bring if cash inflows and outflows are as projected. Organizations set a minimum rate of return that investment in the project must achieve to be considered acceptable. If the IRR is greater or equal to the minimum return rate, the proposal may be accepted. Otherwise, the investment proposal should be rejected. To calculate IRR, solve for the value of i in the formula for NPV that will make NPV equal zero. Solve by successive approximation, graphing, or computing the value of i. Many business calculators have programs for computing the value of i. As an alternative, the next example solves for i by successive approximation.

IRR Calculation Example

In the previous example, NPV = \$40,665 when $i = 17\%$. Since NPV is positive, select a higher value for the discount rate to lower NPV towards zero. If $i = 27\%$:

$$\begin{aligned}\text{NPV} &= 50{,}000(0.7874) + 100{,}000(0.6200) + 200{,}000(0.4882) - 200{,}000\\ &= 39{,}370 + 62{,}000 + 97{,}640 - 200{,}000 = -\$990\end{aligned}$$

Since the NPV is now on the negative side, lower the discount rate to raise the NPV toward zero. If $i = 26\%$,

$$\begin{aligned}\text{NPV} &= 50{,}000(0.7937) + 100{,}000(0.6299) + 200{,}000(0.4999) - 200{,}000\\ &= 39{,}685 + 62{,}990 + 99{,}980 - 200{,}000 = \$2{,}655\end{aligned}$$

Therefore, the IRR is between 26 and 27%.

4.12.6 Further Interest-Based Selection Issues

These interest-based selection methods (e.g., NPV and IRR) can also be adjusted to take into account other factors such as (Gordon and Pinches, 1984):

- Risk: the probability of generating the expected revenues
- Interaction between projects: the extent to which an organization's products compete for the same customer dollars
- Unequal project lives: two proposed projects have different expected lifecycles (e.g., 3 years versus 5 years)
- Capital rationing: the idea that an organization has a fixed amount of money to invest in projects
- Abandonment: the point at which a project is halted
- Inflation: the effects of the expected inflation rate on the calculations
- Different discount rates: the need for higher rates for different groups of projects

See Gordon and Pinches (1984) for a detailed discussion of these topics.

There are associations among the measures that should be noted. Cost-benefit ratio and return factor express the same information; profitability index extends these measures by including the time value of money. Net present value and internal rate of return are related and provide information about different parts of the same formula.

Analysis of Changing Project Circumstances

The following exercises illustrate the use of a sophisticated analysis technique in a project with changing circumstances. Questions are presented and answers are provided after each of the four parts.

Exercise, Part 1. In this exercise, we will analyze the consequences of decisions regarding human factors work on a project and determine the overall contribution of human factors to the project's success.

A new product is under development. It will be completed in 9 months and will be on the market by the end of year 0. There will be no human factors work on the project during its development. The present value of the investment in the project for its entire lifecycle is $230,000. Projected net revenues from the sale of the product are expected to be $100,000, $100,000, and $100,000 for years 1, 2, and 3, respectively. After 3 years, the product will be replaced.

Question 1: What is the internal rate of return for this project without human factors?

Answer to Part 1:

Solution: To calculate IRR, solve for i in NPV so that NPV = 0.

$$\text{NPV} = F_1(1/(1+i))^1 + F_2(1/(1+i))^2 + \cdots F_n(1/(1+i))^n - C$$

For these exercises, work with units of 1000 to simplify cash values, and two decimal points of precision in the PVIF values. See the PVIF chart in Table 4.1. In Part 1, the values are:

F_1 = \$100,000 or 100 units

F_2 = \$100,000 or 100 units

F_3 = \$100,000 or 100 units

C = \$230,000 or 230 units

Try $i = 12\%$, where PVIF = 0.89, 0.80, and 0.71 for years 1 through 3.

$$\begin{aligned}\text{NPV} &= 100(0.89) + 100(0.80) + 100(0.71) - 230 \\ &= 89 + 80 + 71 - 230 = 10\end{aligned}$$

NPV is positive, so raise i to lower NPV towards zero. Try $i = 15\%$, where PVIF = 0.87, 0.76, and 0.66 for years 1 through 3.

$$\begin{aligned}\text{NPV} &= 100(0.87) + 100(0.76) + 100(0.66) - 230 \\ &= 87 + 76 + 66 - 230 = -1\end{aligned}$$

NPV is negative, so lower i to raise NPV towards zero. Try $i = 14\%$, where PVIF = 0.88, 0.77, and 0.68 for years 1 through 3.

$$\begin{aligned}\text{NPV} &= 100(0.88) + 100(0.77) + 100(0.68) - 230 \\ &= 88 + 77 + 68 - 230 = 3\end{aligned}$$

So IRR for the 3 years of the project is between 14% and 15%.

Exercise, Part 2. Now, a business case is made for including human factors in this project. The project manager will allocate \$50,000 for usability engineering activities over the 9-month development period. Usability activities will be included in the project plan being developed now. The project manager would like an overview of the resources and time schedules required for the usability engineering work on the project.

Question 2: What usability activities would you outline for the 9-month development period to achieve the maximum human factors benefit for the $50,000 investment in the project?

Answer to Part 2:

Solution: The human factors activities would be tailored to the specifics of the project and would include a mix of activities to accomplish human factors goals within the resource constraints and project time schedule. Trade-offs regarding resources required for, and benefits to be achieved from the use of various human factors techniques would be analyzed, and decisions would then be made. As already stated, specific activities chosen would depend on the project circumstances. In general, the human factors work would include activities to identify user issues (e.g., benchmarking, user-requirements definition, usability objectives specification, studies of user work context) and iterative usability reviews, walkthroughs, and testing of representations of the user interface (e.g., low- and high-technology prototypes or an integrated system).

Exercise, Part 3. Based on past project data, it is expected that, by including human factors in project development, the project can be delivered on the same schedule and at a present value investment cost of $250,000, including $50,000 for human factors. Including human factors in the project is expected to lower development, training, service, and maintenance costs so that the net investment cost for human factors is $20,000 ($230,000 + $50,000 for human factors − $30,000 in human factors-related project lifecycle cost reduction = $250,000 total present value of investment cost). Projected net revenues are expected to be $100,000, $200,000, and $100,000 for years 1, 2, and 3, respectively.

Question 3: What is the IRR for the project that includes usability engineering in development?

Question 4: What is the IRR on the investment in human factors on this project?

Answers to Part 3. *Question 3:* What is the IRR for the project that includes usability engineering in development?

Solution: To calculate IRR, solve for i in NPV so that NPV = 0.

$$\text{NPV} = F_1(1/(1+i))^1 + F_2(1/(1+i))^2 + \cdots F_n(1/(1+i))^n - C$$

For this exercise, work with units of 1000 to simplify cash values and two decimal points of precision in the PVIF values. For Question 3, the values are:

F_1 = \$100,000 or 100 units

F_2 = \$200,000 or 200 units

F_3 = \$100,000 or 100 units

C = \$250,000 or 250 units

Try $i = 25\%$, where PVIF = 0.80, 0.64, and 0.51 for years 1 through 3.

$$\begin{aligned} \text{NPV} &= 100(0.80) + 200(0.64) + 100(0.51) - 250 \\ &= 80 + 128 + 51 - 250 = 9 \end{aligned}$$

NPV is positive, so raise i to lower NPV towards zero. Try $i = 27\%$, where PVIF = 0.79, 0.62, and 0.49 for years 1 through 3.

$$\begin{aligned} \text{NPV} &= 100(0.79) + 200(0.62) + 100(0.49) - 250 \\ &= 79 + 124 + 49 - 250 = 2 \end{aligned}$$

NPV is positive, so raise i to lower NPV towards zero. Try $i = 28\%$, where PVIF = 0.78, 0.61, and 0.48 for years 1 through 3.

$$\begin{aligned} \text{NPV} &= 100(0.78) + 200(0.61) + 100(0.48) - 250 \\ &= 78 + 122 + 48 - 250 = -2 \end{aligned}$$

So IRR for the 3 years of the project is between 27% and 28%. The IRR for this project without human factors (14–15%) might not provide a sufficient ROI, whereas with human factors the rate of return (27–28%) would probably warrant a decision to invest in the project.

Question 4: What is the IRR on the investment in human factors on this project?

Solution: To calculate IRR, solve for i in NPV so that NPV = 0. For Question 4, the values related to human factors are:

F_1 = \$0 or 0 units

F_2 = \$100,000 or 100 units

F_3 = \$0 or 0 units

C = \$20,000 or 20 units

Try $i = 35\%$, where PVIF = 0.55 for year 2.

$$
\begin{aligned}
\text{NPV} &= 100(0.55) - 20 \\
&= 55 - 20 = 35
\end{aligned}
$$

NPV is positive, so i must be greater than 35%. The answer may be considered complete at this point since the chart ends at 35%.

If you want to know the exact value of i, solve for i so that: F_2 (PVIF)^2 = 20.

$$
\begin{aligned}
100(\text{PVIF})^2 &= 20 \\
\text{PVIF} &= 0.2 \\
(1/1+i)^2 &= 0.2 \\
(1/1+i) &= 0.45 \\
1 &= 0.45(1+i) \\
1 &= 0.45 + 0.45i \\
0.55 &= 0.45i \\
i &= 1.22 \text{ or } 122\%
\end{aligned}
$$

4.13 SUMMARY AND RECOMMENDATIONS

As you begin to try cost-benefit analysis and business case methods:

1. Try one simple test case and then expand to more involved cases.
2. Be conservative in your analysis: Use low-end estimates of benefits and high-end estimates of costs in business cases.
3. Update your cost-benefit data periodically and when major changes occur in a project or new variables are uncovered (e.g., a new module will be included that will allow for a significant boost in user productivity if it is usable, or there will be a change in the types of users employing the application).
4. Use the human factors and project cost-benefit data as feedback to improve the use of human factors resource at individual or organization-wide levels.
5. Report your cost-benefit and business case data to the appropriate audiences, including your project champion or customer, the development team, your management chain, the financial analysts or CFO, the process owner or manager of the live site or deployed application, marketing, and your

professional conferences (as part of external publications and de-identified as necessary).

This chapter explains cost-benefit analysis of usability engineering and encourages professionals in the field to consider this matter as a part of their daily work. Methods and guidelines for these analyses exist and have been used for a number of years to track the cost-benefits of usability engineering in traditional application and hardware development. These methods are now being extended to project planning and tracking of applications for Web sites to develop more accurate measures of the human factors contribution to the success of these development projects and their organizations. Usability engineering of Web applications involves new complexity in designing with an understanding of multi-channel access, trust, and privacy protection for customers and the identification and design for customer value on the Web in this context. Development teams working on either internal or external Web applications as well as other forms of software and hardware can reap many benefits for themselves, their organizations, and the organization's customers, clients, or constituents through attention to usability engineering cost-benefit data and the appropriate use of human factors resource on development projects.

4.14 A LOOK AHEAD

As I mentioned at the beginning of the chapter, the strategy for winning the inclusion of usability in a project goes beyond the quality of the ROI argument that can be developed for the business case. First and foremost, it depends on the people involved. In the technological world in which we live, the heart of business is the business relationships between people. The HCI professional must be able to make a credible presentation to the decision maker who has strategic vision or who is a pragmatist able to consider experimenting with new methods to reach desired business goals. During this presentation the HCI professional must build trust with the decision maker. The HCI professional must also be able to tie the ROI arguments to the business goals and strategies of the organization.

There may be conflicting strategies in play, and the decision about funding usability work may well be determined by a higher order priority that HCI efforts cannot address. In one case, I sat across from a "C" level executive who told me that the logic of the ROI argument for usability was very sound and he was ready to fund the effort except that he knew the effort would fail because of an organizational strategy in place at that time. He was right; he had to deal with that issue first. In another case, I spoke with a senior executive who, based on the HCI business case presented, agreed to fund the project because of the thoroughness and quality of the completed analysis and because the proposed project goal was to go beyond the current business goal for the organization and provide a new competitive advantage in the marketplace. Our team had taken their organizational goals one step further and this showed the organization a whole new level of partnership that they found very valuable.

A final critical factor involves world events that are affecting the business world. Trends regarding the global workforce, the limited free time that families have together, the pervasive nature of technology, the movement of the Baby Boom generation toward retirement, the impending skills shortage, and increasing life expectancy rates in many parts of the world are influencing organizational business strategies, and the products and services that organizations offer.

In light of these factors, I recommend that you start by building a business case for the HCI work for a project you want to advocate. Understand its relationship to organizational business goals and strategies and be able to explain the impact that the HCI work will have on the project, and the impact the project itself may have in achieving business objectives. Listen and adjust your HCI plans and help to guide the project as you learn about new issues and requirements. Determine whether the HCI work addresses an emerging trend that can increase its value to the organization. Use this chapter and the rest of this book as a reference guide and a toolkit to facilitate your communication of the value of your work. Go for it!

REFERENCES

Bosert, J. L. (1991). *Quality Functional Deployment: A Practitioner's Approach.* New York: ASQC Quality Press.

Brown, J. S. (1991). Research that reinvents the corporation. *Harvard Business Review,* 69 (1), 102–111.

Burrill, C., and Ellsworth, L. (1980). *Modern Project Management: Foundations for Quality and Productivity* (pp. 209–223). New York: Burrill-Ellsworth.

Casey, S. (1998). *Set Phasers on Stun: And Other True Tales of Design, Technology, and Human Error.* Santa Barbara, CA: Aegean Publishing Company.

Chapanis, A. (1991). To communicate the human factors message, you have to know what the message is and how to communicate it. *Human Factors Society Bulletin,* 34 (11), 1–4.

Chatham, B. (2004). Online privacy concerns: More than hype. *Forrester Report.* Cambridge, MA: Forrester Research, Inc.

Conklin, P. (1991). Bringing usability effectively into product development. *Human-Computer Interface Design: Success Cases, Emerging Methods, and Real-World Context* (July 24–26). Boulder, CO.

Curtis, B. (1992). Carving a niche in the organizational chart. *IEEE Software,* 9 (1), 78–79.

Due, R. T. (1989). Determining economic feasibility: Four cost/benefit analysis methods. *Journal of Information System Management,* 6 (4), 14–19.

Galegher, J., Kraut, R. E., and Egido, C. (1990). *Intellectual Teamwork: Social and Technological Foundations of Cooperative Work.* New Jersey: Erlbaum Associates.

Gilb, T. (1988). *Principles of Software Engineering Management.* Reading, MA: Addison Wesley.

Gordon, L. A., and Pinches, G. E. (1984). *Improving Capital Budgeting: A Decision Support System Approach.* Reading, MA: Addison-Wesley.

Johnston, M. (2004). Personal communication about her thesis at the City University of London on the return on investment in usability.

Herman, J., and Raju, S. (2004). *The Business of User Experience.* Presentation in Seminar on People, Computers, and Design at Stanford University (Feb 20).

House, C. H., and Price, R. L. (1991). The return map: Tracking product teams. *Harvard Business Review,* 69 (1), 92–100.

Karat, C. (1990). Cost-benefit analysis of usability engineering techniques. *Proceedings of the Human Factors Society* (pp. 839–843). Orlando, FL.

Karat, C. (1991). Cost-benefit and business case analysis of usability engineering. Tutorial presented at the *ACM SIGCHI Conference on Human Factors in Computing Systems* (April 28–May 2). New Orleans, LA.

Karat, C. (1993a). Usability engineering in dollars and cents. *IEEE Software,* 10 (3), 88–89.

Karat, C. (1993b). Cost-benefit and business case analysis of usability engineering. Tutorial presented at the *ACM SIGCHI Conference on Human Factors in Computing Systems* (April 24–29). Amsterdam.

Karat, C. (1994). A business case approach to usability. In R. G. Bias and D. J. Mayhew (Eds.), *Cost-justifying Usability* (pp. 45–70). New York: Academic Press.

Karat, C., Blom, J., and Karat, J. (Eds.) (2004). *Designing Personalized User Experiences in eCommerce.* Amsterdam: Kluwer.

Karat, C., Brodie, C., Karat, J., Vergo, J., and Alpert, S. (2003). Personalizing the User Experience on ibm.com. In K. Vredenburg (Ed.), *IBM Systems Journal,* 42 (4), 686–701.

Karat, C., Campbell, R., and Fiegel, T. (1992). Comparison of empirical testing and walkthrough methods in user interface evaluation. *Proceedings of CHI '92 Human Factors in Computing Systems* (pp. 397–404), May 3–7. Monterey, California.

Karat, C., Karat, J., Vergo, J., Pinhanez, C., Riecken, D., and Cofino, T. (2002). That's entertainment! Designing streaming, multimedia web experiences. *International Journal of Human-Computer Interaction*, 369–385.

Kobsa, A., and Teltzrow, M. (2004). Contextualized Communication of Privacy Practices and Personalization Benefits: Impacts on Users' Data Sharing and Purchase Behavior. To appear in *Proceedings of the Workshop on Privacy Enhancing Technologies*, Springer Verlag, Lecutre Notes in Computer Science. Toronto, Canada.

MacIntyre, F., Estep, K. W., and Sieburth, J. M. (1990). Cost of user-friendly programming. *Journal of Forth Application and Research*, 6 (2), 103–115.

Mantei, M. M., and Teorey, T. J. (1988). Cost/benefit analysis for incorporating human factors in the software lifecycle. *Communications of the ACM*, 31 (4), 428–439.

Martin, J., and McClure, C. (1983). *Software Maintenance: The Problem and Its Solution.* New Jersey: Prentice-Hall.

Mauro, C. (2002). *Professional Usability Testing and Return on Investment as It Applies to User Interface Design for Web-Based Products and Services.* New York: MauroNewMedia.

Pressman, R. S. (1992). *Software Engineering: A Practitioner's Approach.* New York: McGraw-Hill.

Riechenauer, A. (2004). Personal communication about his dissertation at the University of Regensburg, Germany on usability and information architecture of Web sites, including an analysis of the cost-benefit of usability work in this area.

Roach, S. S. (1991). Services under siege—the restructuring imperative. *Harvard Business Review*, 69 (5), 82–91.

Rosenberg, D. (1989). A cost benefit analysis for corporate user interface standards: What price to pay for a consistent look and feel? *Coordinating User Interfaces for Consistency* (J. Nielsen ed.). Boston: Academic Press.

Scerbo, M. W. (1991). Usability engineering approach to software quality. *Annual Quality Congress Transactions*, 45, 726–733.

Schlesinger, L. A., and Heskett, J. L. (1991). The service driven service company. *Harvard Business Review*, 69 (5), 71–81.

Souza, R. (2001a). Get ROI from Design. *Forrester Report.* Cambridge, MA.

Souza, R. (2001b). How to Measure What Matters. *Forrester Report.* Cambridge, MA.

Teltzrow, M., and Kobsa, A. (2004). Impacts of user privacy preferences on personalized systems. In C. Karat, J. Blom, and J. Karat (Eds.), *Designing Personalized User Experiences in eCommerce.* Amsterdam: Kluwer Academic Publishers.

Wasserman, A. (1991). Can research reinvent the corporation? *Harvard Business Review*, 69 (2), 164–175.

Wixon, D., and Jones, S. (1992). Usability for fun and profit: A case study of the design of DEC RALLY version 2. Internal Report, Digital Equipment Corporation.

5 Marketing Usability

CHAPTER

Richard L. Henneman Internet Security Systems

5.1 INTRODUCTION

Justifying why what they do matters is a common activity for usability practitioners. Other chapters in this book can assist in developing such rationale; however, this chapter does not. Starting with the premise that usability matters, this chapter provides some thoughts on how to convince others not only that it does but also that they should exchange something of value (often money) to get it. Here, usability practitioners learn how to market what they do.

This marketing need is common for usability engineers working across a range of organizational contexts. "Big name" usability consultancies with large staffs seek to attract high-end clients through sophisticated marketing strategies based on the reputation of their principals. Lone usability contractors may market their services by word-of-mouth based on their expertise in the usability issues of a particular domain. Internal usability groups in corporations may adopt a strategy of nurturing close relationships with the developers with whom they work daily, while gaining strategic support from top management. In each of these situations, the usability professional must put together a plan for convincing one or more people of the benefit of usability engineering. Questions that should be answered in each of these situations include the following:

- Who is my customer?
- Who are the major stakeholders?
- What problem am I trying to solve for the customer?
- What is my proposed solution?
- How will this solution benefit the customer? How will it benefit other stakeholders? How will it benefit me?

- How will I promote my solution?
- What will my solution cost?
- How will I deliver the solution?
- Who is my competition? What are their strengths and weaknesses?
- What else might my client spend his or her resources on?

The extent to which a usability organization develops well-informed answers to these questions will influence its success. This chapter presents factors to consider when answering these questions. First, however, the remainder of this introductory section lays the groundwork for what follows by defining terminology, describing a survey of usability professionals who market their services, and providing a brief introduction to the field of marketing.

5.1.1 Some Notes on Terminology

One of the characteristics of the usability field is the lack of a common understanding of various terms. To minimize confusion, definitions of some terms used in this chapter follow.

- *Usability services* encompass any activities performed to improve the experience of use of an application or Web site (e.g., minimizing task performance time and number of errors, decreasing time to learn, improving overall level of satisfaction). These activities may include needs-finding analysis, interface design, information architecture, expert evaluation, and user testing.
- *Usability engineer/practitioner/professional* refers to an individual who provides usability services. These terms are used interchangeably and broadly, encompassing human factors engineers, information architects, and user interface designers. Note that this list excludes individuals who may specialize in pure graphic design or software programming.
- *Usability group/consultant/firm/organization* refers to one or more usability professionals working to provide usability services to a customer or client.
- *Customer* or *clients* are the recipients of usability services performed by usability engineers who may be part of a usability group. The customer or client may be in the usability group's parent organization (in the case of an internal usability group) or external to it. Note that the client is almost always not an end user.

Marketing usability will be considered from three organizational perspectives. Each of these organizational situations has particular characteristics that require different marketing approaches.

- An *internal group (IG)* exists under the umbrella of a larger company and is typically responsible for the usability of that company's products.
- A *lone consultant (LC)* is an independent consultant who mostly works on his or her own for other companies.
- An *external group (EG)* is a group of consultants that provides consulting services to other companies. The external group might be large enough to have a support staff responsible for administrative and marketing functions.

5.1.2 Survey of Usability Practitioners

The idea for this chapter had its genesis in my own experiences while working over the past 15 years as a usability practitioner in organizations that typify the three situations described previously—IG, LC, and EG. In each environment, marketing was a significant part of my job responsibility.

To provide some validation for these marketing usability experiences, a survey was distributed to 21 usability engineers who were involved in such efforts around the United States. Along with background questions, respondents were asked to describe their target market, how they promoted their services, the type of services they provided, who their competition was, and suggestions for how to market usability services more effectively. The goal was not to conduct a comprehensive survey, but instead to get a sense for whether my experiences could be generalized.

The survey was distributed to consultants who represented a cross-section of usability professionals in the organizational contexts described previously: IG, LC, and EG. Of the 21 distributed, 13 were returned: 6 from IGs, 3 from LCs, and 4 from EGs. Responses from the questionnaire are used throughout this chapter to offer guidance relative to marketing tools and techniques other practitioners have found useful. The appendix to this chapter contains all the responses for three of the questions on the survey.

5.1.3 Marketing 101

It is not the purpose of this chapter to summarize the extensive popular and academic literature that explores the world of marketing. However, it is worthwhile to note a few basic concepts that provide a framework for discussing approaches to marketing usability. Kotler, Hayes, and Bloom (2002) describe the "Seven P's of Marketing"—factors that define an organization's marketing plan. The way in which usability service organizations implicitly or explicitly address these Seven P's define their marketing strategy. Organizations combine these marketing tools to obtain the results they want in their target markets. The Seven P's include:

- **Product:** The set of services the usability organization offers to its customers.
- **Price:** The compensation the usability organization receives from the client in exchange for its services.
- **Promotion:** Those activities that communicate the benefits of the service with the goal of persuading the customer to purchase it.
- **Place:** The physical location where the usability organization delivers the service.
- **Physical evidence:** The tangible clues that provide an image of the quality of the service received, such as the design and quality of furniture in the office, business cards, and report covers.
- **Processes:** The methods used by the usability group to deliver services.
- **People:** The individuals who comprise the usability group.

This chapter will focus on the first three Ps: Product, Price and Promotion, each of which is especially relevant to marketing usability services. (The remaining four, Place, Physical Evidence, Processes, and People, are important but outside the scope of this chapter.) Implications for each of the organizational types (IG, LC, and EG) will be presented. Where appropriate, supporting evidence from the survey will be provided. First, however, considerations for identifying the most appropriate target market for usability services are addressed in the next section.

5.2 THE MARKET

The first step in developing a marketing plan is to research the internal and external factors that may influence what an organization can and should be doing. Many of the research techniques usability engineers already know are appropriate here, from conducting interviews and surveys to reviewing and analyzing trends in the trade press, university research, and on-line data bases (e.g., LEXIS-NEXIS).

5.2.1 Internal Factors

One of the early important steps in developing a market plan is for the usability consulting organization to perform a realistic assessment of its strengths and weaknesses. Answers to such questions are especially important in determining what services the organization can offer. Factors to consider include the following:

- **People:** What skills do the people in the organization have? In what areas are they truly expert? In what domains do they have experience? What is their attitude toward their work and the organization?
- **Client base/contacts:** Who are the organization's potential customers? In what other organizations does the usability group have contacts? What prior clients can be approached for more work? What is the organization's reputation? Will previous clients serve as references?
- **Facilities:** If the group has an office, where is it located? Does it have sufficient space for growth? Does the group have a usability lab, or access to one? How does the office appear to visitors? Is there suitable space to meet with clients?
- **Equipment:** What sort of computer support does the organization have? Is it unable to or restricted from delivering some types of service because of limitations in the infrastructure?
- **Finances:** What is the state of the organization's finances? What is the group's revenue requirement? Is the organization willing to invest now with the possibility of future gains?

Internal factors are a major determinant of what an organization can or should do. A survey from one LC said, "Almost all of my work is doing usability testing. Although I say I can do anything, that is what I'm good at."

5.2.2 External Factors

In addition to internal factors, the usability organization should also have a good understanding of the external factors that will shape the marketing plan. Kottler, *et al.* (2002) identify three categories of environmental factors that will affect the organization:

- The *public environment* includes groups that may have an impact on the usability organization, such as clients, the usability community, the media, and the general public. For example, increasing awareness of usability as reported in the general media has certainly contributed to a growth in demand of usability services. For internal consultants, Mayhew (1999) suggests several environmental factors that could be exploited to advance the cause of usability: a high-visibility disaster, a perception of competition and market demand, a desire to address general business goals, a need for an objective means of resolving conflicts, a powerful internal advocate, and the need for education.

- *Macro forces* include economic, political, or technological forces that will affect the usability organization. For example, the bursting of the technology "bubble" in 2000–2001 changed the market considerably for usability consultancies. Also, legislation requiring that certain types of systems be made accessible to people with disabilities opened the doors for a new type of service offering from usability organizations.
- *Competition* includes other organizations that may be competing for the same business. Factors to consider regarding competition are explored below.

To position usability services correctly in the marketplace, consultants must have a good understanding of their competition. As the usability field matures, the number and variety of potential competitors is increasing. Some of the competitors are in the same line of work and pursue the same type of customers. Other competitors are nontraditional in the sense that they may not have any formal usability training, yet customers might view them as offering viable solutions. Respondents to the survey listed their competitors, which included:

- Other consulting firms
- Independent usability consultants
- Design firms
- Internal usability groups
- Other internal groups (with no usability training)
- Not doing the work at all

It is valuable to evaluate each competitor's strengths and weaknesses, especially relative to those of the usability group. What are the backgrounds of the principals? What services does the organization offer? What do they charge? What are their limitations? What is their perceived competitive advantage? Various sources exist to research competitors, including company Web sites, customers, publications, and conferences, as well as other marketing material distributed by the competitor.

5.2.3 Identifying the Market

The first question usability professionals are taught to ask is, "Who is the user?" Similarly, the first question usability professionals should ask when developing a marketing plan is, "Who is my customer?" By understanding who the customer is, the consultant can develop a marketing plan that matches that customer's needs.

The types of target market vary among IG, EG, and LC. The customers for external groups and lone consultants can be classified along several dimensions. Respondents to the survey mentioned the following segments they target:

- Geography (e.g., "the Northeast")
- Industry (e.g., "the telecommunications industry," "the government")
- Technology (e.g., "companies that produce software")
- Size (e.g., "medium to large corporate clients")

Most of the respondents, however, qualified their responses by saying they do not explicitly segment the marketplace. A typical comment was, "I'll do anything someone pays me to do."

For IGs, the question of target market differs. Here, the key dimension is organizational function. Bloomer and Croft (1997) identify several market segments that comprise the IG's target market: senior management, potential allies, developers, clients, and users. Another important segment is front-line management, including product managers and engineering directors. For example, although senior management may ultimately decide at what level the IG will be funded, if the development team or front-line management has not bought into having the usability group work on the project, the IG's efforts will be ineffective (see Wilson and Rosenbaum, Chapter 8, for a related discussion). Thus, IGs must base their decisions regarding target market on a careful reading of their own corporate environment. Survey respondents for IGs indicated that when they must prioritize their projects, they target high-profile, high-potential revenue projects.

5.3 PRODUCT

The "product" is the set of services the usability group makes available to the market. In an ideal world, the market needs and the organization capabilities determine the appropriate mix of services. Judging from responses to the survey, most usability organizations offer services that can be classified as follows:

- **Research:** Includes needs-finding analysis, requirements gathering and prioritizing, cognitive modeling, literature reviews, and reviews of competitive products
- **User interface design:** Includes Web site design, traditional software user interface design and development, graphic design, prototyping, and storyboards
- **Usability evaluation:** Includes expert evaluation, heuristic evaluations, and usability testing
- **Training:** Includes short courses, tutorials, lectures, workshops, and lunchtime seminars
- **Strategy:** Includes strategic design planning, consulting with management, and process improvements

Survey respondents were asked to list their services and estimate the percent of their income derived from that service. Because they were using their own terminology to describe their services, I have assigned the responses to one of the five preceding categories. Table 5.1 lists the summarized responses.

Based on these data, several observations concerning service mix can be made. First, the IGs tend to focus more of their effort on design than the external groups. This result is not surprising, as the IGs are more likely to be integrated into product development efforts within their companies. Moreover, because they are dedicated to one particular company, they are more likely to have the domain expertise necessary for good design. Organizational boundaries circumscribe the functions they can perform, however. Note, though, that the small sample size for each group may call into question the appropriateness of making such observations.

Table 5.1 Average Percent of Total Revenue/Budget Derived from Service Categories

	Research	*Design*	*Evaluation*	*Training*	*Strategy*
IG (Internal Group)	7%	61%	33%	—	—
LC (Lone Consultant)	33%	40%	19%	1%	6%
EG (External Group)	10%	30%	53%	5%	2%

The LCs and EGs tend to have a broader range of services, encompassing more research, training, and strategy consulting. The EGs appear to spend more time performing evaluation activities, while the LCs have a more even distribution of time across research, design, and evaluation. This could be because, once established, an LC becomes viewed as a "trusted adviser" who can perform a range of tasks for the client organization.

Several of the external usability groups noted that their strategy was to "get in the door" by offering a relatively inexpensive heuristic evaluation of an application and then expanding this initial activity into high-valued design activities.

5.4 PRICING

One of the most difficult and important elements of marketing usability services is deciding how much to charge the customer. If consultants charge too little they may be overwhelmed with work. (It is also possible that charging too little could actually decrease demand, as customers may associate higher price with higher quality. Such a pricing approach, called premium or prestige pricing, may also attract high-end clients who want to authenticate their success or status by hiring a high-profile consultant.) On the other hand, if the usability organization charges too much, it may find itself with too much free time. Of course, either of these approaches might be valid, depending on the organization's goals. Pricing should be based on what the market will bear; however, understanding what the market will bear requires an understanding of the needs and resources of the customer, the usability organization, and the competition. Some factors that may influence the price point the usability organization selects are as follows:

- **Competition:** Are there competitors for this project? What are they charging? Are they already working with the customer on another project?
- **Consultant:** Is this customer/project important to the organization's long-term strategy? Is the organization in great need of work? Is this work that would improve, extend, or develop a skill set in the organization?
- **Customer:** What is the customer's available budget for this project? Could this work lead to other work? What are the long-term prospects for working with this customer?

Kotler *et al.* (2002) identify four alternative fee-setting objectives. The first, current profit maximization, results in the highest possible profit levels by setting price to maximize the difference between the consultant's expected revenue and the cost of delivering the service. The second, market penetration, typically results in the firm charging fees lower than what the competition charges. This approach makes sense if the market is sensitive to differences in pricing or if the consultant needs experience to attract and retain customers. In the third approach, market skimming, the consultant seeks to make a large amount of profit from a small number of clients by charging very high fees. This approach makes sense when the market associates high fees with high quality or for independent contractors who do not wish to add personnel. In the final approach, satisfaction bases, the consultant charges just enough to cover costs, to live comfortably, and to enjoy a reputation for quality work at a reasonable price.

5.4.1 Pricing Models

Several valid pricing models exist for usability services. In the *fixed-price* model, the consultant provides the customer with a firm price for his or her service based upon the level of effort required or its perceived value. Unless the consultant is able to create a very solid estimate or use a phased approach, this type of pricing model can be risky. If the organization or consultant estimates low, the effective hourly rate may drop significantly; if the organization estimates high, the customer may be reluctant to engage the consultant in the future. Fixed-price models often have a winner and a loser. Nevertheless, customers often prefer fixed-priced fees because they know how much to budget.

In a *time-and-materials* pricing model, the consultant and the customer agree upon an hourly rate. The consultant keeps track of his or her time spent working on the project and then submits a bill at project milestones. Consultants often prefer this pricing approach because it minimizes their risk. Even if unanticipated events or delays arise during the project, the consultant is still compensated for his or her efforts.

Another pricing model that deserves mention is a *performance-based* model. In this model, the consultant receives a percentage of revenue that results from sales of the product or cost savings associated with an improved process. The obvious benefit of this approach is that if the product is successful, the economic gain could be substantial. If the product is not successful, the chances for payment are slim. This model may be most appropriate when working with a startup or a company with limited resources.

A related model is an *equity-based* model, in which the consultant receives an equity-based stake in the company in exchange for services.

These models can be combined for a client and/or combined with various fee-setting tactics, including offering discounts for clients who purchase a large volume of services, premiums for specialized services, or varying rates for different types of interaction (e.g., a lower cost for transcribing audiotapes of interviews, but more for actually conducting the interviews).

Both internal and external survey respondents indicate that they are primarily using a time-and-materials approach, although some well-defined services (e.g., heuristic evaluation) may be offered as a fixed-price service. Hourly rates for the external groups ranged from $85 per hour to $250 per hour. This broad range suggests that some very different pricing strategies are at work within this sample of external consultants. Four of the internal usability groups do not recover their costs when they work with product teams, while the other two work on a time-and-materials basis. Working internally on a time-and-materials basis holds some unique price-setting challenges. For example, sometimes the parent organization may dictate an internal transfer price that may or may not work to the usability group's advantage. If the price is too high, the internal customer could even look for an external consultant to meet its needs more economically. One IG reported that it worked for both internal and external clients. It was required to work at a lower rate internally than what it could get working for external clients, significantly complicating its pricing strategy.

5.5 PROMOTION

The target market learns about what the usability consultant can do for them by means of promotion. Promotion encompasses an organization's marketing communications—the messages sent from the consultant through any medium to the target market. Thus, marketing communications represent a complex set of activities that include not only advertising and sales but also such things as attracting potential employees and establishing or enhancing the firm's brand image. All the messages an organization delivers should work together consistently. At one level, this might mean ensuring the corporate logo is used consistently; at another level, this might mean that the wording used in brochures, presentations, the Web site, and advertising needs to be consistent with the organization's marketing strategy.

Maister provides three principles that should guide the approach the usability organization takes toward marketing promotion. The first is "the raspberry jam rule"—the wider you spread it, the thinner it gets (Weinberg, 1985, as quoted in Maister, 1997). For promotion to be effective, it should be targeted and not spread widely. It makes a lot more sense to spend time and money promoting services to a small number of likely prospects than to a large number of unlikely prospects.

The second principle is that marketing works better when it demonstrates than when it asserts. A usability organization that *shows* prospects what they can do is much more effective than an organization that simply *tells* prospects what they can do, as in brochures, direct mail, or cold calls. Five of the six IGs that responded to the survey said that one of their most effective approaches to creating interest in their services is to conduct a usability test and invite potential internal customers to observe. This approach has some inherent risk, in that if the product does well in the test, the customer may decide that the consultant's services are not needed.

The third principle says that "in-person" promotional tactics should be given preference over the written word. As Maister (1997) points out, the concept of mass marketing is foreign to professional services, in which clients are acquired one at a time based on a highly personal dialogue between the provider and the client. The goal is to get clients talking about their problems. The sooner this can occur, the better.

5.5.1 Promotional Tactics

Survey respondents identified six promotional tactics upon which they depend to develop business. Looking first at the external consultants, the following tactics are listed in order of frequency mentioned: repeat business, word of mouth, tutorials, industry event speeches, publishing, and public relations (e.g., getting quoted in the media as an "expert" source).

The least expensive business to acquire is repeat business, although sometimes significant effort must be spent maintaining the relationship with the client. Word of mouth is also inexpensive and desirable; the problem with it, however, is that in times of economic instability, this tactic is really outside the consultant's control.

Maister (1997) mentions the remaining four (tutorials, speeches, publishing, and public relations) in his list of marketing tactics. Based on his experience in advising professional services firms, Maister lists these tactics in order of descending effectiveness.

- Seminars (small-scale)
- Speeches at client industry meetings

- Articles in client-oriented (trade) press
- Proprietary research (self-funded)
- Community/civic activities
- Networking with potential referral sources
- Newsletters
- Publicity
- Brochures
- Seminars (ballroom scale)
- Direct mail
- Cold calls
- Sponsorship of cultural/sports events
- Advertising
- Video brochures

Maister argues that the tactics near the top of the list adhere to his principles, while the ones toward the bottom do not. Unfortunately, the ones toward the bottom often tend to be the easiest ones to produce.

5.5.2 The Internet

The Internet has developed into an essential tool for promoting usability services (although it probably would not appear toward the top of the Maister's list). At the very least, customers expect that a company will have a Web site on which basic information about it can be found, such as a description of services, a list and/or biographies of the principles, contact information, a list of clients, and some work samples. Some Web sites also provide a secure area to support client-consultant interaction. Intended or not, customers will look at the organization's Web site as an example of the group's work. Because customers will probably be unable to distinguish between the quality of the information design, the graphic design, and the implementation, it is essential that the group's Web site reflect principles of good user interface design.

An important consideration is a customer's ability to find the organization's Web site. Even if the site is an accurate reflection of the organization's capabilities, it will be ineffective if potential customers are unable to locate it. Strategies for ensuring that customers can find the site include the following:

- Select a memorable URL.
- Include the URL on all communications from the organization.
- Work with other organizations to place a link to the site on their sites and vice versa.
- Ensure the site is ranked highly by search engines.

To ensure that users visit a usability group's site often, provide meaningful content that is updated on a regular basis. A good example of a site that does is this Jakob Nielsen's site, www.useit.com. On this site, Nielsen provides access to a biweekly column, "Alertbox," which should be of interest to the site's audience (presumably other usability professionals and potential clients). Columns from 1995 to the current issue are archived and searchable for free. Moreover, the design of the site supports many of the principles articulated by Nielsen within the site's content. Thus, not only are users encouraged to return to the site (helped further by the meaningful URL, www.useit.com) but the site reminds the user of Nielsen's expertise.

Other sites that illustrate many of these same principles include Creative Good (www.creativegood.com) and Human Factors International (www.humanfactors.com). In both of these cases, not only is original content updated on a regular basis but old content is archived for future use, thereby drawing customers to the site on a regular basis.

5.5.3 Internal Promotion

IGs require a different approach to promotion. An important factor that will affect the choice of promotion type is how the internal group must account for its expenses. If the internal consultants are responsible for recovering their costs via a budget transfer from the product groups, the promotion could be similar to that required by an external consultant. If the internal consultant's expenses do not have to be recovered, a different promotion approach is necessary.

Approaches to promoting usability within organizations have been well documented. Chapters in this book provide several perspectives on techniques for internal marketing efforts, such as Chapter 8 by Wilson and Rosenbaum. Wiklund (1994) contains several chapters describing approaches to promoting usability within a company. Kuniavsky (2003), Norman (1998), and Cooper (1999) describe some of the organizational hurdles facing human-centered development. Finally, in a chapter in the first edition of this book, Mayhew and Bias (1994), provide eight "success factors" that could be applied to both internal and external promotions:

- Establish credibility.
- Communicate effectively.
- Get "buy-in."
- Be an engineer, not an artist.
- Produce well-defined work products.
- Manage expectations.
- Clarify value added.
- Conduct usability tests whenever possible.

As Siegel (2003) points out, effective promotion of usability requires a good understanding of the context in which key decision-makers are operating. Although a rational cost-benefit analysis may clearly indicate why it would make sense to invest in usability, a number of other factors may influence the decision, such as a real outlay of money now versus hypothetical future savings, an internal reward structure that values current over future performance, and difficulty in attributing product performance to actual usability changes. Siegel goes on to identify several specific concepts that can make usability promotion more effective, including taking a conservative approach to analyzing the cost-benefit tradeoff, addressing hard-to-measure impacts of usability that are not captured in the cost-benefit analysis, and tailoring the promotion to the specific concerns of the decision-maker.

5.6 CONCLUSIONS

This chapter has considered the activity of marketing usability services from several perspectives, including that of the IG, LC, and EG. Each of these groups has characteristics that call for differing marketing approaches. Successful marketing requires the usability organization initially to identify its target market(s) based on an assessment of its own strengths and weaknesses, the needs of the marketplace, and its competition. In response to this assessment, the usability organization can develop a marketing strategy comprised of the services it offers (product), the amount it charges (price), and the tactics it uses to communicate with its target market (promotion), among other factors. A usability organization will be more or less successful depending on the decisions it makes about its marketing strategy.

This approach should, of course, be applied to any organization that markets its professional services. However, usability engineers face a different set of challenges than those faced by many other professional service organizations, such as lawyers, doctors, or accountants. The market for those types of professional service already has an appreciation for the way in which the service can help meet one or more of its needs. Attorneys apply their knowledge of the law to solve legal problems. Physicians apply their knowledge of the human body to cure disease or alleviate pain. Accountants apply their knowledge of the tax code to prepare tax filings. In each case, the marketplace already has a good understanding of what its needs are. In the case of usability professionals, however, the target market may not realize it has a problem. For example, one of the respondents to the survey wrote: "Systems are developed without usability professionals all the time that work well. Users often blame themselves, not the design of the product, and succeed by working a little harder. People learn to use the difficult systems, and the products are considered 'successful.' Showing the incremental value becomes challenging; changing something already fielded becomes harder to justify." Thus, usability professionals marketing their services not only have to convince potential customers that they can meet a need, but they also have to demonstrate that a need exists. These characteristics of marketing usability services make the activity particularly challenging. Cost-benefit arguments need to be made not only to customers who must decide whether or not to invest current resources to reap potential future usability-related benefits, but also within the usability organization itself as it evaluates the merits of alternative marketing strategies.

ACKNOWLEDGMENTS

I want to thank the 13 usability professionals who took the time out of their busy schedules to complete the survey described in this chapter. I would also like to thank Dr. Janet Fath, long-time usability consultant (IC) and wife (also long-time), for reviewing this chapter.

REFERENCES

Bloomer, S., and Croft, R. (November + December 1997). "Pitching Usability to Your Organization." *Interactions*, v. 4, n. 6, 18–26.

Cooper, A. (1999). *The Inmates Are Running the Asylum.* Indianapolis: SAMS.

Kotler, P., Hayes, T., and Bloom, P. (2002). *Marketing Professional Services.* New Jersey: Prentice-Hall.

Kuniavsky, M. (2003). *Observing the User Experience.* San Francisco: Elsevier Science.

Maister, D. H. (1997). *Managing the Professional Service Firm.* New York: Simon and Schuster.

Mayhew, D. J. (September + October 1999). "Strategic Development of the Usability Engineering Function." *Interactions,* v. 6, n. 5, 27–33.

Mayhew, D. J., and Bias, R. G. (1994). "Organizational Inhibitors and Facilitators." In R. G. Bias and D. J. Mayhew (Eds.), *Cost Justifying Usability* (pp. 287–318). Boston: Academic Press.

Norman, D. A. (1998). *The Invisible Computer.* Cambridge: The MIT Press.

Siegel, D. A. (May–June 2003). "The Business Case for User-Centered Design: Increasing Your Power of Persuasion." *Interactions,* v. 10, n. 3, 30–36.

Weinberg, G. M. (1985). *The Secrets of Consulting: A Guide to Giving and Getting Advice Successfully.* New York: Dorset House Publishing.

Wiklund, M. E. (Ed.) (1994). *Usability in Practice: How Companies Develop User-Friendly Products.* Boston: Academic Press.

APPENDIX: SURVEY QUOTES

Survey participants were asked several open-ended questions regarding their experience in and recommendations for marketing usability. Many of these comments have been integrated into the chapter, but for the sake of completeness, all are included here. References to specific companies have been removed.

Do You Have Any Advice for Someone Interested in Marketing Usability Services?

Internal Groups

The biggest challenge is straddling the line between adding value and just delaying shipment of the product to make it perfect.

This needs to be a dedicated effort, not something that gets done when there is a lull.

Start to do heuristic evaluations for free, or without being requested to do so. That was one of the best ways for our small group to get exposure. Initiate guerrilla usability tactics within the building. We built our own prototypes and asked the admins in our building to take 5 minutes to evaluate them. They were happy to help, and we could take anecdotal evidence back to project teams.

Once usability lab testing becomes available, invite anyone and everyone to attend sessions. Sometimes our observation rooms are crowded, but all of the groups that we work with love to observe and are anxious to continue with research.

Educate upper management and key executives on the value of usability and get their buy-in!

Lone Consultants

Get involved in local and national events focused on your area of consultation. This supports networking among professionals in highly related fields. If you're a "smaller" consultant, also get involved in local community activities where others can learn of your expertise.

No. Do not consider myself to have any real marketing skill—have been lucky to have established a market and a reputation long before there was any competition to speak of.

Provide easy-to-understand, cost-benefit analyses in your area of consultation.

Create a top-notch Web site. Many companies, especially larger ones, now want to see a Web site; it seems to indicate you're a serious professional. A Web site is also useful in general for describing your skills and what you can accomplish for a client.

Provide training courses. Training courses are a great way to make contact with people who may become clients when they realize the usability work is more difficult than they thought it would be.

Network with other usability professionals. Team up. Share information and leads.

External Groups

The job is tedious, exhausting, and unending. Be patient.

Be prepared to do a lot of education regarding benefits, particularly business related benefits.

Yes—please hurry. Actually, I STILL believe (with perhaps no supporting data) that the way to do it is with cost-benefit data.

What Do You Think Are the Biggest Challenges with Respect to Marketing Usability Services?

Internal Groups

Find out what the company culture values and then add value (perceived or actual) through your services. Areas in which usability can add value can include reduced time to ship, reductions in the number of quality issues and less tech calls.

Many of our projects start with doing splash screen graphics. In doing the graphics, we point out to the Engineers or PMs how their product can be even better if they made specific recommended changes. The changes should be small enough to make a difference but never large enough to greatly impact the shipping schedule. As the relationship develops, the Engineers will involve you more and more up front in the specification of the products.

Knowing the technical aspects of how devices and software works helps a lot as well. If you know the details of how to implement your suggestions, you can easily determine the feasibility of the changes before you suggest them, thereby avoiding making unreasonable changes that will just alienate you from the engineers and PMs.

Keeping up with the organizational changes that keep happening.

Being able to justify the benefits without sounding self-serving.

Getting to the right people in an organization is probably a challenge. People more often look for designers and producers than usability testers. They assume testing is testing is testing and don't differentiate usability testing from other forms.

You often don't have a comparison to use; organizations don't often have baseline data upon which to show improvements. Compounding this, it is impractical to have two identical developments, one with usability services and one without.

Systems are developed without usability professionals all the time that work well. Users often blame themselves, not the design of the product, and succeed by working a little harder. People learn to use the difficult systems and the products are considered "successful." Showing the incremental value becomes challenging; changing something already fielded becomes harder to justify.

Determining where a usability team should fit within the organization. I am not convinced we are in the right organization.

Lone Consultants

Helping the huge untapped market to realize they need usability. I virtually always work for people who have already figured out they need usability and are just choosing the best consultant. There is a huge untapped market out there of organizations who need us but don't know it. I would love to know how to wake them up and have them join the market!

Convincing potential clients that usability is important relative to all of the other issues they're confronting.

Finding potential clients.

All of the time and effort that can be required for talking with potential clients, preparing proposals, even doing preliminary work—time and effort that often does NOT result in paid work. Determining when to do the unpaid prep work and when not to do it is a major judgment call. I've learned a lot of interesting content by preparing for projects that never came through!

Identifying the right people to target at a potential client.

Getting repeat business.

Staying current with new technologies.

External Groups

Clients don't consider usability a key factor. They don't understand usability. Other things are creeping in, like branding, usefulness, appeal that water down usability's impact.

Everyone and his/her dog—and ESPECIALLY every software developer and his/her perhaps software-developer dog—thinks that because he/she is a user, his/her intuitions about what is and is not usable might have some FREAKIN' value. Alas, they are wrong.

What Do You Think Are the Biggest Opportunities with Respect to Marketing Usability Services?

Internal Groups

Defining very tangible ways in which usability professionals can add value to the development cycle.

Potential for growth if handled well.

The majority of 3rd party vendors we have worked with have not had their own usability experts. We have found many of these vendors to be very open to receiving our feedback, and many have incorporated it into their products.

Probably going after people like me who supplement their internal organizations with consultants.

Lone Consultants

Convincing potential clients that usability is important relative to all of the other issues they're confronting—demonstrating the benefits relative to the cost.

Moving usability from being a niche market to something in mainstream development.

Not having to justify your existence.

External Groups

Usability for mobile device products and services.

The state of the (usability) art is SO modest and behind the times, that any purse-holder who gets it, who invests just a little dang money in systematic, professional, user-centered design, is likely to beat his/her competitors.

6 CHAPTER Valuing Usability for Startups

David Crow Brant Street Industries

6.1 INTRODUCTION

Could user-centered design have saved us from the dot-com crash? Probably not. The dot-com crash resulted from a complex relationship of many contributing factors. Poor customer experiences contributed to the eventual decline of many dot-com companies, but they were not the sole reason. Many of the dot-com companies had the perfect solution to a problem that did not exist or for a market where no customers existed. The abundance of venture capital, the rush to bring products to market, and the focus on building wealth—these were bad situations that allowed startups to spend and grow rather than to "learn to execute properly, listen to their customers and build their businesses" (Perkins and Perkins, 2001, p. 83). The usability of many products and services provided by dot-com firms could be questioned, but it was only one of the factors that contributed to the eventual downfall of so many. Building a business that can operate profitably is the goal for any successful venture; it lessens or eliminates the need for further use of investment cash (Gurley, 2003). To do this, startups need to build products and services that their customers will buy and use. The quantity of purchase orders for the products eventually determines whether the company will be successful (Bell, 1991). Donoghue (2002) posits that the ultimate success measures for the user experience will be financial. For many startups these financial measures may be irrelevant early in their corporate development. Startups and investors focus on different corporate and product development metrics during the different stages of startup. By understanding these metrics, we can understand both where user-centered design fits and the value it brings at each stage of startup development.

The Web has made the user experience one of the critical success factors for new ventures. As user experience professionals, we need to understand the business value and impact of our work (see Chapter 22). By analyzing the organizational values of the entrepreneurial and venture-capital communities we can begin to understand why usability was not perceived as being important during startup development. The Bell-Mason Diagnostic provides a tool for user-centered design professionals to understand the competing priorities for startups during each stage of development. In addition, user-centered methods provide entrepreneurs and venture capitalists with a tool to better understand their customers and their customers' behavior. The combination of these approaches can help venture capitalists manage risk and entrepreneurs build more successful startups.

6.2 NOTES ON DATA ABOUT STARTUPS

Gathering data about the role of usability at a startup is extremely difficult. Much of the data used comes from companies that have gone public. This is because companies are required to disclose corporate information when they file with the Securities and Exchange Commission for public sale of stock. Most startups are not public companies; they are private enterprises with no legal requirements to share or disclose business practices or financial information. Unsuccessful startups tend to evaporate; the employees and investors do not want to discuss their failures. Even less incentive exists for successful startups to discuss their investment or product development practices. A significant risk for successful startups is that they may disclose a design practice, business decision, or technology that may reduce their success in the market by giving corporate secrets to potential competitors. Much of the analysis is based upon my personal experience working with software startups from 1999 to 2001.

6.3 RESISTANCE TO USABILITY IN STARTUPS

Even with overwhelming evidence that user-centered design practices can help a company develop more successful products (Mauro, 2002), the adoption of usability practices in many startups has been difficult. The importance and value of the customer experience have not been lost on startups; many startups invested heavily in the user experience and the design of their software and sites (Mauro, 2002). However, many Internet startups spent more than 300 times as much money on advertising as they spent on usability (Nielsen, 1999). The focus was on unique, interesting, and compelling visual design to bring new visitors to their sites with very little effort to convert these visitors to customers. Online retailers and content providers measured success in terms of bringing users to their sites; this was measured through page-views and unique Web server hits (Trueman *et al.*, 2000). Advertising was used to keep a steady stream of "eyeballs" (Stone, 2000) coming to Web sites. Bell (2003) discussed the fundamental reasons for the failure of dot-coms including the following:

- The myth of advertising—although the "first-mover" advantage is important in many markets, too many dot-coms put undue faith in the belief that advertising would get them market share quickly
- Inappropriate business models
- Selling at a loss

Many early startups had a business model in which time-to-market and a focus on increasing valuation shifted the focus away from the traditional product development, market identification, and customer sales efforts. The concentration on these different values reduced the emphasis on profitability and why customers buy.

6.4 DIFFERENT VALUES DURING THE INTERNET BUBBLE

The Internet Bubble years of 1994 to 2000 were a unique period. The intersection of unprecedented amounts of investment capital and an explosion of new Internet-related business ideas fostered a large number of entrepreneurs and venture capitalists, who were not focused on building companies and products that would last. "Built to flip" was a model that many startups followed: no need to build a company of enduring value, just pull together a good story to implement a rough draft of an idea and, bingo, instant wealth occurred (Collins, 2000). The built-to-flip model changed the focus of new ventures from the identification of a product or technology, the evaluation of a market and competition, and the plan to execute, to a "gold rush" to create immediate wealth, which outstripped the need to build lasting value through products and companies (Table 6.1).

The race to create new companies and new products and to bring new ideas to market was referred to as "Net Speed." Collins (2000) summarizes the concept as "Develop a good idea, raise venture capital, grow rapidly, and then go public or sell out." The pressure to do it fast outweighed the need to build lasting value. The boom years were a time when companies would rush to market to take advantage of the market opportunity for an exit event (an acquisition or initial

Table 6.1 Ten Largest First Day Stock Price Increases 1973 to 2002

	Company	*Date of IPO*	*Opening Day Percentage Increase in Price*
1	VA Linux	Dec. 9, 1999	697.5%
2	Globe.com	Nov. 13, 1998	606.0%
3	Foundry Networks	Sept. 28, 1999	525.0%
4	Webmethods	Feb. 11, 2000	507.5%
5	Free Markets	Dec. 10, 1999	483.3%
6	Cobalt Networks	Nov. 05, 1999	482.0%
7	MarketWatch.com	Jan. 15, 1999	474.0%
8	Akamai Technologies	Oct. 29, 1999	458.0%
9	Cacheflow	Nov. 19, 1999	426.5%
10	Sycamore Networks	Oct. 22, 1999	386.0%

From Ritter, 2002.

public offering [IPO]). The market created more personal wealth for the founders and investors rather than a profitable business or a sustainable organization (Table 6.2).

The focus for many entrepreneurs during the dot-com boom years then became building companies that could be "flipped" quickly to many investors, thus building value and wealth for the initial investors rather than building lasting products or services. In 1999, 117 IPOs doubled in value on their first day (Ritter, 2002) compared with only 34 IPOs doubling on their first day during the prior 24 years combined. Creating a company and quickly moving it to an IPO was a way to generate personal wealth for founders; for instance, after the Netscape IPO, James Clark (formerly of Netscape) had a personal net worth that was estimated to be $550 million. (Steinert-Threlkeld, 1995; Lewis, 1999).

Table 6.2 IPO Data 1997 to 2000

	1997	*1998*	*1999*	*2000*
Number of deals	390	247	286	406
Total proceeds (billions)	$52	$45	$93	$97
Average deal size (millions)	$133	$181	$191	$240

From Renaissance Capital, 2000.

Development of a company with the potential to quickly generate a large return on investment for the founders and venture capitalists became increasingly important. The markets are extremely efficient at identifying participants that do not produce real results (Collins, 2000).

6.5 TIME-TO-MARKET AND THE "FIRST-MOVER" ADVANTAGE

During the dot-com boom being first mattered. Being first was perceived as having a significant advantage. Many entrepreneurs raced their ideas and products to market with the goal of being the "first-mover"—the company that enters a particular market segment before any competitors. Being the first-mover in a marketplace provided companies advantages including the following (Lieberman, 2002):

- **Proprietary technology**—The development of superior technology provides a temporary advantage to a firm by keeping it ahead of competitors; however, in the absence of patents the ability to sustain this advantage for an extended period of time is questionable.
- **Preemption of scarce resources**—Early entrants into a market are able to preempt superior physical assets, a better domain name, and positions in the customer perceptual space. This includes the ability to build brand recognition and name recognition in their market space.
- **Customer switching costs**—Early entrants enjoy greater opportunities to capture customers through switching costs. For software products that require a large initial investment by the buyer (e.g., enterprise resource planning systems) switching costs arise from the fixed-cost nature of the initial investment and incentives to maintain compatibility over time.
- **Network effects**—Network effects arise when the value of a product or service to a given user increases with the number of other users. The first entrant into the market has the opportunity to exploit the network effect; for example, the successful coordination of buyers and sellers has allowed eBay to become the dominant Internet auction site.

If it were only as simple as being first! Later entrants are able to leverage the first-mover's investments and may benefit by waiting until key technological and market uncertainties have been resolved. Lieberman (2002) concludes that there are significant first-mover advantages for "pioneers in environments with network effects, and firms with key patents." Even for firms with these key attributes, on the Internet the effect of the first-mover advantage was not as great as many entrepreneurs and investors expected.

Weiss (2000) provides historical information that market pioneers fail—47% of companies that are first to market fail. Only 11% maintain a market leader position several years later. First-movers have to educate the market. It can be extremely difficult for first-movers with a new technology offering; customers often do not understand revolutionary concepts. Although many of the products and services developed during the dot-com boom were not difficult for customers to understand, it did take time for customers to develop the trust and confidence in the novel services and products. It was important for early entrants in the Internet market to invest heavily in advertising to develop brand recognition and to develop customer trust (Leiberman, 2002). Although successful companies such as Amazon and eBay invested heavily to build brand recognition and customer trust, spending heavily on advertising did not guarantee brand recognition, trust, or corporate success. Online retailers Pets.com and eToys were very well-recognized online brands that failed to leverage their first-mover advantages (Lorek, 2002).

Webvan invested more than $1.2 billion in building warehouses in 26 cities to enable customers to buy groceries online (Reingold, 2004). For Webvan, the problem wasn't that customers didn't want to buy groceries online, it was just that the cost of the technical infrastructure was too great and the time necessary to change customer behavior was too long. Webvan could not generate enough revenue to overcome the rate at which it was spending money.

6.6 FOCUS ON VALUATION—WHAT'S A STARTUP WORTH?

Determining the value of a startup company is more complex than determining the valuation of an established business (Levine, 2001). Generally accepted business valuation approaches often leave investors with little insight or meaningful information about the value of a startup. Most startups, such as the dot-com companies, have limited operating histories, have never generated a profit, and project significant growth. This combination of characteristics makes it difficult for investors to establish the fair-market value (valuation) of the company. The difficulty in assessing the valuation of the dot-com startups coupled with an abundance of capital to be invested allowed companies to create a false sense of importance and value. Built-to-flip companies changed their customer focus from the end-users of the products and services to the venture capitalists and "new-economy" journalists. Venture capitalists are early adopters and visionaries (Winblad and Roizen, 2004). They are looking for technologies that provide companies with an advantage against competitors.

Moore (1999) offered the idea that the core goal of a new product or service is a business goal, not a technology goal, and it involves taking a quantum leap forward in how business is conducted in an industry or by customers. Visionaries are not looking for an improvement; they are looking for a fundamental breakthrough. A key quality of visionaries is that they are in a hurry. Does this sound familiar? We're back to a self-fulfilling cycle of looking for a breakthrough. The breakthrough turned out to be built to flip—it was a new business model that focused on being first, and increasing valuations at each funding round that was the revolutionary idea. For startups to grow and become successful they need to deliver products and services that are valuable to their customers and provide a concrete return on investment for their customers and investors. With startups treating venture capitalists and journalists as customers, the easiest way to provide a return on investment during the dot-com boom years was to look for an exit strategy that increased value for existing shareholders and investors.

Venture capitalists are looking for technologies—technologies that provide companies with an advantage, technology that is defensible (needs expertise to replicate and can be patented), and technology that affords differentiation in the marketplace and provides a possible exit strategy (Winblad and Roizen, 2004). After the Bubble burst, the potential customers and market segments for most startups changed from the innovators and early adopters to the early majority (Mayfield, 2002). This is the "chasm" (Fig. 6.1) described by Moore (1999), where organizations have to move from marketing and selling to the mass market—the pragmatists. The change in the market segment of potential customers for startups decreased the tolerance for technology risk, investment, and strategic advantage for customers. It changed the metrics and values that corporations used to evaluate potential technology purchases (investments). However, you can't sell a technology to customers; customers buy solutions (or maybe products). If no one understands the technology, that's good; if no one understands the application, that's bad.

6.7 SUCCESSFUL STARTUPS VALUE THE USER EXPERIENCE

Usability offers startups a sustainable competitive advantage. Successful firms such as Amazon have understood that the user experience is a core value for

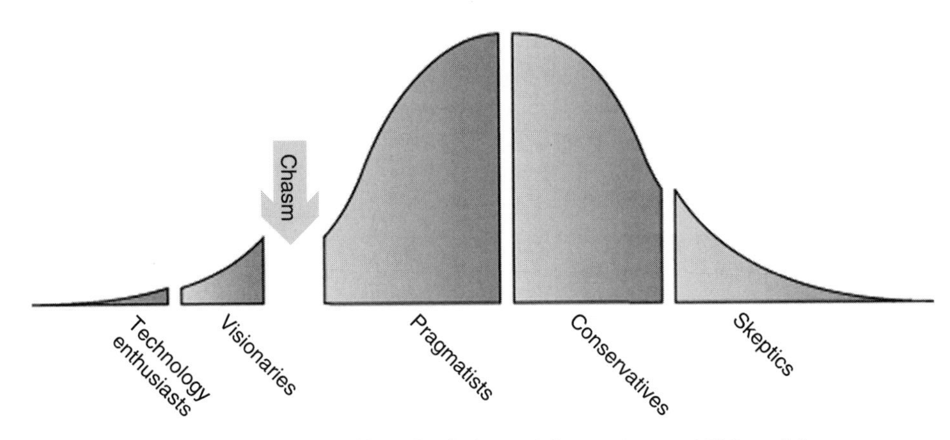

FIGURE 6.1 The technology adoption lifecycle. (Adapted from Moore, 1999 and from *Wikipedia*, en.wikipedia.org/wiki/Crossing_the_Chasm.)

their organization (Hurst, 2002b). At Amazon, the importance of the customer experience was championed at the highest levels—Jeff Bezos, founder and CEO, has promoted the importance of understanding and improving the customer experience. At Amazon.com everything starts with the customer, says Bezos, "we want to start with a customer problem and then invent to a solution. That's how we approach everything we do" (*BusinessWeek*, 2004). Google has built one of the most successful brands and business models on the Internet. Google opened in September 1998 with an initial investment of almost $1 million (Google Inc., n.d.) and has risen to be the leading search provider on the Internet with an estimated market valuation of approximately $23 billion (Google Inc., 2004; Joyce and Shread, 2004). Drummond (as cited in Hornik, 2003) points to four key factors for Google's success: technology, business model innovation, brand, and focus on the user experience. The focus on user experience has allowed Google to prioritize product decisions. The user experience is directly tied to the technology and the business model. By providing better search results and more targeted ads, Google is able to keep users returning, which in turn generates revenue from targeted advertising. Google provides an uncluttered user experience that allows a user to get "what you want, when you want it" (Hurst, 2002a). "The flashy Web designs that were the hallmark of nose-ring New York firms such as Razorfish are now seen as slow and confusing; Yahoo!'s credo of fast, functional and boring has won the day" (*The Economist*, Dec. 7, 2000). By providing a focused search and directory user experience, Yahoo was able to usurp customers and marketing dollars from the leading search technology vendor, AltaVista (Forbes, 2001). By providing the information users need to complete their tasks firms such as Amazon and Google have been able to leverage their focus on the user experience to business success.

6.8 WHAT USABILITY OFFERS STARTUPS

"Getting it right before getting it out there matters more" (Anderson and Braiterman, 2001). An emphasis on usability can help reduce the importance of increasing valuation by changing the focus from funding and technology defensibility to customer needs and solution value. The product development lifecycle for startups forces executive activities to be focused on getting funding. The sales and marketing costs are front loaded; that is, you have to pay for them first to get benefit later. Execution and hiring are predicated on business plan hypotheses of customer acquisition and revenue goals. A heavy spending hit occurs if the product launch is wrong. You don't know whether you are wrong until you are out of money (Blank, 2004). A systematic approach for identifying the product and market for a startup with the greatest likelihood of creating growth is needed (Anthony *et al.*, 2004). By observing customer behavior and tasks, usability methods can be used to identify the important jobs that customers are seeking to get done but can't adequately address with current products or solutions. User-centered design methods can help entrepreneurs identify well-defined customer problems and limit the product complexity to only essential features (as defined by the customer). Technological innovation and defensibility as a sustainable competitive advantage will continue to be a driving factor in venture investing (Winblad and Roizen, 2004). However, it is difficult to sell a technology to a customer. Customers are looking for solutions to problems and a short-term return-on-investment (ROI). Usability methods and early customer involvement will help startups provide value to customers and to investors. User-centered methods can be used to systematically identify customer-oriented, high-potential opportunities. Then a preliminary business case can be built.

6.9 A TOOL FOR UNDERSTANDING AND EVALUATING STARTUPS

Gordon Bell stated, "You don't have to understand the technology to ask the right business questions" (Bell, 2000). The Bell-Mason Diagnostic was developed as a method for understanding new ventures. It provides a common set of ground rules and checklists for analysis and diagnosis of startups and an implied prescription for entrepreneurs, engineers, marketers, the financial community, and the general business community to evaluate startups. It is an alternative to business cases and tries to encapsulate venture capital wisdom to improve the startup process (Bell, 2003/2004). The ultimate goal is to ensure that an entrepreneurial venture is started up and run with a very high (greater than 50%) likelihood of success. And it works. According to data from Nanyang Management Pty Ltd. in Australia, companies that scored an average of 75 on each of the 12 Bell-Mason Diagnostic measures had a business success rate of 95% (Bell, 2000).

The Bell-Mason Diagnostic assesses the health of a startup at four critical stages of organizational development (which are closely related to similar stages in the product development cycle) (Bell, 2003/2004). Customer input and user-centric methods are extremely valuable in reducing the risk of product and corporate failure (Franke and Schreier, 2002; MacCormack, 2001). Using a tool like the Bell-Mason Diagnostic can help user-centered design professionals understand the key business drivers throughout a startup's development. The Bell-Mason Diagnostic is a systematic, rules-based approach to assessing the health and viability of a company in its early stage based on a development model for measuring risk, predicting the course, tracking progress, and improving the success of high-technology, high-growth, early-stage ventures (Bell, 1991). The startup is compared to the Diagnostic's ideal company using a set of rules that are applied by asking a series of questions at each stage of growth. The Diagnostic was developed on the basis of Bell and Mason's experience with hundreds of early-stage ventures (Bell, 2003/2004).

Despite the variety and difference between startups, all healthy startups in the information technology field must pass through four predictable, measurable, sequential growth stages on their way to a fifth stage known as steady state—a mature but still growing stage at which companies are considered to be stable, sustainable organizations. The four stages are Stage 1, Concept; Stage 2, Seed; Stage 3, Product Development; and Stage 4, Market Development. These four stages correspond to key product, market, and corporate development milestones that are intentionally distinct from a definition based on the infusion of capital (i.e., rounds of funding) (Fig. 6.2). During each stage a startup is analyzed using a set of rules and heuristics divided along 12 dimensions. The 12 dimensions are designed to cover every aspect of a startup's operations including input (people, cash, financeability, and technology), output (product and service and the ability to produce and deliver products), balance sheets, the organization and the people who run the company, and the processes

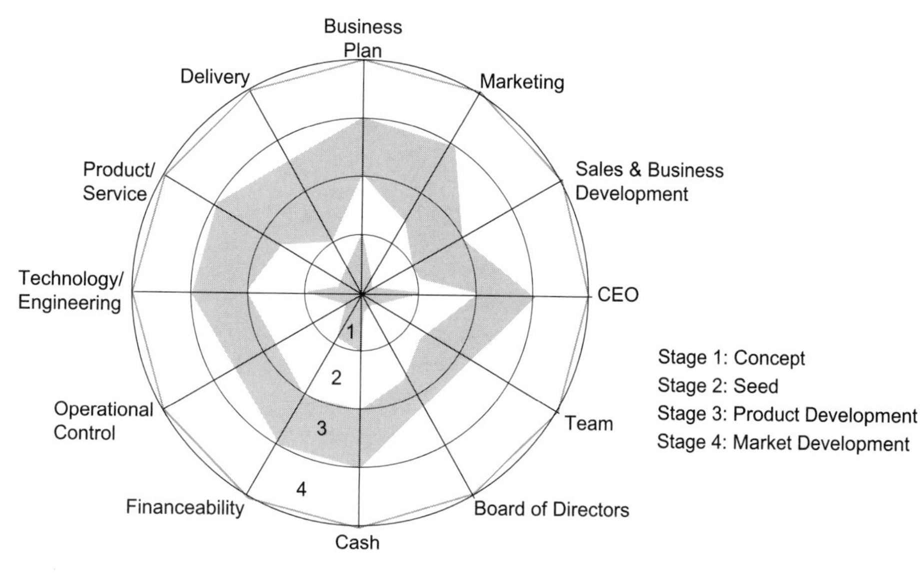

FIGURE 6.2 The Bell-Mason graph of an ideal startup. (Adapted from Bell, 1991, 2003.)

Table 6.3 The 12 Dimensions of the Bell-Mason Diagnostic

Technology/Product	*Marketing/Sales*	*People*	*Finance/Control*
Technology/ engineering	Business plan and vision	CEO	Cash
Product/Service	Marketing	Team	Financeability
Manufacturing/ product delivery	Sales and product support	Board of directors	Operational control

From Bell, 1991, p. 264.

(Table 6.3). The heuristics applied are specific and measurable (Bell, 2003). The diagnosis is carried out by answering questions that come from the heuristics or rules that define an ideal company.

At each stage of start-up development for a company, the Bell-Mason Diagnostic can be used by user-centered design professionals to determine and assess the health of and answer critical questions about a startup. Other research (Nielsen, 1993; Hix and Hartson, 1993) has shown which usability methods can provide value during each of the different stages of product development. The most important factor is that a startup needs to create profitable products. By understanding the stage a startup is currently in, we can begin to understand the values and appropriate measure to be applied.

Stage 1: Concept—The concept stage is the company's starting point. During the concept stage, startups usually have a small number of founders who want to develop an idea they have for converting some technology into a product (Bell, 1991). Some of the questions asked as part of the Bell-Mason Diagnostic include "Has the company translated its technology uniqueness into a relatively concrete product concept (what) that is also self-sustaining (i.e., that provides for the evolution of future generations of the product)?" and "Does the startup have an initial outline of channel-of-distribution alternatives, their typical requirements (e.g., sales cycles, cost of sale, and a first model of sale)?" (Bell, 1991). User-centered design methods offer entrepreneurs a reliable method for identifying customers and their specific needs that can be addressed by a new product. The key to identifying potential markets and start-up opportunities is never to compete against the customer manifest priorities, but instead to facilitate them (Silverthorne, 2003). Many organizations can continue to develop and refine their products through costly and time-consuming iterations or versions until the company delivers value to customers instead of a product to market (Drucker, 2001). The risk for startups is that they will run out of funds before they can deliver a successful product. Instead of designing products and services that dictate consumers' behavior, let the tasks people are trying to get done inform the design (Christensen *et al.*, 2003).

Stage 2: Seed—The purpose of the seed stage is to ensure that any critical technology is under control so that Stage 3 (product development) can be planned and scheduled. Stage 2 is where an initial product definition is created so that the market can be assessed (Bell, 1991). Questions asked during the seed stage include "Does a product definition or functional specification exist for the product being designed?" and "Have sets of customers and their applications been identified for use during the product development stage?" Faulty market segmentation can help to explain the stunningly high rate of failure of new-product development. Cooper (as cited in Franke and Schreier, 2002) argues that the majority of products do not even make it to market, and those that do face a failure rate of up to 90%. Most companies define markets in terms of product categories and demographics (Christensen *et al.*, 2003). This results in products that are removed from customers' needs and expectations and increases the risk of the product failing (i.e., customers not purchasing the product).

Stage 3: Product Development—The goals of the product development stage are to hire the staff, specify and plan the product, and design and produce an actual working product (Bell, 1991). The keys are knowing that there is a true need for a product and service and being able to respond to competitive pressures in the marketplace. Some of the questions asked as part of the Bell-Mason Diagnostic include "Can it be converted into a product that customers buy—or is it just a feature?" and "Have the real customers been identified for the product?" (Bell, 2003/2004). Usability techniques and tools provide startups with the ability to better understand their customers and the features that matter to customers. There is a strong relationship between the success of software companies and customer involvement in the development process. Unsuccessful companies often had inefficient product development teams, who built features that were important to developers and executives, not the features that were identified by customers and user input. In a study of 29 projects from 17 companies, MacCormack (2001) found that successful products and projects involved building an initial low-functionality version of the product to get it into customers' hands at the earliest possible stage and using customer feedback and an iterative approach to add and refine functionality. Smagalla (2004) cited research that analyzed the financial results of 304 publicly held software companies from 1990 through 1998. He concluded that successful companies obtained customer feedback early and only invested in features that customers needed.

Stage 4: Market Development—The market development stage is where all of the planning performed during stages 1 to 3 is tested and tried out in the market place (Bell, 1991). After the product is introduced to the market, the product, marketing, pricing, and sales planes are modified as needed until a refined plan for profitability is decided on. The market development stage is where the ultimate fate of a company becomes apparent; the product decisions and marketing plans from early stages are tested. The goal is for the company to eventually reach a steady-state operation, where an organization can operate at a profitable steady-state. Market development focuses company activities on producing revenue and adjusting the fixed and variable costs associated with engineering, marketing, manufacturing, and administration.

6.10 HOW DO STARTUPS DETERMINE VALUE?

A large number of high-growth startups fail; statistics show that approximately 19 of 20 startups will be unsuccessful (Cringely, 2001; Cusumano, 2004). Startup organizations that fail are, for the most part, those that are unable to deal with the complexity of technology and the fast pace of technological change while simultaneously growing as organizations. As technologists, our instincts usually lead us to look for design flaws or problems in the underlying technologies when trying to understand the sudden collapse of a company. Although these problems can almost always be found, the real roots of the trouble usually can be traced back to basic human foibles and problematic organizational dynamics (Bell, 2003/2004). The odds of success are against startups from the very beginning. During the dot-com boom many companies had to invest tens of millions of dollars to establish operations and then spend $400 to acquire a new customer (Cusumano, 2004). The challenge becomes convincing entrepreneurs and venture capitalists that there is real value in usability investments when resources are extremely constrained as they are in startups.

Calculating the benefits of usability using ROI can be difficult in startups, particularly if the startup is unable to stay in business when the ROI is zero. ROI is usually calculated by tracking business metrics over time; past projects can be used to estimate the value of potential future projects (Hirsch, 2003a). If usability ROI metrics are not tied directly to business strategy and metrics, it can be difficult to assert usability's direct effect on high-level financial metrics. What is needed is a standardized methodology for calculating a project-specific ROI estimate given the needs of customers and the current and future needs of businesses (Hirsch, 2003a). By looking at the metrics and evaluation methods of other key corporate dimensions, entrepreneurs and usability professionals can begin to evaluate their value to a startup. Marketing personnel can look at a variety of key business metrics including customer lifetime value (CLV), and Customer acquisition costs, and use ROI calculations to evaluate different marketing and business strategies (Lenskold, 2002).

Table 6.4 shows a hypothetical example of the use of CLV to calculate an ROI. The example is simplified and does not include discounting of future customer value. The numbers are based on a reported number for Amazon.com (Seybold, 2000). The simplified example provides a framework for a retail organization to make decisions based on both sides of the financial equation—the costs and the returns. It shows a comparison of two projects: one project will reduce customer acquisition costs by 20%; the second project will increase customer revenues by 20%. An underlying assumption is that investment to complete the hypothetical projects is covered in the annual costs/customer. The example demonstrates that by evaluating project returns over the expected customer lifetime, the financial impact of the two proposed projects can be compared. The analysis shows that a project that increases revenues by 20% per customer has a greater return on investment (514.00%) than a project that reduces customer acquisition costs (46.25%).

Table 6.4 Customer Lifetime Value and Return on Investment Example

	Baseline	*20% Decrease in Acquisition Cost*	*20% Increase in Revenue*
Annual revenue/customer	$125.00	$125.00	$150.00
Annual costs/customer (18.6% + 76.1% = 94.7% from Amazon.com Annual Report 2003)	$118.37	$118.37	$118.37
Annual net profit/customer	$6.63	$6.63	$31.63
Average customer lifetime (years)	3.3	3.3	3.3
Customer lifetime value (total net profit* average customer lifetime)	$19.89	$19.89	$104.38
Acquisition cost	$17.00	$13.60	$17.00
Return (considered net CLV)	$2.89	$6.29	$87.38
Change in return (baseline CLV/ net CLV)		217.65%	3,023.53%
Investment (acquisition cost)	$17.00	$13.60	$17.00
ROI (return/acquisition cost)	17.00%	46.25%	514.00%

Based on Lenskold, 2002.
*Not discounted for simplicity.

6.11 HOW MUCH SHOULD STARTUPS INVEST IN USABILITY?

There is no magic number for the ROI for usability. There is no formula that says, "for every dollar invested in usability, you will see a two dollar return." The value of usability to a project or startup depends on a large number of variables including the team, the stage of the company, the stage of the product development, the type of product or service being offered, and many other factors. Hirsch (2004) suggests that to measure the value of a specific design project a company needs to understand the high-level business needs that are being met by the project. The business needs and values are different during the different stages of a startup. Determining and measuring the value of usability to an organization is addressed more in Karat and Lund's Chapter 10. The values and metrics that are provided for larger companies and Web projects are similar to those for startups. These need to be used for the unique stages of a startup and the issues that are faced by entrepreneurs and investors in assessing the risks at each stage.

Usability, like other product development activities, does require an investment. The investment is in people, process, and alignment of usability to business strategy. Most entrepreneurs and business leaders need to understand their customers and the customers' needs that their product is meeting. Usability is only one of many factors in determining the success of a startup. In Chapter 9, Mauro says that ". . . few have any real concept of how to identify viable expertise in the form of experienced and properly educated usability professionals. This problem has become more acute as the definition of what constitutes acceptable levels of expertise is widely debated in the boom and bust of the Internet where many Web development firms offered usability services often without appropriate technical expertise." Along with the challenge of finding experienced usability personnel, finding usability professionals with the skills and desire to step out of their usability roles and contribute to other aspects of business, product, and organizational development is an even bigger challenge for startups. As usability professionals we need to understand the underlying business models of the industry in which we operate (see Chapter 22). Cusumano (2004) offers an excellent introduction into the three predominant software business models: product-focused company; services-focused company; or a hybrid of the two. The fundamentals of products, markets, strategic positioning, and corporate development are necessary for usability professionals to be successful in startups.

Usability plays an important role in product development and engineering circles. However, going forward, we need to leverage our usability tools and skills in the realm of sales and marketing. We usually don't own the customer contact point; however, we need to work with sales and marketing to involve customers in the design and engineering process. Going forward, the economic climate for start-up companies is changing. Venture capitalists and economic markets will no longer pay for mistakes. Although the addition of usability expertise adds to the expenses column for an organization, the result of incorporating usability in the product development lifecycle is to reduce the investment risk by involving customers early in the process. Usability methods are critical and will provide the tools to understand your customers and the features that are valuable to them and provide a means of course correction and planning for current and future products. Most entrepreneurs and startups "vastly underestimate how difficult it will be to build and deliver an easy-to-use software product to customers" (Cusumano, 2004, p. 207). Usability and user experience professionals need to understand the business of starting companies. Tools such as the Bell-Mason

Diagnostic provide a framework to understand the competing financial and organizational pressures on startups. The focus for usability in startups needs to be on the business drivers and the creation of products and services that deliver compelling value to customers.

REFERENCES

Anderson, R. I., and Braiterman, J. (2001). Strategies to make E-business more customer-centered. In J. Bawa, P. Borazio, and L. Trenner (Eds.), *Usability: Politics and New Media.* London: Springer-Verlag. Retrieved on March 11, 2004, from www.well.com/user/riander/chapter.html.

Anthony, S. D., Johnson, M. W., and Eyring, M. (2004, August 9). A diagnostic for disruptive innovation. *Harvard Business School Working Knowledge.* Retrieved August 20, 2004, from hbswk.hbs.edu/item.jhtml?id=4300&t=innovation.

Bell, C. G. (1991). *High-Tech Ventures: The Guide for Entrepreneurial Success.* Reading, MA: Perseus Books.

Bell, C. G. (2000). *Entrepreneurial Ventures: How Do You Do Them?* Online presentation retrieved August 19, 2004, from research.microsoft.com/~gbell/BMD0002.ppt.

Bell, G. (2003/2004). Sink or swim, know when it's time to bail. *ACM Queue,* 1(9). Retrieved August 19, 2004, from www.acmqueue.com/modules.php?name=Content&pa=showpage&pid=106.

Blank, S. (2004). *Customer Development Model.* Retrieved March 4, 2004, from www.stanford.edu/class/msande273/resources/Blank%20presentation%20101403.pdf.

BusinessWeek (2004, August 2). Jeff Bezos on Word-of-Mouth Power [Interview with Jeff Bezos]. Retrieved August 18, 2004, from www.businessweek.com/magazine/content/04_31/b3894101.htm.

Christensen, C. M., Raynor, M. E., and Anthony, S. D. (2003, March 10). Six keys to building new markets by unleashing disruptive innovation. *Harvard Business School Working Knowledge.* Retrieved August 20, 2004, from hbswk.hbs.edu/item.jhtml?id=3374&t=innovation.

Collins, J. (2000, March). Built to flip. *Fast Company,* Issue 32, 131. Retrieved March 12, 2004, from www.fastcompany.com/online/32/builttoflip.html.

Cringely, R. X. (2001, June 14). *Rules of the Road: High Tech Startups Are Set to Boom Again, so Here Are Some Rules for Getting Rich, Then Getting Out* [I, Cringely column]. Retrieved August 20, 2004, from www.pbs.org/cringely/pulpit/pulpit20010614.html.

Cusumano, M. A. (2004). *The Business of Software: What Every Manager, Programmer, and Entrepreneur Must Know to Thrive and Survive in Good Times and Bad.* New York: Free Press.

Donoghue, K. (2002). *Built for Use: Driving Profitability Through the User Experience.* New York: McGraw-Hill.

Drucker, P. F. (2001). *The Essential Drucker.* New York: HarperBusiness.

The Economist. (2000, December 7). *Consultant, Heal Thyself.* Retrieved August 16, 2004, from www.economist.com/displayStory.cfm?Story_ID=444402.

Forbes, S. (2001, April/May). Upwardly mobile. *Context Magazine.* Retrieved August 19, 2004, from www.contextmag.com/setFrameRedirect.asp?src=/archives/200104/impact.asp.

Franke, N., and Schreier, M. (2002). Entrepreneurial opportunities with toolkits for user innovation and design. *The International Journal on Media Management,* 4 (4), 225–235.

Google Inc. (n.d.). Corporate information: Google history. Retrieved on March 31, 2004, from www.google.com/corporate/history.html.

Google Inc. (2004, August 18). Prospectus: Registration No. 333-114984. Retrieved August 19, 2004, from www.ipo.google.com/data/prospectus.html.

Gurley, J. W. (2000, February 21). The most powerful Internet metric of all. *Above the Crowd Newsletter.* Retrieved August 20, 2004, from news.com.com/2010-1071-281288.html.

Gurley, J. W. (2003, April 23). Dot-com double take. *Above the Crowd Newsletter.* Retrieved August 18, 2004, from www.benchmark.com/cgi-bin/suid/~bcmlp/newsletter.cgi?mode=show&year=2003&date=2003-04-23.

Hirsch, S. (2003a, October 17). The red herring of usability ROI. Review. *BayCHI October Program.* Retrieved April 2, 2004, from netnow.blogspot.com/2003_10_01_netnow_archive.html#106642724548276267.

Hirsch, S. (2003b, November 13). User experience accountability: Assessing your impact on business results. *Adaptive Path Essay Archives.* Retrieved April 3, 2004, from www.adaptivepath.com/publications/essays/archives/000276.php.

Hirsch, S. (2004, July 20). *ROI Is Not a Silver Bullet: Five Actionable Steps for Valuing User Experience Design.* Retrieved August 2, 2004, from www.adaptivepath.com/publications/essays/archives/000338.php.

Hix, D., and Hartson, H. R. (1993). *Developing User Interfaces: Ensuring Usability through Products and Process.* New York: John Wiley & Sons.

Hornik, D. (2003, May 2). *4 Keys to Google's Success.* VentureBlog entry. Retrieved March 31, 2004, from www.ventureblog.com/articles/indiv/2003/000080.html.

Hurst, M. (2002a, October 15). *Interview: Marissa Mayer, Product Manager, Google.* Retrieved August 17, 2004, from www.goodexperience.com/columns/02/1015google.html.

Hurst, M. (2002b, November 21). *Interview: Maryam Mohit, Amazon.com.* Retrieved August 17, 2004, from www.goodexperience.com/columns/02/1121.amazon.html.

Joyce, E., and Shread, P. (2004, August 19). *Google's IPO Opens at $100.* internetnews.com Business. Retrieved August 19, 2004, from www.internetnews.com/bus-news/article.php/3397291.

Lenskold, J. (2002). *Customer Lifetime Value vs. ROI Marketing Measures: Supporting Analysis for Marketing Management Article "Marketing ROI: Playing to Win."* Retrieved April 2, 2004, from www.customerpathing.com/leadership/private/CLV_ROI_whitepaper.pdf.

Levine, S. (2001, March). Business valuation issues related to start-up companies. *Leading Companies Online Magazine.* Retrieved August 17, 2004, from www.beysterinstitute.org/includes/cfbin/output/article_slot_view.cfm?ID=311702.

Lewis, M. (1999). *The New New Thing.* New York: W. W. Horton & Company.

Lieberman, M. B. (2002). *Did First-Mover Advantage Survive the Dot-Com Crash?* Retrieved March 4, 2004, from www.gsb.stanford.edu/facseminars/conferences/strat_conf/pdfs/Lieberman%20InternetFMA.pdf.

Lorek, L. A. (2002, May 19). Internet duds gobbled dollars—The 10 worst dot-coms show how bad ideas fed the Web bust. *San Antonio Express-News.* Retrieved August 17, 2004, from NewsBank database.

Macdonald, N. (2001, November 2). After the fall. *AIGA Gain,* 1(2). Retrieved August 17, 2004, from www.spy.co.uk/Communication/Articles/Gain/AfterTheFall/.

MacCormack, A. (2001). Product-development practices that work: How Internet companies build software. *MIT-Sloan Management Review,* 42 (2), 75–84.

Mauro, C. L. (2002). *Professional Usability Testing and Return on Investment.* MauroNewMedia, Inc. Whitepaper. Retrieved March 1, 2004, from www.taskz.com/pdf/MNMwhitepaper.pdf.

Mayfield, R. (2002). *Timing Your Business Case with the Technology Valuation Lifecycle.* Retrieved April 3, 2004, from radio.weblogs.com/0114726/whitepapers/Timing%20Your%20Business%20Case%20-%20Ross%20Mayfield.pdf.

Moore, G. A. (1999). *Crossing the Chasm: Marketing and Selling High-Tech Products to Mainstream Customers* (rev ed.). New York: HarperCollins.

Nielsen, J. (1993). *Usability Engineering.* Chestnut Hill, MA: Academic Press.

Nielsen, J. (1999, November 28). Usability as Barrier to Entry. *Alertbox.* Retrieved March 11, 2004, from www.useit.com/alertbox/991128.html.

Perkins, A. B., and Perkins, M. C. (2001). *The Internet Bubble: The Inside Story on Why It Burst—And What You Can Do to Profit Now* (rev. ed.). New York: HarperBusiness.

Reingold, J. (2004, March). What we learned in the new economy. *Fast Company,* Issue 80, 56. Retrieved March 12, 2004, from www.fastcompany.com/magazine/80/neweconomy.html.

Renaissance Capital. (2000, December 18). *2000 IPO Year End Review: Going to Extremes.* Retrieved August 16, 2004, from www.ipohome.com/marketwatch/review/2000review.asp.

Ritter, J. (2002). *Big IPO Runups of 1975–December 2002.* Retrieved August 15, 2004, from bear.cba.ufl.edu/ritter/RUNUP750.pdf.

Seybold, P. (2000, September 29). *Don't Count Amazon Out.* Retrieved August 20, 2004, from www.business2.com/b2/web/articles/0,17863,528272,00.html.

Silverthorne, S. (2003, April 23). Interview with Clayton M. Christensen: Are crummy products your next growth opportunity? *Harvard Business School Working Knowledge.* Retrieved August 20, 2004, from hbswk.hbs.edu/item.jhtml?id=3437&t=innovation.

Smagalla, D. (2004, Winter). The truth about software startups. *MIT-Sloan Management Review*, 45 (2), 7.

Steinert-Threlkeld, T. (1995, November). Can you work in Netscape time? *Fast-Company*, 1, 86. Retrieved August 16, 2004, from www.fastcompany.com/magazine/01/netscape.html.

Stone, A. (2000, December 19). Crawling from the dot-com wreckage. *BusinessWeek Online.* Retrieved August 21, 2004, from www.businessweek.com/bwdaily/dnflash/dec2000/nf20001219_800.htm.

Trueman, B., Wong, M. H. F., and Zhang, X. (2000). *The Eyeballs Have It: Searching for the Value in Internet Stocks.* Retrieved August 21, 2004, from faculty.haas.berkeley.edu/trueman/valuation59.pdf.

Weiss, A. (2000, July 6). *Trailblazers on the Internet.* Retrieved March 11, 2004, from www.marketingprofs.com/2/fma.asp.

Winblad, A., and Roizen, H. (2004, March 3). *2004 New Era of Optimism in Software.* Retrieved March 11, 2004, from etl.stanford.edu/handouts/0304_win/heidiandann.ppt.

7 CHAPTER Cost-Justifying Usability in Vendor Companies

Janice A. Rohn World Savings Bank

7.1 INTRODUCTION

The popularity and adoption of the Web over the past 10 years have benefited user experience (UE) professionals in two ways: a wider and less technical user population has adopted computers, and users are more likely to select products based on ease of use, rather than solely on features (Black, 2002; Souza *et al.*, 2001). As a result, companies recognized the importance of hiring UE professionals, and the popularity of UE groups in vendor companies was at a peak in the late 1990s and early in the new millennium. The demand for UE professionals at that time was greater than the supply, resulting in increases in compensation, fast hiring, and on-the-job training for individuals from other fields. In 2000, competition for UE professionals was so strong that people were poaching candidates from competitors' recruiting parties at the Association for Computing Machinery Special Interest Group on Computer-Human Interaction conference, and cash sign-on bonuses were being offered. During this period, most high-tech companies had such high revenue or expectations of high revenue (in the case of startups funded by venture capital), with explicit customer demands for usability, that cost-justifying usability was not a strong focus.

Unfortunately, following the cyclical nature of economics, the high-tech boom ended. Initially, unlike in prior economic downturns, UE professionals were not the first to be downsized (personal communications). Companies had learned a valuable lesson about customers' expectations of usable products. However, as the downturn turned from months into years, companies started downsizing their UE groups once again. Some companies reduced their groups radically, down to a small percentage of the original group size, and some downsized their entire UE departments (Sanders, 2004; personal communications). A few vendor companies were able to keep their groups, and fewer still actually grew their UE departments over the past few years, but these companies were in the minority (for example, Oracle, PeopleSoft, and World Savings Bank were all able to grow their UE departments while many other companies laid off UE professionals).

These cyclical economic upturns and downturns have demonstrated that although vendor companies recognized the importance of UE professionals more than in previous years, understanding the economic benefits of usability and educating companies about them are always important goals. When economic downturns occur, educated companies can make better decisions about the importance of UE professionals, and when economic upturns occur, the companies that have invested in UE analysis are better positioned to take advantage of these upturns. These companies are also able to retain the loyalty of top UE professionals for their stable investment in this important job function.

In this chapter we first examine how to analyze your own company's culture and then look at the costs and benefits for vendor companies, their customers, and other stakeholders in vendor product and service selection. The chapter concludes with some strategies for maximizing the effectiveness of the UE department.

7.2 UNDERSTANDING YOUR COMPANY

One of the ironies of the role of UE professionals is that the field of usability focuses on understanding users and their requirements, yet very few UE professionals practice this principle on one of the most important influences on their work: their employer. (See Chapter 8 for a discussion of understanding and persuading stakeholders of value.) Many UE professionals enter the field because they want to improve products and services for people and concentrate on this aspect of their work. However, understanding the culture, value system, and business goals of the company should be the first order of business for a UE department and its members.

Every company has a unique culture, goals, organization, and core competencies. Understanding these factors is key to doing a job effectively. Ideally, you are able to perform at least some of this analysis before you join the company, so that you can ensure at least a certain threshold of efficacy and job security. For example, because high-level support is important, you can find out from the hiring manager what levels of executives support UE. You can also verify this information by asking to meet with a reportedly supportive executive during the interview process. If the executive is too busy to meet with you, the level of support is probably not there. Because the UE field is a communicative community, you can probably find someone who used to work at that company and get his or her assessment. You can also try to obtain "back-door" references from people who work in other groups at the company to get their assessment of the UE group and conditions. Another gauge is whether the company has displayed a proactive or reactive attitude toward UE—is it innovating in design and usability, or is it following? If reactive, the company probably does not support or invest in innovation, which would translate to a lack of support for field studies and other methods that would enable it to take a leadership position in UE. You can also examine the company's historical behavior: has the company had cyclical or consistent support of UE? If support is cyclical, with a pattern of building up and laying off, it is probably not a good long-term choice.

Let's examine some common company cultures and how they affect a UE professional's strategies and influence.

7.2.1 Customer-Focused Company

A customer-focused company has an advantage in its understanding that knowing and satisfying the needs of customers are important business strategies. This type of company has obvious advantages from a UE professional's perspective. Access to customers is typically easier, and UE professionals are often included in meetings with customers. The most common downside in these types of companies is that the organization can sometimes adopt a reactive, rather than proactive, response to customer issues. A number of companies will prioritize the needs of a few key customers without verifying whether the requirements are shared across customers or are specific to those customers. Thus, the latest customer issue can have an inordinate influence on the current project priorities, which can change daily if companies don't observe best practices in gathering data, prioritizing projects, and executing this prioritization over a period of time (at least weeks or months) before making changes to this prioritization. By observing this best-practice strategy, a company can analyze whether the requirement of a key customer should be in the general product, whether it should be in a customer-specific implementation, or whether it is even a real requirement. On numerous occasions a requirement, as stated by a customer, turned out to be completely different from what was described once the UE group investigated the request.

7.2.2 Product-Focused Company

A product-focused company also has an advantage over some other types of companies in that the focus is on the products. Therefore, any factors that influence product requirements, such as user requirements, are typically taken into consideration. The most common downside in this type of organization is that other factors, such as functionality, may occasionally be given a higher priority than usefulness or utility. If the company is focusing on meeting a functionality checklist rather than user requirements, one strategy that UE professionals can employ is to ensure that usability and utility requirements are also considered during the requirements phase. If data and explicit goals can be provided, this can be an easier sell. For example, by performing competitive usability studies, a UE professional can communicate a usability goal, such as to have a higher task-completion rate or a lower time-on-task rate for key tasks, compared with that of a competitor. This is a particularly effective strategy if the competitor is disliked by executives and has better usability numbers, because executives often dislike "losing" in any way to that competitor.

7.2.3 Technology-Focused Company

One of the most difficult types of companies in which a UE professional can be effective is a technology-focused company. This type of company is ruled by technology and focused on applications for the technology. The obvious downside is that customers and users are often a lesser focus. On rare occasions this type of company can have a product that appeals to the masses, but more often these types of companies struggle.

At best, it can be a challenge for a UE professional to work in this type of company; at worst, it can be a frustrating experience with little empowerment. UE professionals, no matter how talented, will never be able to change companies at their core cultural level unless they become the CEO, president, or other top-level executive. They can certainly influence, make incremental changes, and make contributions to processes and products, but the level of effort required to make significant changes may not be worth expending or they may not wish to stay at this type of company.

7.2.4 Executive-Focused Company

An executive-focused company is any company in which the founder, CEO, or other key executives have a very strong influence over product decisions at a more detailed level than would be typically warranted. Company cultures are often strongly influenced by the founder or CEO. In some cases, this can be a positive factor for UE professionals when the executives are proponents of UE methods. In other cases, executives sometimes start to believe that they have a "golden touch" in all areas, including those outside of their skills and training. These can be some of the most difficult organizations to work in, because sometimes the founders dictate product details on opinion rather than on customer data. Sometimes these executives can be influenced when they hear customers' usability issues or they repeatedly see strong data demonstrating usability issues.

7.2.5 Optimal Data-Driven Company

Ideally, companies practice data-driven decision-making. This means that data for product requirements are gathered from a variety of large- and small-sample-size sources, such as field studies, usability lab studies, market research, surveys, Web metrics, and online customer panels. These data are collated and prioritized according to importance and frequency and then are used to drive requirements decisions. Data-driven also means that design decisions are based on data from usability lab studies, online customer panels, and other sources, rather than from stakeholders' opinions.

A data-driven company would be an ideal environment for UE professionals. Unfortunately, in reality, there is no pure data-driven company. However, some companies at least practice this approach frequently, and the more you can influence your company to do so, the more successful you and the company will be. For example, by collecting and prioritizing requirements data, a UE group can become more influential just by having information and presenting it in an organized fashion. Because most companies have reduced development schedules, many teams have very little time for the requirements phase. This situation results in a requirements phase that in reality extends across the entire lifecycle, with key requirements discovered late in the development and deployment phases. If data appear credible and result from sound analysis, a prioritized list from the UE group is sometimes welcomed by teams with very little time.

7.3 COST JUSTIFICATION IN VENDOR COMPANIES

A vendor company is defined as a for-profit company that obtains its revenues by selling products or services. UE departments within vendor companies may work on the products that are the core business, such as software or hardware products, or may work on the delivery mechanism for products or services, such as the e-commerce site for the company. UE departments also often work on the internal business software, company intranet sites, and improving the development processes to be more efficient and data-driven. UE departments are now found in major retail companies, financial companies, educational companies, and others, in addition to the major computer companies where they have been for more than 10 years.

Cost-justifying usability can be both easier and more complex in a vendor company. Vendor products typically affect both a large number and a wide variety of users, thus acting as a multiplier for any usability improvements. For example, a usability improvement in an application can save companies thousands of dollars per week in reduced support call costs. When quantitative data are available, this type of cost benefit is clear.

However, with business-to-business (B-to-B) products, multiple tiers of decision-makers who all have influence over vendor product selection are often present. The varied requirements of the different stakeholders can make the identification and prioritization of usability enhancements more complex than with business-to-consumer (B-to-C) products, in which the consumer is both the purchaser and the end user. Let's look at an example.

A typical product sold to consumers must meet the requirements of a wide variety of users, who often range from novices to experts. Designing a product to meet their needs is a challenge. For products sold to other businesses (B-to-B), not only are the requirements of the customers' businesses highly varied, but the requirements within a single customer company are also highly varied. For example, a call center application must typically meet one set of requirements for one set of actual end users (the tier 1 call center agents), another set for other sets of actual end users (tier 2 call center agents, etc.), and additional sets for the managers of the agents (who are using the product for analyses and resource management), the information technology (IT) department who must install, deploy, maintain, and upgrade the applications, the third-party vendors who configure the product, and the executives who approve the product selection and budget. Each set of stakeholders' requirements must be considered and prioritized, leading to even more complexity for UE professionals.

One benefit of cost-justifying usability in a B-to-B context is that quantitative usability measures—such as successful task completion, time on task, number and severity of errors, and number of assists (the number of times a user can not proceed in completing the task without some assistance)—can become important sales tools for a vendor company, thus increasing the importance of UE methods. A small reduction in time on task in an application can result in a huge cost savings for a customer company using that product. Usability in B-to-B products can be successfully cost-justified in terms of productivity gains.

In contrast, other usability measures are often more important for B-to-C products, because consumers are less (although still somewhat) aware of productivity improvements. For example, B-to-C products typically must be easy to learn without the user having to read a manual. Usability improvements in ease of learning can be cost-justified in terms of easier marketing of the product and in better product reviews.

7.4 COST JUSTIFICATION FOR WEB APPLICATIONS

In the earlier years of the Web, a common practice was to treat the world as a large usability test. An often-untested design would be posted, and the site or application would either partially work or fail. If a retail company's e-commerce site had few purchases, the root cause of the problems was often unknown. Rather than investigate the root cause, companies would often make some changes, post them quickly, and see whether purchases then increased or decreased. Also contributing to the poor usability of Web sites was the fact that many companies hired offline advertising professionals to design their Web sites and applications rather than software UE professionals. This practice resulted in a large number of sites with heavy use of high-concept graphics but poor usability and performance. For example, the original home page for many sites had an animated design with audio, with little to no text. The user had to then click to the next page to get even a hint as to the purpose of the site.

Web applications also cover a wide range of products, ranging from highly complex enterprise applications that are software products delivered via browsers, to an e-commerce product selection and checkout system, to a hotel reservation system, to an online dictionary. Although complexity varies, cost-justification can still be made based on the business goals, including increases in successful purchases and reductions in abandonments.

Another important cost-justification is the low barrier to leaving a site; the user can go "next door" with a single mouse click. If a business requires employees to use a certain type of software, the employees often have no choice: They are expected to use the software as part of their job, regardless of how usable it is. Only rarely can employees exert influence during the product selection process—usually they can't decide to simply switch to different software. Or imagine a college student going online to register for courses. It is not as though he or she has the option to try another university's online registration application. Conversely, users on the Web can try one travel reservation system, and if it has poor usability, they can switch to another system within seconds. Improved usability helps a company's site attract and retain more users. The cost-justification of Web-based applications is addressed in more detail in Chapter 10.

7.5 COSTS

One of the most problematic shortcomings in business is that companies measure and track what is easy to measure and not necessarily what is important to measure. As a result, some costs, including employees, equipment, buildings, and construction, are measured and tracked, whereas other costs, such as project overruns, inefficient or missing processes, poor product decisions, reorganizations, and conflicting or unclear communication, are typically not tracked. Surprisingly, years after revealing publications such as *The Mythical Man-Month* (Brooks, 1995) and the Standish Group's CHAOS report (Standish Group, 1995), businesses have not demonstrated significant improvements in how costs and benefits are measured and how decisions are made.

Another effect of performing cost-benefit analyses is a change in how companies invest for the future. Most companies have become more short-term focused in attempts to increase their stock prices. Investments made for longer-term payoff have become increasingly difficult, with reduced cycle times and quicker product releases desired to demonstrate quarterly increases in revenues. Many companies have downsized their research groups to reduce short-term costs that have less known effect on near-term revenue. As a result, some of the longer-term cost-benefit justifications that would have worked in the past are no longer of interest to many companies. To exert influence with decision-makers, any costs now should produce nearer-term beneficial effects on projections of savings or revenue increases.

The total costs of UE practices can be difficult to assess. (Because companies sometimes desire confidentiality about investments and costs, some of the data contained in this chapter will appear without a reference.) Unfortunately, the savings produced by UE methods are even more difficult to measure. Another factor that contributes to the difficulty of cost-justifying UE expenses is that the costs come out of the current budget, whereas depreciation costs for capital equipment regularly come out of future years' budgets. Even though current costs applied to UE practices can translate into exponential future savings, the lag time and the fact that most organizations don't typically track

UE benefits can make cost-justification a challenge. For example, usability enhancements in the core architecture of enterprise software can often take years to release into the market, thus making it more difficult to track benefits in several years from today's costs. However, these same vendor companies typically value customer loyalty and satisfaction, which can often be improved by providing customers with a usability roadmap. This roadmap not only demonstrates the value placed on usability but also reassures the customer that it will be worthwhile to stay with the vendor rather than go through the expense and trouble to change vendors.

Actual costs spent on usability can vary greatly, depending on the importance of data validity, thoroughness, efficiency, and acceptance of the results (Rohn, 1994). During the late 1990s, when many companies had record revenues, a large amount of money was often spent on usability. Some companies spent tens of thousands of dollars on usability lab equipment and hundreds of thousands of dollars on salaries of UE professionals annually. Although many companies are still investing in UE group salaries, many labs are built more economically now.

Despite the recent rise and fall in expenditures, usability costs can range from a small expense of a cubicle or an office and the cost of the employee's time, to well over $1 million for multiple high-quality labs, equipment, and employees. Fortunately, some costs, such as video recording and editing, have decreased significantly over the past 10 years.

7.5.1 Initial Costs of Building Usability Laboratories

The initial costs may include construction of the facility, purchase and installation of video and audio equipment, tools, furniture, and the products, such as computers and software.

Laboratory Construction

A dedicated space is an important component for performing efficient usability evaluations and providing a tangible presence for the UE team. Nakamura (1990) estimated that the number of usability labs was fewer than 10 in 1985 and increased to more than 100 by 1990. Although there are no recent formal counts of the number of labs, currently hundreds of them exist if the informal and portable labs used in smaller companies are included. When UE analysis is introduced into a company, performing usability evaluations that can be viewed by project team members is one of the most effective ways to convince management of some of the benefits of practicing UE methods (Rosenbaum *et al.*, 2000). If the budget to start a usability lab is limited, a single room or even a cubicle can suffice, with the observer sitting in the same room as the participant. This arrangement can work quite well, but it does have a number of disadvantages, including potential distraction of the participant and a need to limit the number of observers. To circumvent this last disadvantage, test sessions can be viewed from colleagues' computers by using Web casts and providing the stakeholders with password-protected access to the real-time streaming video. Currently available software enables colleagues to view test sessions from their offices across a local-area network.

Because this first-hand observation is one of the most effective ways to educate teams about problems with the design and the benefits of UE methods, most companies that can afford the cost have two rooms with a one-way mirror between them. The one-way mirror obviates the problem of the participant feeling reticent to critique the product or feeling intimidated by the presence of the developers in the same room. This also enables the usability engineer to take notes, control video equipment, log events, and perform other tasks that might otherwise be distracting to the participant. In addition, the development team can watch and contribute valuable information to the usability engineer that couldn't be discussed in front of the participant.

The cost of creating usability labs can range from a few thousand dollars to hundreds of thousands of dollars. Costs can include building new rooms or remodeling existing rooms; installing the one-way mirror, special lighting, video and audio equipment, and special air-conditioning and heating equipment; and soundproofing the walls, windows, and doors.

Video and Audio Equipment

Fortunately, the costs of video equipment, along with its complexity and training to use it, have decreased significantly over the past 10 years. What used to require racks of expensive and specialized hardware can now be accomplished through mostly software and minimal hardware. In most labs video cameras are still used to capture the expressions and movements of the participants, but increasingly screen-capture software is used to record the screens and actions on the computer. Video editing has also become quicker and easier through the use of less expensive software programs. A lab can be outfitted with equipment for less than 20% of the cost of 10 years ago, although construction costs have continued to increase. Intel spent well over $20,000 to equip their former lab with analog equipment, but more recently spent less than $5,000 for a digital solution (Jones and Bullara, 2003).

Portable usability labs have also become more available and affordable. One advantage of portable labs is that they can be used both in-house and for field studies.

Tools

The costs of prototyping and data logging and analysis tools are sometimes overlooked. Software for prototyping is highly cost-effective: it enables designs to be evaluated in less time than it takes to write actual code, and the design is more easily altered (although this is truer for enterprise software than, for example, for Web sites). The code from some prototyping software is improving to be closer to production-quality code, so that in some cases portions of the code from the prototyping tool can be reused in the product.

One of the most cost-effective tools is a data logger, which enables the usability engineer to record the times and types of events that occurred during a usability evaluation. These logging tools can save hundreds of thousands of dollars over time by making data analysis four to eight times more efficient (Hammontree, 1992). Companies with larger UE departments typically have developed their own versions in-house to better meet their needs, whereas smaller UE departments often use commercially available logging tools. Data loggers can also be quite effective in efficiently enabling the recording of objective data, such as the time required to complete a task and the number of errors, and supporting reporting and analysis functions, so that charts, graphs, trends, averages, and other analysis tools are quickly generated from the raw data. Most companies embrace and support activities that can be communicated as a measurable science rather than as an opinion-based art, and data-logging tools can help to make UE departments more credible and more effectively represented in release criteria and other product lifecycle milestones.

A third category of tools is software that automatically logs what the participant does on a computer, such as keyboard input and mouse movements. This "keystroke logger" sounds like it would be a money saver but it often can be costly to produce because it must be customized to the hardware platform and the operating system and thus is typically written for in-house use. In addition, unless the software is written with filters that reduce the amount of data to be analyzed either before or after data are recorded, the logs are typically too cumbersome to analyze. The adoption of these keystroke loggers has been low because of their low return on investment (ROI). These loggers can also present additional privacy and security issues: the participants must be informed if logging software is being used, and the capture and storage of data can be problematic if personal information is being entered.

Furniture

Costs of furniture can also vary greatly in cost, ranging from virtually nothing, if existing furniture within the company is used, to tens of thousands of dollars, if special ergonomic furniture that can be customized by each participant is purchased. For example, tables with adjustable heights, breakdown tables that store flat, and fully adjustable ergonomic chairs are all beneficial for performing usability evaluations if budgeted money is available and the use of the lab and furniture is high.

Products

A sometimes overlooked cost is that of the products themselves. Both internal products under development and external products for performing competitive evaluations can cost money. For example, creating a hardware model for usability evaluations can cost up to thousands of dollars, but ultimately save hundreds of thousands of dollars. The cost of products can range from the cost of a computer (public Web sites, for example, cost nothing more than the computer to access them) to more than $100,000, depending on the types and numbers of products being evaluated.

More challenging than cost, however, is the fact that sometimes competitive products are not available for acquisition by any reasonable means. For example, enterprise software can be expensive to license, install, and run, and is not easily attained for competitive studies. In addition, some companies have strict licensing agreements that disallow customers from publicly revealing the results of evaluations. Sometimes other approaches are mandated, such as examining product demos, interviewing participants who have experience with competitors' products, or using a third-party evaluator who has access to multiple products.

7.5.2 Sustaining Costs

Like initial costs, sustaining costs can vary significantly, depending on the need for timely, accurate, detailed, and easily accessed information. The sustaining costs may include employees, contractors, recruitment of and compensation for participants, travel for field studies, videotapes or digital storage, equipment maintenance, and upgrades to video and audio equipment, computer equipment, and software.

Employees

The most important investment that a company can make is hiring high-quality and trained UE professionals. These individuals can perform cost-efficient usability evaluations not only by knowing which methods to use, how, and when, but also by knowing principles for performing usability inspection methods, such as heuristic evaluations. Nielsen (1992) found there was a correlation between the number of usability problems discovered in the interface and the quality of the methodology used.

To obtain more accurate feedback on designs, usability professionals who can examine the project more objectively because of their training and who know how to design evaluations are required. Given the number of projects and the benefits that performing iterative testing provide, ideally a company should staff for full UE staff coverage of all projects. Although many companies have increased their investment in UE professionals, with a few companies having more than 100 on staff, few, if any, companies have full coverage for their products. Because companies do not have enough usability engineers on staff to cover their product development sufficiently (Nielsen, 1993), trade-offs typically must be made.

As a result, most companies prioritize their projects, ensuring that at least some projects receive the minimally acceptable level of UE involvement. UE professionals and upper management typically examine several factors when prioritizing projects, which include the following:

- How much effect will this product have on the success of the company?
- Is this product in a new market for the company?
- Is a competitor coming out with a similar product that may reduce the company's market share?
- Is this a brand-new product or a new version of an existing product?
- If this is a new version of an existing product, did any or most of the previous versions have UE professional involvement?
- Are the changes in this version major or minor?
- Is this a product on which other variations are based, such as an industry-specific or different-language version?
- What is the perceived current level of usability of this product?

These questions and others help to focus the limited UE resources on the most critical projects.

Evaluation of projects can range from requiring multiple dedicated UE professionals to requiring only a percentage of one UE professional. For example, a company has 12 projects that will combine into four products to ship this year. Eight of the projects are deemed to be top priority (P1), with the other four deemed to be next in priority (P2). Given the types of projects, the company decides that the minimum number of usability evaluations (before the release date) on the P1 projects is four. The decision is to perform at least three usability evaluations on the P2 projects. Thus, the minimum number of evaluations would be 8 P1 projects × 4 evaluations = 32 evaluations, plus 4 P2 projects × 3 evaluations = 12 evaluations. The total then would be 44 evaluations of various types that need to be completed during the year.

Usability lab studies typically last from a few days to six weeks (including preparing for the study, running the study, and analyzing the results), with the average study taking 3 weeks; thus, each usability engineer could perform an average of 12 usability evaluations per year, allowing time for vacation and other work-related activities. This estimate also assumes that prototypes and products for evaluations are delivered on time, which they often aren't, and that evaluations don't overlap, which they often do. This is also a simplified example, because UE professionals typically are working on multiple projects and methods at once. For example, an efficient UE professional can be preparing for a usability lab evaluation and working on preparations for a survey or field study at the same time. From this simple example, a minimum of four full-time usability engineers and a part-time contractor would be needed to evaluate the products. With more typical amounts of other methods and activities, a minimum of five full-time UE professionals would be warranted.

Contractors

Many companies rely on contractors in addition to full-time staff to perform evaluations and field studies. In a survey of six vendor companies, five of them regularly used contractors (Rohn, 2004). The use of contractors can be beneficial for a few reasons, including variable workload and skill-set augmentation. When the workload is variable, contractors can be brought in to address the short-term need for additional resources and peak workload requirements. Contractors can also be helpful when they have expertise in an area not yet found within the UE group. For example, when field studies are performed in other countries, contractors who are bilingual can help with the studies.

Full-time employees are typically the better solution for addressing UE needs for a number of reasons. An employee can learn the domain and work on the same products over time, so that the learning curve does not have to be repeated for each iteration of evaluation and the increased knowledge can be applied to create better design solutions. Efficient performance of usability evaluations depends on knowing the company and group processes. Employees are already familiar with the processes and how to use them effectively, whereas contractors need to be educated about them. Another disadvantage of the use of contractors is that they are often working on multiple projects for multiple companies, so their attention is divided and turnaround time can be longer. Four of the five companies surveyed that used contractors mentioned high overhead costs for training, support, and management of the contractors. However, companies typically are able to hire contractors more easily and quickly than employees, so often necessity dictates the use of contractors.

Participants

Recruiting participants from outside the company who are representative of the target users is critical to providing valid data. Coworkers are influenced by company knowledge, technology, terminology, and motivation. Obtaining external participants who match the proper profiles is more expensive and time-consuming than obtaining coworkers as participants, however. How well companies match the target profiles for their participants ranges from not at all to quite accurately. One of the most beneficial aspects of recruiting for the target profiles is feedback on the profiles. For example, it is not uncommon to receive a profile from another department, such as marketing, only to find that when the people are screened for participation, they are not identical to the profile, and the profile is not accurate.

Using representative customers as participants adds three costs: designing the screening questionnaire, recruiting the participants, and compensating the participants. Most UE departments do not pay for travel, but some perform remote usability studies to obtain feedback from others cities, states, and countries. The UE professional should work with the product marketing and product teams to gain information about the types of customers who will be using the product and design a questionnaire to be used to screen for users who have the desired profiles. Locating and screening users requires either the time of an employee, such as a usability participant recruiter, or the use of a market-research company or a temporary agency. This can be a very time-consuming process, especially if a database of users and their backgrounds is not readily available. Without the aid of an existing pool of participants, recruiting often consumes from 1 to 6 hours per participant, with an average of 3.4 hours, depending on the particular profiles warranted by the test (Rohn, 2004).

Outside agencies typically have more infrastructure and more leads for recruiting some types of participants, but sometimes charge high rates and fail to provide participants who match the requested profiles. Outside agencies are also typically not good sources for finding current customers of the company. For this reason, some companies use a hybrid approach of a combination of an internal resource, such as a participant recruiter, to recruit customers, and an outside agency to recruit prospects. Agencies typically charge between $25 and $100 per participant, depending on the agency and the difficulty of finding the type of participant.

The compensation for external participants can be either money or gifts. Compensation for participants has risen over the years, and companies commonly pay between $50 and $150 per participant per session, depending on the type of participant and the length of the session. Some companies are able to give either their products, such as consumer software, or other gifts, such as clothes and backpacks, as compensation for participants. With many evaluations using between 6 and 12 participants, a company can spend more than $1,000 per study on a single evaluation. This is a small cost, however, and can be recouped through the reduction of a few dozen support calls.

Equipment, Supplies, and Upgrades

In addition to the typical office supplies that any project consumes, usability studies typically require either videotapes and/or digital storage. Because both computer and video technologies are constantly changing, it is not uncommon to upgrade the computers, video equipment, and software used in the lab and in the field every few years. Sometimes a single computer or operating-system upgrade can necessitate an upgrade in software tools and video equipment.

7.6 BENEFITS

Incorporating UE into the product development cycle is a win-win situation for both the company and its direct and indirect customers. The company benefits in two key ways: reduced costs and increased revenues. The company saves time and money by investing its resources more wisely and reducing the likelihood of canceled projects. It also benefits from increased sales and lower support costs. The customers clearly benefit by having a product that is useful and usable, requires less training, and increases productivity.

Benefits should also be examined for all affected organizations. For example, for a B-to-B product, usability improvements can have an impact on the vendor, the direct customer (another company), the third-party integrator who works to customize and deploy the product for the customer company, the customer's IT department, and the customer's internal users and their customers. The business processes should be examined and considered for all impacted parties when benefits are assessed, because a small usability improvement can become significant when multiplied by all users and involved groups.

7.6.1 Benefits to the Vendor Company: Increased Revenues

Increased Sales

Many vendor companies are seeing a benefit from UE analysis in their sales. Although it may not be possible to determine the exact number of sales that result from improving the usability of a product, an increasing number of customer companies and end users are basing their purchase decisions, at least partially, on the ease of use of the product. Fortunately, analysts and the press have become more aware of usability; 10 years ago references to usability in business publications were rarely found, whereas today they are commonplace. Some analysts now specialize in UE, helping to ensure that competitive pressure on vendors remains.

Whereas products in the past competed primarily on features and prices, the usability of a product has become a significant selling point. More usable products are easier to sell. Sales and marketing representatives from several companies believe that most people decide how usable a product is in less than 1 hour.

More companies are leveraging UE professionals as part of the sales process for key customers. For example, in enterprise software, where the business customer will be deploying the product to sometimes hundreds or thousands of people, usability is an important factor. It is not uncommon for these same customers to have their own UE departments and run usability evaluations on the vendor's software. When usability issues are identified, these are communicated between the customer and the vendor, and the addressing of these issues is a factor in consideration of future sales. When customers use UE as a selection criterion, vendors pay strong attention to usability. There have been cases in which deals for hundreds of thousands of dollars have been made at least partially because of the involvement of the UE department.

Measuring the impact UE has on increasing sales can be challenging. One way to measure the impact is to survey customers to discover the reasons that they purchased a product. Another way is to use sales projections to assess the increased sales attributable to producing more usable products. For instance, companies typically have a way to determine sales projections. If a new product or a new version of a product has had significant usability enhancements, the difference between the actual sales and the projected sales can be at least partially attributable to an increase in the usability and, therefore, the salability of the product. When this activity is combined with a survey for customer feedback on the reasons for purchasing the product, the estimates for sales resulting from increased usability become more concrete.

Increased usability can affect both sales of vendors' products and sales from vendors' Web sites. IBM invested millions of dollars in a site redesign, and in the first week after the new site was launched, use of the "help" button decreased 84%, and sales increased 400% (Tedeschi, 1999). Skechers achieved an increase in sales of more than 400% by moving its product selection closer to the home page (Oreskovic, 2001). After Dell applied usability principles to its e-commerce Web site in 1999, its Web sales increased dramatically: online purchases rose from $1 million per day in September 1998 to $34 million per day in March 2000. Nielsen states that e-commerce sites lose nearly half of their potential sales because users can't figure out how to use them. According to studies by NetRaker, a 5% improvement in usability could increase revenues by 10 to 35% (Black, 2002). A study by Forrester (Souza *et al.*, 2001) found that 42% of purchasing customers made their most recent online purchase because of a previous positive experience with the vendor. Forrester also states that financial services executives rated usability as the most important contributor to the success of a bank or brokerage site.

Customer confidence is another factor that increases sales: customers want to be assured that the vendor is stable and credible. Stanford University's *Web Credibility Project* showed that "ease of use" was the second highest factor contributing to a customer's overall perception of a credible Web site (Bisant Interactive, 2002). In 1998, Oracle increased investment in UE after losing sales to potential customers who frequently cited an inability to figure out how to use Oracle's products as a reason for not purchasing (Black, 2002).

For example, in one company, 20 of the most serious usability problems were fixed in the second release of an application-generator product. The revenues for the second release grew by 80%. This revenue increase was 66% higher than sales projections. Although the impact of UE could not be proven precisely, the field test customers repeatedly pointed to improved usability as one of the most significant changes in the product (Wixon and Jones, 1991).

To sell products in Europe, companies have to meet European Community (EC) standards. The EC has already passed a directive stating that, for all display screen workstations put into service in the EC, "software must be suitable for the task," "software must be easy to use," and "the principles of software ergonomics must be applied" (European Economic Community, 1990).

Increased Customer Satisfaction and Loyalty

An increased number of companies are measuring customer satisfaction and loyalty. Companies are tracking these numbers, measuring themselves against competitors, and including them in their corporate goals and bonus programs. Customer satisfaction and loyalty are important because companies can maximize revenue only by growing, not churning, customers, so they rely on keeping current customers while adding new customers. The cost of selling to a current customer is less than that of selling to a new customer, because selling to a new customer requires more marketing and advertising dollars. Satisfied customers not only have brand loyalty but also are much more likely to buy the same brand in the future with less researching of the particular product. Dissatisfied customers are even less likely to consider a brand in the future, even if marked improvements are incorporated into the new version or product.

Customers also influence their friends and families. The numbers cited in internal corporate studies have varied, but conservative estimates are that satisfied customers influence 4 other people to buy the same brand, and dissatisfied customers influence 10 other people to avoid the brand. Internal studies have demonstrated a correlation between increased usability and increased customer satisfaction. The U.S. Better Business Bureau estimates that for every 100 customers that have problems with a site, 50 will tell 8 to 16 other people about their bad experiences. In addition less easily measured costs of a negative experience include brand erosion, lost customers from other channels, and damaged corporate image (Bisant Interactive, 2002).

Increased Customer References

Companies have become increasingly dependent on customer references over the past 5 years. This situation is partly attributable to the fact that analysts and customers are less likely to make a decision based solely on a company's own reports and are thus seeking outside references. For example, analysts typically want either to evaluate and verify the usability and features themselves or to verify them with other customers, rather than relying solely on the vendor's assessment. Prospective customers also increasingly want to talk to existing customers to verify the claims made by companies. More advertising is utilizing quotes from satisfied customers as an effective strategy. Companies are finding that they can remain successful only by ensuring that their customers find their products useful and usable.

Increased Favorable Reviews

In addition to customer references, favorable reviews from analysts and the press are increasingly important to companies' sales. Several projects at one company were improved through usability evaluations and redesigns, resulting in awards citing usability as a key feature. Awards such as these can then be used by the sales force to help with customer sales. Telles (1990) states that the "interface has become an important element in garnering good reviews."

7.6.2 Benefits to the Vendor Company and Customers: Decreased Costs

Many usability benefits decrease costs for both the vendor company and its customers. Because the vendor company often uses its own products (sometimes known as "eating our own dog food" or "drinking our own champagne"), any improvements in usefulness and usability apply not only to customers but also to the vendor itself. Cost reductions within the vendor company can also be easier to measure, which benefits cost-justification activities.

Increased Productivity

Productivity can be increased in two major ways: time to become productive and daily use. Products that are usable take less time to learn, and users can also reach a higher productivity level more quickly. For daily use, usable products support a higher sustainable productivity level. Productivity is one of the best ways to demonstrate the cost benefit of usability. For example, if a usability study demonstrates a 1-minute reduction in the time required to complete a task through more efficient screen layout, keyboard shortcuts, and better navigation, the time saved can be multiplied by the number of users and the number of times per day they complete the task. The savings can then be expressed in dollars (when their salaries or fully-burdened salaries are taken into account) or in more tasks per day. Thousands of dollars of savings per week can be demonstrated for time-sensitive applications, such as those used in call centers, through improved usability. For example, design changes from usability work at IBM resulted in an average reduction of 9.6 minutes per task, with a projected internal savings at IBM of $6.8 million in 1991 alone (Karat, 1990). After UE improvements in Oracle's database manager, database administrators were able to perform their duties 20% faster (Black, 2002).

Poor usability also contributed to the fact that companies did not see the increases in productivity they expected when introducing computers to their employees. An analysis of IT budgets from 138 large U.S. firms between 1988 and 1994 indicated that the gain in IT budgets of 67.4% was much more rapid than the increase in revenue, which was 29.6%, and in profit, which was 39.7%. These differences were at least partially attributable to lost productivity because of poor usability (Strassmann, 1996).

One study found that the average software program has 40 design flaws that impair employees' productivity. The cost in lost productivity is up to 720% (Landauer, 1995). Roach (1992) states that although $500 billion is spent annually in the United States on computers, networks, and information technology, productivity has decreased.

Decreased Training

Improved usability has been demonstrated to decrease training and also to be more effective for learning than relying upon training. A study of management information systems managers found that the training time for new users of a standard computer was 21 hours compared with only 11 hours for users of a more usable computer (Diagnostic Research, 1990). Chief information officers have tightened their budgets and are more wary of software that requires lengthy and expensive training for use (Black, 2002).

However, using training for cost-justification can be a sensitive issue if the vendor's organization has revenue income from training services. Although the customer will benefit from a reduction in training, the vendor company may not. If this is the case, it is best to assess whether this cost-justification will work in favor of the UE department or will instead cause some internal friction if the company is not ready to address a cost reduction in training.

Decreased Errors

Increasing usability can decrease the number of both major and minor errors. These errors not only can cost the company productivity time but can also cause issues with the validity of the data. For example, users of a particular application did not realize that their data were being saved after they entered them into a field, even though they had not explicitly indicated to the application that the data should be saved. When the users abandoned that task, they did not realize that their erroneous data had been submitted, causing problems for the employees receiving the data. In addition, the records in the database contained a significant percentage of mistakes, which caused further problems in the future. By reducing the number of errors, productivity, satisfaction, and data integrity all increase.

Decreased Support

Companies have both direct and indirect support costs. These can be tracked both in technical support calls and in the "hidden" cost of coworkers helping each other. One study estimated that this extra cost is between $6,000 and $15,000 every year for every computer (Bulkeley, 1992). This cost is often borne by computer "gurus" who help their colleagues and is considerably more than the $2,000 to $6,500 that is typically budgeted for the up-front cost of buying the computers and networks and employing a support staff. If products were developed to be more usable, less time would be wasted.

Improvements in descriptions and images on www.lucy.com product pages resulted in more than a 20% reduction in product-related inquiries (Souza *et al.*, 2001). Another company stated that 60 to 80% of support calls were usability-related (personal communication). At an average cost of $250 per support call, the company was spending millions of dollars every month on usability issues. Microsoft also tracks its support call costs and has seen a significant cost savings resulting from improving the usability of its products, such as Microsoft Word (Reed, 1992).

Decreased Development Costs

Significant overruns in cost and schedules occur for most product development projects. Many projects are canceled because of inadequately researched user requirements. The Standish Group (1995) states that more than 30% of software development projects are canceled before completion, primarily because of inadequate user input. The result is a loss of approximately $80 billion annually to the economy. Nussbaum and Neff (1991) state that 46% of all new product development costs are spent on failures. Lederer and Prassad (1992) found that costs for 63% of software projects exceeded estimates, with the top four reasons all being related to product requirements and usability: frequent requests for changes by users, overlooked tasks, users' lack of understanding of their own requirements, and insufficient user-analyst communication and understanding.

By better definition of user requirements at the beginning of a project, development of features can be better prioritized, and feature creep of costs can be reduced. Because the cost of changes increases greatly through the development cycle, identifying and prioritizing development of features early in the process greatly reduces costs. The cost of change is 1 unit in the definition phase, 1.5 to 6 units during the development phase, and 60 to 100 units after release (Pressman, 1992).

Via early identification of user requirements through field studies, task analyses, and product usability comparisons, more informed decisions can be made about which features to implement, reducing the likelihood that additional maintenance releases are needed to address unforeseen customer requirements. Many companies are gaining increased awareness that customer input is beneficial to their success, but don't have the processes or in-house skills to incorporate the data into project funding, definition, and design decisions. The implementation of usability engineering techniques has demonstrated reductions in the product development cycle by 33 to 50% (Bosert, 1991).

Decreased Installation, Configuration, and Deployment Costs

Some less often examined effects of poor usability relate to installation, configuration, and deployment of products throughout the customer company. However, these are critical phases that can have a great impact on costs. For example, at one company a major upgrade of an existing product was planned. Goals were set to improve the ease of installation and configuration, areas in which the previous version had received poor press and generated many service calls. Usability evaluations of the existing and competitive products were performed, followed by iterative design work and usability testing. As a result, usability-related service call rates recorded during the first six months of the product's availability were decreased by approximately 40% compared with those for the previous version during a similar stage in its lifecycle, normalized to the number of licenses shipped.

Decreased Maintenance

Most IT costs occur in the maintenance phase. Although much attention is spent on reducing bugs, only 20% of maintenance is due to bugs or reliability problems, whereas 80% of maintenance is due to unmet or unforeseen user requirements (Martin and McClure, 1983; Pressman, 1992). By identifying these requirements earlier in the development cycle, additional releases and customer satisfaction issues are not as problematic.

7.7 COST-BENEFIT EXAMPLE

The following is a hypothetical, realistic example of a detailed cost-benefit analysis.

The Calculation of the Cost Includes the Following:

- In-house usability staff
- Time spent by staff (UE professionals, developers) × wage rate (same units)
- Additional variable costs (contractors, participants, videotapes, travel)
- Percentage of fixed costs (lab, equipment)

In-House Costs Include the Following:

- Average loaded head count per employee (salary, benefits, vacation time, office space, phones, equipment) = $120,000/year
- Hours per work year = 40 hours/week × 48 = 1,920 hours/year
- Hourly wage = $120,000/1,920 = $62.50/hour
- Time spent on usability evaluation by UE professional (planning, implementation, analysis, recommendations) = 160 hours
- Time spent by interface designer on redesign = 60 hours
- Time spent by development engineer for usability activities = 22 hours
- Staff cost = 160 + 60 + 22 = 242 × $62.50 = $15,125

Laboratory Costs Include the Following:

- Participant recruiting at $100/participant: 9 participants × $100 = $900
- Participant compensation at $50/participant = 9 × $50 = $450
- Videotapes at $5/each = 9 × $5 = $45
- Percentage of lab and equipment costs = amortized cost of lab/hour = $50/hour × 20 hours = $1,000
- Total lab costs = $2,395
- Total usability costs = staff cost + lab costs = $15,125 + $2,395 = $17,520

Benefit 1—Support Call Reduction:

- Support call = $200/call
- 200,000 product version 1 sold

- Support calls due to usability problems = 580,000 × $200/call = $116 million
- Support calls/product sold = 2.9 calls/product
- UE analysis done on version 2
- 300,000 product version 2 sold
- Support calls due to usability problems = 260,000 × $200 = $52 million
- Support calls/product sold = 0.87 calls/product
- Reduction in support calls = 2.03/product
- Support call cost savings due to increased usability = 2.03 calls/product × 300,000 × $250/call = $152.25 million

Benefit 2—Increased Productivity:

- Task A improved by 3 minutes
- Task A performed 5 times/day
- 200 users perform Task A
- Hourly wage (from loaded head count) = $62.50/hour
- 200 users × 3 minutes × 5 = 3,000 minutes saved/day = 50 hours saved/day
- 50 hours × $62.50/hour = $3,125
- Annual amount saved through increased productivity = $3,125 × 240 work days/year = $750,000/year

ROI:

- Return on investment = cost/savings per period, expressed in time period units
- Total cost of usability engineering activities = $17,520
- Total savings from usability engineering = $152.25 million + $750,000/year = $153 million/year = $2,942,308/week
    - Payback period is less than 1 week
- With only direct support call costs of $750,000/year = $14,423/week
    - Payback period = $17,520/$14,423/week = 1.21 weeks

Even if the costs of usability in this example are two to three times as high, the costs are still a small fraction of even one of the benefits.

7.8 STRATEGIES FOR MAXIMIZING EFFECTIVENESS

7.8.1 Understand and Align with Business Goals and Values

Regardless of the type of company, UE professionals can become more effective by understanding the business goals and values of the company and analyzing how to align themselves with those goals and values. Often a change is as simple as adoption of terminology, such as stating benefits in terms of ROI, decreased costs, customer satisfaction, customer loyalty, and quality. In companies that focus on customers, it can mean quoting customers' communications of usability issues back to the company. Some organizations place more weight on issues they hear from key customers than from a usability evaluation. If the company is technology focused, a UE professional can assist in the identification of useful applications of the technology through UE methods.

Another issue is how employees are measured and rewarded in companies. Including usability measures in goals can become a double-edged sword. On the one hand, it would appear to be an accomplishment to have usability measures included in the goals of employees outside of the UE department, so that the organization is focused on making usability improvements. On the other hand, however, executives and other employees also are aware of ways to manipulate the system to ensure a bonus. For example, most professionals have seen people declare success, whether or not the goal was, in reality, met. In addition, when people outside of the UE department are allowed to affect the implementation of the goals, for example, by creating usability tasks that are so unrealistically easy that a high percentage of task completion is guaranteed, then the point of the goal is lost. This situation is not unique to UE goals; the same issues occur with engineering, sales, and other goals. When UE-related goals are created, UE professionals need to be involved in every aspect of planning, implementing, and measuring the goals.

UE professionals are successful when they learn the economic, cultural, and social values of their companies, and leverage these to increase their influence on the products and processes.

7.8.2 Perform Baseline and Ongoing Measurements

Among the most important activities for a UE department to perform are baseline and iterative measurements. To objectively track any type of change, initial measurements must be made before any additional UE work is performed, including metrics such as task time, support calls, satisfaction scores, or other measures. Many UE departments make the mistake of working for several years without measuring and tracking quantitative data. By continually translating usability improvements into tangible cost reductions and revenue increases, a UE department demonstrates its economic benefits to the business. In addition, quantitative data can enable UE to be included in the release criteria reviewed by executives, so that UE metrics are reviewed along with performance, quality, and other release metrics.

7.8.3 Practice Proactive Public Relations

Many skills in business are underrated, and one of these is how to handle internal public relations. A key to successful cost-justification is found in this area, because some costs and benefits are easier to recognize and track than others. Unfortunately, most projects do not stay on schedule, nor do results typically meet initial expectations. A truism of product development is that when planning quality, features, and schedule, you can only pick two of the three. You can set a quality goal and a schedule, but then features may have to be dropped. Or you can create a product that meets the quality goals with the planned features, but the schedule will be longer than expected. Despite this, most people still try to plan projects, which typically don't meet the goals, and then try to determine why.

To minimize the chance that UE will be identified as a cause of a problem, UE management should engage in proactive public relations. For example, most companies do not dedicate sufficient time to planning, and any time spent before development can be viewed as an unnecessary delay. If a UE department does not publicize the benefits of planning and performing UE analysis, the time spent on these activities could be viewed as reasons that product development did not stay on schedule. A UE department can be successful by paying attention to the culture of the company and promoting public relations for the benefits of UE, communicated in terms of the values of the organization. (See Chapters 5 and 8.)

7.9 SUMMARY

By first assessing the company, putting together a strategy for aligning UE with business goals and values, and placing metrics in place to track improvements,

UE departments can demonstrate their value to companies in terms understood by the companies' executives. Although many challenges exist, as with any great accomplishments, UE professionals can be successful and effective in their companies. UE professionals have an opportunity to assist companies in measuring and tracking important metrics, such as reductions in costs due to usability improvements. In addition, cost-justifying usability with near-term benefits is increasingly important in today's business climate. Many of the benefits cited in this chapter can be gauged in one development cycle and by examining the multiple types of beneficiaries, including the vendor company, the customer company, the third-party integrator, and the end user.

REFERENCES

Bisant Interactive (2002). ROI—Usability, Customers and the Business. www.bisant.com.

Black, J. (2002, December 4). Usability is next to profitability. *BusinessWeek online.*

Bosert, J. L. (1991). *Quality Functional Deployment: A Practitioner's Approach.* New York: ASQC Quality Press.

Brooks, F. P., Jr. (1995). *The Mythical ManMonth.* Reading, MA: Addison-Wesley.

Bulkeley, W. M. (1992, November 2). Study finds hidden costs of computing. *The Wall Street Journal,* B4.

Diagnostic Research (1990). *Macintosh, MS-DOS, or Windows: A Synopsis of What MIS Managers and Business Computer Users Had to Say.*

European Economic Community (1990). Council Directive of May 29, 1990 (90/270/EEC). *Official Journal of the European Communities,* L 156(21.6.1990), 14–18.

Hammontree, M. L. (1992). Tools for integrated data capture and analysis tools for research and testing on graphical user interfaces. Presentation at *CHI '92,* Monterey, CA.

Jones, J., and Bullara, F. (2003). Building blocks for a digital solution to usability testing. In *Usability Professionals Association 2003 Proceedings,* Scottsdale, AZ.

Karat, C. M. (1990). Cost-benefit analysis of usability engineering techniques. In D. Diaper (Ed.), *Human Computer Interaction—Interact 90 (351–356).* Amsterdam: Elsevier.

Landauer, T. (1995). *The Trouble with Computers.* Cambridge, MA: MIT Press.

Lederer, A. L., and Prassad, J. (1992). Nine management guidelines for better cost estimating. *Communications of the ACM* 35(2), 51–59.

Martin, J., and McClure, C. (1983). *Software Maintenance: The Problem and Its Solution.* Englewood Cliffs, NJ: Prentice-Hall.

Nakamura, R. (1990, November 19). The X Factor. *Infoworld,* 51–55.

Nielsen, J. (1992). Evaluating the thinking aloud technique for use by computer scientists. In H. R. Hartson and D. Hix (Eds.), *Advances in Human-Computer Interaction* (Vol. 3, pp. 69–82). Norwood, NJ: Ablex.

Nielsen, J. (1993). *Usability Engineering.* Boston: Academic Press.

Nussbaum, B., and Neff, R. (1991, April 29). I can't work this thing. *BusinessWeek,* 58–66.

Oreskovic, A. (2001, March 5). Testing 1-2-3. *The Industry Standard.*

Pressman, R. S. (1992). *Software Engineering: A Practitioner's Approach.* New York: McGraw-Hill.

Reed, S. (1992, December). Who defines usability? You do! *PC/Computing,* 220–232.

Rohn, J. A. (1994). The usability engineering centers at Sun Microsystems. *Behaviour & Information Technology,* 13 (1,2), 25–35.

Rohn, J. A. (2004). *Survey of User Experience Organizations.* Unpublished study.

Rosenbaum, S., Rohn, J. A., and Humburg, J. (2000, April). A toolkit for strategic usability: Results from workshops, panels, and surveys. In *Proceedings of the ACM CHI 2000 Conference.* The Hague, The Netherlands.

Sanders, B. M. (2004, May). HF/E Managers Speak Out. *HFES Bulletin,* 1–2.

Souza, R., Manning, H., Sonderegger, P., Roshan, S., and Dorsey, M (2001, June). Get ROI from Design. *The Forrester Report.*

Standish Group (1995). *The CHAOS Report.*

Strassmann, P. A. (1996). Spending without results? *Computerworld,* 16 (30), 88.

Tedeschi, B. (1999, August 30). Good web site design can lead to healthy sales. *The New York Times on the Web.*

Telles, M. (1990, April). Updating an older interface. In *Proceedings of the ACM CHI '90 Conference* (pp. 243–247). Seattle2.

Wixon, D., and Jones, S. (1991, July). Usability for fun and profit: A case study of the redesign of the VAX RALLY. In *Proceedings of the Workshop on Human-Computer Interface Design: Success Cases, Emerging Methods and Real-World Context.* University of Colorado.

8 CHAPTER Categories of Return on Investment and Their Practical Implications

Chauncey E. Wilson WilDesign Consulting
Stephanie Rosenbaum Tec-Ed, Inc.

8.1 INTRODUCTION

The first edition of *Cost-Justifying Usability* (Bias and Mayhew, 1994) fostered a decade-long discussion about the value of usability and, more broadly, user-centered design (UCD) in product development. Concerns about cost-justifying usability activities waned somewhat during the economic boom years of the late 1990s, but return-on-investment (ROI) concerns resurfaced with a vengeance as the dot-com Bubble burst in 2001, recession loomed, and corporations started looking more closely at UCD activities and their impact on ROI. The tight business market and corporate emphasis on understanding the contribution of UCD to ROI led to pleas in usability forums for hard data, clear examples, and best practices for cost-justifying UCD activities.

Practitioners can now find many books, articles, and links on the impact of UCD on ROI (Landauer, 1996; Nielsen and Gilutz, 2003; Sinha, 2003). However, few details are available on the practical issues that confront UCD practitioners, whose activities are often only part of a large project and whose credibility may depend as much on perceptions of ROI as on actual ROI metrics.

We believe that UCD practitioners should consider both perceived ROI (the belief that UCD adds value to a product) and measured ROI. In some cases, perceived ROI will sustain us when factors beyond our control, such as economic conditions or system reliability, counteract improvements that were predicted from usability activities and reduce corporate ROI metrics. In a recent workshop on ROI (Rosenbaum and Shroyer, 2003), colleagues from major software companies told us stories about products that had undergone iterations of UCD which yielded improved usability (based on test results) and predictions of improved ROI over previous versions of the product. The reality after the product was released was not always so pleasant. Sometimes products showed less ROI than expected after a round of UCD because of the following:

- **History:** The improved site went online just before the terrorist acts of September 11, 2001, and the disruption on the world economy resulted in reduced sales and profits.
- **Marketing strategy:** A more usable site or application was unveiled, but the advertising budget had been cut just before release and customers weren't made aware of the usability enhancements.
- **Unexpected code issues:** When a usability bug was "fixed," a serious new problem was introduced that was not discovered until the product was released. The new bug resulted in customer complaints and a dip in sales.
- **Competition:** A competitor released a major site upgrade that was lauded in the press for its usability and that site "stole" market share.

Like negative results in psychology, medicine, and other fields (Fallik, 2002; Sterling *et al.*, 1995), the failures of UCD to contribute, directly or indirectly, to improved ROI generally go unreported in the literature, although they may be quietly discussed in workshops or private conversations. In this chapter we address some issues that have not been discussed in previous publications as well as other activities that can affect ROI. Later in this chapter, we discuss how even well-conducted UCD activities do not guarantee a positive ROI, but first, we will describe three major ROI categories that all UCD practitioners need to consider in their quest to justify significant involvement in product design.

8.2 CATEGORIES OF RETURN ON INVESTMENT

Usability ROI can be divided into three major categories or components.

Internal ROI. The first component we call "internal ROI." Internal ROI focuses on perceived or actual efficiencies that occur during the development of a product or service that can be attributed to the UCD staff. Perceived internal ROI is the belief that UCD activities are improving the development process. Actual internal ROI is a measured improvement in the development process ("we saved 2 weeks of programming because the paper prototype testing revealed a very serious problem with the proposed user interface [UI] architecture"). UCD activities that promote clear product requirements, eliminate major problems early in the development process, improve communication among the product team, support the re-use of design components, and reduce the cost of development contribute to perceived or actual internal ROI.

External ROI. The second category we call "external ROI." External ROI occurs when UCD practitioners make products or services more profitable for the company, as a result of making them better for customers and user communities. The external ROI comes from more profitable sales to customers, not from development efficiencies. External ROI research cited in this book shows that UCD activities improve direct measures such as sales conversions, revenue, reduced support calls, and number of hits, as well as indirect measures such as customer satisfaction or brand awareness that are viewed as potential contributors to current and future profit.

Social ROI. A final category that UCD practitioners need to consider is "social ROI." Social ROI is the perception of internal stakeholders—managers, developers, and other members of the product team—that UCD provides a return on investment, even when there are no hard data to justify that perception. Increasing social ROI in an organization requires persuasive skills, recognition of what is important to other product team members, and the development of support networks that perceive value in UCD. We will discuss how social ROI affects both internal and external ROI.

In this chapter, we focus on the practical issues associated with internal, external, and social ROI. The first part of the chapter presents specific techniques for improving internal ROI (or the perception of internal ROI). The second part of the chapter addresses how we measure the impact of UCD on external ROI.

8.3 INTERNAL RETURN ON INVESTMENT

Internal ROI focuses on improvements to product development that reduce costs, eliminate rework, and improve efficiency for the development team and other internal stakeholders. Does the work of UCD practitioners have a positive effect on development? Do we, in our daily interactions with product teams:

- Reduce the overall cost of development and maintenance?
- Improve communication among group and individual stakeholders?

- Understand how our UCD activities affect other product team members whose performance is not necessarily based on usability?
- Promote usability as an equal in the trade-off among product attributes (for example, maintainability, reliability, and performance)?
- Support the goals of others (developers, product management, documentation, training, support, sales, and marketing)?

We contend that UCD practitioners must be perceived as contributing to internal ROI before they can contribute to measurable, external ROI. If our colleagues on a product team think that we make their work difficult or inefficient (a common concern of development and product managers) or that we promote usability to the exclusion of other business goals, we may not be able to affect the external ROI.

One solution for getting product teams to believe in the value of UCD is to conduct the first UCD activity with no direct costs to the product team and demonstrate improvement to the development process (for example, reduce bugs, cut down on rework resulting from ambiguity or poor requirements, shorten development, or even cut down the number of meetings). Doing an "on-spec" (speculative at no cost to the development team) UCD project to show how internal ROI is improved has both benefits and risks. The benefit is that an on-spec project shows how UCD can fit smoothly in the development process, cut costs, and reveal design issues or problems. The risks with doing an on-spec project include the following:

- The revelation that the product has some very bad problems that were not overtly recognized (or that were considered "taboo"). To reduce this risk, choose a product that you believe is not fatally flawed but has some usability or design issues that are not obvious but that would affect product quality and are fixable, given the point in the schedule at which you conducted your on-spec UCD activity.
- Getting involved too late in development. Conducting a "free" usability test late in development is risky for a UCD practitioner new to the product team because any negative results may put the development and product managers in a bad light and create a subtle, but lasting, resentment (negative social ROI). To reduce this risk, it is better to plan an on-spec activity early in the development process, such as field visits to obtain requirements or evaluation of prototype design concepts. As a UCD team establishes social and actual ROI, the risks of "late-breaking" results may decrease.

8.4 INTERNAL SOCIAL RETURN ON INVESTMENT

From an internal perspective, social ROI deals with the perceptions of stakeholders in an organization that UCD practitioners add value to the development process, even in the absence of specific supporting data. Social ROI is an important concept in the discussion of both internal and external ROI issues; we discuss external social ROI later in this chapter.

Internal social ROI is important for UCD practitioners in an organization because the impact of UCD on overall product ROI is often difficult to measure, takes time to establish (Conklin, 1995; Ostrander, 2000; Ward, 2002), and may not adequately capture the impact of activities such as task analysis, requirements definition, UI inspections, and paper prototyping. Some groups may track the hours they spend on UCD, but not track all the time that other stakeholders spend collaborating with them or performing activities that could be considered UCD. The move toward collaborative design, in which UCD practitioners serve more as facilitators than designers, complicates the calculation of actual internal ROI.

Research from the software engineering literature provides evidence that it is easier and more cost-effective to change a product early rather than late in development (Pressman, 1992; Hoo *et al.*, 2001), but we rarely have data to support that hypothesis in day-to-day (or even month-to-month) development work. Managers, directors, and members of product teams have to *believe* that UCD activities provide benefits, because tracking the specific input of UCD teams is often difficult and time-consuming.

Moreover, in many environments, design is so interwoven among collaborators that it is difficult to identify the contribution of any particular group or activity and proclaim that "our persona work contributed 5% to the ROI for product X," or "our UCD team's work led to 10% more profit." Of course, if you can perform ongoing data collection and have the services of a statistician, you can tease out these internal ROI data using multivariate statistical techniques, but this approach is beyond the resources and training of most practitioners.

In addition to the difficulties of collecting data on the contribution of UCD activities, other events can overshadow the contributions of UCD practitioners. Consider the impact that the tragedy of September 11, 2001, had on hotel occupancy. A UCD team working on requirements and UI specifications for a Web reservation system in late 2001 might have designed a more effective site that needed less development time than a prior version without UCD contributions. However, the effect of 9/11 on travel would easily have obscured whatever contributions to internal development ROI were made by the UCD team. The reduced sales could lead to the perception that the UCD efforts added no special value.

The 9/11 travel example is a case in which sales and profits would fall despite the greater efficiency and improved quality of the development team's efforts. However, internal social ROI, the product team's belief that the UCD team did a good job, could carry them through until sales improved and the value of the redesign was proven. In the next section, we will provide practical advice on methods for increasing internal social ROI.

8.5 IMPROVING INTERNAL SOCIAL RETURN ON INVESTMENT

Internal social ROI is a function of the number of colleagues with whom we have connections, the types of exchanges we have with those colleagues, and our perceived value in those exchanges (Brown, 1986). When UCD practitioners start in a new position, a primary task is to create the perception—as well as the reality—that we are adding value to product development and improving potential sales.

To measure social ROI, a dynamic phenomenon, we need to consider iterative measures of our perceived value to development, such as the internal equivalent of a customer-satisfaction survey in which our customers are members of the product team. Consider enlisting a third party to conduct a semiannual, anonymous survey that measures perceived effectiveness and value of UCD to the organization and collects suggestions for improvements. Like any other survey, an internal social ROI survey (please don't call it that!) should go through iterative design and usability evaluation.

One author of the chapter received valuable feedback from such a survey of product team members in a corporate environment. Although the ratings of his effectiveness, value, sensitivity, and communication skills were quite positive, the survey also indicated that he sometimes gave clients too much data and did not spend enough time helping them understand how to turn the data into concrete design recommendations.

On one project, for example, a client team had collaborated on a large number of contextual interviews, created affinity diagrams (Beyer and Holtzblatt, 1998) with the UCD team, and were then left to ponder the mystery of how to translate the huge affinity diagrams into UI designs. The clients were too embarrassed to admit they didn't know what to do with all the data and so didn't mention their confusion. The author was unaware of the struggle to turn the user data into useful design requirements until 6 months later, when the internal satisfaction survey was conducted. This anecdote highlights the value of an internal satisfaction survey and also points out that follow-up on projects can increase communication effectiveness, an attribute of social ROI.

As a result of the internal satisfaction survey, a decision was made to plan more time to work on design recommendations directly with developers, and the next survey showed that the author was perceived as more supportive and effective. His social ROI improved, and he was invited to all major design meetings for input. Over time, the client became facile with the analysis and interpretation of data. The improved social ROI led to a more efficient development process and eventually measurable improvements in usability.

Other, less formal, measures of social ROI include the following:

- The number of invitations that UCD practitioners receive to planning and management meetings
- The number of project teams requesting our help
- The number of people who visit or call to ask questions
- The number of times we are mentioned in senior management meetings and announcements
- The number of product team members who are involved in UCD
- The number of people who believe that our work contributes to a more efficient development process and more salable product (Conklin, 1995)

All of these measures are related to the number of connections that we make in an organization. We recommend that UCD teams explicitly consider and collect these "soft" metrics, as well as hard metrics such as hours spent on project work.

8.5.1 Making Connections

Establishing internal social ROI begins when we make connections with stakeholders in development and other product groups. The number of successful connections is an informal, but important, measure of social ROI. One useful private tool for practitioners is a "stakeholder-connection" diagram that tracks personal connections with stakeholders, the potential value of those connections to the practitioner and the stakeholder, and the goals and rewards that are most important for those strengthening the connections.

The stakeholder-connection diagram shown in Figure 8.1 can be used as a planning tool for building social ROI in an organization, as well as a visual representation of the UCD team's connectedness and type of relationship with stakeholders. The concept of the stakeholder-connection diagram is loosely based on rich picture (Monk and Howard, 1998) and sociometric diagrams (Forse and Degenne, 1999), which show the types and direction of relationships among individuals and groups.

These diagrams highlight the high-level issues and connections among stakeholders involved in product design and delivery. The relative size of the circles indicates organizational power. The circle for a particular individual contributor may be larger than a formal title would indicate if that individual is a guru who has the ear of senior management. Arrows show whether relationships between the UCD team and the various stakeholders are nonexistent, one way, or reciprocal. The type of line can show whether the connection is positive or negative. In Figure 8.1, solid lines indicate neutral or positive connections and dashed lines indicate negative connections. You could also use colors to indicate connections. Green could represent positive connections, red could represent negative connections, and gray could represent neutral connections. The diagram can serve as a rough measure of UCD visibility, credibility, and connectedness in an organization.

Our practical advice about making connections is to:

- Identify the leaders, key experts, and other influential people in an organization. They are often the people who can provide top-down support for UCD.
- Set up general information meetings with these people and determine others who will be involved. Keep the initial meetings short and nonthreatening; listen carefully to the business and professional issues that are key motivators for each stakeholder.
- Develop a method to track the development of UCD connections with stakeholders. The stakeholder-connection diagram is one example of a visual tracking method, but you could develop other visualization methods such as a matrix or scatterplot.
- Set goals for increasing both tactical and strategic connections. Invitations to requirements meetings would be an important tactical connection given the significance of good requirements for product success; invitations to senior management meetings where the CEO, vice presidents, and directors gather would be strategic connections. Make tactical and strategic connection goals a formal part of your UCD project plan. You might create a plan to meet with several new tactical and strategic leaders each month.

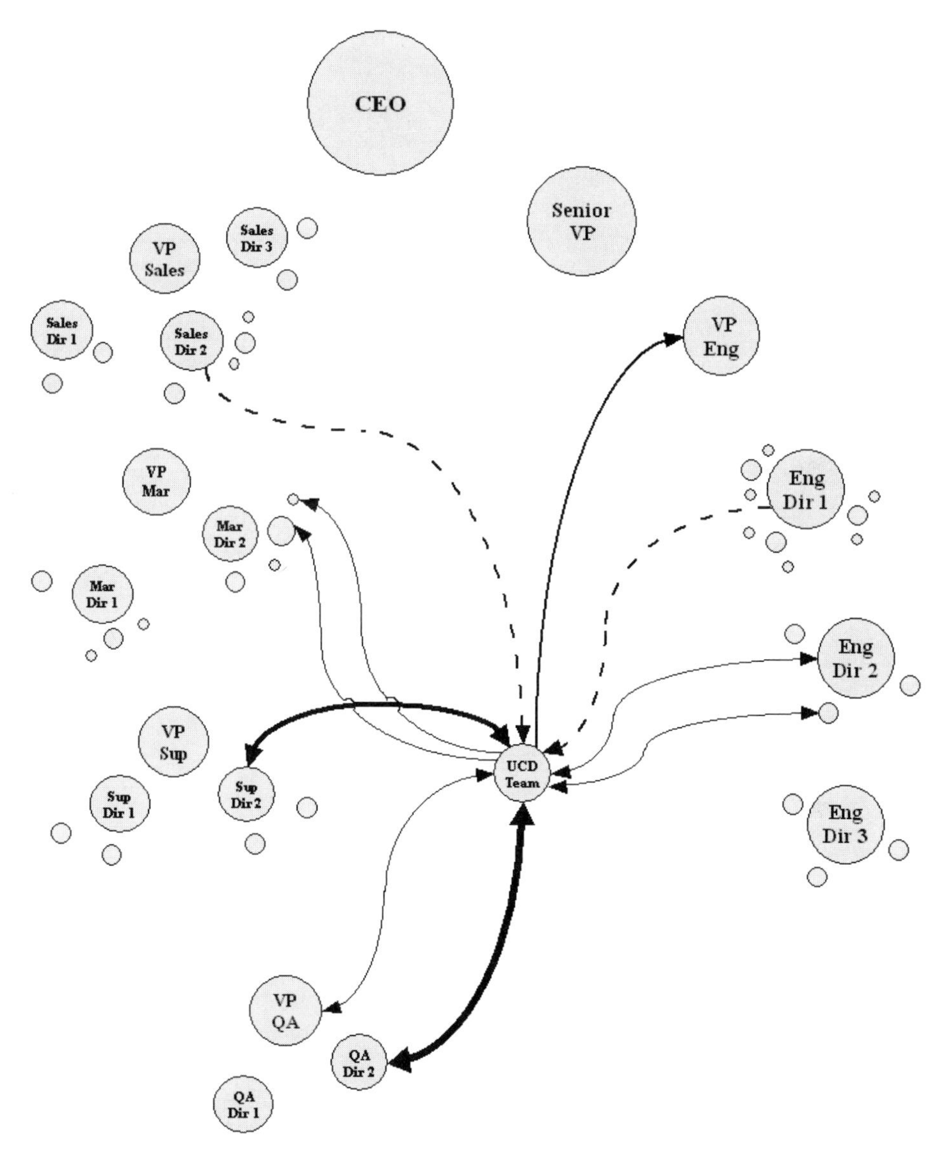

FIGURE 8.1 Sample stakeholder-connection diagram. Connections with senior management don't exist yet. There is some opposition from an engineering and sales director and a variety of weak to strong links with sales, marketing, quality assurance, and engineering. For a new group, this might illustrate connections after one to six months. Some connections are bidirectional and show reciprocity, whereas others show mostly one-way relationships. The dashed line from Sales Dir 2 to the UCD team indicates that the sales director has had negative interactions with the UCD team, perhaps in the form of criticizing the team or keeping them away from customers.

- Because the detailed activities of design can involve working with individual members of the product team, track your connections with those individuals as well as the tactical and strategic leaders. Getting buy-in from developers, writers, and quality assurance (QA) engineers is important and provides a bottom-up approach to complement the top-down approach that involves tactical and strategic contacts.

8.5.2 Communicating with Product Teams

Another method for assessing and improving social ROI involves the tools we use to communicate with product teams. The perceived value of our reports and presentations can have a substantial impact on our social ROI. Therefore, we should request feedback on the various documents, databases, and presentation methods we use to communicate the results of UCD. Consider the following concrete activities with regard to reports and presentation materials that can improve internal social ROI:

- When beginning work with a product team, interview them about their reporting requirements (Molich, 2001). Find out how much detail they need (or can handle). Does the team want just the top 20 problems or the 50-page report with all 100 problems? How would the team like you to rate the severity of the problems? Does the team want you to send an e-mail summary right after a study, with a more detailed report later? The UCD community is debating about how to present UCD data, with some practitioners sticking to a particular approach ("I give them what I believe are the top 20 problems and nothing else!") whereas other practitioners ask how much detail is needed and tailor the report to the situation. In our experience, the requirements for usability data vary considerably among product teams, with some wanting every detail and others wanting only the most serious issues. Determine the requirements for each of your clients and tailor your data presentation to meet those requirements.
- Provide product teams with sample documents that show how you present data. The sample documents should be based on a product that is not being built by any of the teams with whom you will be working. Consider preparing a report on a competitive product, if possible, because that provides both a work example and useful information for the team. One caution here is to verify that doing a competitive review is not seen as an infringement on the role of marketing or a competitive analysis group in a corporation (this could produce negative social ROI).

- Ask product teams to provide feedback on the first report you give them. Have them do a usability edit in which they mark up the document as they read it, and then tell you how the report could be improved. Your willingness to accept criticism may increase your social ROI.
- Make sure that your recommendations are clear, unambiguous, and detailed enough for the rest of the product team to understand and use as the basis for actual changes to the product.

8.5.3 Respecting the Rest of the Product Team

We have encountered colleagues who take the role of "usability evangelist" or "user advocate" to the extreme and alienate their colleagues on the product team, causing negative social ROI. Our practical view is that usability should have the same level of importance as other product attributes such as maintainability, reliability, installability, performance, and scheduling. Product development involves complex trade-offs, and usability is not the only critical attribute; our social ROI will go up when we strive for equality rather than superiority and eschew arrogance for understanding.

As an example of arrogance that can harm our social ROI, consider the common practice of showing videotape highlights of the most egregious problems faced by participants in usability tests. These highlights, popular at professional conferences and seminars, often make members of the audience laugh and wonder just how "those developers could be so stupid to design something this bad." Showing only the most extreme cases of user suffering creates an adversarial relationship with our product team colleagues and reduces our credibility.

We can do many things to avoid the perception of arrogance, but the most important thing is to acknowledge a person who does good design work or invests resources in UCD. As a profession, we are strongly biased toward finding the bad in designs, and we need to recognize that others can do good design or contribute to positive user experiences as well. This acknowledgment can be either from the bottom up, starting with one-on-one relationships with developers and other product team members, or from the top down, by telling senior management about positive as well as negative findings (Molich, 2001). What specific things can we do to create a collaborative environment in which we are not viewed as the "GUI police"?

- *Praise even when suggesting major changes.* When we work with product team members, we can add subtle praise to our suggestions about how to improve design. Our social ROI with a developer is more likely to go up when we say "The layout of the Search results is generally good, but I wonder if we should put a bit more separation between the columns of information so users can scan it more easily," rather than, "These columns are too close together and users will find it hard to read the information." The first statement acknowledges that your colleague has done something right and suggests a way to make it "even better," whereas the second statement just implies bad judgment on the part of the developer.

- *In meetings, acknowledge the positive contributions of others, even if the main reason for the meeting is to critique a design artifact.* When conducting group inspections, for example, make it a policy to explicitly acknowledge positive features. This practice can have a benefit on external as well as internal ROI, as the authors have seen cases in which a good design is altered (with no clear justification) and the revision leads to diminished usability, delays for fixes, reduced customer satisfaction—and ultimately—reduced profits and a negative impact on overall product ROI.
- *In reports of UCD activities, acknowledge good design as well as areas that need improvement.* It is important that praise be legitimate and timely (Molich, 2001).
- *Increase your likability.* Cialdini (2001) presents compelling evidence that liking is an important factor in persuasion. If we want the product team to take our recommendations to heart, we should consider how to help them to like us more. One important factor in getting colleagues to like (in addition to respect) us is similarity. People tend to like people who are similar to them in opinions, personality, traits, background, and lifestyle (Byrne, 1971; Myers, 2002) and the more they like us, the more they will agree with us. There are some simple ways to invoke the similarity principle that can benefit the UCD practitioner, increase liking, and lead to more effective interaction. For example, one author took a course on a development tool that he did not expect to use much, but felt that knowing its limitations would help him understand what developers experience when building Web applications. The developers noticed this increased similarity and were more positive toward his design suggestions. Consider how you might increase social ROI by being more similar to your stakeholders (without being disingenuous, of course).
- *Increase your familiarity* (Cialdini, 2001; Zajonc, 1968). A simple example of the "repeated interaction" principle is attending development or management meetings when you are not explicitly on the agenda. The return here may not occur immediately, but attending meetings with teams even when

the focus is not on usability increases familiarity, liking, and social ROI. If team members see UCD practitioners more often, they are more likely to view us as part of their team and ask us questions early that might prevent significant problems later. It is especially important for usability professionals who are organizationally remote to demonstrate that they are personally involved with the product being developed (Bias and Reitmeyer, 1995).

- *Foster mutual and successful cooperation.* Avoid activities that are one-sided and adversarial, especially early in relationships with product teams. A mistake that some talented practitioners make when they start a new job is to list the usability problems with a product without knowing the product's history. Several years ago, during an initial meeting with a senior architect (who had influence with senior management), one author made a comment about how childish some icons looked. The icons had been designed by the architect's 10-year old daughter! It took about 2 years to recover from this inadvertent adversarial comment—and two more versions of the product to get rid of the bad icons drawn by the daughter.

8.5.4 Understanding How Others Are Judged

To amass social ROI in an organization, we should find out how others are judged and respect those criteria, even as we try to change them. For example, developers are often judged on meeting ambitious schedules, fixing serious bugs in a given time period, and solving complex coding problems. A good user experience is generally not part of the formal performance goals that affect raises, bonuses, and respect or whether a developer has a cube or an office (Wilson, 2002). UCD goals might be ones that create a good user experience for our Web users and UCD practitioners may be graded on achieving them, but our other colleagues on the product team may have quite different goals that we have to understand.

A strategic objective for a UCD team should be to have usability or user experience goals included as part of the formal performance plan for development and products managers, developers, writers, and other members of the product team. When one author was a development manager, he worked with his developers to revise their performance plans to include goals related to UI consistency, usability, and participation in UCD activities. The developers were then quite willing to expend more effort making the interface better when UCD was an explicit part of their performance plan. Another approach is to persuade the product team members that UCD activities will reduce future problems (isn't that what we say paper prototyping will do?) and allow them to focus more on those things that bring them respect (and bonuses), such as adding new features, optimizing performance, or creating robust code.

8.5.5 Supporting Groups Outside Mainstream Product Development

A general method for increasing social ROI (and perhaps even overall product ROI) is to support and involve groups outside development—marketing, sales, sales support, training, the legal office, and even human resources. By "support" we mean doing small things to help make those groups successful. Although this may seem like common sense, it is easy in a busy environment to neglect those with whom we don't interact daily. Small favors invoke the norm of reciprocity (Gouldner, 1960; Webster *et al.*, 1999), which states that people will try to repay favors out of sense of obligation. Examples of support and involvement that can increase our social ROI by helping others achieve their goals include the following:

- Conducting early training sessions that highlight issues found in reviews, testing, and other UCD activities. These training sessions help those who are not directly involved in development—such as quality assurance, technical support, and documentation teams—understand the product's strengths and weaknesses. The sessions can be short and informal, but have many benefits for UCD practitioners including the following:
  - Contacts and relationships with people who know customers and users
  - Increased visibility within the organization
  - Feedback from attendees about new products that can be incorporated in future design iterations
  - Colleagues who will alert us to problems or support future UCD activities
  - A better understanding of the product
- Compiling a readable summary of new features and usability enhancements that sales and marketing can use in their literature and presentations to customers. A nontraditional deliverable that highlights the positive aspects of a product can create positive social ROI and also affect the actual product ROI, by providing sales and marketing with early tips that help them explain and demonstrate the product to customers more effectively (Conklin, 1996; Wilson, 2002). Helping sales and marketing understand new products can also provide increased access to users. A side benefit of helping sales, marketing, training, and customer support is that these groups may become great sources of usability data (Bias, 1997).

8.5.6 Enlisting Quality Assurance Teams as Partners in User Interface Quality

UCD teams generally evaluate limited portions of any large Web site or application. Usability testing, for example, often focuses on a relatively small set of frequent or important tasks that people can perform in a few hours. UI inspections can cover broader issues such as branding and consistency, but even UI inspections may touch only a small number of pages in a large Web site or application.

One method for improving the overall quality of a product is to train QA teams how to spot UI problems in features and pages that will not be tested or evaluated (Wilson, 2002). QA teams are often the very first users of features and have to test them in more depth than is possible in usability testing. QA testers can alert UCD practitioners to common inconsistencies, awkward task sequences, confusing documentation, standards violations, bad tab order, accessibility problems, poor error messages, and places where simple errors could be prevented. Enlisting QA teams and training them about UI and usability problems involves the following steps:

1. Befriend and make the QA director and QA teams your allies. QA directors often have final sign-off authority on new versions of a product even when usability managers do not, so their support can be helpful if a product is not meeting usability and UI design specifications. We can make allies of QA team members by inviting them to observe usability sessions and participate in UI inspections. UCD practitioners can support QA teams by reviewing their test plans, which invokes the norm of reciprocity (Gouldner, 1960) and creates an obligation to support the UCD team in return for the test plan review.
2. Devise a training session for the QA team that describes common categories of UI and usability bugs that could be caught during QA testing and reported in the bug-tracking system. A half- or 1-day training session can serve as the catalyst for enlisting the QA team as an extension of the UCD team. One of the authors did this several years ago for a Fortune 500 company by creating a half-day training session on "Usability Bugs." The session explained various categories of bugs, introduced a severity scale for usability bugs that paralleled the severity scale for technical bugs, provided examples of how to rate the severity of bugs, and elicited the QA team's input on how usability bugs should be reported in the official tracking system.

3. Before delivering the QA training, brief the product development managers about the QA group's involvement in finding and reporting a wide range of usability bugs (Wilson and Coyne, 2001). One author made the mistake of reversing this sequence. The day after the training, the QA team (who previously had not been required to report many categories of usability problems) felt empowered and reported hundreds of new bugs for several products in late stages of development. Hostile calls and e-mails poured in from development managers who suddenly saw their release dates and revenues slipping.

   After these managers received the same training on usability bugs, they agreed that usability problems should be reported like any other bug, but they recommended that reporting by QA of usability bugs should be started at the beginning of a development cycle so developers are not stunned by seeing many new bugs just before the product is ready for release.

4. Set up a mechanism to provide feedback on usability bug tracking. This could be a weekly review to assess the consistency of ratings, the level of detail needed to describe a problem, and the reactions of developers to the problem reports. Ongoing feedback can also help us to develop explicit guidelines for usability bug tracking. The UCD practitioner will probably need to look for patterns of common problems and address these global issues through a Web style guide, management meetings, and additional UI training for designers and developers.

5. Collaborate with the QA team on how to include usability issues in their test plans. For example, the QA testers can develop test scripts that examine issues such as error handling, and the UCD team can provide general guidelines for reporting issues such as inconsistent form design and awkward task flows that might have previously gone unreported. QA teams can become close allies of UCD practitioners and be a vanguard for usability improvements as well as for the technical quality of products.

8.6 REUSE AND INTERNAL RETURN ON INVESTMENT

Reuse of code has long been a rallying cry in product development. To the extent that we can reuse content, major features (such as search boxes, registration forms, log-in pages, and list builders) and UI code, we can reduce the costs of development, maintenance, and support, as well as improve time to market, quality, and cost-effectiveness. UCD practitioners have many opportunities for identifying areas of Web products or services where reuse is possible, including the following:

- Requirements gathering
- Consistency inspections, with which we compare different products across an organization to identify inconsistencies in common features (e.g., do we have 12 different log-in UIs or different methods for creating lists of objects?)
- Style guide development and institutionalization
- Development of UI patterns that show solutions to common problems

In the next section we will describe how UCD practitioners can promote reuse and affect the internal ROI of a product in development.

8.6.1 Methods for Raising Awareness of Reuse

Product managers in many Web companies are focused so intently on their own product that they are not aware that other groups may have better solutions to similar problems. In large Web projects, we have often found that teams working on different products, services, or parts of Web sites fail to reuse features, content, and consistent UI patterns. Cross-team collaboration among different product or component teams is hindered by organizational, political, and financial barriers.

UCD practitioners can be a link across these barriers and highlight areas in which everyone can gain from reuse. One approach to raising awareness is to generate a comparative catalog of similar features. For example, let's say that in a large Web site the UCD team discovers through testing, inspections, or other evaluation methods, that there are multiple—and very different—registration forms that generally require the same information, as well as multiple search engines, calendar widgets, or methods for building a list of desired items.

The implication for both internal and external ROI is clear. Many people are solving the same problem with no awareness that someone else may already have a better, faster, cheaper, or more flexible solution. Furthermore, these different solutions to the same problem increase maintenance costs and probably reduce user satisfaction of people who visit more than one part of the site.

Some other approaches for raising awareness about reuse are also relatively cost-effective. Even in small development groups, it is useful for team members working on different areas of a Web site or application to show their designs to each other early in development, something that seems obvious, but is often not done until it is too late. In larger organizations, UCD teams can host demo days or Web UI fairs (Good, 1989) during which different groups present sketches, prototypes, or working products and see what other groups are doing.

Another method is to create design posters showing segments from a style guide or UI pattern catalog. Placing these posters in conference and meeting rooms can activate the "familiarity breeds liking" principle (Cialdini, 2001; Zajonc, 1968) and be a gentle persuasive technique for consistency and reuse.

8.6.2 Within- and Between-Product Consistency Inspections

The first step in establishing reuse involves an assessment of where reuse of UI features and designs is possible. The basic method for assessment is a comparative consistency inspection to identify how similar features or content are presented within and between products. The results of the consistency inspection should include the following information:

- The components and product names
- A name for the feature (e.g., required field indicator, search, or registration form)
- Side-by-side pictures of the different implementations of the same feature
- Notes under each picture that describe consistency and usability issues related to that feature (These notes should include both positive and negative aspects of the features.)
- A summary of the inconsistencies that emerge from the examples

When UCD practitioners start in a new position or on a new project, consistency and usability inspections are often part of "coming up to speed." Doing a semi-formal consistency inspection when you are in a new position or consulting opportunity is also a method for reducing time to understanding and increasing credibility among the product team, because we can learn a lot about related projects quickly. Keep in mind that it might not be politic to circulate this report to management until we learn more about the organization (or, as noted earlier, whose child designed the icons).

8.6.3 User Interface Patterns as a Tool to Promote Reuse

The work by Alexander and his colleagues (Alexander *et al.*, 1977) on architectural patterns prompted the development of programming patterns (Gamma *et al.*, 1994) and UI patterns (Tidwell, 2003; Van Duyne *et al.*, 2003). Alexander's patterns were structural features that could be applied during design to improve the "habitability" of a town, street, or building. In the UI design area, patterns are potential solutions to design problems that can increase the "habitability" of virtual environments. The components of a UI pattern can include the following:

- A name for the pattern
- An illustration of the pattern
- An indication of when you would use the pattern (what problem is it solving?)
- A description of how the pattern works (a description of the key objects and interactions in the pattern)

For example, related patterns (higher-level patterns) can be built up from lower-level patterns. The use of patterns imposes regularity across designs, thus promoting UI reuse. Patterns can suggest solutions to problems such as how to present a list of selected items (e.g., a shopping cart pattern), how to deal with hierarchical information (e.g., an expandable tree), or how to step a person through a task (e.g., what many call a wizard). Examples of UI patterns on the Web can be found at www.welie.com/patterns/ and time-tripper.com/uipatterns/index.php?page=About_Patterns/.

The value of patterns is that they can provide members of the product team who code UIs with ideas both for high-level design problems such as what a "community site" home page should contain and lower-level problems such as how to represent an input prompt (Tidwell, 2003). The patterns for various UI problems or features can serve as a starting point for design and usability requirements and reduce the time needed to develop a design. UI patterns can be institutionalized into style guides that recommend approaches to particular design problems. However, patterns are flexible, and the particular pattern or details in a pattern must fit the users, tasks, and domain of the product. So, the use of design patterns can impact development by the following:

- Reducing the time required to formulate solutions to design problems
- Providing developers who also design UIs with a tool for learning the rationale behind particular design solutions (thus generally enhancing the developer's sensitivity to design issues that can affect usability)
- Improving consistency across an organization by establishing patterns that can be reused

The development and use of UI patterns is not yet a mainstream activity for most UCD practitioners and developers, and many questions remain about how patterns are developed and validated for specific domains and applications; but the general concept shows much promise for reducing the costs of development and raising the general level of design awareness. What practical advice can we offer practitioners about UI patterns?

- Review current collections of patterns, looking for existing solutions to problems we face in our UCD work.
- Examine common problems identified through usability testing, heuristic reviews, consistency inspections, and other design and evaluation techniques and consider generating patterns for those problems. Validate the patterns by reviewing them against user, task, domain, and technical requirements.
- Publicize the validated patterns as a way of reducing internal development costs, because many people can use similar designs.
- Use the patterns as a way to teach design rationale.

8.6.4 Group User Interface Inspections

A wide variety of techniques for conducting UI inspections are available. Most techniques permit some form of group review, some after individual reviews. A simple but potent method for increasing reuse and improving internal ROI involves inviting managers, architects, and other stakeholders from different product teams to participate in UI inspections. The inclusion of stakeholders from different product teams is likely to reveal unneeded redundancy in UI code and make other groups aware of possibilities for code-sharing, improved maintainability, and cost savings.

One author conducted an inspection with representatives from several product lines. During the inspection, we learned that one group had purchased a "date-picker" widget to address a complaint that the product under review required users to type in a day for a financial transaction, rather than choose it from a calendar (which was important because some transactions were not possible on weekends and holidays). That development manager offered the date-picker widget to other groups, thus improving consistency and usability, as well as reducing costs and maintenance. If the managers had not met at this UI inspection, they might have purchased different date-pickers and directly reduced the internal ROI for both products, as well as increased the potential for customer complaints.

8.6.5 Graphic User Interface Rolls for Workflow Assessment and Comparison

Several years ago, one author worked as a human-computer interface consultant on a suite of Internet products that were designed to be used together. However, as the senior manager for the product discovered, the specs and prototypes for the individual products seemed to come from different planets. The manager wanted the product teams to consider reuse of code and designs to reduce costs and improve the user experience. The author's job was to evaluate the set of tools for UI consistency and make recommendations on how to improve the overall consistency of the entire product suite, including specific suggestions about where features, designs, and code could be reused. Some of the products were already released, some were in beta testing, and some were in the functional/UI specification stage.

As part of the review, we created a visual task flow using screen shots of the products or graphics from the functional/UI specification. The task flows were created with Visio and printed out on long rolls of paper with a large-format plotter. The main flow was one path, and all the dialogs, pop-ups, and common error messages were shown below the main flow. These graphic user interface (GUI) rolls were as long as 20 feet, and we created them for eight separate products.

Then developers and product managers from the different product groups reviewed several rolls at a time, looking for places where the products did similar things differently or were different in other ways. Laying out task flows on a GUI roll using the actual UI objects was much more revealing than seeing a single page of objects or even walking through prototype code. The result of the review was an overall improvement in consistency, increased reuse of common components, and an artifact that developers could take home and show to their children (the fact that the developers could show their children what they did in a visual way provided unexpected social ROI).

8.7 USER-CENTERED DESIGN INFRASTRUCTURE AND RETURN ON INVESTMENT

UCD infrastructure involves the skills, technology, processes, templates, and strategies that an individual or UCD team uses to support the business goals of the company, organization, or the UCD team. The goals of a strategic infrastructure are to promote efficient UCD within a team, the business group, and the company or organization as a whole (McElroy and Wilson, 2003).

An effective UCD infrastructure would reduce the expense of creating and modifying Web products and services and reduce the cost of development. We believe that UCD teams should invest heavily in an infrastructure to facilitate reuse, reduce the time required for usability activities, and eliminate common design mistakes.

The basic component of a UCD infrastructure is skill. Do the UCD practitioners have the necessary social, technical, and domain skills to provide efficient support for a particular product development environment? As in many new professions, extensive debate over the core skills for a UCD practitioner is ongoing. Organizations such as the Usability Professionals' Association (UPA) are trying to define the core body of knowledge for UCD practitioners. Several general skills are at the heart of an effective UCD infrastructure.

The first skill is an ability to consider the business (and sometimes personal) goals of all the internal stakeholders in a product. We have heard colleagues make proclamations such as "I am the user advocate" and "Usability is the most important attribute of a product." The people who stand on a usability pulpit need to remember that usability should be considered at the same level as features, reliability, maintainability, evolvability, and other "ilities." Although evangelism is vital, design is largely a social activity that is driven by personal, business, and life goals, and unbridled evangelism can result in isolation rather than collaboration.

The second general skill is the ability to formulate and ask good questions and then listen to what others are saying about a product. This skill requires practitioners to reflect on how their background and the current situation may bias what they ask and what they hear. It also requires experience in asking open-ended questions that don't "lead the witness," or we will hear only what we expect, rather than gain insights from others.

The third general skill that UCD practitioners should possess is continuous self-assessment of our skills and knowledge. We believe these are the most basic skills required by UCD practitioners; without them, we can have little impact on the ROI for a product.

Once we have the necessary skills to support a strong UCD infrastructure, the next step is to develop efficient ways to plan and conduct UCD activities, including the following:

- A system for recruiting and tracking participants. Recruiting is time-consuming and can tax the resources of a UCD team, so a robust and efficient system can contribute significantly to internal ROI. A recruiting infrastructure could include the following:
  - Recruiting personnel (internal or agencies)
  - Recruiting "ads" on Web sites
  - Templates for the recruiter's script and screening questions that can be easily modified for different studies
  - Recruiting guidelines
  - A database that can track past and potential participants. The recruiting database should be designed for a wide variety of activities including testing, requirements gathering, diary studies, and other UCD activities. If participants can sign up from the Web, the sign-up data should feed directly into the recruiting database to minimize data entry.
  - A clear statement of how people are to be treated during recruiting and the study itself, to provide guidance on potential recruiting problems and ethical dilemmas
- A semiformal method for conducting in-house usability testing sessions. This methodology is generally an important aspect of the UCD infrastructure, because a session really begins when the participant enters the facility and only concludes when the person leaves the building. The primary textbooks on usability (Nielsen, 1994; Rubin, 1994; Dumas and Redish, 1999; Barnum, 2002) provide many suggestions that are useful for creating a test session infrastructure, which should consist of the following:
  - A method for pilot testing and making modifications to the test plan and script
  - A method for separating participants from observers and briefing observers on proper etiquette. Expedited testing methods often lead to invitations for developers and other stakeholders to "drop by" and watch

sessions. Depending on the layout of the testing environment, the observers can be located in the same room as the participant, in an adjacent or nearby observation room, or at an entirely different location. A protocol should be developed for each setting; failure to have a smooth process for observers, wherever they are located, reduces credibility (social ROI) as well as increases set-up costs. For example, a group at a remote site may observe a session using a remote conference tool such as WebEx or streaming video. That group needs procedures for how to ask questions; they could transmit questions via Instant Messenger to the facilitator unbeknownst to the participant or by phone conferences between sessions. Remote observers must also receive the task script, so they have a context for their observations.

  - ✦ Guidelines for greeting and bringing participants to the testing room. Many UCD practitioners fail to realize that the first few moments when the facilitator meets the participant set the stage for the rest of the session. A nervous or harried facilitator could easily transmit anxiety to the participant.

- ✦ Processes for collecting, analyzing, and distributing data, both for individual studies and across studies, such as the following:
  - ✦ A set of processes for collecting qualitative and quantitative data. The process should include training for new notetakers on what to look for and how to code the information. The procedures may be different for in-person and remote observers or for UCD observers and observers from development teams.
  - ✦ A process for distributing the data collected during the sessions. Methods range from e-mail notes that observers send to the product team to daily e-mails that summarize the results from testing sessions, to automated reports generated with logging systems.
  - ✦ Guidelines for managing the product team's desire to have near-real-time reports on usability tests, the hazards of which include overzealous observers who take notes and send them along after each session, as well as a focus on minor visual problems (e.g., the user does not like an icon) rather than global issues (e.g., the navigation architecture is poor). For example, one author ran a Web usability test in which the first three participants had virtually no problems. The product manager watched those sessions and sent a "no problems" e-mail to the product team and vice president. Unfortunately, the next 10 participants had much more trouble, and some serious (and consistent) problems emerged. When all the data were analyzed, and the overall results were found to be quite

negative, numerous diplomatic engagements were required to explain to the angry stakeholders the discrepancy between the first e-mail message and the formal report.

- And finally, a process for examining patterns that emerge across usability studies. In environments in which many products are being tested, the use of a database or visualization method for uncovering meta-problems—problems that occur repeatedly across products—can reveal general problems that are more serious than any single study might indicate. For example, if you run 20 studies in a year and find that in each study, several participants encountered a similar medium-severity problem, that may indicate a more fundamental problem such as a common GUI object (e.g., a calendar or graph widget) with an inherent design flaw used across products.

8.7.1 Templates and Common Documents

One relatively easy and often neglected way to reduce costs is to create a set of UCD templates and common documents that can be reused and are easy to access by UCD and product team members. The initial creation of templates and common documents requires some upfront investment in branding, design, and evaluation (all documents should be reviewed and tested). The templates and common documents should be designed with several goals in mind:

- *Establish a brand for the UCD team.* Brand recognition is important in these tough economic times; UCD teams need to establish their presence with a brand that both fits in with corporate or organizational culture and highlights the unique contribution of the UCD team.
- *Explain the process.* UCD practitioners repeatedly have to describe how their methods fit into development. Create a one-page summary document and a more detailed document that explain how UCD fits cost effectively into development. These can save considerable time, because each group or client you work with is likely to ask you to explain the process. If you have a lab that is the centerpiece of a corporate UCD team, having brochures always available will reduce the need to send special information to visitors.
- *Increase reuse of common materials.* For example, scripts for Web testing will have many common features (an introduction to the method, the informed consent issues, the general "rules" for testing, etc.). Investing in a template that includes the common features, with an easy way to include project-specific information, can be a tremendous time saver.

- *Have sample documents ready for stakeholders and potential clients.* A common request in the consulting and corporate worlds is a preview of what a usability inspection or contextual inquiry report would look like. Sample reports that are not proprietary can save time and increase the social ROI with internal or external clients. Sample reports should be based as much as possible on real data from a known Web site and should be reviewed for clarity, readability, branding, and language that is direct and clear, especially in the executive summary. For a set of sample reports, be sure to apply similar branding to each document (for example, consistency in the cover page and general format).
- *Provide additional information about UCD.* Provide a brief reading list (one to two pages suffice for most situations) of articles, books, and Web sites that would be useful for internal and external stakeholders. Over the years, we have found that many people want to read about UCD methods, case studies, or its impact on ROI. Update the list several times a year and make sure that the entries are consistent with your own UCD philosophy. If you have published articles, books, or online materials, include them on the list to enhance your credibility.

The next sections of this chapter discuss external ROI. First we describe the different kinds of metrics for external ROI: historical, predictive, and simultaneous; direct and indirect. Then we review approaches and variables that can be used to measure improved user experience. Finally, we address the challenges practitioners face in measuring external ROI.

8.8 EXTERNAL RETURN ON INVESTMENT

External ROI for usability occurs when we make products and services better for our company or organization, as a result of making them better for our customers and user community. The return comes from the goods and services we sell, rather than from our efficiency in producing them.

There are two levels of external ROI. In the first level, usability engineering and UCD address the behavior and perceptions of customers and users. External usability ROI is tied to how much we can improve the user experience or how well we can achieve explicit usability goals. For the first level of external ROI, the question we want to answer about UCD activities is, does the work UCD teams do have a positive effect on what we want to accomplish? For example, does UCD work make Web sites and products:

- Easier for customers to use (more usable)?
- More effective at helping customers achieve their goals (more useful)?
- More satisfying to use?

Most UCD practitioners would agree that the answer is a resounding "Yes!" (Bias and Mayhew, 1994; Donoghue, 2002; Donahue *et al.*, 1999; Usability Professionals' Association, 2003). The related ROI question is, do the benefits of accomplishing these goals exceed the costs associated with UCD? Our customers clearly benefit from UCD activities that improve product quality, but how does our own organization benefit?

In many companies, improved user experience itself does not constitute ROI. To produce ROI, usability enhancements must also increase profits or improve branding (or goodwill). Increased profits are the traditional financial measure for ROI, and improved branding and goodwill are among the ways to measure the social ROI of companies or external social ROI.

Thus for the second, broader, level of external ROI, we need to show that improved user experience increases sales, reduces post-sales expenses, or produces enough external social ROI that the organization believes there will eventually be a positive effect on the bottom line. (The other way to increase profits, by reducing development expenses, is addressed earlier under "Internal Return on Investment.")

8.9 HISTORICAL, PREDICTIVE, AND SIMULTANEOUS MEASURES

External ROI often entails comparing before-and-after measurements of usability. Sometimes we assess how usable a product or service was in the past, compared with a current product developed with UCD activities; these are called historical measures. Alternatively, we can measure the usability of a current product and predict (and then measure) how usability will be improved in the successor product that our efforts help create; this is called predictive testing or pro forma analysis (Brealey and Myers, 2003).

Not every comparison has to be across time or before and after. We can make concurrent or simultaneous measures, where the comparison is usually with a competitor's product. For example, a usability goal might be for shoppers to go through the checkout procedure 10% faster than on a competitive site.

Each of these approaches has strengths and weaknesses. In a recent workshop on usability ROI (Rosenbaum and Shroyer, 2003), several participants commented on the difficulty of collecting historical data about pre-UCD products. At one company, the usability group was rarely involved in projects early enough to capture real "before" numbers and then was usually deployed elsewhere before they could capture "after" numbers.

Also, the cost of measuring a current product's usability (establishing a baseline) when extensive changes are already planned may itself be difficult to justify, especially when the new features will require the UCD team to focus on the new product. When the usability of competitive sites or products is assessed, skill and resources may be needed to define comparable user tasks, and some features or functionality may not be comparable.

Measuring external ROI becomes more important as practitioners are faced with more usability work than they can perform with current budgets and as they compete for resources with other parts of the organization. Even though collecting metrics for external ROI can be challenging, the process provides not only data but also insights into social ROI: how UCD teams can work together with other groups in an organization to improve product quality, customer satisfaction, and the bottom line.

8.10 DIRECT AND INDIRECT MEASURES

Direct measures of external ROI are improvements to the user experience that increase profits directly; these improvements either increase revenue or lower the cost of supporting released products. Indirect measures of external ROI are improvements to the user experience that increase user productivity or customer satisfaction and thus increase profits indirectly. In financial discussions, these are often called "hard dollars" and "soft dollars," respectively.

We find direct measures of external ROI when we can tie user goals to business goals. Most of the first-level metrics for external ROI address how well the product meets user goals. To measure the contribution of usability to profit, it's critical to consider related business goals. For example, a user goal for an Internet service provider might be to have adequate storage for messages in e-mail boxes. The related business goal might be to increase the adoption rates for an extra-cost storage option, something we can measure.

Similarly, if our business goal is to have more people make purchases on our e-commerce site, and we assume that there are people who want to do so, then we can usefully measure the completion or drop-out rates for a registration process required before visitors can make purchases on the site. Increasing registration completion rates can improve both the user experience and sales.

However, some improvements to the user experience don't affect sales so directly. Thus the social aspects of external ROI are needed to convince stakeholders that a positive user experience will, in the long run, improve the financial health of the organization. This thesis is comparable to justifying the ROI of public relations or branding campaigns. All are indirect measures, but we can nevertheless demonstrate ROI.

For example, we can publicize improved user experience to increase sales and then measure sales, even though other efforts and occurrences ("confounds") can have a concurrent effect on sales. By confounds, we mean interfering situations, such as the effect of September 11, 2001, on the hotel reservation Web site described earlier. In psychology, confounds are defined as variables that change with the independent variable being measured, so their effects cannot be meaningfully assigned to either the confound or the independent variable. Perhaps more simply, confounding occurs when we do not control extraneous variables that may influence the results of our research.

Any ROI we can demonstrate will help UCD teams cost-justify their usability activities. However, corporate management usually finds direct measures more compelling (social ROI is needed to help support indirect measures of ROI). Therefore, as our UCD teams look at improving the user experience and measuring its impact on ROI, we should give priority to efforts that have direct measures—or if the measures are indirect, the ROI should be dramatically high.

For example, when redesigning the architecture of an e-commerce site, you can measure how easily users navigate the current site and complete purchases. Then, design the new architecture both to improve the usability of the site and to make it easier for shoppers to buy high-profit items. The improvement to overall site usability is an indirect measure, and the increase in sales of high-profit items is a direct measure of external ROI.

8.11 MEASURING IMPROVED USER EXPERIENCE

What should we measure to demonstrate external ROI? The major categories of improved user experience are the following:

- First-time use
- The learning experience
- User performance by experienced users ("continued use")

- Need for customer support and service
- Customer satisfaction and attitudes

If we look at the ways in which usability practitioners measure improved user experience, most of the categories have both behavior and perception measures. The perception measures are social ROI; it's harder to tie them directly to profits, but they contribute to an organization's long-range success or to its key strategies. In the 2003 UPA workshop on ROI (UPA, 2003), practitioners compiled an extensive list of factors or variables to consider as we built guidelines for measuring ROI and justifying investments in UCD. In this section we expand on the workshop list. For each of the categories of improved user experience, we list variables, describe methods for measurement (with examples), and discuss the likely confounds specific to that category. Confounds that affect all external ROI measures are described later under "Challenges to Measuring External ROI."

8.11.1 First-Time Use

For some products, Web sites, and services, the business *and* the usability goal is that the target audience can "walk up and use" the product, without training. For these "kiosk-like" UIs, variables to measure can include the following (Whiteside *et al.*, 1988):

- Number of errors made during initial use
- Success rate in achieving stated goals with the product
- Types and severity of errors (consequences and ability to recover from errors)
- Time spent in errors, compared with productive time
- Users' perception of their success (or failure)

Depending on the nature of the application, the "initial use" may begin with installation or registration, in which case these tasks are usually a major area of concern. All these variables can apply to the installation process, and successful installation is critical because it must precede all other use of the product. Registration is a comparable mission-critical task for sites that require it before access to key features.

Classic "out of the box" usability testing is extensively used and highly successful at measuring these variables. Other data-collection methods can also measure the initial-use variables:

- Field studies for public Web sites and for products for which we can observe installation in the actual context of use
- Remote and online usability testing using a variety of screen-sharing software tools (Bartek and Cheatham, 2003; Chen *et al.*, 1999; Perkins, 2001)
- Automated data gathering of variables such as drop-out rates during registration or payment, although making effective design changes is more difficult without knowledge of user goals

An example of external usability ROI for first-time users is eBay's Sell Your Item (SYI) form, for which UCD consulting and usability testing directly improved eBay revenue (Braun, 2002). UCD team members from eBay looked at page statistics for the SYI form and hypothesized that the complexity of the form contributed to high drop-out rates. They estimated that a 1% improvement in the drop-out rate would increase listings correspondingly. At 1.1 million listings a day, the increase would be 11,000 listings a day. If each eBay customer paid only the minimum listing fee of 30 cents, revenue would increase more than $1.2 million a year, exclusive of final value fees for items sold or customer support costs for people calling with questions about the form. To improve the SYI form, an eBay usability engineer worked with the design team for three months of usability testing and consulting (additional costs included participant incentives and an assistant during two testing weeks). Redesign after usability testing required an additional week—the total was a small investment for the predicted revenue increase.

What confounds affect our measurement of first-time use? Both usability improvements such as the eBay example and technological improvements can have a positive impact on Web sites and products intended for use without training. The current trend toward "autonomic computing," in which self-managing systems function without user intervention, can increase user success without significant changes in the UI (Helland, 2001; Milewski and Lewis, 1997; Rosenbaum, 2003).

Today most autonomic features manage administrative functions such as updating software or virus protection or compiling an inventory of network devices; but end-user features can also be at least partially self-managing. For example, in registration forms, a zip or postal code entered by the visitor could cause the site to generate (editable) city and state fields, reducing data-entry errors and increasing user input efficiency. UCD teams can provide affordances that improve user acceptance of cost-effective autonomic systems, thus increasing the ROI of both usability and technology contributions.

8.11.2 The Learning Experience

For more complex products and sites, some kind of user learning or education is a requirement for satisfactory performance. User training has both internal and external ROI components. The internal component is the savings in budget and schedule when learning materials (courses, online training, online help, and documentation) are developed for products better designed to meet user goals.

The external usability ROI component of user education can be measured either by the reduction in the length and cost of learning needed to achieve a specified level of user performance or by the improvement in user performance gained by a given investment in training. For reduction in length and cost of learning, the variables include the following:

- ✦ Users' time away from job tasks to seek online help, to consult documentation, or to call customer support
- ✦ Time spent by colleagues to provide informal training or coaching
- ✦ Direct costs of training courses, materials, or documentation
- ✦ Users' time away from the job to take formal training classes
- ✦ Number of calls to customer support during first *X* months of use

For improvement in performance after training, the same variables apply as for first-time use (listed above), with the addition of the following efficiency measures:

- ✦ Length of time to perform specific tasks
- ✦ Number of requests for assistance or calls to customer support during specific tasks

The confounds that affect measuring the external usability ROI of training are compounded by the before-and-after nature of measuring training itself. The ROI of training is usually calculated by measuring user performance before and after training. If we want to measure the contribution of usability activities to training, then the "before" measure is the user performance improvement (after training) for the previous product version or predecessor product, and the "after" measure is the user performance improvement for the current product.

For example, the training department of a company building Web-based call center applications found that customer call center representatives handled 20% more calls per hour after taking the company's training course for Version A of the product. Publicizing these measurements as part of the company's marketing program could either increase sales of the call center product itself, sales of billable training, or both, thus demonstrating the ROI of training.

If we want to measure the usability ROI associated with training, we need to compare the improvement after training on Version A with the improvement after training on Version B of the call center application, which we supported with usability engineering. Suppose that in Version B, call center representatives handled 30% more calls per hour after training. Some of that gain comes from UCD activities that improve the application UI. But the training department probably believes the improvement is due to the following:

- Creating higher-quality training materials and online help
- Hiring more experienced training delivery staff
- Using more effective authoring tools for Web-based training or online help, so the same amount of training development time produced better results

Both teams are correct; improvements in ROI are often the result of joint efforts, even when the work wasn't explicitly collaborative. Reducing the need for training and improving the effectiveness of training both enhance external ROI, and the UCD team helped produce the gains. From a social and organizational standpoint, the UCD team need not be the sole owner of an ROI increase. Recognition as part of the team that improved profitability is a similarly valuable contribution.

8.11.3 User Performance by Experienced Users

Measuring the ROI of usability for experienced users of an application is a major challenge. Most usability evaluations and assessments are performed with new or inexperienced users of a Web site or application, and we are usually concerned with the initial experience of the target audience. However, there are Web applications such as order-entry and human resources systems in which the target audience consists primarily of experienced users who might resist changes to the UI.

The scenario that arises most frequently with experienced users is that they have already invested time and effort in learning to use a site or product; they don't want to give up the productivity tied to their current skill level. A redesigned site or a major new version of a product is not necessarily desirable to experienced users. Rather, they ask: are the new features and benefits worth enough to them to justify the new learning curve? As members of a UCD team, we need to address the trade-off of changing the design to broaden the user base versus keeping the installed user base happy by making minimal changes.

When the LexisNexis online legal information service first became available, it had a steep learning curve; typically one person in a law office (often a paralegal or a junior attorney) became the "LexisNexis guru" for the entire office. During development of an easier-to-use Windows® version, LexisNexis developers were surprised to encounter resistance from the gurus in their customer law firms. The people who had made the investment to learn the old system didn't want to give up their expert status for a service anyone in the office could use.

In another example, a major computer company had developed an internal software tool for tracking defects, which was used by thousands of development and QA engineers in many locations. The company planned to replace this tool with a customized Web application that managed defects and problems throughout the product lifecycle and was integrated with their customer relationship management (CRM) system to improve company-wide business practices.

The company conducted contextual inquiries with the current users and learned about the many ways engineers used their internal system. Despite inconsistent use among different business units and virtually no training or documentation, experienced users were fairly productive and satisfied with their existing tool. After iterative exploratory usability testing of the new Web application, with measurement of usability improvements as development continued, the user community had one primary question: could engineers report and track defects as quickly with the new system as with the old tool? The target audience of engineers was more interested in maintaining their personal productivity than in usability or the larger business goals the new system helped the company achieve.

Therefore, the usability practitioners needed to design and conduct a quantitative usability test comparing users' productivity with the two defect systems (a complex study involving controlled training sessions on the new application before the test sessions and statistical analysis of the data). If the stakeholders in the target audience were shown they would not lose productivity, they would be more willing to cooperate in deploying the new tool. In this case, the ROI was tied to the improvement to the computer company's CRM and business processes. The development team and the UCD team both contributed to the better user experience of the application, the former by reducing processing time and the latter by improving the UI.

In the 2003 UPA workshop (UPA, 2003), one participant described how an Internet company provides a "preview" feature so users can click and "see how this page will look in a redesign," then have the opportunity to provide feedback. Even though the majority of responses from experienced customers are typically in favor of the existing page, the process sets customer expectations that change is coming.

8.11.4 Need for Customer Support and Service

For an ideal Web site or product, the target audience uses the site or product successfully to achieve their personal or business goals, either with or without learning aids. Once the product is shipped or the site goes public, the company simply collects revenue and develops the next version. Of course, in the real business world, users have problems that the company needs to solve.

Thus almost all companies offer some kind of customer support, customer service, telephone "help" lines, maintenance contracts, e-mail support, or "live chats" on their Web sites. Most companies offer several types or levels of support.

The business model for customer support varies according to the organization and its products and services. Whether or not customer support groups are designed to be profit centers, they are an operating expense to the company and reducing the cost of calls to customer support improves profitability. Even when a company sells maintenance contracts or service agreements (or charges by the minute for service calls), that revenue creates profit only when customer requests for service don't exceed the expectations on which pricing was based.

If UCD activities improve Web sites and products so they require less customer support, those "self-reliant" products make a direct contribution to external ROI. The 2003 UPA workshop (UPA, 2003) identified the following factors or variables for measuring the contribution of UCD to reducing the need for customer support:

- Number of calls
- Length of calls
- Length of time customer waits "on hold"

- Percentage of calls escalated
- Frequency of the same topic occurring in questions or problems
- Use of frequently asked questions

Much of this information can be collected using content analysis (Neuendorf, 2001; Rockley *et al.*, 2003) of customer call logs. Interviews with support staff can provide insight into the data and may even lead to improving the quality of customer call data collected. Remote surveys of customers who called for support can be used to collect data about customers' perception of product quality; or UCD teams can conduct critical incident surveys (Flanagan, 1954) with customers.

In one example cited at the 2003 UPA workshop (UPA, 2003), total calls to customer support did not decrease, so the UCD team examined the categories of calls, looking specifically at usability-oriented issues. This approach was only partially successful; under the pressure of fielding calls, the customer support representatives didn't worry much about categorizing, so the data had poor inter-rater reliability. Nevertheless, some improvements were significant, such as product installation, for which the total number of calls was reduced by almost half. It would have been even better to measure the reduction in calls per license or calls per visitor; unfortunately those data were not available.

Another UPA 2003 workshop participant described the challenge of showing usability ROI for large enterprise applications for which revenue is also generated by company services such as training and support. The UCD group hoped to show that revenue increases from licensing fees would bring more ROI in the long run than a reliance on revenue from services that increase the total cost of ownership for customers. However, at the time of the workshop, multidepartmental tracking to support this thesis was not taking place.

Some concerns discussed in the workshop were that customer support groups might see call data analysis as a threat to job security or resist UCD work because support revenue would diminish. However, from a company-wide view, the same support groups could handle a larger customer base if each customer required less support. The issue of support revenue is comparable to that of training revenue; if a company's business model depends heavily on revenue from these sources, then they should not be our first choice for measuring usability ROI.

Customer support is also similar to training in that reductions in the need for customer support can have more causes than UCD activities. A new release might have fewer bugs than its predecessor or might be marketed to a more technically sophisticated audience.

8.11.5 Customer Satisfaction and Attitudes

The ISO 9241-11 Guidance on Usability standard (International Organization for Standardization, 1998) defines usability as "the extent to which a product can be used by specified users to achieve specified goals with effectiveness, efficiency, and satisfaction in a specified context of use." Although traditional human factors research focuses on behavior, UCD practitioners in commercial environments recognize the critical importance of measuring customer satisfaction.

This aspect of user experience is often measured in parallel with others. While we measure first-time use, the learning experience, or continued use of products and Web sites, we can concurrently ask questions about customer satisfaction. Many usability research designs include a post-task participant "activity" consisting of a satisfaction questionnaire. We can also conduct specific data-collection projects focused on customer satisfaction:

- Surveys
- Ethnographic interviews
- Competitive analysis

The UPA 2003 workshop (UPA, 2003) discussed several kinds of "usability report cards" for both internal and external customers. For example, an internal ROI measure was how many of the recommended steps in the usability process the development team followed. However, customer satisfaction measures are usually applied to external customers, and measures such as the Gómez Scorecard (Gómez, 2003) and other industry ratings were discussed. Because such heuristic evaluations lack primary data from customers, they should always be combined with or supplemented by direct customer surveys.

Customer satisfaction surveys are one of the few methods we have to demonstrate the correlation of improved user experience with sales. A generally accepted management principle is that increased customer satisfaction improves profitability; companies with satisfied customers either sell more products or can set higher prices for their products and services. If we couple our user experience measurements with before-and-after customer satisfaction measurements, we can correlate improved user experience with greater customer satisfaction and thus demonstrate external ROI.

For example, one UPA 2003 workshop (UPA, 2003) participant described how her group developed a customer survey and report card. They planned to administer the survey to usability test participants as well as to postlaunch customers and then generate scores on the report card for products. However, corporate management was concerned that research and development staff would overwhelm customers with requests for information, so it was difficult to obtain approval to distribute the questionnaire. This example also illustrates the interrelationship of social and external ROI; most UCD practitioners observe that customers like to have us ask their opinions, but management may be unaware of this goodwill (see "ROI from Communicating Improved User Experience" later in this chapter).

Sophisticated methods now exist for collecting customer satisfaction data on the Internet. For example, at least one U.S. vendor, Foresee Results, uses an online version of the American Customer Satisfaction Index (ACSI), produced by the University of Michigan since 1994 and published quarterly in the *Wall Street Journal*. Their online "customer satisfaction dashboard" continuously analyzes customer feedback, and the ASCI directly links customer satisfaction measurement with financial returns.

Another potential measure of customer satisfaction is reduced frustration. A University of Washington researcher invented a "frustration meter" that usability test participants could manipulate during task performance, creating a record of frustration levels at various steps of product use (Ramey, 1999). Some online survey questions address user frustration, but when the question is not concurrent with the causal activity, such summative reporting may not be reliable (for example, numerical self-ratings of physical pain used in clinical settings are always immediate measures, not historical ones). Although we believe that UCD activities reduce user frustration, reliable contemporaneous measurement techniques are still to come.

Branding is yet another factor relating to customer satisfaction. Without straying too far from our topic of external usability ROI, we should be aware that branding programs can have both positive and negative effects on customer satisfaction. When it fulfills its intended purpose, branding provides reassurance to site visitors and customers that an organization they trust is influencing their experience with a product or service. However, branding can also contribute to clutter (discussed later in the sections on confounds), which tends to reduce customer satisfaction.

Because so many experiences can influence customer satisfaction, it's especially important to collect customer satisfaction data whenever possible, both during other usability activities and as separately designed projects. Also, we should remember that the correlation between user satisfaction and user performance is less than 1.0. Many studies report inconsistencies between users' performance and their perception of success (Bailey, 1993; Nielsen and Levy,

1994), another reason—in addition to social ROI—to collaborate with stakeholders in setting usability goals.

8.12 CHALLENGES TO MEASURING EXTERNAL RETURN ON INVESTMENT

Many factors affect our ability to measure the external ROI of usability or influence the impact of our measurements. The primary challenge is that most of us are working in the real world of commercial products and services, not performing well-controlled experiments. As a result, no before-and-after measurements are "pure."

Yes, we can collect historical data about a Web site developed with little usability engineering and limited customer data and then compare its usability with that of the redesigned site produced with UCD methodology. Inevitably, because of wider business goals and the resulting design, development, and marketing decisions, the new site will be different in many ways from the old one—and some of those differences will be unrelated to the UCD contribution.

8.12.1 User Experience Confounds Outside the User-Centered Design Team

In most organizations, the UCD team is one of many groups contributing to the business decisions that affect how the user audience experiences a Web site or application. The result is that user experience—and subsequent sales—may change for reasons we didn't anticipate or couldn't avoid.

For example, consider a product that tests "100% usable for e-commerce customers." The improvements go public, and the UCD team expects an upturn in sales or service, but that doesn't happen (in fact, sales go down slightly) because at the last minute, several major recommended changes were not compatible with a database update. Or the product team had market research data saying 60% of the company's primary target audience used broadband service, so they specified a site redesign that required a high-speed connection to display complex pages quickly. Meanwhile, in the geographic region where most of the company's customers live, one broadband supplier went out of business and another promised service "next month" for six months without delivering it.

On sites that include advertising, "clutter" can be a major confound. Studies of ad clutter (e.g., Broussard *et al.*, 2001) show that too many ads cause visitors to leave a site, despite the usability of its other features. A consortium of Internet companies is working with the Interactive Advertising Bureau on research into what constitutes clutter for different audiences, including number and size of ads, audio elements, and length and type of animation.

The Internet community is encountering the same dilemma that traditional media have faced for years: the advertising that finances publishing must appeal to the target audience—or at least be minimally acceptable—or circulation will decline. Niche media such as fashion, sports, and culinary magazines have learned this skill; their advertisements increase readership.

Most UCD teams are not involved in setting advertising policy for the Web site, yet the level of ad clutter can either attract customers or cause them to abandon the site. A recent example of Web advertising on a popular news site showed a cheerful ad for spring clothing in the middle of a chilling description of explosions in Baghdad and Karbala that killed more than 143 people, an absurd juxtaposition that could easily produce negative reactions to the entire site.

8.12.2 Addressing Confounds with Controlled Testing

Many marketing and development decisions affect the user experience, yet the UCD team doesn't "own" them. How can we handle such confounds? Basically, we need to take confounds into account and control for them. A major benefit of the Internet Age is that we don't have to be huge organizations to run controlled market tests. For Web sites and Web applications, it's possible to expose a UCD improvement to a small segment of the user community and monitor the results, making a comparison between old and new UIs with the same ad clutter or incompatible database.

One UPA 2003 workshop (UPA, 2003) participant described an example of a "beta launch" used as a controlled market test in an Internet company. The company chooses either a product category or a country, launches the new design for this limited market, and then also conducts usability testing on the beta site.

Another workshop participant discussed his "live testing" program, in which a new page design is launched on a special server that shows the new interface to a small percentage of the user population, but a large enough sample for statistical treatment. This live testing can produce results in as little as a few hours or up to a month if there are significant cyclic variations (e.g., page views for entertainment topics increase on Thursdays and Fridays).

Conducting such live market tests is even more valuable when the confounds are unrelated to user experience, as described next.

8.12.3 Confounds Unrelated to User Experience

Our measurements of the success of a new Web site or application are also affected by many confounds unrelated to user experience:

- *Seasonality.* Bricks-and-mortar businesses have long considered seasonality in their strategic planning. Obvious seasonality confounds, especially for consumer e-commerce, are the peaks that occur before Christmas, Mother's Day, Valentine's Day, and other major holidays. A redesigned consumer site released in time for Christmas shopping may bring in much more revenue than an equivalent site released in March.

  Seasonality doesn't only affect consumer sites. Most organizations' budgeting and purchasing decisions are tied to their fiscal calendars. However, sales management in major companies knows how quarter-end, year-end, and holidays usually affect sales, so it's possible to adjust for this confound when external ROI is measured.
- *Economic climate.* Not surprisingly, sales of many products and services decline during a recession and increase during boom times. However, some market segments grow during difficult economic times. For example, home remodeling in the United States blossomed during the recent economic downturn, because many families who might have moved to a more expensive house instead chose to improve their current homes. Google returns more than two million results for a search on "home remodeling."
- *Availability of resources.* For hardware products, problems in obtaining parts from vendors can reduce the amount of product manufactured for sale and thus sales revenue. A shortage of software engineers, database architects, or network administrators can cause an application to be released with bugs in critical functionality or without key features.
- *Distribution channels.* Transportation strikes can affect how quickly products reach retailers and customers. A new Internet virus or worm can affect e-commerce sales. Or a "disconnect" can occur between the e-commerce and bricks-and-mortar sides of a business, for example, when a site features a toll-

free phone number for customer service, and the call center staff access a database with different pricing or delivery options from those on the Web site.

- *Sales efforts.* Revenue can increase—or decrease—due to sales activities unrelated to the user experience. The company may hire more or better sales staff or provide better sales training to existing staff; conversely, a competitor may just have hired the company's three best salespeople. Management may change the sales commission structure to favor the business unit whose products were designed with—or without—UCD activities.
- *Marketing programs.* Revenue can increase when a company conducts a larger or more effective advertising campaign. On the other hand, the authors have worked on projects for which a new marketing campaign is launched that alienates customers, who then migrate to another site.
- *Competitive activity.* If a competitor announces or releases a new product or launches an effective marketing campaign, these events are likely to have an impact on the sales of other companies in the field. Other competitor activities can decrease or increase sales as well; for example, a competitor may undergo a merger or an internal reorganization that affects its position in the marketplace.
- *Public relations and branding.* Public relations and branding usually have long-term effects on revenue, rather than immediate ones. Their goal is to improve the image of the company to its target market and to increase goodwill. For example, the dual gas/electric Prius hybrid car has improved Toyota's branding and generated goodwill among ecology-conscious drivers, although it contributes little to the immediate profitability of Toyota Motor Corporation. In the short term, if a company receives sudden publicity, revenue may unexpectedly rise or fall. A stock split or an employee receiving a Nobel or Pulitzer Prize will probably increase sales, whereas a key employee being indicted for fraud may have the opposite effect.

Although we can't control such confounds, any effort to measure usability ROI should include an inventory or audit of confounds, so that we can communicate the overall picture to colleagues. The contribution of usability professionals improves the user experience, but to address ROI, we also need to show the relationship between the user experience and the rest of the business environment.

As described earlier, market testing is the best way to isolate the contribution of UCD work, and the Internet environment ameliorates its cost. Nevertheless, any kind of ROI measurement requires an investment. We can't spend $50,000 to justify the ROI of a $30,000 contextual inquiry project. Market tests should measure as much as possible with the least risk. For example, we might run a market test:

- After an iterative UCD process of design consulting, exploratory testing, field studies, and validation testing has resulted in extensive changes to the UI.
- During a relatively level sales period; in e-commerce it's not recommended to test new ideas in November and risk Christmas sales, even if only part of the target audience experiences a new site design.

Measuring external ROI is usually more difficult than measuring internal ROI, because so much of the improvement takes place outside the UCD group. Nevertheless, careful observation of confounds, supplemented by controlled market testing whenever possible, will help these measurements, and working with other corporate groups seeking to measure ROI can demonstrate external ROI resulting from joint or collaborative efforts. One of the most straightforward ways to cost-justify usability is to reduce the customer support burden, as described earlier, which can produce dramatic external ROI.

8.13 RETURN ON INVESTMENT FROM COMMUNICATING IMPROVED USER EXPERIENCE

"Send me a memo!" It's human to want praise, and UCD teams make significant contributions that deserve appreciation. But it's not enough to improve the usability of a product or Web site; our job isn't done until we communicate the improvement to two key stakeholder groups—customers and management:

- Showing customers the benefits and value of UCD is one way to increase sales, a contribution to direct external ROI.
- Communicating the benefits of UCD to management increases social ROI and improves the strategic value of usability within organizations.

Rhetoric, the art of expressive or persuasive speech, is valuable for achieving ROI. The rhetorical aspect of external ROI is that making the user experience of our products and Web sites better is only the beginning—we have to communicate that it's better. We have to tell success stories, and we have to write the stories—in the business world, if something isn't in writing, it didn't happen.

Some of the many ways that writing stories about UCD helps increase sales are the following:

- Case histories in journals
- Reviews in media
- Descriptions in marketing literature
- Sales training to help the sales staff to describe the benefits of UCD

Because usability is inextricably tied to users, many of the most dramatic external measures of usability ROI will probably always be indirect, although reducing the need for customer support and service is a strong direct measure. Most of our UCD efforts produce improvements in user productivity, which don't become ROI measures until we translate them into sales or profits. If we care about ROI, we can't rest on our user experience laurels, but have to take the next step of communicating how our usability gains contribute to the bottom line.

8.14 SUMMARY: WHY WE CATEGORIZE USABILITY RETURN ON INVESTMENT

When UCD practitioners conduct usability research, we normally observe and measure both user behavior and user perceptions of a product, system, or Web site. User behavior is usually convincing evidence, although we gain insight, supporting data, explanations, rationale, and new ideas from users' perceptions and opinions. Usability research findings have the most effect when user behavior and perceptions both point to the same conclusions or lead us in the same design directions. Even when user behavior and perceptions differ, we must consider both, and we must create design solutions that address the *gestalt* or unified whole of what we observe.

The categories of usability ROI have a similar model. Measuring direct ROI, whether internal or external, is like observing user behavior—and just as in usability research, confounds can make our measurements more challenging. In internal ROI, we can measure the amount of reused code or the time to market saved by an effective UCD infrastructure. In external ROI, we can measure the decreases in drop-out rate for user registrations, increases in sales, or the reduction in customer support costs.

However, because we *can* make such measurements doesn't mean we should always do so—on the contrary. Just as we tell development teams that it's never practical to conduct usability tests for every product feature, the difficulty and cost of direct ROI measurements mean they are not always cost-effective themselves. Social ROI justifies UCD efforts when direct ROI measurements are impractical or incomplete—or when they are confounded by the collaborative activities that actually increase UCD teams' contributions to the organization's goals.

If we observe or measure that our users enjoy our Web site—they display satisfaction (not frustration) during usability testing, smile when showing us the site during field research, and give the site high ratings in customer satisfaction surveys—then we are likely to be less concerned about a "time on task" that is somewhat longer than our formal usability goal. Certainly we will continue to improve the registration form, but the positive user perceptions will help convince stakeholders that UCD practitioners add value when hard behavioral data are unavailable. Similarly, social ROI helps the rest of our organization understand and believe in what we do.

In this chapter, we have talked about internal, external, and social ROI; and we have suggested methods for achieving and measuring these three components of usability ROI. Our underlying message is that UCD practitioners and teams *can* convince stakeholders that our efforts produce ROI. The methods, guidelines, and tips described here capture the authors' (and our colleagues') practical experience in demonstrating the ROI of usability.

ACKNOWLEDGMENTS

The authors acknowledge the substantial contribution made by the panelists from the UPA 2002 Panel on "Measuring ROI for Usability" (Randolph Bias, Kelly Braun, J. O. Bugental, and Ed See) and the participants in the UPA 2003 Workshop on "Measuring Usability ROI to Justify Usability Investments" (Eugenie Bertus, Laura Borns, Donna Cooper, Mike Katz, Se-Hoon Kim, Scott Kincaid, Madhuri Kolhatkar, Christian Rohrer, and Roberta Shroyer).

REFERENCES

Alexander, C., Ishikawa, S., and Silverstein, M. (1977). *A Pattern Language: Towns, Buildings, Construction.* New York: Oxford University Press.

Bailey, R. W. (1993). Performance vs. preference. In *Proceedings of the Human Factors and Ergonomics Society, 37th Annual Meeting* (pp. 282–286).

Barnum, C. M. (2002). *Usability Testing and Research.* New York: Pearson Education.

Bartek, V., and Cheatham, D. (2003, January). *Experience Remote Usability Testing, Part 1: Examine Study Results on the Benefits and Downside of Remote Usability Testing.* Retrieved December 22, 2003, from www-106.ibm.com/developerworks/web/library/wa-rmustsl/.

Beyer, H., and Holtzblatt, K. (1998). *Defining Customer-Centered Systems.* San Francisco: Morgan Kaufmann.

Bias, R. G. (1997, July). Free usability data. Presented at *Interact 97, the 5th IFIP Conference on Human-Computer Interaction.* Sydney, Australia.

Bias, R. G., and Mayhew, D. J. (Eds.) (1994). *Cost-Justifying Usability.* Boston: Academic Press.

Bias, R. G., and Reitmeyer, P. B. (1995). Usability support inside and out. *interactions, 2*(2), 29–32.

Braun, K. (2002). Measuring usability ROI at eBay. Panel on measuring return on investment for usability (S. Rosenbaum, Chair). Panel presentation at the *Usability Professionals' Association 2002 Conference.* Orlando, FL.

Brealey, R. A., and Myers, S. C. (2003). *Principles of Corporate Finance* (7th ed.). New York: McGraw-Hill.

Broussard, G., Graham, J., Reichig, D., and Ryan, M. (2001, October). *Visual Noise: The Role of Site Clutter in Advertising Branding Effectiveness.* White paper by Dynamic Logic, Inc. Retrieved December 12, 2004, from www.dynamiclogic.com.

Brown, R. (1986). *Social Psychology* (2nd ed.). New York: Free Press.

Byrne, D. (1971). *The Attraction Paradigm.* New York: Academic Press.

Chen, B., Mitsock, M., Coronado, J., and Salvendy, G. (1999). Remote usability testing through the Internet. In *Proceedings of the Eighth International Conference on Human-Computer Interaction 1* (pp. 1108–1113).

Cialdini, R. (2001). *Influence: Science and Practice* (4th ed.). Boston: Allyn & Bacon.

Conklin, P. (1995). Bring usability effectively into product development. In M. Rudisill, C. Lewis, P. G. Poulson, and T. D. McKay (Eds.), *Human-Computer Interface Design: Success Stories, Emerging Methods, and Real-World Context* (pp. 367–374). San Francisco: Morgan Kaufmann.

Donahue, G., Weinschenk, S., and Nowicki, J. (1999, July). *Usability is Good Business.* Compuware White Paper. Retrieved December 12, 2004, from http://interface.free.fr/Archives/Usability_Is_Good_Business.pdf.

Donoghue, K. (2002). *Built for Use: Driving Profitability through the User Experience.* New York: McGraw-Hill.

Dumas, J., and Redish, J. (1999). *A Practical Guide to Usability Testing* (rev. ed.). Exeter, UK: Intellect.

Fallik, D. (2002, August 2). Science makes much ado about nothing. *Detroit Free Press.* Retrieved December 21, 2003, from www.freep.com/news/nw/journ2_20020802.htm.

Flanagan, J. C. (1954). The critical incident technique. *Psychological Bulletin,* 51 (28), 28–35.

Forse, M., and Degenne, A. (1999). *Introducing Social Networks.* Thousand Oaks, CA: Sage Publications.

Gamma, E., Helm, R., Johnson, R., and Vlissides, J. (1994). *Design Patterns: Elements of Reusable Object-Oriented Software.* New York: Addison-Wesley.

Gómez (2003). *Scorecards: Banker.* Retrieved December 22, 2003, from www.gomez.com/main.aspx?m=5&s=2&tc=1.

Good, M. (1989). Developing the XUI style. In J. Nielsen (Ed.), *Coordinating User Interfaces for Consistency* (pp. 75–88). New York: Academic Press.

Gouldner, A. W. (1960). The norm of reciprocity: A preliminary statement. *American Sociological Review,* 25, 161–178.

Helland, P. (2001). Autonomous computing. Presented at the *Ninth International Workshop on High Performance Transaction Systems (HPTS).* Pacific Grove, CA.

Hoo, K. S., Sudbury, A W., and Jaquith, A. R. (2001). Tangible ROI through secure software engineering. *Secure Business Quarterly,* 2(1), 1–5. Retrieved August 14, 2004, from www.sbq.com/sbq/rosi/sbq_rosi_software_engineering.pdf.

International Organization for Standardization (ISO). (1998). *Ergonomic Requirements for Office Work with Visual Display Terminals (VDTs). Part 11: Guidance on usability.* ISO 9241–11.

Landauer, T. K. (1996). *The Trouble with Computers: Usefulness, Usability, and Productivity.* Cambridge, MA: MIT Press.

McElroy, J., and Wilson, C. E. (2003). *UPA 2003 Workshop Proposal: Building and Sustaining a Usability Infrastructure: The Framework behind an Efficient and Effective Usability Team.* Available from C. Wilson at chaunsee@aol.com.

Milewski, A. E., and Steven H. L. (1997). Delegating to software agents. *International Journal of Human-Computer Studies,* 46(4), 485–500.

Molich, R. (2001). *230 Tips and Tricks for a Better Usability Test.* Fremont, CA: Nielsen Norman Group.

Monk, A., and Howard, S. (1998, March/April). The rich picture: A tool for reasoning about work context. *interactions,* 5(2), 21–30.

Myers, D. G. (2002). *Social Psychology* (7th ed.). New York: McGraw-Hill.

Neuendorf, K. A. (2001). *The Content Analysis Guidebook.* Thousand Oaks, CA: Sage Publications.

Nielsen, J. (1994). *Usability Engineering.* San Francisco: Morgan Kaufmann.

Nielsen, J., and Gilutz, S. (2003). *Usability Return on Investment.* Fremont, CA: Nielsen Norman Group.

Nielsen, J., and Levy, J. (1994). Measuring usability: Preference vs. performance. *Communications of the ACM,* 37(4), 66–75.

Ostrander, E. (2000). *Usability Testing of Documentation Has Many Benefits of Unknown Value.* Retrieved August 14, 2004, from www.stcsig.org/usability/newsletter/0010-pilotstudy.html.

Perkins, R. (2001). Remote usability evaluation over the Internet. In R. Branaghan (Ed.), *Design by people for People: Essays on Usability* (pp. 143–179). Chicago: Usability Professionals' Association.

Pressman, R. S. (1992). *Software Engineering: A Practitioner's Approach.* New York: McGraw-Hill.

Ramey, J. (1999). Oral communication with S. Rosenbaum.

Rockley, A., Kostur, P., and Manning, S, (2003). *Managing Enterprise Content: A Unified Content Strategy* (1st ed.). Indianapolis: New Riders Publishing.

Rosenbaum, S. (2003). *Future Trends in Usability: Standards and Innovation.* Presentation at *IBM's Make It Easy 2003 Conference,* San Jose, CA.

Rosenbaum, S., and Shroyer, R. (2003). *Measuring Usability ROI to Justify Usability Investments.* Presentation at the *UPA 2003 Workshop.* Available from S. Rosenbaum at stephanie@teced.com.

Rubin, J. (1994). *Handbook of Usability Testing: How to Plan, Design, and Conduct Effective Tests.* New York, NY: Wiley.

Sinha, R. (2003, February 22). *ROI of Usability: A Collection of Links.* Retrieved December 28, 2003, from www.rashmisinha.com/useroi.html.

Sterling, T. D., Rosenbaum, W. L., and Weinkam, J. J. (1995). Publication decisions revisited: The effect of the outcome of statistical tests on the decision to publish and vice versa. *The American Statistician,* 49, 108–112.

Tidwell, J. (2003). *UI patterns and techniques.* Retrieved December 21, 2003, from http://time-tripper.com/uipatterns/index.php.

Usability Professionals Association (UPA). (2003). Business benefits of usability. *UPA Website: Resources: Usability in the Real World.* Retrieved December 22, 2003, from www.upassoc.org/usability_resources/usability_in_the_real_world/benefits_of_usability.html.

Van Duyne, D. K., Landay, J. A., and Hong, J. I. (2003). *The Design of Sites: Patterns, Principles, and Processes for Crafting a Customer-Centered Web Experience.* Boston: Addison-Wesley.

Ward, T. (2002). *Measuring the Dollar Value of Intranets.* Retrieved August 14, 2004, from www.intranetjournal.com/articles/200104/pii_04_25_01a.html.

Webster, J. M., Smith, R. H., and Rhodes, A. (1999). The effect of a favor on public and private compliance: How internalized is the norm of reciprocity? *Basic and Applied Social Psychology,* 21(3), 251–259.

Whiteside, J., Bennett, J., and Holtzblatt, K. (1988). Usability engineering: Our experience and evolution. In M. Helander, (Ed.), *Handbook of Human-Computer Interaction* (pp. 791–817). Amsterdam: North-Holland.

Wilson, C. E. (2002). Lessons learned: Looking at both sides. *User Experience,* 1(2), 8–13.

Wilson, C., and Coyne, K. (2001, May/June). The whiteboard: Tracking usability issues: To bug or not to bug? *interactions,* 8, 15–19.

Zajonc, R. B. (1968). The attitudinal effects of mere exposure. *Journal of Personality and Social Psychology Monographs,* 9, (2, Part 2).

9 CHAPTER

Usability Science: Tactical and Strategic Cost Justifications in Large Corporate Applications

Charles L. Mauro MauroNewMedia

9.1 INTRODUCTION

In my role as founder of a professional consulting firm delivering formal usability science to large corporations and government agencies for more than 25 years, it has become increasingly clear to me that usability science can be applied in large corporate projects as a tactical development tool or in a more comprehensive manner as a critical and valuable corporate strategic asset. The most effective implementation of formal usability science in these large and complex projects depends on several factors including the sponsoring organization's willingness to adopt innovations and the skill set of the team charged with integration of formal usability science.

It has also become increasingly clear that the integration of formal usability science in large complex projects follows quite closely principles of "innovation diffusion," (Rogers, 2003) which predict in a meaningful manner the resistance those who do this work encounter in such projects. This resistance is based on three factors: usability science is disruptive, expensive, and time consuming when it is first encountered by high-level corporate decision-makers. However, the cost benefit (cost justification) of successfully integrating usability science as a tactical tool or strategic asset is overwhelmingly positive. Usability science saved the New York Stock Exchange billions of dollars, explained the failure of major infrastructure systems (Three-Mile Island), and makes possible our most advanced and effective weapons systems. But realistically usability science is in the very early stages of diffusion with respect to wide adoption by large corporations and government agencies. For those who deliver formal usability science as consultants and colleagues who strive to implement such expertise on large internal development teams, this is simply a fact of life that only time will cure.

In this chapter I discuss both the tactical and strategic implications of formal usability science as it apples to large and complex corporate development projects. The views taken in this material can be useful both to those who wish to execute large-scale usability science projects on a consulting basis and to those executives within large corporations who wish to implement formal usability science for tactical or strategic benefits.

9.2 BENCHMARKING AWARENESS AND ISSUES OF IMPLEMENTATION

To those who practice formal usability science on a professional basis, this book represents an opportunity to put into a formal context that which we know to be self-evident: *improving usability of technology-based systems at any level is a benefit.* However, not everyone in industry is so enlightened. This lack of insight is not the fault of industry at large but is linked to issues with formal usability science as a development discipline and its relatively new entry into complex and highly stratified corporate decision-making models. For a discipline that has as its central tenet increasing the effectiveness of all manner of technology-based systems, usability science as a professional discipline has had difficulty packaging this powerful new expertise into a form that corporations and government agencies can understand and benefit from. Some of the primary issues that have impeded the widespread integration of formal usability science into large, corporate product development process models include the following:

- Lack of a formal definition of what usability science actually constitutes
- Limited professional certification of qualified usability experts
- Professional societies that lack rigorous membership requirements
- Poorly defined and sometimes widely varying development methodologies
- Lack of objective and rigorous cost-benefit models
- Disregard for the impact that usability science has on group dynamics
- Poor integration with other critical corporate research disciplines
- Disregard for the politics of corporate decision making

- Widely varying costs and fee structures
- Lack of consideration for corporate funding cycles
- Failure to frame usability science as a strategic asset

Although this list paints a complex picture of the delivery and use of professional usability science in large corporate settings, the actual benefits when such issues are overcome produce stunning levels of success and return on investment (ROI).[1] The resistance encountered with the use of formal usability science is not unique but is similar to problems associated with the early entry of other new development disciplines such as market research, econometric modeling, and "just-in-time" manufacturing methods. All of these disciplines and their underlying methods found significant resistance breaking into the main body of product development methodologies, and all are now profoundly and permanently embedded in development processes of leading corporations. No executive in a position of responsibility at the high corporate level would consider making critical development decisions without the benefit of these disciplines.

9.3 A PROBLEM OF EXPERTISE

Although it is clear that most development organizations can benefit from usability science, few have a meaningful concept of how to identify viable expertise in the form of experienced and properly educated usability professionals. This problem has become more acute as the definition of what constitutes acceptable levels of expertise is widely debated in the boom and bust of the Internet where many Web development firms offered usability science often without appropriate technical expertise. In this chapter, references made to *usability science* will adhere to the following definition:

> Professional usability science is defined as a formal research and development discipline that adheres to the processes and rules of scientific investigation as developed and taught in formal graduate level programs in the cognitive and biomechanical sciences. Practitioners of this type of research hold advanced degrees in usability science, ergonomics, or other relevant cognitive science fields. This field of expertise is also known under the

[1] Many of our most effective military systems would not have been possible without the rigorous application of formal usability science as an active element in systems development.

professional terms of usability science, usability engineering, human factors engineering, ergonomics and cognitive ergonomics.[2]

As awareness reaches critical decision-makers usability science will play a critical role in successful product development programs. Later in this chapter I discuss factors that effect the diffusion of usability science into large, complex corporate development programs. Currently the cost-benefit models for integration of usability science into large corporate settings are based on optimization of narrowly focused problems related to products and systems already in production or use. This approach for cost-justifying usability science is important but will never lead to widespread adoption by large corporate development organizations. The examples in some of the other chapters of this book are, for the most part, focused on "spot" optimization of critical usability problems. But this model and approach produce a dramatic mismatch between how corporate executives think about problems and what usability science can do to become appropriately aligned with the mental models of high level corporate executives. (See the case study of the Sprint experience in Chapter 21 for an example of a *good* match.)

9.4 THE OBVIOUS COST-BENEFIT FACTORS

The overall net loss to industry from poor usability is literally in the hundreds of billions of dollars a year. As you parse this book you will encounter many examples in which usability science has been applied to the resolution of problems resulting in a significant cost benefit. All of these are valid in the context of discreet segments or factors that map to corporate cost-benefit models such as formal "return on investment," "net present value," and others (Bias and Mayhew, 1994). Examples include the following categories of costs that are directly and measurably reduced by improved usability:

- Reduction in training costs
- Reduction in operator-induced errors
- Reduction in service and maintenance costs

---

[2] This definition adheres tightly to established expertise and experience profiles provided by the Human Factors and Ergonomics Society and other professional organizations that deal with professional usability research as a formal development discipline. For an interactive definition of user-centered design, an important aspect of formal usability science, visit www.taskz.com/definitions.

- Reduction in workplace injuries and lost time
- Reduction in employee turnover
- Reduction in product development costs and better user acceptance
- Reduction in project lead times for new product development
- Reduction in product liability costs and associated insurance

Even though these factors are the primary focus of the material in other chapters of this book, surprisingly these are not the reasons that ultimately lead corporations to retain professional usability science as a centerpiece of product development. Certainly these are valid and important factors, but professional usability science at the large corporate level is simply about two factors: 1) *improving competitive position* and 2) *reducing risk.* Reducing risk can apply to the risk of a failed product, a failed system, or a failed career. These factors are not determined by what costs can be saved (the cost reduction model) but by what profits can be made proactively from producing more competitive and successful global products. The question of how to employ usability science is not self-evident even to the brightest development team. Formal usability science must be marketed and sold in a manner consistent with the real needs and limitations of the corporate executive. These limitations and needs include 1) concern for objective project funding cycles and methods and 2) proof that usability science is a viable and critical asset worth investing in. These are not trivial factors.

9.5 HOW USABILITY SCIENCE CAN BE UTILIZED IN A LARGE CORPORATE SETTING

Cost-justifying the use of formal usability science can be parsed into two primary use patterns that are tightly aligned with the decision-making models of top executives. In large, complex development programs usability science can be utilized either as a tactical methodology for making "spot enhancements" to a system or in a much broader way as a "strategic asset" comprising knowledge and experience that become part of the intellectual capital of the corporation. These two approaches are well aligned with leading research from the field of group problem solving derived from the social-cultural psychological literature. For example, various widely respected researchers including Getzels and Csikszentmihalyi (1976) divided the approaches to problem resolution into two basic forms: the first being "problem finding" and the second being "problem solving." This research suggests that creative individuals or teams approach problems in one of two ways as defined above. Rarely do teams employ both cognitive models when solving problems. This model is useful for identifying how usability science as a professional product development discipline can be utilized in large, complex projects. This approach is consistent with professional experience gained from participation in many large projects over the past 25 years.

How you employ usability science as either a tactical or a strategic asset is, to a significant extent, determined by the development culture of the corporation itself and its willingness to adopt new methods and practices. If your product development teams tend to approach projects from the "problem-finding" perspective, you are well served to retain usability science as a tactical methodology of significant power. This means simply that usability science should be a methodology you use to solve problems with existing products and services that are already under development or are currently in the hands of important customers. The overwhelming application of usability science in this context focuses on "usability testing"[3] methods. This component of formal usability science can be applied at any point during development or after product launch. The cost-justification for this approach can be considerable, but in reality such modeling is almost always unnecessary because assigning a value (monetary) to a problem is at best highly subjective. This leaves ROI models based on usability testing often weak and subject to debate. Still such models can be useful.

On the other hand, if your methods and teams tend to approach development from the "problem-solving" point of view, usability science can become a powerful strategic asset. Like other major forms of innovation usability science follows basic principles of innovation diffusion. But how do large development teams know when usability science is required or important?

9.5.1 Usability Science as a Tactical Asset

Today, many corporations and government agencies have little or no knowledge of what constitutes a usability science problem and, furthermore, have less knowl-

---

[3] *Definition—Usability testing:* Professional usability testing is defined as a formal research methodology that adheres to the processes and rules of scientific investigation as developed and taught in formal graduate level programs in the cognitive sciences. Practitioners of this type of research hold advanced degrees in human factors engineering, ergonomics, or other relevant cognitive science fields. This field of expertise is also known under the professional terms of human factors engineering, usability engineering, human-computer interaction, and cognitive ergonomics. It is important to note that, within the overall field of formal usability science, "human-computer interaction" (HCI) has become a well-recognized subspecialty. The primary focus of this chapter is on the application of professional usability engineering and testing methods to screen-based products and services. A large and active body of work is taking place outside the HCI field.

edge of how to solve such problems. This general lack of awareness on the part of industry at large dramatically complicates the sales, delivery, and adoption of professional usability science. Even the most aggressive and targeted marketing of professional usability services goes without success unless executives of the sponsoring organization are "primed" with a specific problem outside the experience sphere of their development team. These events have essentially three primary stimuli:

- Liability threats or claims
- Customer complaints
- Competitive pressures

One can see immediately that under this view the primary stimuli for seeking usability science expertise are *reactive*. Executives are looking back at problems from the view of a product in the marketplace. Taken from the standpoint of cost benefit this reactive utilization of usability science can have a dramatic, relatively short-term, impact on the corporate bottom line. Formal usability science used in this manner is overwhelmingly "diagnostic" in nature and can be a powerful "tactical" development tool.

Tactical utilization of usability science is discussed in detail in the case studies section of this chapter. However, an example is helpful at this point. At the core of one aspect of usability science is usability testing. When properly and professionally executed such studies produce reliable and highly focused research on the usability of products and services. These types of analysis focus on an evaluation of the "efficiency" of the interface between the customer or user and the system itself. The metrics or dimensions of this analysis generally focus on two types of data: subjective impressions and objective performance. When usability science experts are employed to execute such studies, large and complex usability issues can be identified in significant detail and in a relatively short time.

Example: Transfer Functions and Usability Science

For example, customer complaints indicating that customers cannot effectively use the funds-transfer function on a new home banking application may be filtering in from call centers. This problem is resulting in customers making double transfers of funds between accounts. By utilizing formal usability science the exact nature of the customer's problems can be defined in terms of time and errors (objective measures). The impact these problems are having on customer satisfaction and retention (subjective factors) can also be determined. By conducting formal usability studies executives can pinpoint both the structure and extent of the problem. With data from this type of research it is then a short step to produce an economic model of the impact such issues are having on the bottom line. This actual example showed that the cost of fixing the Web-based interface (approximately $150,000) was more than balanced by the cost of reacquiring lost customers and damage to the underlying brand attributes of the bank. In this example customers were high net-worth individuals. It is important to point out that in many tactical applications usability science is a powerful and cost effective means of *diagnosing the problem.* It is *not* a means of creating a design solution. Such expertise falls generally into the discipline of *user-centered design,*[4] which is an allied field that draws heavily on formal usability science as a design methodology.

9.5.2 Usability Science as a Strategic Asset

In the previous example usability science was used to diagnose an existing problem that in the end proved to be related to poor interface design of the software itself. The cost benefit of fixing such problems is critical. However, a more advanced and cost-effective means of applying usability science in large complex development efforts is to employ it as a strategic asset. This approach produces the largest ROI because it makes use of usability science in the *design* of the system. Many months or even years before the first customer interacts with your system usability science can be employed to create a system that is easy to learn, is error resistant, and reinforces the brand attributes of your corporation. In the example of the funds-transfer functions of a home banking application, formal usability science working during early stages of development would have shown that the Web interface being developed compromised the users' short-term memory, did not provide adequate feedback, and left the customer with undetected errors. These findings were possible before a single line of code was written or detailed product engineering was undertaken by applying formal usability science analytical methods such as task analysis and interface concept testing. When viewed from this perspective, it is clear that formal usability science essentially reduces the risk of a design defect, but, more importantly, it also guarantees the enhancement of customer satisfaction. These are the objective and

---

[4] *Definition—The user-centered design:* User-centered design (UCD) is a comprehensive, development methodology driven by 1) clearly specified, task-oriented business objectives, and 2) recognition of user needs, limitations, and preferences. Information collected using UCD analysis is scientifically applied in the design, testing, and implementation of products and services. When applied correctly, a UCD approach meets both user needs and the business objectives of the sponsoring organization.

subjective benefits of employing an innovation such as usability science as a strategic asset and not as a tactical methodology. One can see immediately that the downstream effects of solving the funds-transfer problem would have had an impact on many more areas of the screen-based system than employing usability science as a tactical methodology.

9.6 THE ACTUAL DOWNSTREAM IMPACT

For example, by solving the funds-transfer problem before detailed engineering begins you can see that the downstream impact of usability science is extensive. By solving this problem (and a wide range of other critical problems) usability science, as a risk-reduction methodology, directly touches all of the major cost centers in the following list, which in turn have an impact on the corporate bottom line. But the more important point is that when usability science is applied as a strategic asset, it often leads to increased customer acquisition, retention, and migration rates. These factors are the cornerstone of corporate profit on a proactive basis. This is in direct contrast to the utilization of usability science to solve tactical problems as discussed previously in this chapter.

Usability science employed as a strategic asset affects the following:

Cost reduction factors with downstream impact

- Project scheduling and allocation of capital
- The cost of capital and utilization of production facilities
- The cost of high-level product engineering
- The cost of infrastructure design and engineering
- The cost of detailed engineering design
- The cost of engineering and planning staff
- The cost of product prototype development
- The cost of product testing and certification
- The cost of product production
- The cost of product marketing and promotion
- The cost of product support literature and employee training
- The cost of customer support at launch
- The cost of customer support into system maturity

- The cost of customer relationship management systems
- The cost of product support including routine maintenance
- The cost of product support including replacement
- The cost of product marketing and sales

Profit-making factors

- Increased customer acquisition rates
- Increased customer retention rates
- Increased customer migration rates

This approach results in strategic advantages over your competition, but, more important, downstream effects on other cost centers of the corporation are dramatic reductions in costs and improvements and, most important, these come without further expenditure of human resources and actual capital. For example, the creation of a funds-transfer function that is fast, easy, and error resistant means that call center staffing will be reduced, customer relationship management system complexity will be dramatically reduced, and customer acquisition, retention, and migration costs will be reduced. Cost-justification of usability science at this level is the virtual tip of the iceberg. In the earlier example the downstream savings from use of advanced usability science and user-centered design was effectively more than $10,000,000 when amortized over a 5-year development cycle. The tactical approach resulted in minor savings in costs but solved a critical customer relationship problem and reduced customer defection percentages, both significant effects but trivial compared with employing usability science as a strategic asset. The cost of both programs was similar for resolution of the funds-transfer problem.

9.7 CURRENT TRENDS IN THE APPLICATION OF USABILITY SCIENCE

As can be seen from Figure 9.1 there is much more to be gained in terms of strict ROI by employing formal usability science early in development as opposed to after the product or site is launched. However in most applications usability science is currently employed late in the product lifecycle. For more detailed discussion of this issue see Mauro (2003).

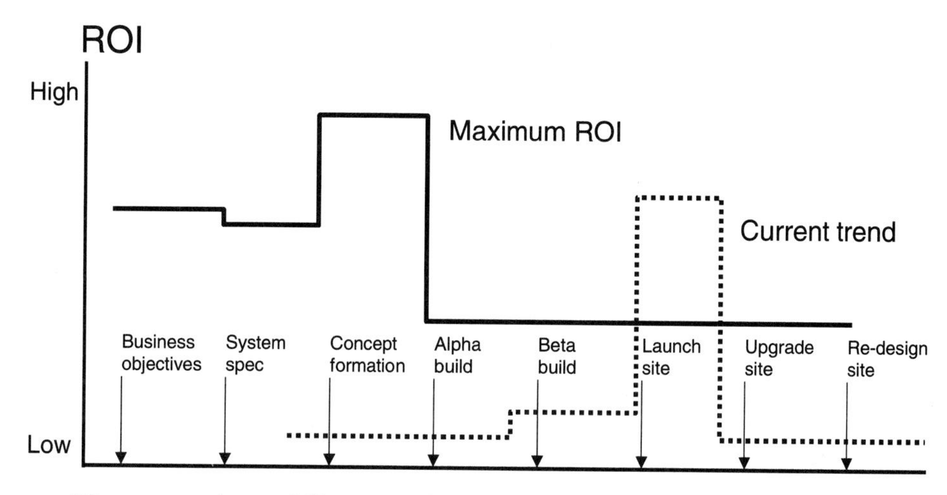

FIGURE 9.1 When to employ usability research.

9.7.1 The Rate of Dispersion of Formal Usability Science into Large Corporate Projects

Usability science as it applies to strategic issues in the large corporate setting is in a relatively early stage of diffusion. As the general research literature on innovation diffusion shows, the pathway to adoption of an innovation such as usability science follows the widely accepted S-curve adoption cycle.[5] This form of diffusion starts with a relatively flat initial phase of adoption followed by an eventual steep rise in adoption rate followed finally by a general flattening of the adoption curve as innovations reach high levels of penetration. The rapid and continued propagation of usability science (and other major innovations) begins to increase as the innovation reaches a "critical mass." As stated in Rogers (2003) critical mass in the context of innovation diffusion is that point at which an innovative adoption becomes self-sustaining. It is clear that formal usability science is just beginning dispersion into large corporate projects. It is in the early part of the cycle, which means simply that for both those who sell usability science and those who intend to adopt this powerful innovation for the purpose of enhancing strategic performance scant real experience is available to give

[5] The S-curve adoption cycle is covered in the text by Rogers (2003, p. 344, Fig. 8.5).

comfort. This does not mean that usability science will not be widely adopted in the future. Certainly given the cost and risk of developing increasingly complex products and services, usability science will become a major factor in strategic decision-making.

In large corporate settings innovation diffusion has been shown to take several years. It is my personal experience that successful integration of high-quality, long-lasting usability science in large, complex development projects requires years of effort, certainly not months. The adoption cycle will shorten over the coming years, but the rate of adoption will only quicken as a critical mass effect takes place on a broad corporate level. Based on established research from the field of innovation diffusion there are several attributes that are highly correlated with the likelihood that a large corporation will adopt usability science. These factors include the following (summarized from Rogers, 2004, p. 412):

1. *Centralization* of decision-making is negatively correlated with adoption of innovations such as usability science. If all decisions are made in a centralized manner, usability science is less likely to be effectively adopted within the organization.
2. *Complexity* of the organization's staff background and educations can be positively correlated with adoption of innovations. If an organization has many highly qualified and professional individuals from varied disciplines, it is more likely to adopt innovations such as usability science. It is especially important that no one discipline dominate product development decision-making.
3. *Formalization* of rules and procedures is negatively correlated with adoption of innovations. The more structured the organization's development rules and procedures are, the less likely the corporation will be to adopt an innovation.
4. *Interconnectedness* is highly correlated with adoption of innovations such as usability science. This means the level of network utilization and interpersonal communications that take place between and across divisions and departments and even within individual teams of corporate employees and executives are strong predictors of willingness to adopt innovation.
5. *Organizational slack* is simply the amount of uncommitted resources that is available to use in the adoption of new and powerful ideas. It is important to remember that all corporations run on a yearly development budget and *never* is an innovation part of that budget. Therefore the resistance to an innovation is always determined by the funds available.

9.7.2 Summary of Usability Science and Innovation Diffusion

The previous five factors are highly correlated with the acceptance and propagation of innovations such as usability science into large, complex corporate development programs. Therefore, it is likely that the extent to which such factors are present in your corporation will be a predictor of how likely usability science is to be adopted. This not a hard and fast rule, and certainly exceptions exist. Looking back over the past 25 years of experience with projects of this type, I see clearly that these attributes are, in fact, highly correlated with personal experience.

9.7.3 Factors That Slow or Completely Inhibit Diffusion of Usability Science in Large Corporate Settings

Three primary factors, which are listed in the following paragraphs, help the corporate CFO/CIO/CEO make a decision to fund a strategic multimillion-dollar usability science program, and, as one might expect, these issues map directly to the bottom line. However, one of the most important problems with the diffusion of usability science is the need for practitioners to make clear the impact that usability science has on existing product development programs. Attempts to adopt usability science as strategic assets will probably end in confusion and eventual rejection unless top management understands that usability science affects their ongoing development efforts in significant ways. The three factors must be made clear to the funding organization before usability science programs are adopted in large, complex development efforts. How to evaluate what type of organization you work in was discussed in Chapter 7 of this text.

Factor 1: Usability science is disruptive. Clayton Christensen in his widely read text, *The Innovators Dilemma: When New Technologies Cause Great Firms to Fail* (1997), discusses the impact that new technologies can have on existing corporate products and services and the eventual success or failure of the corporation at large. His thesis is that new technologies that are not properly understood and integrated into the development decision-making of existing corporations can be highly disruptive if not outright disastrous. Although the focus of his thesis is on changes in hardware and software technologies, the same holds true for the application of new "process" methods. In this way usability science, as a core factor in product development, is no different from other new technologies. In fact it might be argued that major changes to product development "process models" are more disruptive than technology advances because process changes go to the heart of creative decision-making by rearranging the milestones and the weight of key decision variables (user needs and limitations, not technology, drive the design). Integration of formal usability science as a core development discipline initially causes disruption in most standard product development process models. The primary reason for this disruption is the requirement that "user performance" must be stated as the primary design specification for system design. This is a 180-degree shift from the most widely accepted process model, which places technology requirements at the center of complex product development specifications. The metrics change, as do the decision models. A discussion of the exact nature of these changes is well beyond the scope of the chapter, but it is clear that the effective mapping of usability science onto existing product development process models requires serious levels of experience with usability science and the existing development methodologies as strategic assets. The key concept is "go slow" and pick your entry point carefully. Properly done, adoption of usability science will follow normal rules of innovation diffusion as discussed later in the chapter.

Factor 2: Usability science is expensive. One of the most common problems with the delivery and propagation of usability science within large corporate projects is the lack of reliable and accurate cost-setting methods. Usability science is a significant development cost factor; to say otherwise is a mistake. Usability science requires highly educated and creative expertise, testing labs, and access to customers. This fact cannot be denied and must be accounted for in the initial planning for the integration of usability science into complex projects. As we have discussed before, the return on this investment can far exceed the costs but simply hiring expensive usability experts on a spot consulting basis is not the answer. Integration of usability science as a formal component of product development requires 1 to 2 years in the best case. As all experienced corporate executives know only too well, significant increases in development costs must adhere to yearly funding cycles. This funding takes a minimum of 2 years in an up economy and 3 to 5 years in a down market.

Factor 3: Usability science is time consuming. If employed for strategic benefit usability science is initially time-consuming and complex. This is no surprise when one considers that existing product development methods must be modified, resulting in new decision models. There are two factors that must be considered when the time required to implement formal usability science is evaluated. These factors are 1) the methodology issues and 2) the people issues. Taken together these are complex problems that must be carefully managed and monitored. The following is a brief discussion of these two variables.

Variable 1: The methodology issues. As any successful corporate executive will be quick to tell you, product development is a difficult and complex business, which in small projects is risky but in large complex projects is career threatening. A large and expanding research literature on product innovation and development methods is available. Many of these methods focus on controlling the quality of the end product. These methods, including Six-Sigma, Balanced Score Card, and a host of other structured approaches, are aimed at reducing the risk of creating an unsuccessful product. None of these widely adopted methods integrates formal usability science as a critical component of their decision-making models. Some of these methods have given minor acknowledgment to ergonomic standards, but the most widely adopted product development methods are technology centered not user centered. From the outset this creates a conflict between usability science and these methods. These issues can be overcome with the resulting benefits far outweighing the issues of integration. At the heart of this issue is a fundamental shift in what key variables form the decision-making framework of product development when usability science is employed as a strategic asset.

Variable 2: The people issues. The research literature (sociocultural psychology) (Csikszentmihalyi, 1996, and others) dealing with group dynamics in creative problem solving gives structure to the issues that arise when an innovation or new methodology is being integrated into an existing framework. Product development at its core involves creative problem solving and brainstorming, which occurs not only at the outset of projects but also over the entire lifecycle of product development. There are several factors that have an impact on people issues related to adoption of usability science, but the most important is the requirement that existing development team members learn and adopt a new knowledge "domain" that is not part of their existing educational or professional experience profile. Unfortunately, current educational frameworks at both the undergraduate and graduate level do not adequately introduce usability science as a critical component of product development. This leaves the development of domain knowledge as a key technical requirement for the usability professional involved in large complex projects. The concept of retention of a new knowledge "domain" is central to the adoption of usability science in large corporate settings. This means simply that a significant "educational component" is often essential. This *does not* mean employing a usability consulting firm to give seminars to the engineering team. Domain knowledge is gained in only one way—by hands-on solution of a relevant problem using high-quality usability expertise. Attempting to teach usability science on a seminar basis to staff with little or no graduate level psychology background not only wastes funds but also dramatically underplays the impact and complexity of formal usability science as a corporate asset. (See Bias, 2003, on the dangers of amateur usability engineering.) Not everyone within large corporate development teams believes that migrating to a development methodology that places user needs and limitations ahead of technological innovation is a good approach. This situation often leaves pockets of believers and pockets of development team members who will grudgingly go along with the new process but who, at a moment's notice, will return to the traditional engineering decision model. This leads me to the role of the usability science "champion" within the corporate setting. Without a highly placed champion of the approach, usability science is a waste of time and valuable dollars. Do not even consider usability science as a strategic asset without a staunch and highly placed "C" level champion (Mayhew and Bias, 1994). It is also important to point out that this position must be ongoing. If the usability champion moves to a new corporation he or she must be replaced immediately. Over the years I have developed and participated in numerous programs that brilliantly employed formal usability science to great benefit. All of these efforts involved millions of dollars in funding and untold hours of product development reconfiguration. However, I have then seen these same programs (including staff and facilities) being completely dismantled in a matter of months once the usability science champion left for another position. Such champions take their expertise with them, and the loss to the initial corporation is immense. Products that previously benefited from major innovations in terms of user acceptance may return to pre–usability science levels of customer dissatisfaction and associated costs.

9.8 WHERE YOU WILL CURRENTLY NOT FIND USABILITY SCIENCE

The problem of development team members lacking appropriate levels of domain knowledge is significant to those selling usability science and those individuals within a corporation attempting to create asset value from this new science. Based on the previous discussion of innovation diffusion, it is no surprise that usability science has not yet reached into the core of new product development programs, business schools, and software/hardware engineering programs. At a series of yearly lectures given in leading master of business administration (MBA) programs I have found that less than 5% of the students had any prior knowledge of usability science as a viable discipline for solving critical business problems. Many of these graduate students later told me that they had no idea that the formal discipline of usability science existed. When reviewing the dominant product development texts and related case studies used in leading engineering, marketing, and MBA programs, we find virtually no mention of formal usability science as a critical component of product development. In many, if not most, of the academic computer science programs in the world, students can earn a master's degree or a doctorate and never take one course in usability or user-centered design. This omission leaves maturing executives with the impression that formal usability science is not a critical component of product development. The net effect is that most products under development today will not have the benefit of professional usability design, testing, and certification. This confluence of factors dramatically affects the marketing and sales of formal usability science as a development discipline both from the internal perspective and from the standpoint of a service provider (consulting service). How then, can formal usability science begin effective diffusion into large and complex corporate development projects where the benefits will be strategic as opposed to simply tactical in nature?

9.9 MOVING TO THE USE OF USABILITY SCIENCE AS A STRATEGIC ASSET

Based on many years of experience with both highly successful usability science implementations and observations of numerous stunning failures, I have identified five key factors that all corporate executives need to bear in mind when considering implementation of formal usability science in their most important projects. These are the following:

1. *Go slowly.* All major innovations that are widely adopted in large corporate development projects started with small programs that focused on solving relevant and meaningful problems. Usability science as a strategic asset and key component of your development programs should begin with a single, small project that can tolerate some fits and starts. This is essential. Do not put usability science front and center on your largest most visible project because failure is almost assured.
2. *Start with a real problem.* A common and understandable approach for executives to begin integration of formal usability science is by employing consulting firms to conduct "seminars" on this important new discipline. This will not result in significant adoption of usability science. The best process

for initial projects in usability science is to assign an important (but not mission-critical) problem to a team that includes formal usability science expertise, and let the team experience the usability science development model and benefits in the process of solving a real problem.

3. *Fund efforts appropriately*. This seems self-evident, but usability must be properly funded. If the budgeted cost is too high, redefine the problem but do not simply reduce the funding and retain the same problem. As discussed in this chapter usability science is a real science and as such needs proper funding.

4. *Retain those with the best expertise*. This is not only difficult but is also complex. As previously discussed, many firms and individuals offer expertise in usability science. Look for those with expertise that includes many years of experience, advanced degrees in formal usability science, and domain experience in your specific product category. Finally, check references with high-level executives to verify past performance. You will probably start by retaining a consulting firm to assist in integrating usability science into your development programs. When efforts go well consider internalizing some of your usability science expertise. When you do this, work with your consultant, as he or she will be enormously helpful with staff and planning. Keeping the consultant involved over time is strategically important because he or she works in much broader fields and can be a critical asset with new and complex problems.

5. *Let usability science sell itself*. It is hard to resist pushing usability science out to other development projects and divisions. Proactive propagation of usability science on a broad corporate level cannot generally be achieved without first proving its benefits on smaller projects. In a corporation that is highly networked (as discussed earlier) the successful application of formal usability science will spread without stimulus. I have worked on many large projects for which new divisions became clients without any active marketing of usability science. Good work begets more good work, and an active and inquiring workforce ensures propagation of new methods and innovations.

9.10 CASE STUDIES: TACTICAL AND STRATEGIC

Manufacturers seek assistance in the resolution of usability problems for a variety of reasons that appear obvious to them at the time they request help. However, experience has shown that most clients do not have a clear understanding of the scope of their existing problems or the fact that other problems, of which they are not aware, may exist. One of the most important contributions that a comprehensive usability science program can bring to product development is a demonstration of the wide impact that improved usability can have on the financial health of the corporation. The usability science professional is often faced with the problem of determining the exact nature of the client's problems by seeking information in areas not commonly linked in the product development process. By systematically exploring a few key topics, the usability science professional can generally define within useful limits both the nature and scope of the client's usability problems. The following concerns commonly lead the usability science team to establish a comprehensive overview of the problems to be addressed:

- Liability claims
- Service and maintenance costs
- Customer complaints
- Increased complexity in new products
- Market competition

9.10.1 Liability Claims

In the early 1960s, after a series of legal precedents, the U.S. courts became pro plaintiff on issues related to product safety. For the first time, plaintiffs were allowed to attack the design decision-making process of manufacturers. The concept of "design for reasonable use and misuse" became the leading conceptual framework for determining a manufacturer's ability to meet the user's needs. Under the legal construct of strict product liability, the courts allowed usability to take center stage in the determination of legal liability. As a result, the bar, both plaintiff and defense, focused on usability as a driving force in litigation. Lawyers immediately understood the importance of well-documented and well-presented explanations of usability.

In many cases, manufacturers had very poor defense against claims of defective design based on usability. Often, products involved in these claims had never been designed, tested, or certified for usability. At best, a safety committee for compliance with existing safety standards and regulations may have reviewed these products. As any corporate counsel now knows, compliance with industry standards is only one test that the product will face in the courtroom. The legal system holds manufacturers to a far higher standard when the question of usability is at hand. In many cases, the requirement for the manufacturer to "design for reasonable use and misuse" is the overriding requirement. Disproving such a claim is complex and difficult if the product has not been created with professional usability methods that will stand up under cross-examination.

As a result, plaintiffs' attorneys have often retained professional usability science engineers for their expertise in evaluating the usability of the product. These evaluations often proved devastating and provided the jury with tangible information and analysis on how the product should have been designed for usability. This sort of reverse engineering was often the first exposure that manufacturers had to usability science. In cases in which design for usability has been a formal part of the development process, manufacturers have a much better chance of reducing the size of claims, if not winning outright. The cost benefit of this approach is staggering.

The Costs

To bring to trial a case involving serious injury or damage resulting from claims of defective design will cost a minimum of $2 million. These costs include legal fees, experts, discovery, documentation, and more. This first $2 million does not cover the cost of person-hours for corporate staff preparing to testify, producing documents, and meeting with counsel or expenses for travel and other expenses. In a complex case, these costs may easily exceed the original $2 million estimate. Add to the previous investment in legal fees and corporate person-hours the long-term effect of not having management, engineering, and design staff working on critical new product development projects, but instead working to support litigation. This last cost has a direct impact on ROI and may be the largest expense affecting the long-term health of the corporation.

Costs related to litigation are difficult to predict. Taken together, these three sources of costs may easily total $3 million as a minimum. And they are not the full story. Many cases cost far more than this amount. Jury awards are added to litigation costs. Other punitive damages may also apply. Even after these high fees, costly internal expenses, and a sizable jury award, the problem is not over. When a product is found defective by design, a whole new issue arises: repeat litigation. Certain products have actually spawned entire industries of attorneys, experts, and documentation specialists. The cost to corporations in these cases is enormous because they must self-insure after a certain point and may have to carry the cost of litigation themselves.

The cost benefit of designing for usability is so overwhelming when seen from this perspective that it is hard to imagine why more corporations do not adopt a comprehensive program for addressing the usability problem. The answer probably lies in the accounting and budgeting methods of corporations. Of all the costs just discussed, only one shows up as a line-item expense: insurance premiums. All other costs are either not documented, paid out of reserve accounts, or channeled through corporate legal expenses, which often do not have strict reporting or budgeting requirements. As a result, many corporations never see the true cost of litigation. If they did, they would be shocked, as would their shareholders. By comparison, the cost of a thorough usability design program for a product would be at worst only a small fraction of the cost of even a normal litigation proceeding. In many cases, corporations could easily hire a top-quality usability science staff, construct a testing lab, rewrite their product development procedures, and educate their product development managers for a small fraction of the potential cost of litigation. The long-term benefit of ROI would be hard to equal in any other area of corporate development.

The Benefits

Certainly, there is no absolute guarantee that designing for better usability will dramatically reduce litigation costs, but the probability that it will is very high. For example, plaintiffs' attorneys always assess the strength of a defendant based on what they discover by way of formal test results, design alternative documentation, and the background of those responsible for usability design and testing. It is hard to imagine that a plaintiff's attorney would push to trial knowing that a corporation had employed professional usability engineering methods in the design of their products. The attorney would certainly look for an easier case that would involve less costly opposing experts and easier evaluation of the product. In this way, usability becomes a deterrent to litigation and a source for improving the safety of the product. Both of these factors mean a very large potential savings for the corporation.

9.10.2 Service and Maintenance Costs

Typical calls to a software support line are estimated to cost the manufacturer between $12.00 and $28.00 per call. Indeed, in Chapter 10, Karat and Lund report support costs of up to $250 per call. This cost includes the salary and benefits of the employees, facilities, insurance, computers, training and updating, and maintaining the customer database, user registration cards and correspondence, telephones, and more. Furthermore, once customers call a first time, they will be much more likely to become repeat or, worst of all, frequent callers. The cost of these seemingly necessary calls can drive the profitability of a product down so far that it is impossible to break even on development costs, let alone show a profit.

With the wide proliferation of user support lines, these costs are becoming increasingly critical to the long-term success of software and hardware manufacturers. In many cases, the use of customer support lines could be reduced dramatically by designing the software from the beginning for better usability. It is often possible for software under development to be subjected to usability evaluation before it reaches alpha or beta testing so that critical incidents can be identified and reduced or eliminated. The net effect on profitability can be significant.

In the simplest case, it is possible to conduct a critical incident analysis of the software for the specific purpose of identifying usability problems that can lead the user to make that first, important call. Such a study can be highly cost effective when viewed against the cost of customer service calls. This type of usability study does not strive to identify in a comprehensive manner all aspects of the user interface that can be improved. Instead, the purpose of this type of usability study is to identify in as direct a manner as possible the *critical* problems that can lead to overuse or misuse of the customer support lines. Often, the usability science consultant must recommend this approach based on the client's lead-time constraints, budgets, and overall product development objectives.

Case Study: The Case of the Problem Printer Driver. A leading manufacturer of printers released a product that had only one serious usability problem. The difficulty was present in the installation and operation of the printer driver software. However, this problem was so difficult to solve that more than 50% of the first 100,000 users called the customer service line. The cost was nearly $0.5 million *per month.* The secondary and equally important problem was the poor reputation created for the manufacturer as a result of overburdened phone systems. Many users had to call over several days to reach a customer service person. The manufacturer was also forced to correct the problem by sending the customers new diskettes with a letter of apology at a cost of $3.00 each. By this time, the distribution of the printer was well in excess of 200,000 units.

The total cost of replacing the printer drivers was $900,000. It was very clear upon analysis that the problem could have been identified and corrected at a fraction of the cost if the product had been subjected to even the simplest usability testing. It is interesting that the printer's usability was originally tested internally by the engineering group responsible for development of the product; of course, they did not have a problem with the driver software.

Increasingly, manufacturers are discovering the benefits of reduced service and maintenance costs brought about by improved usability. However, in many cases, manufacturers and developers are not aware how dramatically such costs can be reduced. It is critically important that the usability team seek detailed information on the client's service and maintenance records before attempting to identify the scope of the problem or propose methods or budgets. Waiting to identify usability problems by monitoring customer support calls is like fixing a leak in the hull of a boat that has already sunk. Once the product is in the hands of the user, the cost of fixing the problem will always exceed the cost of a comprehensive usability design and verification program.

9.10.3 Customer Complaints

Manufacturers of most products maintain a file of customer complaints. Some files are very comprehensive; others are quite informal. The identification of usability problems should always include a review of customer complaints. Occasionally, manufacturers will seek assistance in resolving usability problems based on these types of complaints. In these cases, the usability problems are generally so overwhelming that complaints are a small part of the problem. The most interesting aspect of customer complaints is that they are nearly impossible to measure in conventional terms. The cost to the corporation is qualitative and indirect, showing up in terms of low repeat customer ratios, poor peer recommendations, and generally poor customer relations. These factors have their greatest impact on the long-term profitability of the corporation. Rarely are there any mechanisms set up to measure these variables in a manner that is useful to product management.

The general assumption is that customers' complaints are a problem to be dealt with by replacing the product or making an occasional follow-up call. If the product has in some manner failed mechanically or is otherwise defective in operational terms, this approach is acceptable. However, rarely does a manufacturer receive a customer complaint that focuses on usability. Most consumers will not file a complaint on poor usability. The reasons for this are the subject of much debate. However, it is clear that customers do not yet return products because they are difficult to use, but that trend may not continue forever as consumers generally become aware that products do not have to be so difficult to use or maintain.

Case Study: The Case of the Digital Watch under the Couch. A leading manufacturer of digital watches produced and sold a watch that was so difficult to operate that the end user perceived it as a mechanical failure. When faced with the possibility of changing the watch's time, the user was referred to a miniature operator's manual of some 30 pages. The task was so complex and procedure-specific that most users could not change the time on their watches even after persisting for more than 15 minutes, using the manual as support. When faced with the task of setting the alarm function, they had a similar experience. In a study conducted to identify usability problems, respondents stated that they simply could not set the alarm or, more importantly, turn it off when it rang. In several cases, respondents stated that they actually put the watch under the living room couch cushions so that it would not wake them up when it went off at night. They could not reliably set the alarm or manage its operation.

The manufacturer did receive a significant number of watches back with claims that they were broken. However, when the watches were tested, they were operationally sound. This is clearly an extreme case of poor usability design. However, it does demonstrate that users must often be driven by poor usability design to the point of believing that the product is actually mechanically defective before they will log a customer complaint or return the product for refund or repair.

Usability problems of this magnitude will usually kill a product and sometimes its manufacturer within one or two product lifecycles. This product went to market without any usability testing except that undertaken by the product development team itself. The general rule in usability science is that if usability problems are being mentioned in customer complaints then the problems are probably very serious. Examination of customer complaints can be a useful means of identifying critical usability problems. However, they can often be deceptive.

Although the benefit of reduced customer complaints seems self-evident, the real message is that manufacturers should never refer to a lack of customer complaints related to usability as a measure of the quality of their product's user interface. Furthermore, if customer complaints do begin to show up, manufacturers should pay attention and act immediately to identify problems through the use of usability science testing. If they don't act, the next correspondence they receive may be from a plaintiff's attorney. They can also expect sales of the product to drop precipitously as customers approach the phase of secondary purchase decisions. In markets in which peer recommendations are a critical factor in purchase decisions, problems of this scale will kill a product faster than most manufacturers understand.

It is the responsibility of the usability science consultant to identify and analyze problems for the client by examining customer complaints. Clearly, this should never be the only criterion reviewed to determine usability problems, but only one of several defined in this and other chapters in this book.

9.10.4 Increased Complexity in New Products

Increasing product complexity is a very real predictor of usability problems that usability science consultants must examine if they are to provide the client with critical support in developing an effective usability policy for new product development. The development team rarely understands increasing complexity as they upgrade software packages or create new products. In many instances, the product development team is too close to the product to realize how seemingly minor changes in operational structure combined with new features affect users' ability to use the product effectively.

A critical component in identifying the scope and nature of the complexity problem is examining in detail the product's previous generation while reviewing the proposed new user interface. This is an especially critical issue for products that have a large installed user base or to which significant new features are being added. If the product meets both criteria, considerations of negative transfer and access to new features are absolutely critical. This kind of cross-correlated task analysis is essential in determining what aspects of the user interface must be retained in the new version. Unfortunately, the findings of such studies often conflict with the desires of the marketing team to create a new and exciting user interface.

The training equity that users have invested in a software package cannot be discounted in the design of a product upgrade or a new product. Major corporations have lost significant market share by ignoring or discounting the equity that a user group has established in a software package or user interface structure. Regardless of all the theoretical discussion surrounding the definition of operational complexity, it is clear, based on practical experience, that adding significant new features to a user interface while also dramatically changing the existing interface is courting disaster in the marketplace. This does not mean that every software manufacturer should copy the user interface of the market leader, but it does mean that the design of an effective, highly usable interface is a complex problem that must consider the users' capabilities and limitations as the starting point, not an afterthought. The main theme is to capitalize on the users' previous experience combined with new features that have real functional value. Although this approach seems self-evident, it is rarely followed.

Case Study: Dusting off Old Reliable. A number of years ago, a world leader in sewing machines came to my firm for a usability analysis of a new microprocessor-based sewing machine. The new device offered the user a highly attractive user interface, including touch selection of stitch patterns. The engineering team had used the newest technology to automate some aspects of the sewing process, mainly stitch selection. Other operational requirements, such as threading the machine, making buttonholes, and other frequently used features, were also changed slightly but were not automated. When we recruited the first round of users and conducted tests, we discovered a very interesting fact. Many of the users stated that they frequently kept their old machines out and in working order so that they could switch to the old machine if they needed to do something quickly. The more time users had with the new machine, the more likely they were to abandon the new machine altogether. In fact, several users had actually put the new machine away and returned to using their old machines exclusively. This transfer to the old machine usually took place after many hours of attempting to use the new machine to do what the old machine did supremely well, namely, sew a few simple stitches, make buttonholes, sew on a button, and other high-frequency tasks.

Upon the execution of a professional usability analysis, we discovered that the development team had chosen to automate aspects of the device that were operationally insignificant. They left the key functionality buried behind a facade of slick covers and difficult operating procedures. By poor usability design, they dramatically increased the operational complexity of the product at a time when women (their target market) had less and less time to learn to sew because of increased demands on the family structure for a second income.

For example, an error analysis of a critical operational task, threading the machine, produced very interesting and insightful results. Novice sewers tended to make the same kinds of errors in threading the machine as did more experienced users. However, the novices left a much higher percentage of their errors uncorrected. At one critical point in the thread path, these uncorrected errors caused the machine to jam. Because of the design of the machine's thread path, the jam was very difficult to clear. Even when prompted to unclear the jam using the troubleshooting portion of the operator's manual, the novice users could not fix the machine. The user interface had failed; the user was stuck.

The benefit of designing for reduced operational complexity can have tremendous impact that can be directly measured in terms of profit and loss. In the case of the sewing machine's jammed thread path, the usability science team did not stop at identification of the problem. In addition, they examined in detail the impact that such an operational failure might have on the success of the product and more broadly on the corporation. By examining the warranty for the product, they discovered that the manufacturer would come to the user's home to fix the machine any time in the first 90 days of ownership (i.e., novice user prime time). By gathering service and maintenance records from field offices, they found that more than 50% of the in-home service calls were attributed to jammed thread paths. The estimated cost to the manufacturer was more than $1 million a year in North America alone. The usability study resulted in the redesign of the threading procedures and related thread path, which reduced the number of jammed machines by more than 90%. The cost saving in the U.S. market was several million dollars a year. The savings on a worldwide basis were much larger.

If the product had been subjected to a usability testing program before production, the problem would have been identified and corrected. Furthermore, the effort that went into creating a new user interface for insignificant features would have been redirected, and millions of dollars in development and production costs would have been far better utilized. This product never produced a profit and may have cost the manufacturer a world leadership position in this product category. Simply knowing how to evaluate the relationship between new features, past experience, and basic operational requirements of a product is one of the greatest contributions that a usability science consultant can bring to the product development process. The overall ROI can be measured in strategic, not only tactical, terms.

One important aspect of the new sewing machine was that at first impression (the most important impression when it comes to purchase decisions) it actually *looked* easier to use than the older version. This "designed to *look* easy to use" factor is a critical problem for the designers of complex user interfaces. In fact, much of the usability science literature on screen design and layout is incredibly narrow and accepting of the belief that simple is better.

Experience shows that screen designs need to be different based on the skill base and operational goals of the user. I have seen visually complex screens that were highly error resistant, easy for training, fast, and operationally satisfying to use. At first glance, none of these screens would have won awards for graphic design. That does not mean that the designs were bad but simply that the designer did not follow the naive modernist dictum that simple is better. One of the most difficult issues to resolve in the design of a highly "usable" interface is the notion that graphic layout can and does allow the manufacturer to create the *impression* of operational simplicity. As with the case of the sewing machine, in the long run, these half-truths do not leave the customer with a warm and fuzzy feeling toward future generations of the product.

9.10.5 Market Competition

Frequently, a corporate client may not be aware that significant improvements in usability can lead to increased sales and market share. The basic perspective of typical corporate clients is to view usability enhancements as a necessary evil that must be addressed only if they have very real safety or use problems.

However, an increasing number of corporate development groups are being made aware that enhancing usability can and often does produce directly measurable results in terms of market share and sales.

The most interesting aspect of this issue is that a product development group rarely identifies improved usability as a major project objective when seeking assistance from an outside usability science consultant. The role of the usability expert is to identify and propose improved market share and sales as a primary benefit of the program. Clients are often actually surprised that these absolutely critical variables can be directly affected by usability enhancement.

Cost implications of improved usability depend on a number of factors, including, but not limited to the following:

- The maturity of the market for the product or system
- The competition's emphasis on usability as an aspect of competitive edge
- The presence of an industry standard interface (very rare)
- The costs of training and updating
- Demands for significant ROI based on automation
- Demands for significant ROI based on job elimination

Any one of these factors can signify an opportunity for the client to sell usability as a significant product feature. The difficulty is convincing the client that he or she should fund a larger project so that the user interface performance of the primary competition can be documented as a benchmark for program enhancement. The development of a detailed understanding of the usability performance of the client's products plus that of the primary competition is absolutely essential if the client wants to realize a significant improvement in market share and sales as a result of enhanced usability.

Often the detailed examination of the competitive products can add 40 to 60% to the cost of early phases of a usability program. However, these costs are often insignificant when viewed against the opportunity to improve market share or sales by even a very small percentage. There are many secondary benefits as well, including the opportunity to understand how users actually interact with the competitor's product and why they have purchased the product in the past. Such information is rarely part of a marketing plan or product development objectives.

Case Study: Market Share in Process Control. In the early 1970s, a group of young engineers left a leading manufacturer (Company A) of process control computers to form a new business (Company B) that focused on providing computer automation to the paper-making industry. The first generation of Company B's product was very successful, becoming, within a few years, the market leader in terms of market share, gross sales, and profits. This first product line focused on automating a few basic aspects of the paper-making process, most of which did not require direct human interaction on a real-time basis. Company B grew rapidly and made the original development team and management wealthy.

As is always the case, a successful product launches competitive products, and a third company (Company C) was formed by engineers who left Company B. They created a new system to compete head to head with the main product of Company B. However, Company C offered the industry a new user interface that was marginally easier to use, for a lower price and with more automation. They began to take market share from Company B.

During the following 2-year period, Company B designed and developed a new system that offered far more automation but maintained much of the old interface design, which was very difficult to use. As the new systems began to come online, customers noticed an actual drop in the productivity of their work crews. This, of course, did not make them happy. During this phase Company B began to lose more market share and to develop a questionable reputation in a very tightly controlled industry. In a last-ditch effort to identify the problem with their new system and with the hope of creating a rapid fix, Company B sought assistance from a usability science consulting firm.

After only two site visits and a preliminary task analysis, it became obvious that the new system was exceptionally difficult to use. Furthermore, the new computer took the user out of the decision-making process at critical points when the machine crew had to make rapid decisions about the status of the process if the paper being produced was to remain within specification. As a result, the machine crew would often purposely take the computer offline and record the event as a computer malfunction. In fact, nothing was wrong with the computer itself except that the user interface was totally unusable. The net effect of this problem was a miserable reliability record and a series of very unhappy customers who did not see any reason why their considerable investment was not returning a better product at a lower price.

Manufacturers must realize that it is frequently possible to clearly identify important usability problems in a relatively short period of time. On the other hand, much more effort is often required to convince the client that a new user interface is the answer. In this case study, the client immediately understood the need for a new interface and adopted a user-centered design approach in creating new code for the next-generation system. The program received support from the highest levels of management and was given top corporate priority. Information gathered from field interviews and related observations identified many additional problems and areas of innovation for the client's system and for the overall process of successfully integrating advanced computer control systems with human intelligence to produce a better product at a lower cost.

There are many excellent reasons to integrate usability engineering into the product development process. However, none is as powerful or will be so easily understood by the executive suite as the benefit of creating a marketable product that will increase market share and profits. In any product development effort, the consultant will serve the client best by positioning usability improvements in ways that can be leveraged by the client to increase the sales and profits of the product.

In the paper industry example, the benefit to the client was immediate and significant. From the earliest stages of development, the new user interface was seen as a means for the client to regain market share and return to profitability. The client's advertising agency frequently sat in on presentations by the usability science team. Improved usability became the centerpiece of the new product marketing strategy. Formal usability testing and design became central to the product development methods of the client, thus reducing product development lead time.

The new product, with a completely redesigned hardware and software interface, was introduced with improved usability as its main marketing strategy. Within one year, the product had recaptured all lost market share and gathered new percentage points. The user interface became widely regarded as the best in the industry and is now a *de facto* standard. The client returned to high levels of profitability and was acquired by a world leader in process control computers.

It is safe to say that improved usability was the central driving force in this important case study. However, none of the benefits listed here would have been realized if the management of the corporation had ignored the importance of usability science in the design of their new products. It is unfortunately more common for manufacturers to discount the importance of usability and focus instead on cost, features, or time to market.

The market share of the product discussed here increased while a high profit margin was maintained. Dramatically improved usability allowed the client to maintain its position as the premium producer of process control computers for the paper-manufacturing industry.

9.11 SUMMARY

The preceding case studies have been drawn from my more than 25 years' experience as an independent usability science consultant. During that time, staggering improvements have occurred in the design of hardware and software. The relative cost of computing power has dropped so fast that it is now hard to imagine where it will all lead.

On the other hand, advances in user interface design have not been nearly so dramatic. The reasons for this are beyond the scope and objectives of this chapter. However, it is safe to say that what improvements have been made in user interface design have benefited users and manufacturers in substantial and even dramatic ways, both tangible and intangible. As the next generation of computing power emerges from development labs, we can hope that some significant portion of the new computing power will be channeled toward improved usability for all products, large and small, complex and simple. The key to this may well be the successful application of cost-benefit analyses at opportune moments in product development organizations. The various case studies cited in this and other chapters in this book are the beginning of a substantial and professional literature on cost justification for usability science as a new and critical corporate asset.

REFERENCES

Bias, R. G. (2003, October). The dangers of amateur usability engineering. In S. Hirsch (chair), Usability in Practice: Avoiding Pitfalls and Seizing Opportunities. Presentation at the *Annual Meeting of the American Society of Information Science and Technology.* Long Beach, CA.

Bias R. G., and Mayhew, D. J. (1994). *Cost-Justifying Usability.* Boston: Academic Press.

Christensen, C. M. (1997). *The Innovator's Dilemma: When New Technologies Cause Great Firms to Fail.* Boston: Harvard Business School Press.

Csikszentmihalyi, M. (1988). *Society, Culture, and Person: A Systems View of Creativity.* In R. J. Sternberg (Ed.), *The Nature of Creativity.* New York: Cambridge University Press.

Csikszentmihalyi, M. (1996). *Creativity: Flow and the Psychology of Discovery and Invention.* New York: HarperCollins.

Getzels and Csikszentmihalyi (1976)

International Organization for Standardization (ISO) (2003)

Mauro, C. L. (1994). Cost-justifying usability. In *A Contractor Company* (pp. 123–141). Boston: Academic Press.

Mauro, C. L. (2003). Professional Usability Testing and Return on Investment as It Applies to User Interface Design for Web-Based Products and Services. A review of online lab-based approaches. Retrieved February 18, 2004, from www.taskz.com/ucd_testing_roi_ summary.php.

Mayhew, D. J., and Bias, R. G. (1994). Organizational inhibitors and facilitators. In R. G. Bias and D. J. Mayhew (Eds.), *Cost-Justifying Usability.* Boston: Academic Press.

Norman, D. A. (2001). *Applying the Behavioral, Cognitive, and Social Sciences to Products.* Retrieved February 18, 2004, from www.jnd.org/dn.mss/BCCSandProducts.html.

Rogers, E. M. (2003). *Diffusion of Innovations* (5th ed.). New York: Free Press.

Sawyer, K. R. (2003). *Group Creativity: Music, Theater, Collaboration.* Mahwah, NJ: Lawrence Erlbaum Associates.

10 CHAPTER The Return on Investment in Usability of Web Applications

Clare-Marie Karat IBM TJ Watson Research Center
Arnold Lund Microsoft Moble PC

10.1 INTRODUCTION

In this chapter we provide context for the discussion of the return on investment (ROI) in usability for Web applications and a brief review of the research completed on the topic of the cost benefit of usability. We discuss examples of ROI in usability for the Web, take a "big picture" view of the attributes of value, and identify at a high level how to create and measure value. (For a more detailed business case approach to usability, please see Karat's Chapter 4.) We illustrate how to sell the value of ROI in usability for Web applications and conclude by looking to the future. In our examination of this topic, we draw on usability research and consulting experiences with customers across a range of professional settings in recent years and cite de-identified case study information that has been made available to us or is in the public domain.

This book is focused on cost-justifying usability, with particular emphasis on activities for Web applications and services. The lessons discussed in this chapter come from Web and non-Web contexts and have been applied successfully in Web environments. A wealth of knowledge about human-computer interaction (HCI) has accumulated in the field over the last 25 years (over the last 50 years if the fields of behavioral and computer science, on which HCI is based, are included). In this chapter we provide an overview of the ROI for human factors activities in the development of Web applications and services. Web-specific factors that extend the knowledge gained in the HCI field about ROI techniques for usability of traditional applications are identified. Some lessons from the Web can be applied more broadly as well. In the networked world that is rapidly emerging, the focus is on the Internet, and business processes and applications are converging there.

10.2 RETURN ON INVESTMENT IN USABILITY OF WEB APPLICATIONS

The literature on the impact of applying human factors to designing applications is filled with examples of the remarkable ROI that can be realized (Bias and Mayhew, 1994; Lund, 1997; Donahue, 2001; Souza, 2001a, 2001b) and the business impact it can have (see Chapter 21). The impact of applying human factors to Web design is similarly spectacular. For example, Sapient was contracted to help with the redesign of the United Airlines site. This work began by the teams' developing a deep understanding of travelers and modeling the traveler's experience to inform a new design. The result was a 200% increase in online ticketing and an increase in the satisfaction of United's most profitable customers (Internal Sapient Sales Materials, 2000). The new design doubled the number of daily sessions, and the number of users going deeper into the site increased to roughly 65%. Gómez Advisors (an Internet quality measurement firm) lauded United Airlines as having one of the Top 25 Web Sites. They ranked it as the #1 Airline Web Site, and the site was recognized as providing leadership in customer relationship management.

The Sapient redesign of Wal-Mart's online sales channel resulted in a 214% increase in the number of unique visitors (Internal Sapient Sales Materials, 2000). Wal-Mart faced significant challenges with the redesign of their online store. The previous version of the online channel was performing well below targets and received poor press. According to Gómez Advisors the new release moved Wal-Mart ahead of its competitors and scored particularly well in ease of use and in stimulating consumer confidence. The new design supports the consumer's ability to find the products they are looking for, as well as Wal-Mart's desire to educate people about the range of products available and to cross-sell other products as appropriate. A strong internal design and usability organization has grown within Wal-Mart to ensure that the company can continue to harvest the value of effective design.

Karat *et al.* (2002) researched the usability and entertainment value of streaming, multimedia Web experiences. This HCI research determined that traditional interaction guidelines for task-oriented interfaces do not necessarily apply when the goal of the site is to be entertaining and engaging. The cost-benefit analysis of the HCI research highlighted the interaction methods that would motivate target users to return to the site and determined that use of the site would augment users' enjoyment of brick and mortar cultural institutions around the world, rather than be a substitute for them. The HCI research with target users identified this value and the means to attract and retain the users at the site. The financial value of the usability and user experience (UE) on the site was calculated to be a very substantial amount of money (due to confidentiality agreements, no further details may be provided).

Karat *et al.* (2003) developed the personalization design and strategy for the ibm.com site. The HCI team iteratively designed and tested a set of personalization features and user scenarios with target customers over the course of seven months and identified a core set of 12 personalization features that provided highly valued UEs on the site. As part of this analysis, the team was asked to collect data for the business case for personalization and the related HCI activities necessary to accomplish the design and implementation of the UE. In the course of the research, the HCI team collected customer information on the number of additional visits to the site and additional purchases that would be made if the site provided the type of personalized experience that the customers realized in working hands-on with mid-fidelity prototypes to complete critical work tasks in highly efficient, effective, and pleasant ways. The resulting confidential data showed a very significant ROI in the usability of the personalized UE on the e-commerce site.

There are a variety of other examples. A classic success story, of course, is Dell's ongoing use of its online channel and the continuing evolution of the usability of that channel. Dell has worked to drive significant improvements in the value of the site by systematically tracking and improving their UE (Dell, 1998). The iterative evolution of the US WEST Web site (before it was acquired by Qwest) resulted in a dramatic increase in traffic over 3 years, and the cost of ongoing development support was more than compensated for by the cost savings arising from shifting customers from the use of live customer support to online support (Lund, 1999). A redesign of the customer service site of a large mutual fund company by one usability- and design-oriented firm resulted in a dramatic reduction in live customer service costs, and more than 20% of new accounts were subsequently set up through the online channel (this example is a de-identified case study based on previous research by Lund).

Souza (2001a) reported that 42% of U.S. consumers made their most recent purchase because of a previous good experience with the online retailer. Yet 65% of online shopping attempts end in failure because customers can't find what they are looking for; and according to one analyst, 62% of shoppers have given up on online ordering because of a bad experience (Souza 2001a). New users at one site spent an average of $127 per purchase. Repeat customers spend almost twice as much. It is also well known that poorly designed applications are a major reason users call support lines, and such calls can easily cost $12 to $250 per call (de-identified case study data from Karat). An improved experience for Schneider Automation resulted in $2 million saved in call-center support costs over the first 10 months after the change. There are other benefits as well. One early study of e-commerce trust (Lund, de-identified case study data) found that good navigation and presentation were essential in creating trust, and one of the authors of this chapter, in testing the usability of Web sites regularly, found a link in users' expectations between the quality of the design and trust in the channel and the company's brand.

The benefits are not just experienced in business-to-customer applications on the Web; they are also realized in internal applications. AT Kearny (Internal Sapient Sales Materials, 2000), for example, has reported that poorly designed knowledge management systems and intranets cost organizations more than $750 billion annually. According to *CIO BusinessWeb* magazine, "On a corporate intranet, poor usability means poor employee productivity; investments in making an intranet easier to use can pay off by a factor of 10 or more, especially at large companies" (Kalin, 1999). Forty-one percent of employees at one information technology company couldn't find the information they needed for their jobs on their intranet, and at one oil company it has been reported that 35% of an employee's time is wasted looking for information. Some estimates of time wasted finding information on corporate intranets are as high as 50%. However, improvements in productivity and decreases in learning time of 30% or more can be realized when human factors are applied to the redesign of enterprise Web applications (Kalin, 1999).

10.3 LESSONS LEARNED FROM ANALYSIS OF RETURN ON INVESTMENT IN USABILITY ON THE WEB

The most immediate challenge in demonstrating the ROI in Web applications is to determine what to measure and how to measure it. Common metrics (e.g., hits, visitors, click-streams, and so on) certainly correlate with the value of a site and therefore provide some value in monitoring the impact of design changes. Tracking sales or support activity over the site is typically of greater interest to most managers because it relates directly to the bottom line. Both, however, are similar to taking a patient's pulse and blood pressure. They allow you to monitor health, but not to easily diagnose the "why" behind the changes. They also are difficult to convert into ROI numbers. Being able to track the activity of individual users (recognizing the security and privacy issues implied) provides the clearest picture of both the ROI of the Web as a delivery channel and the factors that influence the ROI. With a user focus, collecting the richer set of data typically used to demonstrate ROI for other products and applications (e.g., speed of learning, user support costs, and so on) and/or confirming in-lab data against field data should be easier.

Whereas most companies seem to focus on the ROI of the channel itself (e.g., "Are we increasing the number of people coming to the site?"), they tend to miss the bigger and more important picture. Just as products can be cross-elastic, meaning that a product can serve many purposes, so can channels. The ROI that should be of interest is the user-focused ROI that recognizes that the user employs the Web along with other ways of meeting his or her goals. A shopper may shop by browsing and collecting information on the Web but then buying in a store, or may shop in a store and buy on the Web. Customers may solve a problem on the Web (and therefore not require any live support), or they may be getting background information on the Web, which makes live support more efficient and enables the live support to sell products (a common telecommunications company customer-support activity). The ROI of interest therefore needs to be user- or customer-focused. Regarding customer activity, the questions of interest include the following:

- How much are they buying from the company?
- How often do they return?
- How much overall does it cost to support them?
- How satisfied and loyal are they?
- What attributes do they associate with the brand?
- How do the various channels influence that ROI?

User-centered design, therefore, could and should include user-centered metrics, which in turn implies a contextual understanding of users to drive coordination across product and channel silos in companies.

Because capturing compelling ROI data requires having the right data collection mechanisms or "hooks" in the right places, another lesson is that the HCI field has far fewer ROI success stories than it might otherwise have. Rarely do Web application development efforts include building in the hooks from the beginning to demonstrate the ROI of an application when it is launched. Human factors professionals might benefit by being accountable for the value of their efforts and arguing for inclusion of the appropriate hooks as requirements. Future Web services may have these hooks because application developers and providers can use them to track total cost of ownership.

Most major technology companies are working toward a Web services world in which the Internet resides on a backbone network with virtually unlimited capacity to carry data and voice; information, images, multimedia, and virtually every other type of content are just data. The data will be coded in different ways to enable richer applications and UEs using XML and in the future taking advantage of the semantic Web. Users will connect to this network through narrowband, broadband, and wireless interfaces. As Moore's law continues to hold sway, in ubiquitous computing virtually everything could have a processor and network connection within it, and the network will also connect to servers with ever-increasing processing and storage capacity. Thus small devices in the hands of users will have the virtual capabilities of large mainframe computers. This is the dream of what is currently known as Web services. Although there will certainly continue to be traditional standalone applications for many years, increasingly Web services will look and feel like graphic user interfaces (GUIs). The Web, therefore, over time will be less and less dominated by hypertext documents and increasingly dominated by interactive applications. Design principles that have evolved over the years for GUIs will have a natural extension to the Web, and research on designing for the Web has caused these GUI design principles to evolve further in application to the Web. Taking the richer contextual view of user behavior to demonstrate ROI is therefore likely to apply more broadly, as ROI needs to be demonstrated for other applications as well.

10.4 THE ATTRIBUTES OF VALUE

Typical benefits arising from applying user-centered design to hardware and software include the following:

- Increased sales, market share, and revenue
- Improved customer satisfaction and loyalty
- Reduced customer support costs
- Stronger brand recognition
- Improved employee productivity
- Reduced errors
- Faster learning and better retention of information
- Improved decision-making and effectiveness
- Decreased development and software maintenance costs

See Karat's Chapter 4 for further information.

It is not surprising, therefore, that these are the same benefits found when user-centered design methods are applied to Web sites. In many ways, the Web experience is the brand. The business value that a Web site is intended to deliver may include helping define a brand, supporting users or employees with a lower cost channel, providing an automated or semiautomated way for customers or businesses to share information, and so on. But most often the business case for Web projects is justified based either on cost savings arising from getting users to do something in a different way (e.g., shifting from the use of live support representatives to online service), increasing revenue (e.g., from new sales achieved by providing a new and better shopping environment for customers), or adding new revenue (e.g., from a new service offered over the Internet).

Achieving the objectives of the business case typically requires users to do something differently. They need to adopt and use the Web application. At the heart of adoption of a new solution is user satisfaction. Users need to find value in the solution and usually that value needs to be significantly greater than the alternatives they are already employing. They also need to find it easy to get at that value, that usefulness. If the usefulness is sufficiently great, ease of use could be relatively less important in the short run, but when there are competitive solutions, making it easy for users to get at the underlying functionality or product is required. If the underlying functionality isn't dramatically greater than alternatives, then significant improvements in ease of use can provide a competitive advantage. Interestingly, satisfaction, ease of use, and usefulness map nicely onto the International Organization for Standardization (ISO) 9241[1] definition of usable software. For many users, this interaction of ease of use and usefulness in large measure defines the quality of an application and a UE, and when combined with perceived financial benefits and costs, results in a sense of value.

An attractive design invites users to experience the value of the application and its ease of use. A well-designed application may even provide some signals about the ease of use of the application (because many principles of good design are identical to the principles of easy-to-use designs). For some domains and applications, the quality and nature of the *experience* are also important to users. Entertainment applications are an obvious example, but our experience is that for many users Web applications are also judged on the basis of the quality of the design. The quality of the design may even drive the trust that the users have in the application (Studio Archetype/Sapient and Cheskin, 1999).

---

[1] ISO 9241-11, Guidance on Usability: Usability is "the extent to which a product can be used by specified users to achieve specified goals with effectiveness, efficiency, and satisfaction in a specified context of use."

The characteristics of the Web application itself, however, are typically not sufficient to drive value. The technology diffusion literature identifies several other characteristics of the experience that can be important. The value needs to be observable and observed. When people feel they are seeing others getting value, they consider using the valued product themselves. Cell phones and VCRs have benefited from being observable, and the "buzz" around certain Web sites such as Amazon and Yahoo have helped them attract first-time users who then can decide if the site works for them. Therefore, in the design and launch of a new UE, it might be beneficial to define a strategy for educating the potential users about its value and about how to begin obtaining that value, to design in feedback mechanisms to determine how well the educational initiatives are working, and to design in features to promote such education. At US WEST we could track usage of the site in sufficient detail that we could monitor whether advertising about the site had an impact, whether the features highlighted in the advertising were actually purchased and used, and whether usage levels eventually dropped all the way back to preadvertising levels or whether we had "converted" some of the potential users into regular users.

Another attribute of a successful application is that it needs to integrate into people's lives. Just as it is difficult to change the person to fit the technology, it is also hard to change the way people fit within the ongoing context in which they live while using the application. The application may seem great in isolation, whereas in practice it quickly gets dropped because it doesn't fit into real people's lives.

How do variables like social context relate to the ROI of Web design? There are dependent variables impacted by the properties of the design and the way the design is presented to users as well as independent variables that can be tracked to diagnose where and how to improve scores for the relevant dependent variables. Usefulness, ease of use, and satisfaction can be measured. User awareness and belief in what is presented can be measured as well. HCI practitioners can measure actual and predicted integration. And practitioners can observe how these measures change as the design of the experience and the various factors that influence the experience are changed.

HCI professionals can also look to the scenarios of use that were identified as part of the initial experience models the design was based on—the scenarios that form the basis of the anticipated user and business value. Say that a mobility scenario is being supported. This scenario involves portfolio managers reviewing the current state of the market from their desks and requesting that brokers place orders as the managers leave to visit a company in which they are investing. The portfolio managers will want to be notified when the orders are completed and will want to track whether events they expect to impact the investments take place. Here the result should be that the portfolio managers are more successful in managing their portfolios (a portfolio performance measure), but it should also be possible to look at metrics that correlate with the success that mobility enables (e.g., the number of companies visited might increase, if that fact is considered important based on the experience models). Some useful examples and articles related to ROI on the Web are listed in the reference section at the end of the chapter (see Karat, 1994; Lund, 1997; Mauro, 2002; Nielsen, 2003; Souza, 2001a).

10.5 CREATING VALUE

Having an understanding of the business goals and the specific outcomes anticipated for the application is a start for creating value. In recent years, those goals have tended to focus on cost reduction or revenue increases, the kinds of goals that can be used in a business case to justify a project. Improvements in branding, customer satisfaction, and similar goals are also possible. These high-level goals need to be operationalized as specific objective outcomes. A cost-savings goal might involve fewer or shorter service center calls resulting from a shift of some customers to online self-service. In a business case, specific numbers would be projected for this shift to demonstrate that the expected cost savings justify the cost of the project.

Once explicit objectives are defined and models of the relevant user behaviors and expectations have been created, the next step is to define and model how the objectives will be met by modifying the UE. Again, if an objective is to move the user from the live service center to an online channel, it should be possible to define rewards for making a change and perhaps even "negative reinforcers" for not making the change. The rewards might include improved availability (relative to the live service center), rich information and services, and services not available through the live service center. Negative reinforcers might include reduced availability of the live service representatives, feedback while on hold concerning the time until the call can be answered, or even charges for live support.

10.6 THE COST OF CHANGE

In addition to creating value through the design of the UE, applying human factors to the design process reduces the overall cost of the product (Karat,

1994). Problems identified late in the development process or even after release are far more costly to fix than those identified early in the process, and the kinds of fixes that are practical are typically only the small ones rather than the restructuring that may be needed to add real value to the project (Karat, 1994). Web design is a slightly special case in this analysis. For many Web applications, the expectation is that the design will and should change fairly often. These changes, however, typically are changes in content and in the images on the site. Changes to the more fundamental information architecture and process flows within the site tend to be more like late-stage changes to other kinds of traditional application software. These late-stage and more complex changes may drive increased customer service calls and costs as users become confused, and they may require rewriting significant amounts of documentation. Web services applications, of course, are more like standard applications and should follow the cost-justification model described for traditional GUIs (Karat, 1994).

10.7 REDUCING THE RISK OF FAILURE

Other opportunities to create value through the application of human factors to Web design include various techniques for reducing the risk of failure when the product is introduced. At Ameritech, the organizational belief was that it would take 10 ideas to produce 1 good idea, 10 good ideas to produce 1 product, and 10 products to produce 1 successful product. If the relationship between the various attributes of the user's Web experience and the target business objective(s) are understood, however, the organization can improve its chances of success with a Web site by measuring the attributes. The organization can even create norms that will allow it to predict success. The USE questionnaire developed at Ameritech was built on the model of usefulness, satisfaction, and ease of use, for example, and was used to collect data across a variety of hardware and software products and applications, customer services, and documentation (Lund, 2001). By knowing which were successful and which weren't, it was possible to create norms that could be used to set ease of use and usefulness rating requirements. The data collected showed clearly the range of ratings associated with products destined for failure, and those required for success (realizing that pricing, marketing, and other factors are obviously also critical for success). Although some consulting companies have been reasonably successful in using various screening metrics to predict success, we imagine that choosing metrics based on real user needs and experiences should be even more effective. The usability practitioner can think of these metrics as measures associated with the usability objectives for the Web site or application. For example, a usability objective might be that 95% of new users with a particular product in mind can navigate or use the site search function successfully to find the items they are looking for within 60 seconds. The development team can track the improvement in the UE across iterations in design and evaluation. When the site has met the usability objectives that the team has created on the basis of user feedback, then the team has reduced the risk of failure.

With these same measurements, it is possible to conduct competitive analyses, and using the models of value creation through the new UE, HCI professionals can obtain diagnostic information about the competition. They can identify features for which a site has advantages in the UE and those for which the competition has advantages. Then the organization can focus on improvements to meet or exceed similar features of the competition. This same information can be used to inform advertising and user education, which in turn helps drive adoption and use of the application. At Sapient, Lund used contextualized heuristic evaluations for competitive analyses of various Web sites, for which he and his team applied an experience model that identified the relevant attributes of the users' experiences and needs to structure the collection and analysis of heuristic data. The resulting analysis allowed the organization to prioritize improvements to a site based on the expected impact on the business goals for the site (Lund, 2000).

10.8 MEASURING VALUE

Creating a baseline refers to directly measuring the variables of interest before and after a change. For example, purchases on a Web site can be measured before it is redesigned and again afterward to determine whether a change had the desired impact. The recommendation is that baseline data be collected for the dependent variables most directly related to those identified in the business case (e.g., time on task, satisfaction with the UE, completion rate, and error-free completion rate) and also for behaviors that are likely to be correlated (e.g., number of first-time visits, number of repeat visitors, and so on). In addition, measurements on related dimensions such as perceptions of ease of use, usefulness, and satisfaction are collected. In addition to assessing the impact of changes before and after a new design is introduced, these measures can be tracked over time to determine when the impact of the design has declined sufficiently to warrant another refreshing (and as user expectations, technology, and design conventions change, the impact is nearly certain to decline). These data can also be used (depending on the measure) for competitive analyses to identify opportunities for enhancing the value of the experience. Some external factors can affect baseline measures and future measures. Wilson and Rosenbaum discuss possible confounds in Chapter 8 of this volume. These may include a competitor who beat you with new features, the quality of the site's advertising campaign, and global economic conditions and events.

It is important to note, however, that although many focus on measurements of the Web experience in isolation, the most important information for a business is often the Web channel changes relative to changes in other channels. In other words, do customers move from the live customer service channel (accessed over the telephone) to the Web channel? Is the Web channel actually as inexpensive as expected or do customers have more problems and therefore require more manual intervention? Do total purchases increase, and if so, are there more or fewer returns from unsatisfied customers? Although a rich set of challenges in measuring and interpreting the Web data exist, even tougher challenges are often seen in integration of data across channels to obtain a meaningful picture of the total ROI resulting from design changes.

How is value created? As mentioned previously, it generally comes from people changing their behavior, their expectations, and their attitudes based on their satisfaction—a satisfaction that comes from awareness, experience, and integrating an application into their lives. The experience and integration are enabled when a Web application is easy to use and useful. This means that if the business goals of the application are defined, identifying aspects of the experience to support those goals should be possible. Design in and assess the value of functions that people will find rewarding and that will meet their needs. By specifying the relationship between the business value of the application and the aspects of the experience that would be expected to drive that value, a value model is created for this context. By assessing whether a user's experience is consistent with the design requirements and how well the application is satisfying the business goals, the HCI professional can determine which aspects of the experience have the greatest impact on the value realized.

For example, suppose that the goal for the organization is to move people from obtaining customer support live over a telephone to obtaining it through the Web, because the Web cost is significantly lower, ranging from 20% to 25% of the cost of live support. Users need one or more compelling reasons—that is, fundamental value—to go to the Web versus the live support. On the basis of previous research, we might conclude that the important values for customers are faster response time and completeness and accuracy of the solution to their problems. For the organization, value will also be in the ability to move customers to newer products. Build these attributes into the design of the site.

Measure the probability that a given customer uses the live support channel versus the probability that he or she will use the online support channel, and measure how that probability changes over time. Assess how much time a given customer spends using the live channel versus the online channel, and make inferences about the total time spent in solving a problem (the number of return visits in each case). With these numbers we can project trends in savings resulting from use of the online channel rather than the live channel.

HCI professionals can also measure how well they have accomplished the value proposition that should compel people to change. To do this, measure the response time of each channel and the time to a solution from the first contact that the user experiences. Use a variety of techniques to measure how much and what type of information is gathered, given a problem scenario. Measure the probability of purchase through controlled user evaluations or click-stream analysis and the revenue generated per customer.

The Internet is not static. It evolves as people find new applications for it, as designers try new things, and as technology advances. As a result, expectations change. The UE that was satisfying and delivered value today may not deliver value tomorrow. That's one of the reasons why the best Web sites continually evolve. The organization must continually collect user feedback on satisfaction with the site and a set of usage metrics tied to usability objectives and capture unmet user needs to determine when a small adjustment or larger redesign of the site is necessary. Because updating a Web site can be costly, however, it is important to have the information needed to drive that evolution intelligently and on the basis of business considerations.

One useful technique is to collect baseline information on the site and monitor changes in user satisfaction and usage over time. A variety of tools and companies are available for assessing the baseline situation on a Web site. Mauro (2002) reviewed the online tools available and found that 95% of the tools are "objectively little more than web-based survey systems" and 5% of the tools "contain simplified behavior-tracking capabilities." Mauro, Web Criteria, Vividence, NetRaker, and others provide services that have been used successfully to help measure and improve a site's effectiveness.

We recommend the use of traditional usability testing of a Web site combined with some online testing and regular competitive analyses. HCI professionals and the organizations they are serving require the valuable objective and subjective data and design insights and ideas that emerge during these individual and small-group sessions with target users who experience a prototype version of the site and core user scenarios. These laboratory studies can be complemented by online testing that can validate laboratory data and track changes on the site. Also, periodic in-house competitive evaluations may include baseline metrics (the set that can be collected from public sites) and help to provide information that is useful in prioritizing improvements. Identified best-of-breed elements may provide inspiration for new designs, and the metrics can be used to set usability objectives.

10.9 HOW TO SELL RETURN ON INVESTMENT IN USABILITY

The survival and the effectiveness of Web design groups often depend on their ability to convince stakeholders that the return on their efforts more than justifies the costs (the ROI). If you survey listservs, discussions at conferences, and relevant publications, you notice that this question of how to convince people of the value of user-centered Web design arises again and again. There are several reasons why this sales process is necessary. Our chapter provides a perspective on this topic within the larger topic of ROI in usability of Web applications. (For an additional perspective on marketing usability, see Henneman's Chapter 5.)

The value that effective design brings is typically separated from the time when most design and user-research activities take place, and it is obscured by the collaborative nature of development. For a stand-alone Web project, the user research and design may take place in the first 2 or 3 months of the project, and the implementation of the design and the backend development work may go on for another 9 months before launch. The people working on the research and design may make up less than 10% of the overall budget for the project. What team members typically see is represented in the way design and user research are often introduced. You may hear something like "This is Joe from design. They make things look nice. This is Sue from usability. They make things easy to use."

Furthermore, many of the players on the team often feel they "own" roles that sound very similar to those held by design and usability people. Market research and project management may feel it is their job to understand the users and to ensure that the product meets their needs. The user interface developers themselves have often designed interfaces as part of past applications. All may wonder why design and usability professionals are needed and wonder may turn into ugliness when the team is pressured to deliver against a tight deadline. Why does the creative process have to take that long and why must the artifacts be created in that order? Can't we learn everything we need to know from a focus group or perhaps even by talking to the sponsor of the work (who presumably should know its users)?

Because the ROI story rarely stands on its own and typically does not in itself have a lot of "stickiness," or integration, within the corporate culture, successful UE managers find themselves continually telling the ROI story. At Ameritech, perhaps the most important event that opened the door for enabling the demonstration of user-centered design ROI was an unexpected meeting of the CEO and the manager of the UE group. The CEO's question, "So what do you do?" and the right response started a sponsorship that lasted for years. At Sapient, the question, "Why can't you give us the design so we can start coding, and while we code you can do the information architecture?" could only be dealt with effectively by discussing where the value comes from and how the creative process logically must flow to deliver the value.

Selling the ROI story begins with quickly and clearly describing the business value of the UE work and the user-centered design process. This speech is often known as the "elevator speech." Crafting the phrases that make up the elevator speech is a lot like crafting a good mission statement. They need to be short and to the point, use the language of the audience, and touch relevant "hot buttons."

As an example, we have written an elevator speech for this chapter:

> There should be no question that the potential return on investment (ROI) for investing in Web usability is tremendous. The benefits occur whether designing services, customer support applications, or productivity tools for use within businesses. The key is a user-centered design approach that delivers ease of use, usefulness, and satisfaction in a compelling way—an approach that adds value to each stage of the adoption process. This chapter discusses the ROI in the usability for Web applications and reviews the research on the cost benefit of usability. Examples of ROI in usability are provided to illustrate the conceptual overview of how user value is created and how to measure it. Woven throughout the chapter are recommendations on how to help businesses leverage that value to become more successful.

A key to articulating an effective ROI story is to use the language of the organization. At IBM, Karat found it critical to speak with the product managers in terms of time, people, resources, ROI, and net present value (NPV). She began by working with one product manager in a grassroots usability evangelism effort. She collected the business case data on the value of HCI while completing the HCI work on the project that enabled it to come in under budget, on time, and with high customer satisfaction. This result was unusual at the time and won bonuses for the entire team. Karat was asked to make presentations on the business case for usability to a group of product managers within the larger organization next. Speaking the product manager's language and demonstrating through a case study the value of the HCI work opened doors within that organization. Karat found that she would need to repeat at least some of the experience as she changed organizations and sees the effort as a combination of support of key executives for HCI in a top-down aspect combined with ongoing grassroots evangelism and continual demonstration of the value of HCI. Providing significant support to a development team by heading off design problems and quickly identifying and resolving other usability problems is a continual basis for bonding with these teams, above and beyond the financial value demonstrated to the executives (supporting what Wilson and Rosenbaum say in Chapter 8).

At Ameritech all managers were trained in the concept of a value engine, a model of how their work drove value for the organization. We were able to talk about how easy-to-use and useful Web applications would create satisfaction, which in turn would result in a strong brand and additional revenue. Many managers understand NPV (see Chapter 4 for more detail) and use it when creating their business cases, and where this is true you can draw on the same language in stating the value proposition for your organization. Business cases, in fact, are an excellent place to look for the kinds of arguments that are persuasive to decision-makers. At Sapient the corporate value proposition related to delivering explicit business outcomes, and we were able to talk about how viewing business problems through the lens of experience and then balancing user and business needs along with the available technology enabled us to deliver the outcomes desired.

Along with the elevator speech, it is important to have the facts at hand. Again, ideally the stories you tell to illustrate the points you are making are drawn from the impact you have had for your business on the metrics that matter to the organization. But if you don't have those core stories, or if you need some information to supplement them, you can always use some of the examples cited earlier and you can cite external sources that have credibility with your audience (e.g., Forrester reports).

Quickly and effectively articulating the value of user-centered Web design is important, but increasing impact usually requires a more assertive public relations (PR) campaign. PR departments and internal newsletters are often looking for good stories, and offering examples of effective ROIs can create "buzz" around an organization. Setting up a usability lab as a stop on corporate tours can bring your message to customers and executives, exposing them to the story you want to tell and leveraging the compelling experience of observing users in action. Cold-calling key executives and managers to share what your organization has been doing for the business and perhaps exploring opportunities to lend a hand with initiatives that are strategically important to the executives can create a foundation that pays off in the future. Identifying strategic and visible projects that should have design and usability support but don't, selling your way into the projects, and then becoming a most valuable player on the teams is another way to be more clearly associated with the eventual success of the projects and the impact that success has on the business.

In addition to being able to quickly and easily describe the value of user-centered design in Web design and to provide a high-level view of how it is generated, describing in more detail how that value is derived is also useful. The "how" of the argument is most important when you work with teams to build in the user-centered design activities that actually drive the value.

Looking across many companies, one can argue that support for user-centered design activities in general and the role of user-centered design in Web design tend to be relationship based. Support comes when there are team leaders who insist that the Web experience be designed on the basis of an understanding of users, and when executives "get it" and insist that the Web development process include design and user-research professionals. These relationships often begin with the initial impressions. For people who are unfamiliar with the UE area, their model of usability is often shaped by the understanding they gain in those first conversations.

10.10 SUMMARY AND FUTURE DIRECTIONS

ROI is a measure of business effectiveness and is typically a ratio of profit to the cost of achieving the profit. The profit is often calculated using one of the sophisticated selection techniques outlined by Karat (e.g., NPV) in Chapter 4 of this book. An important point to consider is that projected ROIs are often used by companies to decide in which areas to invest. For example, a company can invest a given amount of money in a new product A or in improving the usability of Web site B. An HCI professional with an understanding of the costs and benefits of usability and the ability to describe them in business terms such as ROI will help executives make more effective decisions for their companies and (knowing how the evidence leans) for their customers as well.

HCI professionals can help to build the knowledge base in cost justifying usability on the Web by collecting cost-benefit data during their projects and sharing them as appropriate with the community, either in a case study or de-identified format. There are a few ROI research questions on the value of human factors on the Web that we challenge our peers to address as well. First, the HCI field needs to improve the model of the relationship between user value and business value. This model may be domain specific or may be general for the impact of some variables and not others. Second, a clear understanding of how the various aspects of the UE impact ROI is needed to prioritize development efforts more effectively. Research data in these areas could help to improve the allocation of resources to development efforts, including the investment in HCI, which is investing in the quality of the applications and services that all of us will use in different facets of our lives.

REFERENCES

Bias, R. G., and Mayhew, D. J. (Eds.) (1994). *Cost-Justifying Usability.* Boston: Academic Press.

Dell (1998). Retrieved ••, 2004, from http://wwwl.ap.dell.com/content/topics/topic.aspx/ap/corporate/en/pressoffice/archived/nz/1998_06_01_nz_000?c=nz&l=en&s=corp. Retrieved September 1, 2004.

Donahue, G. M. (2001, January–February). Usability and the bottom line. *IEEE Software,* 2–8.

Internal Sapient Sales Materials (2000). Sales materials covered deliverables achieved for a variety of customer references. Contact A. Lund for information about these materials.

International Organization on Standardization (1997). *ISO 9241–1: Ergonomic Requirements for Office Work with Visual Display Terminals (VDTs)—Part 1: General Introduction.* Retrieved September 1, 2004, from www.iso.org.

Kalin, S. (1999, April 1). Mazed and Confused. *CIO Business Web Magazine.* Retrieved September 1, 2004, from www.cio.com/archive/.

Karat, C. (1991, April 28–May 2). Cost-benefit and business case analysis of usability engineering. Tutorial presented at the *ACM SIGCHI Conference on Human Factors in Computing Systems.* New Orleans.

Karat, C. (1994). A Business Case Approach to Usability. In R. G. Bias and D. J. Mayhew (Eds.), *Cost-Justifying Usability* (pp. 45–70). Boston: Academic Press.

Karat, C., Brodie, C., Karat, J., Vergo, J., and Alpert, S. (2003). Personalizing the User Experience on ibm.com. *IBM Systems Journal,* 42 (4), 686–701.

Karat, C., Karat, J., Vergo, J., Pinhanez, C., Riecken, D., and Cofino, T. (2002). That's entertainment! Designing streaming, multimedia web experiences. *International Journal of Human-Computer Interaction,* 369–385.

Lund, A. M. (1997). Another approach to justifying the cost of usability. *interactions,* 4 (3), 49–56.

Lund, A. M. (1999). E-commerce by design. In *Proceedings of the Human Factors and Ergonomics Society 43rd Annual Meeting* (pp. 374–378).

Lund, A. M. (2000). *Assessment of Expected Business Impact through Contextualized Heuristic Evaluations.* Unpublished article available from the author.

Lund, A. M. (2001). Measuring usability with the USE questionnaire. *Usability Interface,* 8 (2), 3–6.

Mauro, C. (2002). *Professional Usability Testing and Return on Investment as It Applies to User Interface Design for Web-Based Products and Services.* White paper. Retrieved September 1, 2004, from www.taskz.com/ucd_testing_roi_summary.php.

Nielsen, J. (2003, January 7). Return on investment for usability. *Alertbox.*

Souza, R. (2001a). *Get ROI from Design.* Cambridge, MA: Forrester Report.

Souza, R. (2001b). *How to Measure What Matters.* Cambridge, MA: Forrester Report.

Studio Archetype/Sapient and Cheskin (1999). *ECommerce Trust Study.* Retrieved September 1, 2004, from /www.cheskin.com.

WEB SITES OF INTEREST

www.usabilitynet.org/management/c_benefits.htm

www.rashmisinha.com/useroi.html

www.nngroup.com/reports/roi/

www.upassoc.org/usability_resources/usability_in_the_real_world/benefits_of_usability.html

www.usabilitynet.org/management/c_business.htm

webword.com/moving/savecompany.html

www.amanda.com/resources/ROI/AMA_ROIWhitePaper_28Feb02.pdf

www.taskz.com/ucd_testing_roi_summary.php

advisor.com/doc/12063

www.sotopia.com/roi/roi.htm

www.theomandel.com/resources/returnoninvestment.html

11 CHAPTER Making the Business Case for International User Centered Design

David A. Siegel Dray & Associates, Inc.
Susan M. Dray Dray & Associates, Inc.

11.1 INTRODUCTION

When clients come to us for help with international user studies, they have often already made the case within their company for the need to do such studies, despite the costs, but sometimes they ask for our help in formulating the arguments. In this chapter we outline the rationales that we share with them, and help you think about how to make these arguments within your company. We draw upon our experience in doing user centered design (UCD) projects throughout the world, and our involvement with the small community of professionals devoted to international design. Throughout this chapter we refer to international UCD instead of simply usability, because UCD is a broader term that can include a variety of activities that are important in creating designs that are usable worldwide. We focus on issues you must address to show a critical but receptive listener that UCD is essential to reduce the risks inherent in developing products for the international market.

In a sense, the general arguments for international UCD are the same as the arguments for UCD overall, except that the costs, risks, and benefits are often an order of magnitude greater. The costs of doing international UCD are usually higher than the costs of carrying out UCD activities in your home country, but the risks of failure if UCD is not included are also much greater. Therefore, the potential benefits of doing it well are much greater. The most general arguments for UCD, wherever it takes place, are as follows:

- Design can be successful only when it is based on deep understanding of the mindset of users and of the dynamics and the context of use
- It is impossible for product planners to adequately grasp these things through intuition
- The conventional ways of introducing information about users into design are inadequate

The impact of these principles is even greater in the international context. As Masao Ito and Kumiyo Nakakoji point out, "interacting with a computer system implies asynchronous communication with people who designed and programmed the system. Thus, using an internationally designed system means collaborating with a designer who belongs to a different culture" (Ito and Nakakoji, 1996). The inverse is also true: The designers are attempting to communicate with users from another culture.

11.1.1 Do We Know Our Users?

Although we may fool ourselves into believing that we know the lives and psyches of users (who we think of as similar to ourselves), this belief is especially unfounded in the international context. This fact is illustrated by a perceptual illusion, known as the rotating trapezoid illusion. A trapezoidal frame is mounted on a post with its unequal edges in the vertical dimension, the tall edge on one side of the post and the short edge on the other, and the post centered between them. When the trapezoid rotates, people perceive it as oscillating back and forth, with its shorter edge always behind. You can see a demonstration of the basic illusion at this site: *www.exploratorium.edu/exhibits/trapezoidal_window/trap_window.html.* If you mount a horizontal stick through the trapezoid, perpendicular to the post, the stick appears to rotate a full 360 degrees. Consequently, it appears to slice through the "oscillating" frame on each rotation. This is a very striking illusion. Neither the additional visual cue provided by the stick, nor one's intellectual knowledge of the real situation destroys it. Interestingly, when Allport and Pettigrew (1957) did a classic study of this illusion, they found that Zulus who lived in traditional villages with round huts and few straight lines and rectangles, were not as susceptible to the illusion. This suggests that the illusion is actually based on higher cognitive processing in which the brain draws on its experience of foreshortened rectangles. The illusion reveals both the cultural dependency of what feels to the viewer like a basic perceptual process, and also our difficulty in sharing the perceptions of others, which one might call "cultural blindness."

There are many examples from the realm of international design that reveal the affects of such cultural blindness. For instance, the story of the Mitsubishi automobile, the "Pajero," which means "one who masturbates" in Spanish, is by now familiar to many readers. However, as Elisa del Galdo (1996) pointed out, what is more interesting is the manufacturer's response when the error was discovered. Mitsubishi changed the name, but only in Spain, as if they did not realize that many Europeans speak more than one language, and European drivers frequently cross borders. French drivers probably do not want to be ridiculed when they cross the Pyrenees. Evidently, in addition to their lack of knowledge of Spanish slang, the manufacturers (or at least the department in charge of naming its products) was also unaware of other aspects of the European lifestyle that were especially relevant to their product.

In *Set Phasers on Stun*, Steven Casey (1993) reports another, more tragic example that occurred in 1971. Seed grain shipped to Iraqi farmers was laced with a toxic fungicide. To warn the farmers that the grain was inedible, it was dyed red, and the bags were marked with the skull-and-crossbones symbol. Unfortunately, this symbol, which the manufacturer assumed was universal, was unknown to the farmers. Some of them were told about the poison, but they associated the red dye with the poison and assumed that washing it away would make the grain safe. Many deaths resulted from this communication failure. For our purposes, the important message of this example is that even when people are sensitive to risk and are trying hard to address it, they are bound by their own cultural assumptions, and have difficulty imagining to what degree and in what ways the experience of other people may be different from their own.

Despite the existence of cautionary tales like these, companies frequently seem to act on the assumption that the functionality and design of their products will be so compelling that users around the world will adopt them. There are situations where a product has at least a temporary monopoly on a particular functionality, and users will have no choice but to cope with poor design. However, there are also many situations in which functionality that is poorly conceived and designed for users in other countries, while usually not fatal to the user, is fatal to the product, or at least limits its success.

Some decision makers are well aware that doing business internationally means venturing into areas where they do not even know what they don't know. Once they understand what UCD offers, they may be receptive to adopting it. Others seem not to grasp this at all. One colleague of ours in a company that sold specialized software globally was aware of problems with an important product in a key Asian market. This UCD professional provided many kinds of evidence to get the problems addressed, including customer problems and requests, requests from the local sales office for design changes that would help them sell more effectively, and examples of blatant mismatches between the application's functionalities and common user tasks. None of these arguments inspired measurable change in how the company addressed the needs of its foreign users. The relevant executives felt that sales were adequate and that the company's technological leadership in its domain outweighed the problems. After several years of frustration, things suddenly changed when a reorganization placed the department under a different vice president. When he heard the arguments, he said, "That makes sense," and suddenly there was funding for developing a usability lab in that country. While many of us have encountered executives who seem immovable, and many dream of finding a corporate sponsor who just "gets it," we are writing with the belief that there are decision makers in the midrange who are critical but potentially receptive to persuasive evidence.

11.1.2 Our Focus in this Chapter

The business case for international UCD is a broad topic. It is difficult to cite representative numbers that will be useful to a broad range of readers because of the tremendous variety of situations, the number of factors influencing results, and the varying strategic goals of different products. The issue of cost justification for international UCD encompasses all the issues for cost justification of UCD in general, with the added dimension of the international variables. Arguments that apply for a given product in one country may not work as well in another country, not only because of country-specific factors, but also because of differences in the current business situations across countries that affect the product. Financial examples focusing on one product may not apply well to another. The full matrix of factors to consider would have to include the following:

> (cost factors) × (benefit factors) × (different types of products) × (different types of UCD activities) × (different business goals) × (different countries) × (different business situations in country)

Obviously, it would be neither useful nor feasible to re-evaluate from the international perspective all the factors involved in cost justification that have been covered elsewhere in this book, and a specific case example is unlikely to be analogous to your situation.

Of course, if you have specific cost figures and reasonable projections of gains or cost savings specifically attributable to design changes, you certainly can apply a structured analysis like those described elsewhere in this book. However, many of the benefits of international UCD are extremely difficult to quantify. We surveyed more than 30 UCD professionals from 10 countries in North America, Europe, Asia and Africa, all of whom have done significant amounts of work internationally. Although these people all described significant impacts, *none* of them were able to come up with numbers to quantify the benefit of international UCD efforts. This difficulty of quantifying benefits is partly because so many of the impacts are broad and there are far more intervening variables than for most domestic projects. No two countries are the same, and it is usually a mistake to lump countries together, even in geographic regions, and assume that you can have one design solution that will fit all of them. This makes it difficult to collect relevant data. In many cases, we feel it is more realistic to treat the UCD activities as an essential part of a many-pronged approach to international product design. Therefore, we have chosen to provide the general elements for a persuasive argument to make the case for international UCD, and must leave it to you to translate these ideas into the specifics of your own circumstances.

Our focus for the remainder of this chapter is on how to think about the strategic business case for international UCD. We begin by discussing the difference between strategic and tactical arguments for UCD. We then discuss the importance of international markets, the strategic risks in designing products for these markets, and why attempts to addressing the risks without full-fledged UCD do not provide sufficient solutions. Then we look at the categories of costs for international studies and identify both why these are usually greater than costs for studies in your own country, as well as how to make wise tradeoffs to maintain the integrity of your studies while keeping cost under control. We end by discussing ways to magnify the benefit of international studies, through long-term, strategic thinking.

11.2 THINKING STRATEGICALLY ABOUT INTERNATIONAL UCD

11.2.1 Making Strategic versus Tactical Business Cases

Many return on investment (ROI) arguments for domestic UCD practices tend to be tactical, focusing on incremental near-term payoffs, such as increased sales or decreased support costs (Rosenberg, 2004). This certainly can apply to international UCD if you have a way to identify and quantify this type of benefit. However, we have found that the more effective arguments for international UCD tend to be strategic arguments. Strategic activities require a larger effort involving the coordination of separate initiatives, have to cope with a more complex array of factors, and assume a longer time horizon. Companies may be more willing to invest in adapting products to the international market if they see the market's importance in the long run (Luong *et al.*, 1995). Fujinuma and Risden (2002) point out that Microsoft had to invest in years of research before Microsoft Word was successful in Japan. In our experience, companies tend to be more receptive to doing international studies (and internal champions tend to have an easier time arguing for funding) when the organization is conscious that it is taking a strategic-level initiative and venturing into new territory. This can be because the company is launching a new product, product line, or technology with a global focus; making a major initiative to improve its international competitiveness; or because it is in the early stages of venturing into one or more international markets.

Although there certainly can be good tactical justifications for international UCD, the hurdle of cost justification for tactical efforts can be higher than for domestic UCD for a number of reasons. While a company's international markets may represent a large and growing share of its business, these markets may be dispersed and fragmented. In addition, the number of users you can affect by doing a study in any given locale or market may be smaller (depending, of course, on the size of your home market) and the cost of doing the research may be greater than the cost of similar research in your own country. Therefore, it is often harder to make an ROI argument focusing on short-term gains for a specific product in a particular locale. In contrast, viewing UCD as part of a strategic initiative makes it part of an investment that is acknowledged to be risky but worthwhile and that has a payoff that will develop in the long run.

What this means for people who are trying to argue persuasively for international UCD is that they have to link UCD to the organization's international strategy. Without this strategic focus, international projects are likely to be pushed aside by seemingly more immediate problems that can be addressed more easily and with a more rapid expected payback.

11.2.2 The Strategic Importance of International Markets

Arguments for international UCD will only carry weight if corporate decision makers perceive *both* that international markets are a priority *and* that they need special attention. A company that sees its international sales as peripheral to its business often has little interest in an investment in international UCD. Of course, in this age of global commerce fueled by the Internet and information technology, the number of companies for which international business is a priority is growing. Figures cited for the percentage of business done internationally by many companies range from 35 to 65%. Of course these figures may be even higher if the home market of a company is small. Large companies selling technology for the mass market often have to compete internationally to grow, which means adapting their products to the international market (O'Sullivan *et al.*, 2003). Even producers of specialized technologies who serve narrow audiences have to think globally to have a large enough market. One of our clients that sells to a narrowly targeted industry and that already enjoys an 80% market share in its domain in the States, knows that the only way it can grow significantly at this point is through actively branching out internationally. In addition, at times when sales of computer technology are stagnant in the United States or in Europe, companies have to look to other markets (Fujinuma and Risden, 2002).

Although our focus is not exclusively on the Internet, it is certainly one area where the trends showing international growth are quite striking. In an article in the *New York Times*, Tedeschi (2004) points to evidence of the growing volume of international web traffic. Sites like Washingtonpost.com and Yahoo.com have a growing percent of visitors from outside the United States, and advertisers who use these vehicles are taking notice. There is far more "upside" to the growth potential of the Internet in areas outside of North America and Western Europe. The size of the online market in China is growing rapidly. For example, according to Tedeschi's article, the China Internet Network Information Center reported an increase of 15% in the number of Internet users in the first 6 months of 2003 alone (Tedeschi, 2004). Meringer (2002) reported on projections from Forrester that the percentage of revenue from e-commerce at Global 3,500 companies would more than double in 5 years, reaching close to 30% of total revenue in sectors like Technology/Telecom and Chemicals/Petroleum. The same report projects North American e-commerce revenues increasing by a factor of about 4.45 between 2002 and 2006, with those in the Asia Pacific region increasing by a factor of 7.2 and in Western Europe by a factor of 10.75. Their figures project a drop in the North American share of global e-commerce from about 73% in 2002 to about 58% by 2006. Cyberatlas.com (Greenspan, 2003a,b) reported that, for the 2003 holiday season, the volume of Western European online shopping was, for the first time, nearly equivalent to that in the United States.

Of course, statistics like these can only make the general case for the importance of global markets. If you are evaluating the business case for international

UCD in your own company, you need to find the equivalent statistics for your product category, for projected growth of your own company, and for competitors in international markets. Sometimes, finding information about how a competitor has moved into international markets from trade publications can be a sufficient motivator for doing international UCD.

11.2.3 Linking Strategic Risk Management and International Design

Globalization exposes all participants to more strategic risk and competitive pressure, and companies will be increasingly differentiated from each other based on how effectively they respond to these pressures. At one time, the notion was proposed that global commerce would lead to a single, integrated global market for standardized products benefiting from enormous economies of scale (Hermeking, 2003). History has proven this hypothesis to be false. In fact, standardization that does not fit the product to the user has been revealed to be the antithesis of quality. Therefore, in many domains, rather than expecting customers and users all over the world to adapt themselves to global standards, companies have to cope with and respond to growing heterogeneity among users of their products as they reach broader markets.

Just as global commerce opens new markets to companies, it also brings them into competition with other companies with whom they have no competitive experience. These include both global competitors and local competitors. Companies that are accustomed to a position of dominance in one area may find themselves at a disadvantage when competing with companies that do a better job of providing local users with products that are useful, attractive, usable, and adapted to the local context. Local competitors may have inherent advantages in this regard, but global companies that have developed the considerable skills needed for tailoring their products to local conditions will also have the upper hand. Tony Fernandes (1995) pointed out that software companies outside the United States are exposed to the need to adapt their products to local markets "from day one." As an example, he cited Catena, a Japanese firm that developed a successful program for business tables because the available Western style spreadsheet programs did not fit the Japanese model of "asymmetric" tables, in which cells do not necessarily line up in vertical columns.

As we pointed out earlier, one all-too-common basis for complacency about competitiveness in the global environment is the notion that, even if your product is not tailored to international users, it will still sell based on its technical and functional capabilities. Sometimes this is indeed true, but reliance on it is a poor long-term strategy. Functionality can be copied, and often is. Furthermore, globalization means that users can compare your offerings with the best in the world (DePalma, 2002). If a competitor matches your functionality in a package better adapted to local needs and preferences, you will lose your edge on the market. Day (1999) points out that it can also be self-defeating for vendors to rely too heavily on the idea that overseas users will pride themselves on having the most advanced foreign products, because this mindset can make vendors too dependent on early adopters. In fact, good design for the local market will only enhance the discoverability, usefulness, and usage of features, and therefore increase the perception of the product's richness while making it more accessible to a larger number of people (Fernandes, 1995).

Reaching out to more diverse and unfamiliar markets increases the risk of mismatch between the design and the users (the risks of asynchronous *mis*communication that we referred to earlier). These mismatches can have effects that go beyond simply deterring people from using a specific product. Companies invest tremendous amounts of capital in their images, reputations, and brands. There is evidence to show that providing a poor user experience in a company Web site not only deters people from patronizing that site, but also can have a negative halo effect. For example, Greenspan (2003a) reports on a survey in which 30% of respondents said that they would stop purchasing from a favorite *offline* store if the online experience is poor. If you apply this to the international context, it seems obvious that companies face additional vulnerabilities, particularly where their brands may not already be well established. Also, the international context brings additional risks of creating negative perceptions. These can come not just from bad design in general, but also from markers in the design that identify the site or product as alien or unfamiliar to the user. Shade (1999) uses the term "psychic bruising" to describe the cumulative impact on users of having to adapt to many small mismatches between their own mindsets and the design of a foreign piece of software, and points out that it can raise general suspicion towards the product. In addition, research on trust and "e-loyalty"—both concepts that are likely to reflect not just on a particular product or web site, but also on the company associated with it—suggest that these develop differently in different cultures (Cyr *et al.*, 2004a,b). Studies like these strongly suggest that money spent on brand image in the international arena will be put at risk if cultural variables and other local factors are not considered carefully.

Any assessment of risk has to look not just at the potential of failure, but also at the size of the investment that is at risk. When considering international UCD, companies should recognize that the size of their overall investment in moving into the global market can be enormous. There are investments in exploring markets, setting up overseas branches, transportation and communication costs, costs of marketing in different countries, logistics and infrastructure costs, legal expenses, support costs, and so on. These are both up-front and ongoing costs that companies accept as part of the strategic decision to move into a market. Even if a company is already well established overseas, the costs of launching any kind of new product or initiative globally are high, and therefore the risk is high. As a percentage of these costs, expenses for UCD activities are probably quite small, but UCD can have a significant role in protecting these other investments. The cost of many UCD activities can be thought of as a form of insurance premium that may be easily justifiable in these circumstances. This argument has been one of the strongest that our clients have used in "selling" UCD to their management. Therefore, we suggest you attempt to get data that shows the overall size of your company's stake in its international business and then identify some of the potential sources of failure.

The IBM "rule of 10s'" states that if you find a problem that costs $1.00 to fix before a product goes to market, the cost of that same fix if you don't discover it until the beta test will be 10 times that, or $10. That same fix will cost another 10 times that cost, or $100, if you don't discover it until the product has fielded (IBM, 2004). The exact multiplier has been debated, but whatever it is for domestic design work, it is likely to be even greater in the international context. First, there are greater risks in the international context of making fundamental errors at the level of product concept and value proposition, conceptual design, and major functionality if you do not do early UCD research. Second, the cost of rework may be substantially higher for international products because of the variety of additional costs that are required to get to the stage of international roll-out, such as localization costs. We discuss localization's necessary—but not sufficient—role in more detail in the following section.

11.3 THE DREAM OF "SIMPLE" ANSWERS

When confronted with the complex challenges of designing for international markets, companies all too often seem to rely on partial solutions. In this section, we consider two approaches that some may think will take care of the problem but that are insufficient. These approaches are (1) relying on translation, localization, and internationalization to address cross-cultural markets, and (2) attempting to use cultural models to provide design guidelines for international products.

11.3.1 Translation, Localization, and Internationalization

Translation

Translation of the interface, content, and/or instructional materials is probably the minimal effort to adapt a Web site, application, or other product for an overseas market. Often, companies seem to think that this is the whole task and that it should be relatively straightforward. Unfortunately, both of these ideas are false. Directly translating text does not adapt concepts from one culture to another culture. Translation that attempts to adapt the underlying concepts to another culture is a very complex and subtle task. Furthermore, text is only part of what conveys information in an interface. Luckily, many companies now move beyond simple translation to localization, or even, in more globally-aware companies, to internationalization.

Localization

Innumerable practices and conventions differ from country to country in how textual and other types of visual information are handled. Localization refers to the effort to go beyond mere translation of text, to assuring that all information is presented in a way that is adapted to a particular country or locale (Uren, 1997). Localization addresses both objective and subjective elements (Smith *et al.*, 2004). Objective elements are those that are easily defined standard local practices. These include, in addition to translation itself, a multitude of things like using correct local date formats, handling currency correctly, and observing appropriate rules and conventions for alphabetization. They can also include conventions about hardware. For instance, Swales *et al.* (1999) report that population stereotypes as to the "correct" direction in which to turn a knob to control downward movement on a screen differ between the United States and China. Some of these—but definitely not all—are well-documented. Subjective elements include culturally-appropriate colors and images, making sure that the creative "look" is culturally appropriate, and adapting the product or site in other ways that are expected to affect the perception of local people that the design "fits" their culture.

We sometimes hear the notion that simply having a person from the target country on the development team will take care of the localization issues. This often reflects a serious underestimation of the difficulty of the task, and is akin to the old notion that having one user representative on a development team eliminates the need for other forms of user research or input. For example, localization may have to address things that are quite specific to a particular domain. It is unrealistic to assume that country knowledge alone is sufficient for this. Second, team members who are originally from the target country are likely to have lost some of their acculturation over time and may not be up to date with the most current practices. Additionally, the fact that they left their country of origin makes them different from those who stayed behind. Third, the idea that any member of a culture is conscious of and able to be explicit about all the practices of his or her home country is a basic fallacy. Fourth, there is an unfortunate tendency to lump together people from entire regions or linguistic groups. There are many forms of Spanish, for example, and a representative from one country in a region such as Latin America cannot speak for the whole region.

Good overviews of areas that need to be addressed for localization and that show the complexity of the task can be found in a number of references (e.g., Aykin, 2005; Aykin and Milewski, 2005; DePalma, 2002; Hoft, 1995; Luong *et al.,* 1995). Be aware, however, that these references are only a partial resource, because domain-specific practices and conventions also can vary by region or country.

Internationalization

Internationalization refers to the effort to design the platform and code in a way that facilitates ongoing localization for multiple countries at minimum cost (Luong *et al.,* 1995; Uren, 1997). This can include providing for bidirectional text support and double-byte language support and supporting alternate date and time formats, such as Julian, lunar, or mixed date formats, among other things. The list of issues that have to be addressed is extremely long, and the processes for addressing them are complex and specialized (see Luong *et al.,* 1995 and Uren, 1997 for more details.) Although more and more companies are localizing their products, fewer of them follow a coordinated internationalization strategy that includes a comprehensive organizational strategy of management of content, localization, and redesign as needed (DePalma, 2002).

Limitations of Localization

Both localization and the internationalization that facilitates it are essential to the subsequent usability of a product or site, but they are not sufficient to ensure it. For one thing, in practice, localization is typically left until the end of the design process, even after documentation and training are developed. This causes many problems that are often reflected in the final product or interface. Furthermore, it is difficult at this point in the process to address issues with the fit of the conceptual design or functionality of the product for a particular locale. In fact, companies do not typically involve localization until they have a stable product. Internationalization does not particularly concern itself with the elements that ensure usability *per se*, other than those covered in the localization process itself. It focuses more on the organizational and structural technical components required to make repeated localization possible.

One could argue that there is nothing in the definition of localization that *inherently* prevents it from getting to deeper levels of design for a given country, and there is a convergence of interest between localization and UCD (Dray, 2003, 2004). However, as localization is typically practiced, this tends not to be the case for the reasons mentioned previously. As a result, providing input into the deep levels of functional design tends not to be the main area of expertise for localization professionals. Localization professionals often have to focus on making culturally-appropriate adaptations to a design or product with flawed underlying utility and usability and these issues are typically out of their scope. Of course, many localization professionals have deep cultural understanding of the target locale(s). Therefore, whenever possible, we try to partner with them, because this both helps us to understand the target culture better and allows them to have earlier input into product design. However, as a profession, UCD is more focused on ensuring that conceptual design, including the functionality, the mental model it instantiates, and the metaphors it uses, as well as the logical design or navigation, are consistent with those of users, wherever they may be located. These go beyond the elements of either localization or internationalization.

11.3.2 Cultural Design Guidelines

Differences in Design Preferences across Cultures

Because we know that people's experiences and circumstances differ in different countries, it is not surprising that numerous studies have shown that when localized designs and products are tested in different countries, there are differences in user response. For instance, Cyr and her colleagues point out (Cyr and Trevor-Smith, 2004) that the factors that influence the tendency to return to a particular Web site are different in different cultures (Cyr *et al.,* 2004a,b). Smith *et al.* (2004) suggest that "trust factors" that affect Internet shopping differ in different cultures. In addition, Chavan and Iyer (2003) found a strong preference for Indian sites by Indian users asked to evaluate American and Indian shopping Web sites, even when the specific site identities were disguised by removing the name, logo, and other identifiers. They suggest that the Indian sites had a different "look and feel" that was more comfortable and familiar to their Indian users. Röse and Zuehlke (2000) studied mechanical engineering applications in India, Indonesia, China, South Korea, the United States, and Germany, and found differences in the ways that engineers interpreted pictorial symbols. Lund (2004) found that, while products designed to support group collaboration did support *Western-style* collaboration well, they did not support Asian-style consensus building or organizational styles very well. And finally, Evers (1999) found that interpretations of the metaphors used on a Web site for a Canadian virtual university (DirectEd) varied considerably for students from different countries.

Can We Come Up with Useful Cultural Design Guidelines?

The list of findings such as these, which mostly have to do with the subjective level of design, is nearly endless. This certainly constitutes additional evidence for the fact that good design in one country does not guarantee a positive response elsewhere. It also raises the question of whether these cultural differences can be captured at a higher level to provide guidelines for culturally-tailored design, which might simplify the task, and make it less expensive. Thus, another effort to facilitate international design is the attempt to identify general characteristics of different cultures that may have implications for design. Although we are not aware of any indication that these approaches have yet been adopted in the business world, they are worth discussing here because of the amount of attention they have received in our field. There are two main approaches:

1. Identifying general cultural models and dimensions based on which countries can be described and classified, and attempting to extract from these general implications for design (Harel and Prahhu, 1999; Hoft, 1996; Marcus, 2005; Marcus and Gould, 2000; Marcus *et al.*, 2003)
2. Auditing existing indigenous designs of products, in an effort to abstract the core characteristics of local design (Barber and Badre, 2001; Dunkley and Smith, 2000; Smith *et al.*, 2004)

We will discuss these in the following sections.

Cultural Models

The three primary schools of research on cultural dimensions that are cited most frequently are Hall (1959), Hofstede (1991), and Trompenaars and Hampden-Turner (1998). Several researchers in human-computer interface (HCI) have been working to spread the word about their theories, to integrate them, and to apply them to design of international Web sites (Hoft, 1996; Marcus and Gould, 2000; Smith *et al.*, 2004). For a more in-depth discussion of a variety of cultural models than we can present here, see Hoft (1996) and Gould (2005).

Hall (1959) is often seen as the intellectual precursor of both Hofstede's and Trompenaars's models. He introduced three concepts to describe cultures: context, orientation towards time, and sense of space. Context refers to how much people derive the meaning in a message from the message itself (low context) versus from its context (high context). Orientation towards time can be either monochronic (linear, sequential time, e.g., people attend to things one at a time) or polychronic (simultaneous, e.g., people attend to things concurrently). Sense of space refers to the types of boundaries people draw in their environment (e.g., personal distance and territoriality). This work has served as a fundamental underpinning for subsequent researchers.

The most commonly cited theory of cultural dimensions in our field is that of Hofstede (1991). Hofstede's dimensions of culture were derived from a questionnaire study of 116,000 IBM employees worldwide in the late 1960's and early 1970's. From the questionnaire responses, he derived five dimensions, which include:

1. **Power distance**—a measure of the interpersonal power in society from the viewpoint of the less powerful
2. **Collectivism versus individualism**—a measure of the relationship between the individual and the collective group
3. **Femininity versus masculinity**—a measure of gender role impact on social roles, especially tenderness vs. toughness
4. **Uncertainty avoidance**—a measure of the degree of anxiety caused by uncertain conditions
5. **Time orientation (long term vs. short term)**—a measure of the focus on the past or the future

Trompenaars (Trompernaars and Hampden-Turner, 1998) describes cultures in terms of relationships with people, attitudes towards time and attitudes towards the environment. He further breaks these down and provides a number of recommendations for managing cross-cultural situations based on these attributes. His is the most "action-oriented" model and includes specific guidelines for managers dealing with cross-cultural projects.

While these theories of cultural dimensions were not focused on design, there has been some interest in trying to apply them in HCI. For example, Marcus and Gould (Marcus, 2005; Marcus and Gould, 2000) worked on deriving design implications of these theoretical cultural dimensions, and came up with some early design "guidelines," which have since been applied by Smith (Smith *et al.,* 2004). We discuss what we see as the limitations of these approaches in a subsequent section, but first we consider the second category of attempts to develop cultural design guidelines.

Cultural Audits of Sites

Another approach is to try to derive cultural design guidelines by auditing existing designs from target cultures. Barber and Bader (2001) refer to this as "culturability inspection." Using this approach, they have catalogued a number of cultural preferences for certain patterns of various visual design characteristics that they believe are indications of preferences for at least some of these cultural markers. More recently, Smith and his colleagues (Smith and Chang, 2003; Smith *et al.,* 2004) reported on audits of Indian and Taiwanese e-finance Web sites to identify "cultural attractors." They, too, hope to derive broader guidelines from their work. While interesting, it is too early to tell whether this approach will actually pay off.

Can We Use these Cultural Variables as a Guide *to Design?*

While we think that both approaches to cultural design guidelines are of tremendous intellectual interest, there are reasons for skepticism about the extent to which guidelines derived either from hypothetical cultural dimensions or from audits of indigenous designs help us to design international products and interfaces more cost effectively. Following are some of the possible problems:

1. **Validity**—Hofstede's and Trompenaars's theories are based on self-report questionnaire data, which is generally considered a weak basis for predicting behavior. In addition, Hofstede's data was collected more than 2 decades ago, and was based on a sample that was probably not representative. We must be very careful about using these models to stereotype cultures.

2. **Applicability to design**—Even if the data are valid for the purposes of deriving cultural models, it is not clear that these dimensions are applicable to product and interface design. Broad generalizations about a culture do not tell you how to design a specific interface, much less what your product or application should be trying to do for people in that culture. Design recommendations that have been emerging from this research often seem to rely entirely on face validity and literal semantic associations. For instance, Marcus (2005) talks about using images of groups for collectivist cultures.
3. **Level of abstraction**—The types of guidelines that can conceivably be derived from this work are necessarily general. As a consequence, while they may ultimately help identify factors that can create an overall impression of cultural fit, they are likely to be of limited potential value in guiding detailed international design.
4. **Normal cultural practice versus best practice**—While indigenous designs may indeed reflect cultural design stereotypes, they do not necessarily point the way to what is useful and usable in that society. Unfortunately, bad design is all too common and common practice may not be something to emulate. Furthermore, these guidelines face the same problem of over abstraction as the guidelines derived from cultural models. Of course, if there are specific design practices that are standard in a given country for particular functionalities, this is something you should know. However, the odds are low that the specific design issue you need information on will be found in a general compendium of cultural design information, even if such a thing existed.
5. **Scope**—Finally, most of the work to date has focused on Web site design and does not address the challenges of designing other types of products for international use.

When Are Cultural Models Useful?

We are not claiming that cultural models or cultural design patterns have no validity in our field. We believe that they may be very useful tools for sensitizing project teams to the depth of differences in mindset across cultures. Indeed, models such as Hall's were developed specifically to provide this kind of sensitization for managers. If their application to design is further validated through rigorous research, we can also imagine that they eventually will provide some look and feel guidelines. We also believe that it can be helpful to audit indigenous designs in the specific genre in which you are interested (e.g., retail e-commerce Web sites) to identify common features or functionality, not so that you can simply copy them, but because these may be clues to very specific local circumstances, needs, or behavior patterns that you will need to take into account. After all, designs are artifacts of the culture that produced them, and they can give you insights related to your product just as other relevant artifacts can.

11.4 CUSTOM UCD RESEARCH

Because generic information, such as localization and cultural guidelines, is not sufficient to manage the risks or seize the opportunities of international design, it is essential to go beyond it by doing custom UCD research, specifically targeted on the data needed for each product. This is a key point in making the business case for international UCD. Design is determined at the level of factors specific to a particular application, product, or functionality, to a given user population and to the specific context. Overlooking these specific factors can spell product failure. These factors tend to occur in constellations that reinforce each other, defining the dynamics of different categories of behavior in different societies. Specific constellations of factors are not derivable from general cultural stereotypes or from following standard rules for appropriate design in a given locale. To move to a deeper level of design, it is critical that companies study their actual audience and their tasks in their actual environment.

11.4.1 Identifying International User Factors and Contextual Variables Specifically Relevant to *Your* Domain

In this section, we provide examples that show how crucial contextual variables are to the design of particular products for particular countries. We use the term "contextual variables" to distinguish them from "cultural variables" or dimensions discussed in the previous section. In contrast to cultural variables, contextual variables are much more specific and are pertinent to a particular product domain. They are relevant to determining all levels of design, not simply the visual but also the conceptual design, functionality, and logical design. Because of their specificity, they almost always require targeted, onsite research that focuses on a particular product domain for a particular geographical and social context. The benefit of studying the contextual variables applicable to your product is that the findings are more specifically prescriptive of design. Because they get at deep levels of design, they are different from the concerns of localization as it is typically practiced.

By focusing on contextual variables, we are not exclusively advocating ethnographic and contextual research, although we think these are particularly powerful in the international environment. Usability evaluation can also reveal international differences in user approaches to tasks and mental models, as well as more specific issues regarding the interaction and user interface design.

Many technologies require a network of users and a constellation of different uses to reach critical mass before their benefits become established. The likelihood of a constellation of factors aligning to promote a given technology can depend on complex dynamics within a given country. For example, there have been numerous efforts to promote the use of smart cards (which contain a computer chip) in the United States, together with much speculation about why the technology has not caught on. In the early 1990s, one of the authors (Siegel) participated in a study of early uses of smart cards in the French healthcare system; the uses included storing identification, eligibility, and basic health-status information. It became clear that the potential attractiveness of this application of the technology in France depended on a complex set of very specific circumstances that were extremely discrepant from the situation in the United States. These included the following:

- Consistent government support and subsidy for the technology, which was perceived as an area of technology where France had a lead
- Pervasive use of the cards as stored value cards, motivated by things like:
  - Higher telecommunications costs that inhibited merchants from using telephone lines to get credit card authorizations
  - Higher rates of credit card fraud than in the United States
  - Extensive use of the cards in public telephones (which could be established by top-down directive in a nationalized telephone system)
- Potential economies of scale and simplicity of administration (both for clinics and for the heath administration) because of the existence of:
  - A single national health insurance system in which essentially the whole population was enrolled, making it feasible both to install card readers in clinics and to use a single format for information records
  - A single nationwide network of physicians participating in the public health scheme

It was also clear that in order to understand the dynamics of such a system in France, one would have to understand many other country-specific factors, such as how the uniform national health coverage would interface with a more fragmentary system of private insurance through "mutual" organizations that provided supplementary coverage, the role of local legal principles regarding ownership of medical records, and so on.

There are many other examples from around the world of particular technologies fitting local conditions in particular ways. For example, there is a much greater relative penetration of cell phones compared to Internet-connected computers in some countries, such as South Africa (Marsden, 2003). Telecommunication rates combined with this differential penetration also contributes to the prevalence of short message service (SMS) compared to e-mail and instant messaging. Of course, it can be argued that for any technology to become established, it has to fit the specifics of the local context. Our point is that it is almost impossible to predict the details of these dynamics from outside. Although examples like these make sense after the fact and are consistent with culture (e.g., the fact that France is a more centralized and top-down country than the United States), one cannot know the specifics without direct investigation, and design depends on the deep level of understanding that can only be achieved in this way. In the following sections we provide a few additional examples of country-specific and technology-specific factors that we consider particularly interesting.

Purchasing Dynamics and Financial Transactions

In several of our own studies in South America, we have found interesting patterns that affected people's attitudes towards purchasing online. These had to do with their reliance on relationships with trusted individual vendors rather than anonymous companies, issues with trust in delivery services, and a general culture of making payments in person as opposed to transmitting credit card numbers or even sending checks through the mail.

The prevalence and role of credit cards, cash cards, debit cards, and cash, whether it is customary to give credit card numbers over the telephone or to fax them, and so on, can all differ across countries and can depend on the type of transaction as well. In a panel at CHI 2003 (Roshak *et al.*, 2003), Ann-Byrd Platt described her experience in a usability evaluation of a utility for mobile payments (m-commerce) in Switzerland. This was envisioned as a system that allowed people to use ATMs to transfer funds into special accounts accessible from mobile phones for micro payments. To people from outside Switzerland, this might sound like a strange concept. Why would users bother to go through this convoluted process of downloading cash in advance to special electronic accounts instead of just using credit or debit cards? The positive response of

Swiss users only begins to make sense when one takes into account factors specific to the Swiss context, such as the heavy reliance on cash and the low reliance on credit cards, daily limits on cash withdrawals from bank accounts, and so on.

Social Structure and Service Expectations

In some societies with small middle- and upper-class segments and a large class of lower-paid workers, we have seen evidence of a culture of direct in-home services of all types beyond what persists in the United States and Western Europe, and less of a culture of "do-it-yourself." This appears to have implications for product concepts such as providing online technical support services to consumers, who are likely to be in the upper classes of these societies, at least for the present (Dray *et al.,* 2003).

Mental Models of Geography

Many applications and Web sites must deal with geographical information. Making sure that geographical information, such as addresses, is formatted correctly is clearly a localization issue, and is standardized within a country. However, deciding how to "chunk" units of geography, what regions users will perceive as close to them or distant from them, depends on specific local circumstances and behavior patterns, and has different relevance for different purposes. For example, without studying the issue specifically, a company that has to provide maps of branch locations in a metropolitan area could have a very difficult time determining (or guessing) what a person in another part of the world would consider to be easily accessible from a given place, or how to subdivide the geography into familiar units. Foreign visitors often have the same problem in reverse when interacting with tools that assume local knowledge of the geography. One of our clients, interested in understanding the right level for presentation of geographic information, conducted extensive ethnographic research in several locales in Europe prior to designing the application for use in Europe.

Use of Physical Space

Use of physical space differs from one country to another in specific ways that can affect the form factor of different products. The implications are very specific to the type of product under consideration and to the specific physical context in which the product will be used. For instance, when we did usability evaluations of the Hewlett-Packard Infiniium Digital Oscilloscope (Dray and Rowland, 1998) in Zurich and Japan, we found that there were particular concerns in Japan about using a mouse to interact with the scope, but it was difficult to tell what the source of these concerns actually was. When we saw the actual workbench spaces that the Japanese engineers used, we realized there was simply no room on their engineering benches for an external pointing device such as a mouse. Luckily, this was discovered before the launch of the product. HP delayed the launch in Japan by 6 months while the team devised an alternative pointing device for the Japanese market. Partly as a result of this design change, the Infiniium scope was extremely popular in Japan.

We had a similar experience when doing usability testing of a multifunction printer/scanner/copier/fax machine for a different HP division in Germany and Korea. In both countries, we allowed users to set up the machine wherever they chose to do so. In Germany, they used the desk next to the computer as users had done in the previous tests in the United States. However, in Korea, when given the choice, virtually all users sat on the floor to set up the machine. This had a profound impact on their experience of setting it up, because the designers had assumed the device would be set up on a table or desk, and thus, this finding has clear design implications for future iterations of the form factor (Dray, 2003).

Maintenance Intervals and How Time Is Marked

In another example, Hermeking (2003) describes a maintenance problem that occurred in Columbia. The German manufacturer of hydroelectric equipment had difficulty getting the workers to oil the machinery at the prescribed intervals of 2,500 hours of operation, approximately equivalent to every 4 months. The operation manuals were ineffective, so they installed instructional plates on the machinery, and also added large meters to display the elapsed number of hours. When the problem persisted, a local expert suggested tying the maintenance intervals to major religious holidays that were celebrated at about 4 month intervals. This solved the problem.[1]

---

[1] Hermeking presents this story as evidence that the manufacturer should have been more aware that it was dealing with a culture classified on Hofstede's dimensions as "high-context," and that not reading manuals or paying attention to measuring devices is consistent with such cultures. However, this seems unnecessarily roundabout, and a case of "post-hoc" reasoning. Since many people in Germany really do read instructional manuals, it might have seemed counterintuitive to them that people elsewhere in the world would not do so. However, there is a big difference between *deciding after the fact* that failure to read manuals is consistent with high context cultures and *predicting* that people in high context cultures will not read manuals and those in low context cultures will. Furthermore, information about where Columbia falls on cultural dimensions would not have been likely to suggest the ultimate solution in this case, which had to be specifically tailored based on local knowledge.

Classification Schemes

Broad areas of information are conceptualized and classified differently in different countries, in ways that cannot be predicted from general knowledge of the country. For example, Elisa del Galdo (2004) described to us country-specific observations regarding an international job-recruitment site. She found several international differences in how job information and occupations were classified, for example whether people searched first by industry sector or by profession (e.g., "auto industry" vs. "accounting"). She also found that physical location of jobs (e.g., close to public transport, and/or in a city nearby or one far away) had a higher priority in some places than others. In some countries, she found that the site needed to accommodate additional classifications based on things like required courses and special diplomas.

Pre-Existing Adaptations to Nonlocalized "Legacy" Applications

Lynn Shade (2003) writes about the mismatch between Western word-processing software and the Japanese model of documents. She points out that even though previous applications may not have been localized, work practices in the country may have evolved to some degree based on the workflows assumed in the software. This complicates the task of planning a new generation of better-localized products. The tradeoffs involved in deciding to what degree to support a more "traditional" workflow, or to maintain the adaptations depend on many things that are difficult to assess at a distance. Without doing specifically targeted UCD research, it is difficult to know exactly what adaptations people have had to make to the nonlocalized software, exactly what work arounds they have developed, how successful these are, what indirect influences there may have been on other processes and systems, and so on. It is also important to know something about how uniform or diverse these adaptations are, and whether the net result is perceived as positive or negative. All too often, the fact that existing users have already made adaptations to nonlocalized products becomes an excuse for continuing down that path (just as the idea that an "installed base of users" would be bothered by changes becomes an excuse for not solving usability problems). Good decision-making calls for a much more nuanced analysis than this to really understand the pros and cons.

Climate and Environmental Conditions of Usage

A client company told us about discoveries they had made during their first international user visits. They manufactured sensitive equipment, and their engineers were accustomed to working in antiseptic, climate-controlled, dust-free environments. They were surprised to discover that their equipment was being used in hot, dusty, open-sided tents in desert oil fields. Another company was surprised to discover that their sensitive electronic products were being used in damp, dusty, and generally unsanitary conditions inside a waste-treatment facility where there was not even a door that could be closed to protect the devices.

Travel Patterns

Degen *et al.* (2005) went through a simulated UCD process to design a travel Web site for people in the United States, Germany, and China. Their sample and methodology were quite limited, in that they gathered requirements via a small focus group of only four participants in each country. In the focus groups, participants discussed their travel behavior and use of the Web for planning. Despite the limitations of the approach, their results are suggestive of fundamental differences in travel behavior and in expectations of the Web. There were differences in factors like what kinds of travel opportunities people seek and in their process of using the Web to consider possible destinations. For example, users from different countries differed in whether they approached the Web with a predetermined idea of their destination. The authors showed how these differences among countries might drive differences in design of a travel Web site at the conceptual level. If their initial findings about intercountry differences are validated, they certainly indicate that a design that would work for users in one country might be very poorly adapted for another country.

11.4.2 Developing Your Own List of Contextual Variables

The previous sections are only a few examples, of course. In many projects over the years, we have identified specific local conditions with unanticipated and profound implications for product design at the conceptual or logical levels associated with things like the following:

- Housing patterns
- Religion and dynamics of religion
- Family constellations and dynamics, affecting things like sharing of technology within families and specific uses of technology
- Dynamics of decision making regarding technology within families
- Replacement period and lifecycle for your company's technology
- Turnover in occupations relevant to your company's technology

- Seasonal and climate effects on behavior related to your company's technology
- Organizational structure, authority, and delegation in workplaces relevant to the role structure assumed by applications
- Patterns of population concentration, travel, and commuting
- Social networks and communication patterns

Each product, system, or technology faces its own particular set of contextual factors. We recommend that you think carefully about what domains of human behavior at the individual and social level are most likely to be relevant to your company's product. Be sure to include in your list some of the assumptions, explicit or implicit, that the product concept and design are based on, including the most obvious or seemingly "safe" assumptions. (There will of course be others you do not realize are assumptions until you encounter data that contradict them, but you should be able to generate a long list on an *a priori* basis.) Develop a list of behavioral and social variables that are potentially relevant to determining if those assumptions are true. Then take an inventory of what data your company already has on those or similar variables to support the assumptions. You may well be able to make the case for targeted international research by identifying some unverified pivotal assumptions that design has been based on.

11.5 UNDERSTANDING COSTS OF INTERNATIONAL RESEARCH

The costs of doing international user studies are, predictably, higher than costs for similar studies conducted in your own country. There are a number of reasons for this. Some categories of cost are unique to international studies, and others are analogous to costs in your own country, but may vary depending on the project specifics. Most of the major categories of costs are applicable to different types of research: ethnographic, contextual field research, naturalistic usability, or usability research in a facility. We have written elsewhere (Dray and Siegel, 2004, 2005) on the details of planning and carrying out international user studies, the rationales for different categories of expenditure, and the risk of "false economies." Although we cannot cover all of these in the same depth here, we will give an overview of the primary costs that have to be taken into account and their rationales. It is important to clarify that our focus is primarily on qualitative research that depends on rich interaction with users.

11.5.1 Bilingual Facilitation

One of the common ways that people try to cut corners in international research is to study only people who speak the development team's language. This can be particularly tempting for researchers who are native English speakers and who tell themselves that English is the *lingua franca,* especially with technical user populations. In general, however, this is a poor idea. It restricts your sample to a group that may not be representative. In addition, a person's own estimates of their fluency in a foreign language may be inaccurate. There is also a big difference between being able to make some sense of written material in a foreign language, and engaging in free-flowing communication, as when doing any kind of inquiry or thinking aloud. Therefore, only very rarely is it appropriate to conduct studies in a language that is not the native tongue of the participants.

This means that when you are conducting studies in a country where you cannot do the facilitation in the local language yourself, you will need to hire a bilingual facilitator. Good facilitation is absolutely critical for UCD research, whether in usability evaluations or field studies. Therefore, it is dangerous to compromise on this. The bilingual facilitator needs to be comfortable enough in both the local language and your language to communicate fluently with you and to understand nuances of the directions you give.

11.5.2 Simultaneous Translation

The arguments in favor of interacting with users in their native languages also give rise to a need for translation during the data collection. Only simultaneous translation will allow you to manage the process in real time if you are working in a language other than your own. When sessions are conducted in facilities, simultaneous translators, sometimes called "interpreters," provide a voice-over that, typically, observers hear through earphones and which is recorded as a track on audio and video tapes. For naturalistic studies, such as contextual inquiry or naturalistic usability evaluations, they provide an ongoing translation of the users' words in real-time during the session. To be able to do this, translators not only must be fluent in both your language and the participant's language, but also must be able to maintain concentration for long periods of time. Just because someone is good at translating documents does not mean that he or she will be able to do a good job at simultaneous translation.

This is a specialized skill and therefore an expensive one. Indeed, simultaneous translators can be the most costly single part of an international study, costing as much as or more than facility rental, depending on the country.

However, they are also one of the most critical elements, so it is usually not a good idea to try to economize on this, for instance, by hiring a local student, or using a member of your country's local staff, to do simultaneous translation. The danger is that you may not have any way to know what percentage of the communication you are missing. If you have a poor translator, your investment in the entire study can be undermined.

11.5.3 Written Translation

All of your documents, including screeners, study documents, and consent forms, need to be translated into the target language(s). Typically, this is charged per page, per line, or per word, depending on the translator and the norms in a given country. You can help to minimize this cost by being very judicious in what you have translated, for instance, by waiting until you have the final script or protocol. The challenge here is that your screener and task scripts may have to be adapted as well as translated (we discuss this later). To save on translation costs and the time spent on arranging for updated translations, it is best to do the adaptation first. In rare cases, you and your facilitator may jointly decide that he or she does not need to have a translated version of scripts and probe points, but this can only work if you are confident that the facilitator deeply understands the protocol, and you have confidence in his or her ability to improvise. We have found that most facilitators are more comfortable having a translated protocol, especially for the initial sessions.

It is a good idea to back-translate documents to make sure they have been correctly translated, even though this can add to the cost. This is one time when it may be appropriate to ask people in your local office to compare the original and local versions and describe any discrepancies. However, you must be aware that they are typically not professional translators. Furthermore, they may be too immersed in the jargon of your company and lose sight of the fact that their terminology is not really colloquial in their own countries.

11.5.4 Adapting the Recruiting and Scheduling Strategy

We mentioned previously that screenings may have to be adapted, which can add to the planning and preparation time for an international study. This means adjusting its criteria, decision rules, and quotas so that they make sense in the target country, and defining the sample that is appropriate to you in that context. This is an iterative process, because the concepts by which you define your target audience may themselves be culturally relative. Your local recruiter or field office personnel may be able to help with this, but you may well have limited insight into this before you study the international user population and context. In fact, learning about the most useful way to define your target market segments in another country can be one of the primary benefits of conducting the study.

The process and cost of recruiting can differ significantly from one country to another. Differences in typical no-show rates can require that more backup participants be recruited in some countries. Also, privacy regulations in many European countries can prevent "cold calling."

Recruiting and scheduling are closely related. Your recruiter can sometimes help you identify ways to make an otherwise challenging recruit less difficult, and therefore, less costly. Variations from country to country in when people can attend sessions can necessitate a longer period of time in the country to collect your data. This translates into longer hotel stays and increased costs.

11.5.5 Air-Travel Expenses

International travel expenses are likely to be higher than domestic travel expenses. You may be able to reduce these by combining several activities in one trip, for instance, taking advantage of being at a conference or sales meeting to make a set of ethnographic visits, or do a round of usability evaluations. Alternatively, you may be able to combine all your data collection into one longer trip. Obviously, specific costs can vary widely depending on where you go. Some people advocate simply hiring a consultant in the country where you want to do the study. This suggestion is only valid with many caveats, which we discuss later.

11.5.6 *Per Diem* Expenses

In addition to the amount of time you need to spend in-country to complete data collection, other factors can influence *per diem* costs. Even though it adds to travel costs, it can be important to allow extra time (at least a day) for the team to adjust to time zones if you are traveling over more than four time zones. It is also important to allow time to train the local people who will be working with you. This includes working with the facility and recruiter(s) to make sure they understand the study requirements, as well as working closely with the facilitator, both before you go and after you are in the country. This is especially true if you are working with a new facility and/or facilitator. We not only orient the facilitator to the test plan (or visit protocol for field studies), but also do a hands-on dry run and roleplay of the sessions once we arrive. We also work directly with the simultaneous translator(s) to make sure the requirements of the particular type of research we are doing are understood. This helps to educate as well as to build critical rapport among your collaborators. Unless you have worked with a specific facilitator and/or translator in the past, you will need to allow time for this. Finally, you also have to be sure to allow sufficient debriefing time between sessions and at the conclusion of data collection to review and capture what is likely to be very rich data.

One way of saving on *per diem* costs is to structure your schedule as efficiently as possible. You should begin the preparation of the facility and meeting with your local partners during your "time-zone—adjustment day." Sometimes it helps to go to a second-tier city rather than a first-tier city in a particular country, assuming that both can provide you with the types of users you need for your research. This can often dramatically reduce costs of hotel, food, and ground transport.

11.5.7 Video and Computer Equipment

Rental of specialized video and computer equipment can be a significant cost item, especially if the facility has to outsource it. Bringing your own equipment along (assuming you have checked to make sure it will function in the country you are traveling to) can yield significant savings. Several systems are on the market that capture a high resolution screen image and a face shot, and if their capabilities are appropriate for the type of study you are planning, you should look at the comparative costs for buying one of them versus renting a special video setup. However, if you do need to rent equipment, and if costs seem significantly higher than you expected, be sure that the facility has understood your requirements and is not adding services you do not need, such as a camera operator.

More importantly, ask yourself if you really need the video. Video can be very valuable when you cannot bring a full team along on the study. They are also useful for doing a "data check," and creating edited clips is much easier with digital recordings. However, it is important to weigh the costs carefully. Often, videotapes sit on shelves, never watched.

11.5.8 What Is the Bottom Line?

As we have explained, it would not be useful to give cost examples for a particular project, because so many parameters differ from one project to another.

Even discussing costs as a percentage of domestic costs is difficult, because these depend not only on where you are doing the study, but also on what typical costs are for different kinds of studies in your own country, whether your company has its own facility or you typically rent a facility, and so on. If you work for a company in Japan, the most expensive market in our experience, it is possible that research in China might seem fairly inexpensive, even allowing for the costs of translation and travel. If you work for a German company, the difference between doing a project in France and one in Japan will be much greater than the difference between research in these two markets would be for someone coming from the United States. You also have to remember that costs can vary widely between cities in a country.

With the above caveats in mind, we can offer some rough guidelines about relative costs. Consider a case in which you are conducting a usability study for an American firm in the United States and in another country where you do not speak the language fluently. Let us assume that in the United States, you need to arrange for a facility, recruiting, and video. You do not need to rent a computer, because facilities will usually allow use of a computer free of charge. In the overseas study assume that you will have to contract for a facility, recruiting, bilingual facilitation, translation, video, and computer rental. In our experience the costs for this package of services in the overseas case can be anywhere from 75 to 125% more than the costs for the shorter list of contracted services in the United States. China is on the lower end of this range, and Korea and Latin America are somewhat higher. Countries in the high end of the range include Western European countries. In our experience, Japan is almost in a category by itself, and translation costs make the major contribution to this.

Costs for any market where you do not need translation and can do the facilitation yourself will of course be much less expensive. For researchers from the United States, Singapore and countries in the United Kingdom (where we can conduct the research in English) are not particularly expensive, apart from travel costs. Depending on the nature of what you are studying or testing, the same may be true of India. Research in Anglophone Canada is no more expensive than research in the United States. (People from the United States need to remember that doing research in Canada is indeed "international" research, even when the focus is on Anglophone Canada. In projects that we have done there, we have found significant contextual differences that affected responses to product and software design.)

Remember that these estimates do *not* include costs for travel or for personnel time (whether this is for an internal UCD professional or a consultant). In budgeting for personnel time, you have to keep in mind that international studies take more planning and coordination, dramatically so if you are new to the area or if you do not already have an international network of resources. Fortunately, costs for personnel time are not proportionately as high as costs for services in your own country. Our charges for consulting time are a much smaller fraction of the total project cost for international studies than for projects in the United States.

11.6 ARE THERE LESS EXPENSIVE WAYS TO COLLECT DATA?

It is only natural that when costs for a service are high, people seek less expensive alternatives. When considering less expensive approaches, though, it is very important to consider the tradeoffs. Changes in approach intended to save money are not very wise if they fundamentally compromise the value of the rest of the investment in the project. We have already discussed the arguments against conducting the research in English when it is not the local language. This is probably one of the most commonly considered compromises, because working in the local language drives so many of the additional expenses of international UCD.

11.6.1 International Discount Methods

Other approaches that people consider include international inspections and international discount evaluations (Nielsen, 1996). Expert reviews probably have a similar range of pros and cons whether they are conducted by overseas experts or in your own country. They are more useful for applications where there is a large accumulated base of detailed design knowledge rather than for specialized tools and applications. These methods are unlikely to give you the deep understanding of your international users and their contexts that you need. Perhaps the biggest danger is that you will believe that you know more than you really do. While these methods may have a place as a quality check at certain points in the design, it would be a mistake to think (or to encourage decision makers in your company to think) that by using them, you have covered the need for international research.

11.6.2 Hiring Local Consultants

Some people may think that a simple way to hold down costs is to hire consultants in the countries of interest. Obviously, this may save travel expenses. You also may be able to benefit from their local knowledge of relevant technology and of the design approaches that users are exposed to. Finally, if you rely on them to conduct the study in the local language, but send you the report in your own language, you save money on bilinguual facilitation and translation. Nevertheless, the tradeoffs are complex when considering the approach of contracting out research to local firms. Although there are active communities of UCD professionals in many countries, in many parts of the world it is still primarily an academic profession, and there are few practitioners. Overall, it can still be difficult to find experienced consultants in many parts of the world. On the other hand, there are focus-group facilities all over the world that have learned the usability buzzword. You must be on guard against focus group facilitators whose usability experience consists only of asking focus-group participants if something looks easy to use. Their experience may be difficult to evaluate from afar. This makes it risky to think that you can lower costs just by hiring a local usability person.

Also, do not underestimate the time and effort required for collaboration, communication, and oversight. Just as the design team needs to gain deep understanding of their users, UCD researchers need a deep understanding of the existing mindset or culturally-based assumptions of the design team. Furthermore, we have found that a tremendous amount of the learning in international studies comes from being there in person, from first-hand experience of the data, and from the contact with unfamiliar ways of thinking and working. Therefore, do not delegate your research to overseas vendors in the belief that you will not have to be as personally involved. Do not think that you will be able to send your scenarios and then just await the report.

Finally, delegating to local researchers can mean that you are sacrificing the cross-cultural perspective, which can be as important as local knowledge. Insights come from seeing the contrasts among countries. If you are using local researchers in a multicountry study, some key observations may not be as apparent to the separate teams as they would be if a single team had done the research in several countries, or if at least one key person from the team had been to all of the sites. Finally, you will complicate the task of coordinating across countries and maintaining "calibration" among the teams to facilitate comparing, contrasting and integrating their findings.

11.6.3 Remote Evaluation

Remote evaluation (Hammontree *et al.*, 1994) is another approach that sounds like an appealingly simple way of conducting international research. Remote conferencing and remote-control software may be useful for interviews and focus groups, but when more in-depth behaviorally-focused research is necessary, we are not enthusiastic about remote methods. Elsewhere, we have written in detail on issues in using remote testing methods for international UCD (Dray and Siegel, 2004). While remote testing may be appropriate in some situations, we suggest that it is best reserved for summative tests in locations where you have already had significant experience.

11.7 MAXIMIZING THE VALUE OF INTERNATIONAL UCD RESEARCH

We have already touched on some ways to reduce the costs of international research without compromising the quality (and value) of the research. There is also a long list of ways that you can increase the benefits of the research to make sure you get the most possible value from the investment.

11.7.1 Target the Right Markets for Your Research

Targeting the right markets means more than just going to where you have the largest concentration of users. There are other factors that might determine where your research may have the biggest impact. For example, you may select markets where you have the best indications of interest and buy in from stakeholders who have to act on findings. You may choose to go to a country that is more unfamiliar, even if the market there is smaller. If you have a long list of countries that you would like to visit, you probably have to prioritize. Emphasize diversity of locations over thoroughness of coverage.

11.7.2 Do Background Research

Do as much background research as possible. Study market analyses of the country or countries in question. Familiarize yourself with demographic patterns and with any available data on the type of technology you are interested in, such as its penetration, the platforms that are current in that country, and so on. Study available data on economic and social structures in the country. Even travel guides and books on international etiquette can be useful. Of course, review any information you can obtain about your own company's prior experience in that market. Background research like this should help you generate some preliminary hypotheses about variables that may be of interest and help you target your research focus and design your approach so that it is more likely to address pivotal issues.

11.7.3 Investigate Opportunities for Partnering

Identify internal partners who can collaborate with you on the research. This includes both participants from the local office and representatives of other disciplines from your home country. Local company representatives have a vested interest in participating in the research. They typically have knowledge of the local situation, which can be beneficial for you, and which they would like to see incorporated in the product. Often they are frustrated because of the sense that the home office does not listen to them. As a result, UCD projects can be organizational interventions that help break down the classic "home office/field office schism." This may add to their motivation to collaborate. At the same time, because they tend to be sales and marketing people, their input tends to be based on user requests and self report and consist of feature requests. Therefore partnering can also be an opportunity for them to benefit by learning about UCD research, which tends to be more behaviorally focused. In addition to the practical help they may be able to give the study, which can lower costs, they may also become longer-range allies and local champions for usability and UCD generally.

11.7.4 Do Ethnographic and Exploratory Visits First

As we have already stated, it is important to make every effort to do early international user research. The difference in ROI for early versus late research is greater in the international context. Beyond this, even short exploratory visits can be extremely helpful in ensuring that follow up research is appropriately planned and targeted. For example, basic knowledge of the context and user practices in the country can help you with such things as scenario development for subsequent studies, thus ensuring that time and resources devoted to later, more structured studies is well spent.

11.7.5 Focus Stakeholders on Broader Benefits and Synergies

Because of the increased likelihood that you will discover fundamental issues in functionality or conceptual design when doing international research, some of your most important findings may be actionable only in the longer run. It can be a mistake to attempt to cost justify each individual project independently on the basis of its predicted short-term impact on sales. One of the benefits of doing this research is that you will discover fundamental design issues that are applicable across a family of products, the implications of which will take time to incorporate. You will do well to manage your stakeholders' expectations appropriately in this regard.

11.7.6 Develop In-Country Resources

Use each study as an opportunity to help you develop relationships and resources that can make future research easier and less costly. Each study you do lays the groundwork for future studies in a variety of ways. As a result, planning and preparation time can be reduced. For example, a screener that you have already adjusted for an overseas market may be more easily adapted for the next study in that country.

11.7.7 Build Cumulative Learning

As you continue doing international UCD, you will not only facilitate the logistics and planning process for future studies, but will also add to your knowledge of international issues in your domain. DePalma (2002) reports that Lands' End found that the costs of doing international research on its Web site went down over time. Their initial outlays included a fundamental redesign of their online catalog processes for international markets as part of their global strategy. Now that these have been completed, they estimate that the cost of entering new markets is little more than the cost of translation.

One of the best ways to build cumulative knowledge is to not limit yourself to focusing only on bringing back descriptive information about a particular country, but to also focus on progressively building a model of international user and contextual variables that will help you generalize to other countries. On any given project, you will probably be able to study only a subset of the countries in which you are interested. By keeping track of the variables that you discover to be relevant in one country, you can add them to your model of factors that have to be considered across countries. Thus, things you learn in one country can help you target your research and drill down deeper in other countries, even countries you have not visited yet. This is part of the payoff of accumulating cross-cultural experience. It is also one reason that sampling a diverse subset of countries is important. It is an important benefit that goes beyond near-term ROI for a given project.

SUMMARY

In cases where you already have baseline data on relevant business metrics for a given international product in a country where you have a large concentration of customers, demonstrating ROI or developing credible projections of ROI is similar to what you would do for a domestic product, with the exception that you probably have additional costs to take into account. However, there are many factors in the international business context that suggest that a longer-range view needs to be applied when making the business case for international UCD, and that, in many cases, short-term ROI is not an appropriate criterion to apply. As international markets continue to grow and represent larger concentrations of customers, it will become easier to use short-term ROI. However, because of the lead time needed to develop organizational expertise in international design, and the deep level at which the data need to influence products, UCD efforts need to begin well in advance of the time when benefits are expected.

Organizations need to build their knowledge, skills, and resources in a cumulative manner in order to be poised to take full advantage of international markets. They need to look for payoffs in the form of increased expertise that they can apply across countries and across product lines. This means longer-range, more strategic arguments apply in the international realm. Of course, our field has advocated UCD as a fundament shift in design strategy for all products. The international domain makes these arguments even more compelling. Fortunately, more and more companies are seeing the necessity for these approaches. Thus, for companies that are not yet doing international UCD, the most powerful business rationale may be that they will be left behind.

Special Thanks to

Adam Asnes, Lingoport (USA), Nuray Aykin, Siemens Research (USA), Renato Beninatto, Common Sense Advisory (USA), Stefana Broadbent, COJI Group (France), Apala Lahiri Chavan, Human Factors International (India), Jose Coronado, Hyperion (USA), Dianne Cyr, Simon Fraser University (Canada), Elisa del Galdo, del Galdo Associates (UK), Don DePalma, Common Sense Advisory (USA), Vanessa Evers, University of Amsterdam (The Netherlands), Pia Honold, Siemens (Germany), Jacques Hugo, JP Associates (South Africa), Robin Jeffries, Sun Microsystems (USA), Paula Kotzé, University of South Africa (South Africa), Masaaki Kurosu, National Institute of Multimedia Education (Japan), Arnie Lund, Microsoft (USA), Ann-Byrd Platt, Swisscom (Switzerland), Girish Prahbu, Hewlett-Packard Labs (India), Lynn Shade, Adobe (USA), Denise

Spacinsky, Hewlett-Packard (USA), Andy Smith, Optimum Web (UK), Tom Stewart, System Concepts (UK), Christian Sturm, Siemens (Germany), and a number of additional colleagues who wish to remain anonymous.

REFERENCES

Allport, G., and Pettigrew, T. (1957). Cultural Influence on the Perception of Movement: the Trapezoidal Illusion Among Zulus. *Journal of Abnormal and Social Psychology*, 1 (55), 104–113.

Aykin, N. (2005). Overview: Where to start and what to consider. In N. Aykin (Ed.), *Usability and Internationalisation of Information Technology*. Mahwah, New Jersey: Lawrence Erlbaum Associates (pp. 3–20).

Aykin, N., and Milewski, A. (2005). Practical issues and guidelines for international information display. In N. Aykin (Ed.), *Usability and Internationalisation of Information Technology*. Mahwah, New Jersey: Lawrence Erlbaum Associates (pp. 21–50).

Barber, W., and Badre, A. N. (2001). Culturability: The merging of culture and usability. Retrieved on February 14, 2005. www.research.att.com/conf/hfweb/proceedings/barber/index.html.

Casey, S. (1993). *Set Phasers on Stun and Other True Tales of Design, Technology, and Human Error.* Santa Barbara: Aegean Publishing Company.

Chavan, A., and Iyer, A. (2003). Creating Recipes for Culture Curry—Is that the Road Ahead? In V. Evers, K. Röse, P. Honold, J. Coronado, and D. Day (Eds.), *Designing for Global Markets 5*. Proceedings of the 5th International Workshop on Internationalisation of Products and Services (pp. 301–303). Berlin, Germany: University of Kaiserslautern Press.

Cyr, D., Bonanni, C., Ilsever, J., and Bowes, J. (2004a). Beyond Trust: Website Design Preference Across Cultures. Simon Fraser University Working Paper.

Cyr, D., Ilsever, J., Bonanni, C., and Bowes, J. (2004b). Website Design and Culture: An Empirical Investigation. Presented at IWIPS 2004 and published in the Proceedings for the International Workshop for the Internationalization of Products and Systems Conference.

Cyr, D., and Trevor-Smith, H. (2004). Localization of Web Design: An Empirical Comparison of German, Japanese, and U.S. Website Characteristics. *Journal of the American Society for Information Science and Technology*, 55 (13), 1–10, 2004.

Day, D. (1999). Thinking Differently, Acting Together: A treatise on technology acceptance in the era of internationalisation. In G. Prahbu and E. del Galdo (Eds.), *Designing for Global Markets*. Proceedings of the 1st International Workshop on Internationalisation of Products and Services (pp. 25–32). Rochester, NY: Backhouse Press.

Degen, H., Lubin, K., Pedel, S., and Zheng, J. (2005). Travel Planning on the Web: A Cross-Cultural Case Study. In N. Aykin (Ed.), *Usability and Internationalisation of Information Technology* (pp. 313–343). Mahwah, New Jersey: Lawrence Erlbaum Associates.

del Galdo, E. (2004). Personal communication.

del Galdo, E. (1996). Culture and Design. In E. del Galdo and J. Nielsen (Eds.), *International User Interfaces.* New York: John Wiley & Sons.

DePalma, D. (2002). *Business without Borders.* New York: John Wiley & Sons.

Dray, S. (March, 1996). Designing for the rest of the world: A consultant's observations. *interactions,* 3 (2), 15–18. Available at: www.dray.com/articles.html.

Dray, S. (October, 2003). User-Centered Design and Localization: Partners in making things usable around the world. Talk given at *Localization World.* Seattle, WA.

Dray, S. (2004). Usable in New York, Usable in Nairobi. *MultiLingual Computing & Technology,* #65 Volume 15, Issue 5.

Dray, S., Kiel, A., Siegel, D., Sturm, C., and Wixon, D. (September, 2003). Ethnography in Organizations: Exploring Questions of Validity and Value. In P. Gray, H. Johnson, and E. O'Neill (Eds.), Presented at the 17th British HCI Annual Conference, University of Bath, UK. *Designing for Society* (pp. 227–228).

Dray, S., and Rowland, L. (1998). "Round the world in 18 days: Learnings from an international usability tour. In L. Trenner (Ed.), *The Politics of Usability.* New York: Springer.

Dray, S., and Siegel, D. (2004). Remote possibilities? International usability testing at a distance. *interactions,* 11 (2), 10–17.

Dray, S., and Siegel, D. (2005). "Sunday in Shanghai, Monday in Madrid?!": Key issues and decisions in planning international user studies. In N. Aykin (Ed.), *Usability and Internationalisation of Information Technology* (pp. 189–212). Mahwah, New Jersey: Lawrence Erlbaum Associates.

Dunckley, L., and Smith, A. (2000). Cultural Dichotomies in User Evaluation of International Software. In D. Day, E. del Galdo, and G. Prahbu (Eds.), *Designing for Global Markets 2* (pp. 39–52). Proceedings of the 2nd International Workshop on Internationalisation of Products and Services. Rochester, NY: Backhouse Press.

Evers, V., Kukulska-Hulme, A., and Jones, A. (1999). Cross-Cultural Understanding of Interface Design: A Cross-Cultural Analysis of Icon Recognition. In G. Prahbu and E. del Galdo (Eds.), *Designing for Global Markets* (pp. 173–182). Proceedings of the 1st International Workshop on Internationalisation of Products and Services. Rochester, NY: Backhouse Press.

Fernandes, T. (1995). *Global User Interface Design.* New York: AP Professional.

Fujinuma, M., and Risden, K. (2002). It's a global economy out there: Usability innovation for global market places CHI 2002 (p. 930). Workshop at Conference on Human Factors in Computing Systems. CHI'02 extended abstracts on human factors in computing systems. Minneapolis, MN.

Gould, E. (2005). Synthesizing the literature on cultural values. In N. Aykin (Ed.), *Usability and Internationalisation of Information Technology* (pp. 79–121). Mahwah, New Jersey: Lawrence Erlbaum Associates.

Greenspan, R. (2003a). Shoppers Demand Decent Design. Retrieved June 18, 2003, from CyberAtlas (now ClickzStats). Available at: www.clickz.com/stats/markets/retailing/article.php/2224101

Greenspan, R. (2003b). Western European E-Com to Reach Nearly 100 B Euros. Retrieved December 5, 2003, from CyberAtlas (now ClickzStats). Available at: www.clickz.com/stats/markets/retailing/article.php/3285861

Hall, E. (1959). *The Silent Language.* New York: Doubleday.

Hammontree, M., Weiler, P., and Nayak, N. (July, 1994). Remote usability testing. *interactions,* 21–25.

Harel, D., and Prabhu, G. (1999). Global User Experience (GLUE), Design for Cultural Diversity: Japan, China, and India. In G. Prahbu and E. del Galdo (Eds.), *Designing for Global Markets* (pp. 205–216). Proceedings of the 1st International Workshop on Internationalisation of Products and Services. Rochester, NY: Backhouse Press.

Hermeking, M. (2003). The Cultural Influences on International Product Development. In V. Evers, K. Röse, P. Honold, J. Coronado, and D. Day (Eds.), *Designing for Global Markets 5* (pp. 135–152). Proceedings of the 5th International Workshop on Internationalisation of Products and Services. Berlin, Germany: University of Kaiserslautern Press.

Hofstede, G. (1991). *Cultures and Organizations: Software of the Mind.* London: McGraw-Hill.

Hoft, N. (1995). *International Technical Communication: How to Export Information About High Technology.* New York: John Wiley & Sons.

Hoft, N. (1996). Developing a Cultural Model. In E. del Galdo and J. Nielsen (Eds.), *International User Interfaces.* New York: John Wiley & Sons.

Honold, P. (2000). Intercultural Usability Engineering: Barriers & Challenges from a German point of view. In D. Day, E. del Galdo, and G. Prahbu (Eds.), *Designing for Global Markets 2* (pp. 137–147). Proceedings of the 2nd International Workshop on Internationalisation of Products and Services. Rochester, NY: Backhouse Press.

IBM "Ease of Use" web site. (2004). www-3.ibm.com/ibm/easy/eou_ext.nsf/publish/558 Especially: www-3.ibm.com/ibm/easy/eou_ext.nsf/Publish/23 for the "rule of 10's."

Ito, M., and Nakakoji, K. (1996). Impact of Culture on User Interface Design. In E. del Galdo and J. Nielsen (Eds.), *International User Interfaces* (pp. 105–126). New York: John Wiley & Sons.

Lund, A. (2004). Personal communication.

Luong, T., Lok, J., Taylor, D., and Driscoll, K. (1995). *Internationalisation: Developing Software for Global Markets.* New York: John Wiley & Sons.

Marcus, A. (2005). User Interface Design and Culture. In N. Aykin (Ed.), *Usability and Internationalisation of Information Technology* (pp. 51–78). Mahwah, New Jersey: Lawrence Erlbaum Associates.

Marcus, A., Baumgartner, V., and Chen, E. (2003). User Interface Design vs. Culture. In V. Evers, K. Röse, P. Honold, J. Coronado, and D. Day (Eds.), *Designing for Global Markets*

5 (pp. 67–78). Proceedings of the 5th International Workshop on Internationalisation of Products and Services. Berlin, Germany: University of Kaiserslautern Press.

Marcus, A., and Gould, E. (2000). Crosscurrents: Cultural Dimensions and Global Web User-Interface Design. *interactions*, 7 (4), 32–46.

Marsden, G. (March/April, 2003). Using HCI to leverage communication technology. *interactions*, 2 (2), 48–57.

Meringer, J. (2002). eCommerce Next Wave: Productivity and Innovation. Seminar on revenue implications of ecommerce, WTO, April, 22, 2002. Geneva, Switzerland. Retrieved on February 14, 2005: www.wto.org/english/tratop_e/devel_e/sem05_e/presentation_meringer.ppt

Nielsen, J. (1996). International Usability Engineering. In E. del Galdo and J. Nielsen (Eds.), *International User Interfaces* (pp. 1–19). New York: John Wiley & Sons.

O'Sullivan, P., Wallace, M., and Yousuf, N. (2003). A Software Model Approach to Accomodating Cultural Diversity in Development of Multilingual Applications. In V. Evers, K. Röse, P. Honold, J. Coronado, and D. Day (Eds.), *Designing for Global Markets 5* (pp. 9–28). Proceedings of the 5th International Workshop on Internationalisation of Products and Services. Berlin, Germany: University of Kaiserslautern Press.

Röse, K., and Zuehlke, D. (2000). Method of Culture-Oriented Design for Technical Products. In D. Day, E. del Galdo, and G. Prahbu (Eds.), *Designing for Global Markets 2* (pp. 53–59). Proceedings of the 2nd International Workshop on Internationalisation of Products and Services. Rochester, NY: Backhouse Press.

Rosenberg, D. (2004). The Myths of ROI. *interactions*, 11 (5), 22–29.

Roshak, L., Spool, J., Evers, V., Molich, R., Page, C., and Platt, A. (2003). Evaluating globally: How to conduct international or intercultural usability research. Proceedings, Conference on Human Factors in Computing, CHI 03, Extended abstracts on Human Factors in Computer Systems (pp. 704–705). New York, NY: ACM Press.

Shade, L. (October, 1999). Cultural Discovery: The Process of Cross-Cultural Design. Paper presented at *Vision Plus 7: Diversification of Minds, Conversation in Processes, Design for Communities*. Tokyo.

Shade, L. (2003). Designing In Design-J: A Case Study in cross-cultural interface design. In V. Evers, K. Röse, P. Honold, J. Coronado, and D. Day (Eds.), *Designing for Global Markets 5* (pp. 259–265). Proceedings of the 5th International Workshop on Internationalisation of Products and Services. Berlin, Germany: University of Kaiserslautern Press.

Smith, A. (2004). Personal communication.

Smith, A., and Chang, Y. (2003). Quantifying Hofstede and Developing Cultural Fingerprints for Websites. In V. Evers, K. Röse, P. Honold, J. Coronado, and D. Day (Eds.), *Designing for Global Markets 5* (pp. 89–102). Proceedings of the 5th International Workshop on Internationalisation of Products and Services. Berlin, Germany: University of Kaiserslautern Press.

Smith, A., Dunckley, L., French, T., Minocha, S., and Chang, Y. (January, 2004). A process model for developing usable cross cultural web sites. In *Interacting with Computers* (16 [1], pp. 63–91), Special Issue on Global human-computer systems, cultural determinants of usability.

Swales, M., Strik, M., Dutta, A., and Potvin, J. (1999). Assigning Safety Functions to Products during Design—A Cultural Perspective. In G. Prahbu and E. del Galdo (Eds.), *Designing for Global Markets* (pp. 217–226). Proceedings of the 1st International Workshop on Internationalisation of Products and Services. Rochester, NY: Backhouse Press.

Tedeschi, B. (2004). American Web Sites Speak the Language of Overseas Users. Retrieved from New York Times on January 12, 2004. Available at: www.nytimes.com/2004/01/12/business/12ecom.html?ei=5007&en=85eee9de1e6f6220&ex=1389243600&adxnnl=1&partner=USERLAND&adxnnlx=1078520738–9CmidO9lm0xxEuGuEFDxXw

Trompenaars, F., and Hampden-Turner, C. (1998). *Riding the Waves of Culture: Understanding Diversity in Global Business* (2nd ed.). New York: McGraw-Hill.

Uren, E. (1997). Annotated Bibliography on Internationalisation and Localization. *Journal of Computer Documentation.* (SIGDOC), 21 (4), 26–33.

Vöhringer-Kuhnt, T. (June, 2002). The Culture of Usability. Masters Thesis, Frei Universität. Berlin, Germany. Retrieved on February 14, 2005: userpage.fuberlin.de/~kuhnt/thesis/results.pdf

12 CHAPTER Cost Justification of Usability Engineering for International Web Sites

Deborah J. Mayhew Deborah J. Mayhew & Associates

12.1 INTRODUCTION

This chapter discusses the cost justification of adding usability engineering to projects aimed at developing software for use in different countries and cultures. I have drawn upon the writings of and conversations with colleagues who have significant experience in international design, and extrapolated from our collective experience to explore the cost justification of applying usability engineering in international contexts.

Cost justification of usability engineering in general is addressed in Chapter 3, by Mayhew and Tremaine. Here I focus on what is unique about adding usability engineering to an international development project and how to adapt the general cost-justification technique in this case. To do this, I first provide some discussion and examples of unique aspects of the usability engineering process for international development projects. I then provide some examples of adapting the technique to cross-cultural projects.

12.2 OVERVIEW OF UNIQUE ASPECTS OF USABILITY ENGINEERING FOR INTERNATIONAL USER INTERFACES

Various aspects of The Usability Engineering Lifecycle described by Mayhew and Tremaine (Chapter 3; see also Mayhew, 1999), will need to be modified or added to, and may be more difficult and expensive when designing for an international audience. There will always be a variety of approaches possible for addressing the unique aspects of including usability engineering on an international development project (see Aykin, 2005; Siegel and Dray, Chapter 11). Here I refer to them briefly to provide context.

In the Requirements Analysis phase, fundamental and application-independent cross cultural differences in design, such as use of color, symbols and icons, date and currency formats, and so on—referred to extensively in Aykin (2005)—will need to be addressed. The good news is that a great deal of this information about a culture—language issues, formatting conventions, use of color, and so on—can usually be accurately learned from a single native individual who also speaks very fluent American English, or the language of the designing organization (but see Siegel and Dray's cautions in Chapter 11 regarding the pitfalls of this approach, including the limitations of a sample size of one, the need for domain-specific expertise from a culture, and regional variations within a culture). It may be possible, and would certainly be desirable, to hire such a person to be onsite with the project team full time. Of course, if there are multiple target cultures, multiple native speakers must be hired.

Another alternative is to find a firm that specializes in researching and providing this type of cross-cultural information and partnering with them. Yet another approach is simply to assume that when in a target country doing the application-unique types of requirements analyses that are a normal part of The Usability Engineering Lifecycle (i.e., User Profile and Contextual Task Analyses), your usability engineer will be attuned to these more general cultural issues, and, as a part of other activities, pick up some of these differences.

Depending on which approach is chosen different costs will be incurred. While gathering this type of requirements data is necessary and must be planned for, strictly speaking, it need not be included in the cost justification of a usability engineering program (unless, perhaps, if usability engineers are heavily involved in this activity), because, as explained above, it is really a basic prerequisite for designing for another culture.

Although a User Profile questionnaire can be distributed via a Web site or e-mail, it may need to be translated into the target language, and recruiting appropriate participants to respond may be more complex than in one's native country. It might be most effective to hire a firm in each target country to recruit questionnaire participants and communicate the sampling requirements to them. The services of a translator may be necessary to communicate the recruiting task to the local firm.

An alternative to an end-user questionnaire for user profiling is to conduct interviews with representatives of target users (e.g., managers) in each country and ask them to characterize the target user population. However, just as when using this technique in one's native country, this sort of second-hand data is always less reliable than data solicited directly from end-users. If you do this, you will need to plan on the costs of hiring a local interviewer (including the costs of recruiting and selecting a competent service provider) as well as the cost of translating the local interviewer's report.

Task analyses will be considerably more costly and difficult in other countries. One really needs access to actual end users to conduct an effective task analysis. One approach is for someone in the designing organization to travel to each of the target countries and, if necessary, employ a translator both to translate all task analysis materials beforehand and to translate in real time during the task analysis sessions with potential users. Time should also be planned to coordinate with and train the translators regarding what is important to pick up and what is not. You would need to factor in travel time as well, as it may not only take much longer for your usability engineer to get to the target country than to travel within the United States or the home country, but there may be a day or two required to recover from jet lag and time zone changes as well.

Alternatively, native speakers skilled in task analysis techniques might be recruited and hired in the target countries and communicated with effectively, which may entail recruiting, hiring and translation costs, as well as long distance phone expenses.

It may also take more time to conduct Task Analyses in some countries than in others. For example, in some countries you might not be able to get people to participate during work hours. In others, you might not be able to get them on weekends. If someone is traveling to a target country to conduct task analysis interviews, you should thus plan more or less travel time to get the interviews completed, depending on what you know about the ability to recruit users inside and outside business hours. Also, in some cultures no-shows may be much more likely than in others, especially if they must come to some other facility away from work. Thus you may have to schedule more users than the actual number of data points you hope to get, and correspondingly plan the additional time to run all planned sessions with whomever does show up.

In the Design/Test/Develop phase, it would be helpful during design activities to have a native member of each target country on the design team. Planning needs to be done in this phase to build a global architecture that can be most efficiently translated into each local target country. In later stages of design, design will need to be localized for each target country.

Besides design, evaluation is a key task in this phase. Usability tests should be conducted with local end users. One approach is to hire native speakers in each target country skilled in usability testing and communicate effectively with them, which may entail vendor recruiting and hiring costs, translation costs, and long distance phone expenses.

Alternatively, you might send someone in the design organization to the target countries and hire translators to work with them. Just as with task analyses, it may take more travel time, as well as more or less time in-county to collect enough data points in a test, as the rate of no-shows may be higher or lower than it would be in your own country, and it may be harder or easier to schedule test users.

An alternative to usability testing is hiring skilled native speakers to perform heuristic evaluations of prototype designs. In this case, recruiting and hiring costs will be incurred, as well as costs to translate the report into the native language of the development organization.

Many of the same issues will arise in the Installation phase, in which usability feedback is desired after implementation. Users are remote, and there will be additional time and expense involved in soliciting their direct feedback. Again, alternatives are hiring local skilled usability engineers, or having someone from the design organization travel and work with a translator. Some feedback techniques can be conducted remotely, using the Web or e-mail, and for these techniques it might be best to hire a local firm to recruit appropriate participants based on sampling specifications. Again, there may be additional costs for various translation services, and increased travel time and in-country time.

At this point, to simplify the discussion of cost justification for international software, I will limit the discussion to Web sites. It should be noted, however, that similar analyses can be applied to shrink-wrapped software and software developed for internal use by specific companies. The benefit categories might be slightly different in these cases, but the overall analyses will be highly analogous. Other sources provide the details of analyses in these cases (Bias and Mayhew, 1994; Bias *et al.*, 2003), and could be adapted with the details of international usability engineering provided in this chapter.

12.3 SAMPLE COST-BENEFIT ANALYSES OF USABILITY ENGINEERING ON INTERNATIONAL WEB DEVELOPMENT PROJECTS

As far back as 1991, five of the six major U.S. computer vendors were bringing in over 50% of their income from international sales (Russo and Boor, 1993). Smaller countries had an even larger percentage of their sales come from outside their own country (Nielsen, 1990). Since then, the Internet and the World Wide Web have made it even easier to market and sell in other countries. Siegel and Dray (Chapter 11) cite figures of 35 to 65%, and growth in international sales continues. For example, a June 2004 news report on the ECommerce Times Web site reported that for Amazon.com, "Net international sales, excluding the benefit from changes in foreign exchange rates, grew 38 percent compared with the second quarter in 2003" (www.ecommercetimes.com/story/35302.html). However, understanding cross-cultural differences and incorporating them into both products and marketing channels like the Web has a profound impact on the success of international marketing efforts, just as incorporating an understanding of domestic users and their tasks into products and marketing channels impacts overall success and return on investment in one's own country.

Here we will consider a hypothetical usability engineering plan in the context of two different international Web development projects, and then consider how you could conduct cost-benefit analyses of that plan for each project.

12.3.1 A Cross-Cultural E-Commerce Web Site

Imagine that a Web development organization is planning to redesign an existing e-commerce site that is not producing the return on investment (ROI) hoped for. Traffic and sales statistics are available from the current site. In particular, let's assume the original site was designed in English for an American audience, but is being used by multiple other cultures, and is not performing as well as hoped in any country, but especially not in the international countries. Part of the purpose of the redesign project is to tailor the site for use in a small number of target countries. The usability engineer prepares a proposed usability engineering project plan to be conducted for *each* target country (including the country of the development organization), and then performs a cost-benefit analysis of that plan.

First the final results of cost-benefit analyses for a "base" country (in this example, the United States) and *one* international country are presented below. Then the derivations of those final results are described. The same type of analysis would apply to *each* international country, although the numbers would vary from county to country depending on costs in individual countries. It should be noted that the cost for usability engineering programs would probably be less for a set of countries than the sum of the costs computed separately for each country, as certain parts of the program could be generalized or "amortized" across countries. For example, creating usability testing materials would likely take most time in the first country, but could be a fairly simple matter of translation for other countries, assuming they would use the same test tasks. However, for simplicity's sake, here we will assume and present the analysis for only one international country.

Table 12.1 shows the overall calculation of the *cost* of a usability engineering plan for a project, including a base country (in this example, the United States) and one international country (imagine a country in South America, where the cost of living and pay scales are significantly lower than in the United States). In this table, the first column identifies the overall project phases. The second column identifies the Usability Engineering Lifecycle tasks (see Mayhew and Tremaine, Chapter 3; Mayhew, 1999) that are planned in each phase. This column also breaks down each task according to general categories of costs, first for the base country and then for the international country.

The third, fourth, fifth, sixth, and seventh columns identify the number of hours required by different types of professionals to complete each task. The last column then summarizes the total cost of each category within a task. Note that values in the cells representing hours and costs for the international country are expressed as *differences* relative to the base country ("International Delta"). For example, under User Profile, you will see that relative to the base country, 12 more hours of the usability engineer's time (beyond the 80 hours required for doing a user profile at home) are expected to be required. Grand cost totals for the whole plan are given at the bottom of the table. All dollar amounts are given in the currency of the base country; in this example, U.S. dollars.

Next, the project usability engineer estimates that the usability engineering plans will produce new site designs with the *expected benefits every month* in the two countries shown in Table 12.2.

Comparing these benefits and costs, the project usability engineer argues that the proposed usability engineering plan at home will likely pay for itself in the first 5 months after launch.

Base Country

Benefits per month = **$33,854**

One time cost = **$160,605**

Payoff period = **4.74 months**

Given the higher costs for the international usability engineering program and the different parameters and predictions of benefits, the project usability engineer then argues that the proposed international usability engineering plan will likely pay for itself within the first seven months after launch.

Table 12.1 Cost of Usability Engineering Plan

Phase	*Task*	*UE Hours at $150*	*Dev. Hours at $150*	*User Hours at $40*	*User Hours at $20*	*Language/ Culture Consultant Hours at $75*	*Total Cost*
Requirements Analysis	**User Profile**						
	BASE						
	Hours	80		12.5			$12,500
	Recruitment Fees						$2,000
	INTERNATIONAL DELTA						
	Hours	+12		−12.5	+12.5	+12	+$2,450
	Recruitment Fees						−$1,000
	Materials Translation						+$1,050
	Task analysis						
	BASE						
	Hours	104		30			$16,800
	Recruitment Fees						$800
	Travel/Phone Expenses						$3,275
	INTERNATIONAL DELTA						
	Hours	+80		−30	+30	+32	+$13,800
	Recruitment Fees						−$400
	Materials Translation						+$300
	Local Facilitator						+$2,500
	Simultaneous Translation						+$2,000
	Visa						+$180
	Travel/Phone Expenses						+$1,575
	Platform Constraints						
	BASE						
	Hours	8	8				$2,400

Table 12.1 *Continued*

Phase	*Task*	*UE Hours at $150*	*Dev. Hours at $150*	*User Hours at $40*	*User Hours at $20*	*Language/ Culture Consultant Hours at $75*	*Total Cost*
	Usability Goal Setting						
	BASE						
	Hours	20					$3,000
	INTERNATIONAL DELTA						
	Hours					+8	+$600
Design/Testing/ Development	**Information Architecture**						
	BASE						
	Hours	60					$9,000
	INTERNATIONAL DELTA						
	Hours	+16				+16	+$3,600
	Conceptual Model Design						
	BASE						
	Hours	60					$9,000
	INTERNATIONAL DELTA						
	Hours	+16				+40	+$5,400
	Screen Design Standards						
	BASE						
	Hours	60					$9,000
	INTERNATIONAL DELTA						
	Hours	+16				+40	+$5,400
	Live Prototype Development						
	BASE						
	Hours	20	120				$21,000
	INTERNATIONAL DELTA						
	Hours	+16				+40	+$5,400

Table 12.1 *Continued*

Phase	*Task*	*UE Hours at $150*	*Dev. Hours at $150*	*User Hours at $40*	*User Hours at $20*	*Language/ Culture Consultant Hours at $75*	*Total Cost*
	Usability Test						
	BASE						
	Hours	104		16			$16,240
	Recruitment Fees						$400
	Travel/Phone Expenses						$2,775
	INTERNATIONAL DELTA						
	Hours	+56		–16	+16	+16	$9,280
	Recruitment Fees						–$200
	Materials Translation						+$3,000
	Local Facilitator						+$2,500
	Simultaneous Translation						+$2,000
	Visa						+$180
	Travel/Phone Expenses						+$1,350
	Redesign						
	BASE						
	Hours	40					$6,000
	INTERNATIONAL DELTA						
	Hours	+16				+16	$3,600
	Detailed User Interface Design						
	BASE						
	Hours	60					$9,000
	INTERNATIONAL DELTA						
	Hours	+16				+16	+$3,600
	Live Prototype Development						
	BASE						
	Hours	20	80				$15,000

Table 12.1 *Continued*

Phase	*Task*	*UE Hours at $150*	*Dev. Hours at $150*	*User Hours at $40*	*User Hours at $20*	*Language/ Culture Consultant Hours at $75*	*Total Cost*
	INTERNATIONAL DELTA						
	Hours	+16				+16	+$3,600
	Usability Test						
	BASE						
	Hours	84		16			$13,240
	Recruitment Fees						$400
	Travel/Phone Expenses						$2,775
	INTERNATIONAL DELTA						
	Hours	+40		–16	+16	+16	$8,080
	Recruitment Fees						–$200
	Materials Translation						+$3,000
	Local Facilitator						+$2,500
	Simultaneous Translation						+$2,000
	Visa						+$180
	Travel/Phone Expenses						+$1,350
	Redesign						
	BASE						
	Hours	40					$6,000
	INTERNATIONAL DELTA						
	Hours	+16				+16	+$3,600
	Total Base	**760**	**208**	**74.5**	**0**	**0**	**$160,605**
	International Delta	**+316**	**0**	**–74.5**	**+74.5**	**+284**	**+$92,275**
	Total International	**1,076**	**208**	**0**	**74.5**	**284**	**$252,880**

Table 12.2 Expected Monthly Benefits for an e-Commerce Site

Benefit Category	*Value in Base Country* per Month	*Value in International Country* per Month
Increased buy-to-look ratio	$15,625	$18,750
Decreased abandoned shopping carts	$15,625	$18,750
Decreased use of call back button	$2,604	$3,125
Total monthly benefits	**$33,854**	**$40,625**

International Country

Benefits per month = **$40,625**

One time cost = **$252,880**

Payoff period = **6.23 months**

Because the new sites are expected to have a lifetime of something much longer than 7 months, the project usability engineer expects the plans to be approved *for these two countries in particular*, based on this cost justification.

Note that the analyses offered here do not consider the time value of money. That is, the money for the costs is spent at one point in time, whereas the benefits come later. Also, if the money was *not* spent on the costs of usability engineering, but instead was invested, this money would likely increase in value. This investment could be compared to the benefits of investing in the usability engineering program. Furthermore, both costs and benefits are somewhat simplified and not exhaustive. Some expenses may have been overlooked, but also some benefits. Usually, the benefits of usability are so robust, these more sophisticated and complex financial considerations and details aren't necessary. However, if needed, calculations based on the time value of money are presented in Karat (Chapter 4) as well as in Bias *et al.* (2003).

Finally, as Siegel and Dray (Chapter 11) point out, the significant benefits of usability engineering on an international project are often strategic rather than tactical. These benefits are simply required to protect the huge overall investment of venturing into international markets, and costs need to be amortized over many projects. They argue convincingly that even if a cost-benefit analysis performed in the context of a single project did not show a benefit, one could still make the case for conducting usability engineering that would benefit the strategic position of the company. However, I suspect most cost-benefit analyses conducted on just a single project plan as in the example herein will show a significant benefit, and as such can only help the case for doing usability engineering in the context of international projects.

Below is a step-by-step description of how the project usability engineer arrived at the previously stated final results.

Start with a Usability Engineering Plan

If it has not already been done, this is the first step in conducting a cost-benefit analysis. The usability engineering plan identifies which Usability Engineering Lifecycle tasks and techniques will be employed (see Mayhew and Tremaine, Chapter 3; Mayhew, 1999) and breaks them down into required staff, hours and other expenses. Costs can then be computed for these tasks in the next two steps. The sample usability engineering plan used in these examples and laid out in Table 12.1 includes most lifecycle tasks and fairly rigorous techniques for each task. It assumes that three tasks—the task analysis and two rounds of usability testing—require travel, but that the rest of the work can be done in the base country location of the development organization.

Establish Analysis Parameters

Most of the calculations for both planned costs and estimated benefits are based on project-specific parameters. These should be established and documented before proceeding with the analysis. Sample parameters are given in the following sections for our hypothetical project.

Base (United States) Parameters

- The site is an e-commerce site (vs., for example, a product information site or a site based on an advertising model)
- The *current site* gets an average of 125,000 visitors per month in the base country (vs., for example, 500,000 users or 10,000 users)
- The current buy-to-look ratio in the base country = 2%
- The current average profit margin on each online purchase = $25
- The current rate of usage of the "call back" button (serviced by a local customer service organization—this is a button that allows Web site users to make contact with a live customer support rep, either by phone or "live chat") = 2%
- Average time to service each usage of the call back button = 5 minutes
- Users paid to participate in usability engineering tasks at an hourly rate = $40

- Customer Support fully loaded hourly wage = $50
- Usability Engineer (UE) and Developer fully loaded hourly wage = $150

International (for One Hypothetical Country) Parameters

- The site is an e-commerce site
- The *current site* gets an average of 75,000 visitors per month in the international country
- The current buy-to-look ratio in the international country = 1%
- The current average profit margin on each online purchase = $25
- The current rate of usage of the call back button (serviced by a local customer service organization) = 4%
- Average time to service each usage of the call back button = 10 minutes
- Users paid to participate in usability engineering tasks at an hourly rate = $20 per hour
- Customer Support fully loaded hourly wage = $25
- Usability Engineer and Developer fully loaded hourly wage = $150
- Language/Culture Consultant (a grad student living in the location of the development organization in the base country) paid a flat rate hourly = $7 per hour

Fully loaded hourly wages are calculated by adding together the costs of salary, benefits, office space, equipment, utilities, and other facilities for a type of personnel and dividing this by the number of hours worked each year by that personnel type. If outside consultants or contractors are used, their simple hourly rate would apply.

The hourly rate for users in this cost estimate is $40 in the base country and $20 in the international country. This is not based on a typical user's fully loaded hourly wage at their job, as it would be in the case of a cost justification of traditional software development for internal users or intranet development for internal users. Instead it is based on the assumption that test users of an e-commerce Web site will have to be recruited from the general public to participate in usability engineering tasks/techniques and that they will be paid at a rate of $40 (or $20) an hour for their participation.

The fully loaded hourly rate for usability engineering staff is based roughly on a typical current salary and benefits for a senior level internal usability engineer in the United States. The hourly rate of developers was similarly estimated.

I emphasize that when using the general cost-benefit analysis technique illustrated here, the particular parameter values used in the sample analyses should not be assumed. The particular parameter values of *your* project and organization and *your* base and international countries should be substituted for those stated previously. They will almost certainly be different from the parameters used in this example. In particular, professionals hired in different countries may have very different rates in the designing organization's currency. These unique parameter values are not always easy to obtain, but every attempt should be made to do so to add credibility to your analysis with your relevant audience.

In addition, as Siegel and Dray (Chapter 11) point out, there are many different approaches for conducting usability engineering on international projects, and your cost-benefit analysis needs to reflect the actual techniques you use, not necessarily the ones used in these examples. However, the examples here should serve to provide a general framework that can be applied in the case of international projects.

Calculate the Costs of Implementing the Usability Engineering Plan

Given a usability engineering plan, the number of hours required for each necessary type of staff can be estimated for each task/technique. Once you estimate the number of hours required for each task/technique, simply multiply the total number of hours each staff type requires for each task by their fully loaded hourly wage. Then you calculate the totals of the different staff types. Additional costs, such as equipment, supplies, travel expenses, and special services, should also be estimated and added in for each task. This is how the task costs in Table 12.1 were calculated.

The numbers of hours estimated for each task/technique in the cost summary table were not pulled out of a hat—they are hypothetical, but based on many years of experience. For example, the 104 hours estimated to conduct the first usability test in the base country (in this case, the United States) was derived as follows:

Design/develop test materials = 40

Two days' travel = 16

Run test/collect data (three days, eight test users) = 24

Summarize/interpret data, draw conclusions = 8

Document results = 16

Total = 104

Similar sample breakdowns of the steps required in each Usability Engineering Lifecycle task/technique can be found in Mayhew and Tremaine (Chapter 3) and in Mayhew (1999).

The cost differences (delta) for the international usability engineering program indicated in Table 12.1 are also hypothetical, but again are extrapolated from the actual experience of some of our colleagues conducting international projects. For example, in the case of the same first usability test, the additional 56 hours of the usability engineer's time is based on a number of anticipated factors:

- ✦ More time spent in preparation for the testing because of the complexity of making international travel and support staff arrangements
- ✦ More travel time (actual in-transit time, plus a weekend stay required to be on site the required number of week days)
- ✦ The assumption of more no-shows in the target country, which means more users must be scheduled to meet the goal of eight data points, which in turn means the usability engineer must plan to spend more time at the testing site
- ✦ Time spent training the local facilitator before testing begins
- ✦ Time spent simply picking up basic cross-cultural differences referred to previously (it is assumed that basic data were provided, but that additional data will be pursued and picked up during testing)

Also note that there are cost categories in Table 12.1 that are given as flat fees rather than broken down into staff types, numbers of hours for each staff type, and hourly rates. For example, flat fees are given for services such as recruitment of users, local facilitator, and simultaneous translation. These dollar amounts are based on the actual experience of colleagues in countries where service providers quoted a flat fee for the service agreed to rather than an hourly rate.

The difference in travel expenses for the base and international countries in this example is primarily a result of the expectation of higher airfares for international travel (this of course varies depending on the destination and the fluctuations in airfares over time). While more travel days are required, both in transit and on site to run the same number of test users as in the base country, it is expected that hotel, meal, and rental car costs will be less in the international country, making overall travel expenses comparable in the two countries.

Remember that both the hourly wage figures and the predicted hours and expenses per cost category for each task used to generate the sample analyses here are just hypothetical examples based on specific experiences. Again, you would have to use the actual costs and time of personnel and other expenses in your country and in your target country (or countries), as well as your own specific project plan, in order to carry out your own analysis.

Select Relevant Benefit Categories

Since this is an *e-commerce site,* only certain benefit categories are relevant to the business goals of this site. For another type of site, different benefit categories would be selected (see Mayhew and Tremaine, Chapter 3).

In this hypothetical case, the project usability engineer decides to include the following benefit categories:

- Increased buy-to-look ratio
- Decreased abandoned shopping carts
- Decreased use of call back button

These categories were selected because the usability engineer knew these would be of most relevance to the audience of the analysis: the business sponsors of the site. There may be other potential benefits of the usability engineering plan, but the usability engineer chose these for simplicity and for a conservative estimate of benefits.

As compared to the existing site design, the usability engineer anticipated that in the course of redesign, the usability engineering effort would decrease abandoned shopping carts by insuring that the checkout process is clear, efficient, provides all the right information at the right time, does not violate any cultural expectations or values, and does not bother users with tedious entry of information they do not want or need to provide. He or she expected to improve the buy-to-look ratio by insuring that the right product information is contained in the site, that navigation to find products is efficient and always successful, and that no cultural blunders are made in terms of product names, use of colors, etc. He or she expected to decrease the use of the call back button by making the information architecture match users' expectations by designing and validating a clear conceptual model so that navigation of and interactions with the site are obvious, and by insuring that language translation is not misleading or confusing. Accomplishing all these things depends on conducting the required analysis, design, and testing activities in the proposed plan with an eye toward uncovering and addressing unique requirements in the international country, as well as on applying general user interface design expertise.

Quantify and Estimate Benefits

Next the project usability engineer estimates the magnitude of each benefit that would be realized *compared to the current site* being redesigned *if* the usability engineering plan (with its associated costs) were implemented. Thus for example, he or she estimates how much *higher* the buy-to-look ratio would be on the site if it were re-engineered for greater usability as compared to the existing site.

To estimate each benefit, you must choose a unit of measurement for the benefit, such as the average purchase profit margin in the case of the increased buy-to-look ratio benefit, or the average cost of customer support time spent servicing each usage of the call back button in the case of the reduced use of the call back button benefit. Then—and this is the tricky part—you must make an *assumption* concerning the *magnitude of the benefit* for each unit of measurement; for example, a 1% increase in buy-to-look ratio, or a 1% decrease in the usage rate of the call back button. Tips on how to make these key assumptions are discussed in Mayhew and Tremaine (Chapter 3). Finally, you will calculate the total benefit in each category based on the unit of measurement, key parameters, and your assumptions about magnitudes of benefit. When the unit of measurement is time, benefits can be expressed first in units of time and then converted to dollars, given the value of time.

Remember that our hypothetical project involves development of an e-commerce site. Based in part on in-house experience, our project usability engineer makes the following key assumptions:

- Buy-to-look ratio will increase by 0.5% of total visitors in the United States, and by 1% in the international country (in this example, a hypothetical country in South America)
- Abandoned shopping carts will decrease by 0.5% of total visitors in the United States, and by 1% in the international country
- Usage of the call back button will decrease by 0.5% of total visitors in the United States, and by 1% in the international country

Note that the assumption is that there will be greater benefits in the international country than in the base country. Remember that currently the international country is using the site designed in English for an American audience. Any general usability problems are thus compounded by the fact that the site is not in the user's native language and that it does not take into account other cross-cultural differences such as preferred measurement units, culture-appropriate imagery, culture-appropriate use of color, currency differences, and so on. Therefore, by decreasing the general "alien-ness" of the site for the international audience, as well as improving the general usability of the site, greater benefits are predicted.

Also note that very conservative assumptions regarding predicted benefits are made, in spite of the aggressiveness and thoroughness of the usability engineering project plan. A 0.5% increase in the buy-to-look ratio in the base country is not much. However, making conservative benefits estimates in a cost justification analysis such as this is always wise if you can show an overall benefit even when your costs are high and your claims regarding expected benefits are very conservative, then you have a compelling argument for your plan.

Based on these key assumptions, the project usability engineer then calculates benefits in each of the selected benefit categories as follows:

Increased Buy-to-Look Ratio

BASE COUNTRY:

0.5% more visitors will decide to buy, and will successfully make a purchase each month

125,000 visitors per month × 0.5% = 625 more purchases

625 purchases at profit margin of $25 = **$15,625 per month**

INTERNATIONAL COUNTRY:

1% more visitors will decide to buy, and will successfully make a purchase each month

75,000 visitors per month × 1% = 750 more purchases

750 purchases at profit margin of $25 = **$18,750 per month**

Decreased Abandoned Shopping Carts

BASE COUNTRY:

0.5% more visitors who would have decided to buy anyway will now complete checkout successfully and make a purchase each month

125,000 visitors per month × 0.5% = 625 more purchases

625 purchases at profit margin of $25 = **$15,625 per month**

INTERNATIONAL COUNTRY:

1% more visitors will decide to buy, and will successfully make a purchase each month

75,000 visitors per month × 1% = 750 more purchases

750 purchases at profit margin of $25 = **$18,750 per month**

Decreased Usage of Call Back Button

BASE COUNTRY:

0.5% fewer visitors will need to use the call back button each month

125,000 visitors per month × 0.5% = 625 fewer calls

625 calls at 5 minutes each = 52.08 hours

52.08 hours at $50 = **$2,604 per month**

INTERNATIONAL COUNTRY:

1% fewer visitors will need to use the call back button each month

75,000 visitors per month × 1% = 750 fewer calls

750 calls at 10 minutes each 125 hours

125 hours at $25 = **$3,125 per month**

This is how the benefit predictions summarized in Table 12.2 were calculated. The usability engineer based the assumptions regarding benefits in the base country on statistics available in the literature, such as those discussed in Mayhew and Tremaine (Chapter 3). In particular, he or she began with the often quoted average e-commerce Web site buy-to-look ratio of 2–3% (Souza, 2000). He or she then based the assumption that this rate could be improved by a minimum of 1% (0.5% from improving the product search process and 0.5% from improving the checkout process) through usability engineering techniques on statistics offered as average by a Forrester report called "Get ROI from Design" (Souza, 2001). This report suggested that it would be typical for as many as 5% of online shoppers to fail to find the product and offer they are looking for (other statistics suggest as many as 45% may experience this problem; Souza, 2000). The Forrester report also claimed that more than 50% of shoppers who do find a product they would like to buy bail out during checkout because of the following:

- They are put off by forced registration
- They are put off by being asked to give a credit card number before being shown total costs
- They are not convinced that this is the best available offer

The assumption of reduced usage of the call back button by 0.5% was based on statistics cited earlier suggesting that as many as 20% of site users typically call in to get more information.

Most of us have experienced these problems and would have little argument that they are typical. Given these statistics, the assumptions made for the base country seem modest indeed. The slightly greater benefits assumptions made for the international country were based on the expectation that decreasing the general "alien-ness" of the site for the international audience would provide benefits beyond improving the general usability of the site.

The basic assumption of a cost-benefit analysis of a usability engineering plan is that the improved user interfaces achieved through usability engineering techniques will result in such tangible, measurable benefits as those calculated in this hypothetical example.

The audience for the analysis is asked to accept these *assumptions* of certain estimated, quantified benefits as *reasonable and likely minimum benefits*, rather than as precise, proven, guaranteed benefits. Proof simply does not exist that for each specific Web site an optimal user interface will provide some specific, reliable advantage over some other user interface which would result—or has resulted—in the absence of a usability engineering plan.

How to generate—and convince your audience to accept—the inherent assumptions in the benefits you estimate in a given cost-benefit analysis is discussed in Mayhew and Tremaine (Chapter 3).

Remember, it is usually wise to make *conservative* benefit assumptions. This is because any cost-benefit analysis has an intended audience who must be convinced that benefits will likely outweigh costs. Conservative assumptions are less likely to be challenged by the relevant audience, thus increasing the likelihood of acceptance of the analysis conclusions. In addition, conservative benefits assumptions help to manage expectations. It is always better to achieve greater benefit than was predicted in the cost-benefit analysis than to achieve less benefit, even if the benefits still outweighs the costs. Having underestimated benefits will make future cost-benefit analyses more credible and more readily accepted.

When each relevant benefit has been calculated for a common unit of time (e.g., per month or per year), then it is time to add up all benefit category estimates for a benefit total.

Compare Costs to Benefits

Recall that in our hypothetical project, the project usability engineer compared costs to benefits, and this is what was found:

BASE COUNTRY:

Benefits per month = **$33,854**

One time cost = **$160,605**

Payoff period = **4.74 months**

INTERNATIONAL COUNTRY:

Benefits per month = **$40,625**

One time cost = **$252,880**

Payoff period = **6.23 months**

Our project usability engineer's initial usability engineering plan for both countries appears to be well justified. It is a fairly aggressive plan (in that it includes all lifecycle tasks and moderate to rigorous techniques for each task), and the benefit assumptions are fairly conservative. Given the very clear net benefit, it would be wise to stick with this aggressive plan and submit it to project management for approval. In fact, based on the modest assumptions, you might be well advised to redesign the usability engineering plan to be even more thorough and aggressive, because increased benefits that might be realized by a more rigorous approach will likely be more than compensated for. In this case the usability engineer might consider increasing the level of effort for the requirements analysis tasks, which usually have high payoffs.

If the estimated payoff period had been long, or there was no reasonable payoff period, then it would be wise to go back and rethink the plan, scaling back on the rigorousness of techniques for certain tasks and even eliminating some tasks (for example, collapsing the design process within the Design/Testing/Development phase from two design levels to one; that is, doing only one usability test) to reduce the costs.

12.3.2 A Cross-Cultural Product Information Web Site

This example is based on a hypothetical scenario given in a Forrester report called "Get ROI from Design" (Souza, 2001). It involves an automobile manufacturing company that has put up a Web site to allow customers to get information about the features of the different models of cars they offer and the options available on those cars. It allows users to configure a base model with options of their choice and get sticker price information. Users cannot purchase a car online through this site—it is meant to generate leads and point users to dealerships and salespeople in their area.

Start with a Usability Engineering Plan

In this example, we again start with the same assumed plan as in the e-commerce site example (see Table 12.1).

Establish Analysis Parameters

Sample parameters for this example follow. Again we are assuming there is an existing site with known traffic and sales statistics and the project involves a redesign.

Base Country Parameters

- The site is a product information site
- The *current site* gets an average of 500,000 visitors per month
- Currently 1% of visitors result in a concrete lead
- Currently 10% of leads generate a sale
- The profit on a sale averages $300
- Users are paid to participate in usability engineering tasks at an hourly rate = $40
- Usability Engineer and Developer fully loaded hourly wage = $150

International Country Parameters

- The site is a product information site
- The *current site* gets an average of 250,000 visitors per month
- Currently 0.5% of visitors result in a concrete lead
- Currently 5% of leads generate a sale
- The profit on a sale averages $300
- Users are paid to participate in usability engineering tasks at an hourly rate = $20
- Usability Engineer and Developer fully loaded hourly wage = $150

Calculate the Costs of the Usability Engineering Plan

We can use the same cost calculations as before (shown in Table 12.1).

Select Relevant Benefit Categories

Since this is a *product information site*, only certain benefit categories are relevant to the business goals of this redesign project. The project usability engineer decides to include only the following benefit category:

- Increased lead generation

Again, he or she selected this benefit category because it will be of most relevance to the audience for the analysis: the business sponsors of the site. There may be other potential benefits of the usability engineering plan, but he or she chose this one for simplicity and to make a conservative estimate of benefits (see the following).

As compared to the existing site design, the usability engineer anticipates that in the course of redesign, the usability engineering effort will increase leads by insuring that visitors can find basic information and successfully configure models with options. Accomplishing this will depend on conducting the requirements analysis and testing activities in the proposed plan, as well as on applying general user interface design expertise.

Quantify and Estimate Benefits

Next the project usability engineer estimates the magnitude of the benefit that would be realized *relative to the current site* being redesigned *if* the usability engineering plan (with its associated costs) were implemented. Thus in this case, he or she estimates how much *higher* the lead generation rate would be on the site if it were re-engineered for usability as compared to the existing site. Table 12.3 summarizes these benefits predictions.

The project usability engineer calculated the estimated benefits as follows:

Increased Lead Generation Rate

BASE COUNTRY:

0.5% *more* visitors will generate a lead

500,000 visitors per month × 0.5% = 2,500 more leads

10% of these new leads will result in a sale = 250 more sales

250 more sales at profit margin of $300 = **$75,000 per month**

INTERNATIONAL COUNTRY:

1% *more* visitors will generate a lead

Table 12.3 Expected Monthly Benefits for a Product Information Site

Benefit Category	*Value in Base Country* per Month	*Value in International Country* per Month
Increased lead generation	$75,000	$37,500
Total monthly benefits	**$75,000**	**$37,500**

250,000 visitors per month × 1% = 2,500 more leads

5% of these new leads will result in a sale = 125 more sales

125 more sales at profit margin of $300 = **$37,500 per month**

Compare Costs to Benefits

Next the usability engineer compares benefits and costs to determine the payoff period.

BASE COUNTRY:

Benefits per month = **$75,000**

One time cost = **$160,605**

Payoff period = **2.14 months**

INTERNATIONAL COUNTRY:

Benefits per month = **$37,500**

One time cost = **$252,880**

Payoff period = **6.74 months**

Again, our project usability engineer's initial usability engineering plan appears to be well justified. It was a fairly aggressive plan, in that it included all lifecycle tasks, and moderate to rigorous techniques for each task, and the benefit assumptions were fairly conservative. Given the relatively short estimated payoff period and the expectation of a site lifetime of much longer than 7 months, it would be wise to stick with this aggressive plan and submit it to project management for approval. In fact, based on the very modest assumptions regarding increases in lead generation, he or she might consider redesigning the usability engineering plan to be even more thorough and aggressive because the increased benefits that might be realized by a more rigorous approach will likely be more than compensated for. In this case it might be advisable to increase the level of effort of the requirements analysis tasks, which usually have a high payoff.

SUMMARY AND CONCLUSIONS

Just as in the case of traditional software and Web sites intended only for a single culture, international Web sites can benefit greatly from usability engineering.

The bottom-line benefits of a proposed usability engineering program on a project to develop an international Web site can be calculated just as they can for development projects of other sorts. The difference is primarily in the details of the usability engineering program and in calculating the costs of that program. As in the two examples given previously, it seems likely that significant usability engineering effort will usually be easily cost justifiable on international Web development projects.

(This chapter is a revised excerpt from a chapter by Deborah J. Mayhew and Randolph G. Bias in Aykin, N. (Ed.) (2005). Usability and Internationalization of Information Technology. *Matwah, NJ: Lawrence Erlbaum Associates. Used with permission.)*

REFERENCES

Aykin, N. (Ed.) (2005). *Usability and Internationalization of Information Technology.* Matwah, NJ: Lawrence Erlbaum Associates.

Bias, R. G., and Mayhew, D. J. (Eds.) (1994). *Cost Justifying Usability.* Boston: Academic Press.

Bias, R. G., Mayhew, D. J., and Upmanyu, D. (2003). Cost Justification. In J. Jacko and A. Sears (Eds.), *The Human-Computer Interaction Handbook* Matwah, NJ: Lawrence Erlbaum Assoc., Publishers.

Mayhew, D. J. (1999). *The Usability Engineering Lifecycle.* San Francisco: Morgan Kaufmann Publishers.

Mayhew, D. J. and Bias, R. G. (2003). Cost Justifying Web Usability. In J. Ratner (Ed.), *Human Factors and Web Development* (2nd Edition, pp. 63–87). Matwah, NJ: Lawrence Erlbaum Associates.

Nielsen, J. (Ed.) (1990). *Designing User Interfaces for International Use.* Amsterdam: Elsevier.

Russo, P., and Boor, S. (1993). How Fluent is Your Interface: Designing for International Users. *Proceedings of INTERCHI,* 50–51.

Souza, R. K. (2000). *The Best of Retail Site Design.* Cambridge, MA: Forrester Research.

Souza, R. K. (2001). *Get ROI From Design.* Cambridge, MA: Forrester Research.

13 CHAPTER Return on Goodwill: Return on Investment for Accessibility

Tom Brinck Diamond Bullet Design

13.1 INTRODUCTION

Accessibility means "able to access." Quite frankly, most products cannot be accessed by everyone. Without accessibility, usability isn't even an issue. Once access is possible, designing for usability makes that access meaningful. Thus, usability and accessibility go hand in hand.

While information technology has become ubiquitous, it still remains inaccessible to a large number of people. A wide variety of design assumptions can create usage barriers—designs for a single computer platform, for a single language, for people who are adept with a mouse, for people with strong vision, for those who can easily distinguish colors, and for those who are familiar with specialized terminology or user interface conventions. Designers may fall into the trap of designing for people like themselves or of assuming their target audience is homogeneous. Other designers may be concerned about the costs of designing for special cases.

Recent internet technologies have made a focus on accessibility more valuable than ever. Web sites are *available* to anyone with an Internet connection, and this has been a wonderful opportunity for people to reach information from any country, despite wide differences in knowledge, experience, and abilities. Nevertheless, without a focus on making these Web sites *accessible,* this opportunity is squandered.

This chapter addresses the costs and benefits of creating accessible designs, with a focus on access for people with disabilities. Web design is the running example, although the lessons apply to almost any design domain. In addition, the tradeoffs of choosing a level of effort in achieving accessibility are discussed.

Accessibility isn't something you do or don't do—there are many gradations, with associated costs and benefits, at each level.

13.2 WHAT DOES IT MEAN TO MAKE A WEB SITE ACCESSIBLE?

If you have ever come across a Web site that didn't display correctly, you have had an accessibility problem. Web sites may be inaccessible because of reliability issues—they're broken when you try to access them. Or they may be inaccessible because you don't have the necessary plug-in or because your screen isn't large enough. In this sense, an accessible Web site doesn't necessarily look any different; in fact, what distinguishes accessible Web sites is that they actually work the way you would expect. However, Web sites can create difficulties for people with visual or movement impairments. Hearing impairments are rarely an issue because audio is not used frequently on Web sites (yet), although if audio is used, text captions and transcripts are recommended.

13.2.1 Visual Impairments

There are many types of vision problems; for example, macular degeneration, cataracts, and glaucoma. These result in impairments ranging from common near-sightedness to complete blindness, and may involve blurriness, color distortion, or limitations in the field of view. People with low vision, for example, may benefit from adjusting the text size on the screen, or using a screen magnifier, a piece of software that allows them to zoom in on a small portion of the screen for better visibility.

For a blind person, a Web site is often accessed with a screen reader—software that uses voice synthesis to read the text on a Web page. Without additional design features, any information that isn't available in text, such as an image that isn't described, is unavailable to that person. The first step in improving such a design is to ensure that all the information on the Web site is available in text. This doesn't mean there has to be a text-only version of the site. Indeed, Slatin and Rush (2003) argue convincingly that providing an alternate text-only site is a bad solution. The most common concern about doing so is ensuring that the text-only site is maintained and contains complete and current information matching the non–text-only site.

The typical solution to making images accessible is to mark up images with alternative text descriptions (alt tags). In HTML, this means including a phrase such as alt = "XYZ Corporation" to mark up the image of your company logo. The screen reader can see these labels and read them aloud. Images that do not convey meaning, such as decorative images, should be marked up with empty alternative descriptions such as alt = " ". This makes the text accessible. The second step is to make the text actually *flow* well. Listening to many Web pages with a screen reader is a tedious experience, in which information can be repeated, appear out of context, or have poor organization and emphasis. Thus the text and controls need to be edited and arranged with good headings and in a sensible order, and pages are ideally tested with a screen reader to verify that they are indeed organized and efficient listening experiences. As Slatin and Rush (2003) say, accessible design is good design.

13.2.2 Movement Impairments

The second common challenge is for people who have difficulty using their hands. This may be the result of arthritis, paralysis, amputation, temporary injury (such as a broken arm), or even circumstances, such as the need to hold something in one hand while using the computer. These difficulties can be assisted by reducing the need for large amounts of typing, the need to press multiple keys at once, and the need for fine motor control in pointing a mouse.

For example, most people with Parkinson's Disease have some degree of tremor in their hands. Each person has different symptoms and different means of adapting to them, so some prefer to avoid the mouse altogether and use the keyboard to navigate the Web, which requires that pages be designed so users can tab to each link. Because of difficulty typing, others prefer using a mouse or trackball, but will have difficulty pointing accurately; thus, a design that employs larger and more widely-spaced buttons, minimizes scrolling, and minimizes controls requiring manual dexterity or multiple clicks, such as cascading menus, will be more accessible.

13.2.3 The Role of Guidelines and Browser Compatibility

Designing successfully for a variety of disabilities is made possible by applying standard design guidelines such as the W3C's Web Content Accessibility Guidelines (W3Ca) and the U.S. Federal Government's Section 508 Guidelines (Section 508, USGSA). Guidelines do not capture all the real problems people may encounter in using a specific design, and some guidelines may be contradictory, so user testing is highly recommended, as are other techniques for gathering user requirements, such as focus groups and interviews. However, user testing is limited by the availability of users with any given disability (although some disabilities may be "simulated" in user testing by blindfolding users, for example, these simulations will often be limited by their inability to capture the expert behavior of someone who lives with that disability), by the inherent variability of disabilities, and is not possible for certain issues (for example, it is not possible to test whether a certain type of flashing would stimulate a seizure). For all these reasons, guidelines are a crucial component in ensuring a reasonable level of completeness in addressing a wide variety of disabilities.

Writing Web pages so that they follow standards and are as compatible as possible with a variety of browsers helps accessibility considerably. Cross-platform compatibility helps everyone who uses alternative platforms, including mobile phones, PDAs, and older and lesser-known computer platforms. People with disabilities adopt a wide variety of technologies to help them access computers, and the more robust, standards-compliant, and flexible a Web site is, the more likely it is to work on the equipment they require.

13.3 THE BENEFITS OF ACCESSIBILITY

For many, the idea of accessibility is abstract, and although designing for accessibility is perceived as a good deed, it is not clear how extensive the value might be. Following are some of the most common reasons to design for accessibility, with an argument made for each reason.

Table 13.1 categorizes these reasons according to who will benefit from accessibility and what type of benefit they receive according to each rationale. For users, the types of benefits they receive from more accessible design include the effectiveness, or success, they have at accomplishing what they intend to do, their efficiency at it, and how satisfied they are with the experience. For the business and Web developer creating an accessible product, the benefits are increased sales, reduced costs, or lower risks, all of which are, to some degree, quantifiable financially. Those aspects that aren't directly quantifiable are labeled here as "karma," which are things like positive market perceptions that are viewed as good for a company but may not have any direct, measurable impact on revenue streams (Wilson and Rosenbaum [Chapter 8] refer to this as social return on investment [ROI]). Positive market perceptions in particular are hoped to indirectly improve sales, and good karma in general may lead to better customer retention, as well as better retention of employees in the developer's organization. The last beneficiary listed is society in general, which

Table 13.1 Reasons for Accessibility and The Types of Benefits Each Provides

	Who It Benefits and Type of Benefit Provided					
	User	*Developer/Business*				*Society*
Reason for Accessibility	Effectiveness, Efficiency, and Satisfaction	Increased Sales	Reduced Costs	Lower Risks	Positive Karma (Social ROI)	Public Health, Welfare, Justice, Employment
Enablement	yes				yes	yes
Accessibility helps everyone	yes	yes				yes
Public good and corporate citizenship					yes	yes
Social justice					yes	yes
Market size		yes				
Niche markets		yes		yes	yes	yes
Positive market perceptions					yes	
Legal requirements			yes	yes		yes
Cost savings in service provision			yes			
Cost savings in software development and maintenance			yes			

achieves quantifiable benefits in terms of higher employment rates and lower healthcare costs. Lower healthcare costs, in particular, result from more cost-effective accommodation of those with disabilities and reduced health complications when they are able to live and work effectively.

In Table 13.1, the benefits to society are central to the majority of the arguments. Social benefits are often the most compelling, and this is one reason that, although it may sometimes be difficult for an individual business to justify the cost of accessibility, society can achieve social ends by enforcing laws requiring businesses to supply accessible Web sites, thus tipping the scales in their cost evaluations and lending broad benefits to society as a whole.

13.3.1 Enablement

In the end, the best argument for making technology accessible is to take the perspective of those who are excluded. By being in a demographic that wasn't taken into account, they find themselves unable to fully participate in the benefits of technology. Building accessible hardware, software, and Web sites empowers everyone, and gives people greater opportunities to participate in society, enjoy greater independence, find meaningful employment, and enjoy a rewarding lifestyle. This is the most exciting aspect of accessibility.

13.3.2 Accessibility Helps Everyone

Accessible Web sites are more usable for everyone. Everyone is disabled in some way, and most people are likely to encounter serious disabilities as they age. Injuries make all of us disabled at some point. Most people encounter *situational disabilities*, meaning they can't use their computers as well because they only have one hand free as they hold a phone or a book with the other hand, they misplaced their glasses, or they are feeling more fatigued than usual. Everyone will periodically appreciate a more forgiving user interface that allows the text to be larger when they are showing a Web page to a large group, that enables them to tab through links when their mouse is not working properly, or that can be viewed easily on a mobile phone when they can't get to a computer.

True accessibility requires good design along basic principles that benefit everyone. Follow standards. Build robust software that doesn't crash or produce incorrect results. Be maximally compatible with multiple operating systems, input and output devices, Web browsers, and varying user preferences. Present information in multiple media forms (text, images, and sound). Minimize the need for users to remember information. Use simple, unambiguous language. All these are broad principles of usable design that are even more important when designing for accessibility.

13.3.3 Public Good and Corporate Citizenship

As individuals, we want to help people in our society who might not otherwise be served. This creates a society in which more people are employed and fewer require public support. As companies, this is a basic responsibility of good corporate citizenship. Accessibility addresses the problems of the digital divide, helping those who have been left out because of economic circumstances, education, cultural background, or disability. Nonprofit organizations often consider reaching people with disabilities and other technologically-disenfranchised groups part of their core mission.

13.3.4 Social Justice

Another perspective on the public-good argument is the ethical view on the basic human rights to equal access and nondiscrimination. While we may not be able to design Web sites that work for absolutely everyone, we have a responsibility to make a reasonable effort to design sites to be as widely accessible as possible.

Many industries, such as government, education, and public utilities, have a core mission of serving the *entire* public. These industries, by the nature of who they serve, must look at groups that may form only 1% of the population and determine how they will be served.

13.3.5 Market Size

Many of the benefits of accessibility cannot easily be financially quantified, but one in particular *can* be: market size. If you design a Web site to work for the most common Web browser and operating system, you may reach as many as 90% of users, but designing for and testing the next most common browsers can enable you to reach 95%, and this can be one of the least expensive ways to expand your market by 5%.

13.3.6 Niche Markets

In some industries, a consideration of disabilities just makes sense. Some Web sites are certain to have more target users who will have disabilities, such as hospitals, HMOs, and health insurance organizations. In addition, any target audience with a high percentage of elderly users will have a high percentage of disabled users.

13.3.7 Positive Market Perceptions

Accessible Web sites provide better customer service for site visitors with disabilities. If customers without specific accessibility needs know about your efforts to create an accessible Web site, they'll view it favorably, which is good public relations. On the flip side, if some customers can't access your site, it generates negative perceptions, which can be both embarrassing and damaging to your image. Given this knowledge, if you have made the effort to create an accessible solution, you ought to brag about it—it builds goodwill among all your customers.

13.3.8 Legal Requirements

In the United States, the most clear-cut legal standard is Section 508 of the Rehabilitation Act (U.S. General Services Administration, 1998). Section 508 requires all federal government-sponsored information technology to be accessible according to specific guidelines (with some exceptions such as military technology). Sixteen of these guidelines apply specifically to Web sites (Table 13.2). Thus most federal agencies and federal contractors must follow accessibility guidelines, and a large number of state and local governments are following suit.

The other major legal requirement in the United States is the Americans with Disabilities Act (ADA). The ADA requires that public services be accessible and that employers not discriminate on the basis of disability. Unfortunately, the ADA doesn't have specific guidelines for what is required of a Web site (I suggest the Section 508 standards as a logical starting point). Regardless, if use of Web sites is required for a job, then they must be accessible to employees with disabilities. Beyond the legal requirement, accessibility may lead to greater employee retention, a benefit in itself.

Laws similar to the ADA can be found in Australia, Canada, and Europe. Many countries have adopted some version of the W3C's Web Content Accessibility Guidelines (W3Ca) as their starting point for Web access. Details on the legal requirements in other countries can be found in Thatcher (2002) and Slatin and Rush (2003), as well as some legal cases. In one recent example (CNN, Aug. 19, 2004), Priceline.com and Ramada.com agreed in an out-of-court

Table 13.2 Section 508 Guidelines for Web Accessibility

These are the formal legal standards for U.S. federal government Web sites.

(a) A text equivalent for every nontext element shall be provided (e.g., via "alt," "longdesc," or in element content).
(b) Equivalent alternatives for any multimedia presentation shall be synchronized with the presentation.
(c) Web pages shall be designed so that all information conveyed with color is also available without color, for example, from context or markup.
(d) Documents shall be organized so they are readable without requiring an associated style sheet.
(e) Redundant text links shall be provided for each active region of a server-side image map.
(f) Client-side image maps shall be provided instead of server-side image maps except where the regions cannot be defined with an available geometric shape.
(g) Row and column headers shall be identified for data tables.
(h) Markup shall be used to associate data cells and header cells for data tables that have two or more logical levels of row or column headers.
(i) Frames shall be titled with text that facilitates frame identification and navigation.
(j) Pages shall be designed to avoid causing the screen to flicker with a frequency higher than 2 Hz and lower than 55 Hz.
(k) A text-only page, with equivalent information or functionality, shall be provided to make a Web site comply with the provisions of this part, when compliance cannot be accomplished in any other way. The content of the text-only page shall be updated whenever the primary page changes.
(l) When pages utilize scripting languages to display content or to create interface elements, the information provided by the script shall be identified with functional text that can be read by assistive technology.
(m) When a Web page requires that an applet, plug-in or other application be present on the client system to interpret page content, the page must provide a link to a plug-in or applet that complies with §1194.21(a) through (l).
(n) When electronic forms are designed to be completed online, the form shall allow people using assistive technology to access the information, field elements, and functionality required for completion and submission of the form, including all directions and cues.
(o) A method shall be provided that permits users to skip repetitive navigation links.
(p) When a timed response is required, the user shall be alerted and given sufficient time to indicate more time is required.

settlement to fix their Web sites to make them fully accessible for blind users and others "in one of the first enforcement actions of the Americans with Disabilities Act on the Internet" (CNN).

For many, the legal requirements are the clincher, not just because it's the law, but because the risk and potential high cost of a lawsuit or a failed sale to a large customer is unacceptable. A single employee suing on the basis of discrimination could be an enormous cost by itself, and the potential is high for class action lawsuits when Web sites serve a large audience.

13.3.9 Cost Savings in Service Provision

If you will be working with customers with disabilities, you may achieve cost savings by serving these customers over a Web site versus other means, such as phone support. For example, online banking is not only more convenient for many customers, but it can be dramatically less expensive to a bank than using bank tellers or telephone banking. An accessible bank Web site can significantly reduce the percentage of people who call or visit the bank while at the same time helping people with disabilities who may be inconvenienced if they need to visit the bank (perhaps requiring special transportation, for example).

13.3.10 Cost Savings in Software Development and Maintenance

While designing for accessibility is unlikely to lower your overall development costs if you already have best practices implemented, it may save costs if it moves you toward better development practices. Accessibility requires a move toward software development practices that follow standards, ensure consistency, allow for flexible presentation, and design for maximum platform independence. Some shortcuts around these may be possible when accessibility isn't a primary goal, but in the long term those shortcuts can be costly when the software must be patched and rewritten.

13.3.11 When Is There No Benefit to Designing for Accessibility?

It's hard to argue that there should be absolutely no effort made toward accessibility. One of the critical arguments for designing for accessibility is the value of picking low-hanging fruit; that is, the low cost of achieving some level of accessibility with tremendous value. For example, ensuring that a Web page can be navigated via the keyboard is quite easy as long as you avoid some mistakes that would interfere (such as pop-up windows or embedded media that is not keyboard accessible) and write code that places links in a logical order. Adding alt tags to images is relatively easy as long as it is planned from the start. Testing additional Web browsers is simple, at least for the most common browsers. These small acts can make a big difference even if they don't achieve the widest possible accessibility.

Any additional work has additional cost, and there are certainly cases where accessibility is cost prohibitive. I discuss estimating those costs in a later section. Because of all the benefits already mentioned, neglecting accessibility even when there are costs can be risky, so if cost is the stumbling block, consider whether your investment will succeed without the greater investment needed for accessibility.

Another basis for not considering accessibility is when you have a known target audience without accessibility needs. This is quite rare—even in apparently homogenous populations there is often surprising variation. As an example, I was once developing an intranet for a company in which the IT staff guaranteed that everyone in the organization had at least an 800 × 600 resolution monitor. However, when I was demonstrating the system to one of the vice presidents, he asked me why the design didn't fit on his screen. It turned out that despite having a high-resolution screen, he had set it to a lower resolution of 640 × 480 so that it was easier on his eyes, which is precisely the kind of situation for which we would like to plan.

Nevertheless, there are Web sites designed for small audiences in which you can verify that no one has a specific disability or dictate what Web browser and computing platform will be used. The risk in these circumstances is that the situation will eventually change. People will come and go, and the nature of your target audience will evolve. New technologies will come. If you've designed without flexibility, you risk that the cost of fixing an inaccessible design could be higher than the cost of creating an accessible design in the first place.

13.4 AUDIENCE DIVERSITY AND MARKET SIZE

Of all of the benefits of accessibility described so far, the most straightforward to quantify is market size, which can be done by describing audience diversity numerically. That is, exactly what percentage of people is excluded from a given design? People vary in many ways, making it easy to inadvertently exclude a portion of your market. People vary by disability, nationality, language, education, computer experience, the technology they use (Web browser, operating system, screen size, network speed), and software preferences and settings. In addressing this audience diversity, you need to start with basic statistics about how many people are affected when you choose to create a design for one segment versus another. While many people think of disabilities as fairly rare, the U.S. Census reports that about one in five adult Americans has a disability and half of seniors older than 65 have a disability (Census 1997). With an aging population, the number of people with disabilities is expected to increase. Table 13.3 shows the percentages of some common disabilities that relate to Web design and some examples of how each is accommodated in Web design.

13.4.1 Estimating Benefits

A basic cost-benefit analysis of market size involves looking at how many people are excluded based on a market segment that is not currently supported and how much it would cost to accommodate that market segment. Thus, if 3.7% of people have difficulty seeing, you can ask how much value you'll gain from a

Table 13.3 Rates of Disabilities for U.S. Residents from U.S. Census 1997 Survey (Census01)

	Population (in Millions)	*Percent*	*Example Accommodation*
All ages	**267.7**	**100.0%**	
With a disability	52.6	19.7%	
Severe disability	33.0	12.3%	
Disability in more than one domain (physical, mental, or communication)	17.9	8.6%	
Age 15 years and over	**208.1**	**100.0%**	
Difficulty seeing	7.7	3.7%	High contrast, scalable fonts
Unable to see	1.8	0.8%	Text equivalents for images
Difficulty using hands	6.8	3.2%	Large buttons, keyboard navigation
Difficulty hearing	8.0	3.8%	Volume controls for Flash
Unable to hear	0.8	0.4%	Text equivalents for audio
Mental disability	14.3	6.9%	Short, direct tasks
Mental retardation	1.4	0.7%	Simplified instructions
Learning disability	3.5	1.7%	Image equivalents for text

3.7% market growth when you repair visual accessibility problems. As a running example, consider making sales online, where an increase in users is directly translatable to income (if your Web site derives revenue in a more indirect way, the reasoning remains the same but follows the business case for the Web site).

A detailed benefit calculation based on increased market size is shown in Table 13.4. If your current Web site supports 10,000 customers (say, annually), and the average revenue per customer is $100, then it generates revenue of $1,000,000. If you modify this Web site to be Section 508 compliant, you can estimate the additional market at 5% more customers (this is a rough estimate; a more or less conservative choice can be made at the estimator's discretion). Similarly, modifying the site to be compatible with additional browsers is estimated in this example to reach an additional 3% more customers.

Table 13.4 Estimating the Benefit of Increased Sales from Market Size Increases Resulting from Achieving Section 508 Compliance and Additional Browser Compatibility

	Current Customers	*Estimated % Increase*	*Estimated New Customers*	*Cost of Adaptation*	*Average Revenue per Customer*	*Total Revenue*	*Revenue Minus Cost of Adaptation*
Current	10,000				$100	$1,000,000	
Section 508		5%	500	$13,500	$110	$55,000	$41,500
Additional browser compatibility		3%	300	$11,250	$95	$28,500	$17,250
			Total adaptation cost	$24,750		**Net increase in revenue**	$58,750
						% increase in revenue	5.88%
						ROI (net increase in revenue : total adaptation cost)	2.4 : 1

Revenue for each adaptation is computed as the number of customers multiplied by the average revenue per customer. The final column shows the revenue increase after subtracting the cost of adapting the site. The cost of adaptation is computed in a later section and can be treated as just another estimate for the moment. The average revenue per customer may remain the same after making a Web site more accessible, but this example shows the impact of estimating increased or decreased revenue per customer based on the type of adaptation made. This would be the case if you expected someone with disabilities to buy either more or less than the average customer. For example, people with disabilities may be more likely to make greater purchases online than the average consumer because they may be less mobile.

Sharp readers may have noticed the following problem in how this calculation is made: Because the disability statistics are based on the population as a whole, we ideally ought to be starting the calculation from the overall maximum market size rather than our current market penetration. That is, suppose that the potential market for our product is 100,000 people and we are currently selling to 50,000 people. If our Web site currently doesn't accommodate people who have difficulty seeing (3.7% of the overall population) because it relies heavily on images, then we are excluding 3.7% of our potential target audience, or 3700 people. Thus, improving the site for this audience could increase our market to 53,700 people. If, as in Table 13.4, we start with our current customers rather than working from the potential market size, we will add 3.7% of 50,000, or half the number of customers. In practice, this oversimplification of the calculation is usually necessary because we have data on only our current number of customers and the potential maximum market size isn't well defined. In doing so, we recognize that we are producing a highly conservative estimate. When the total population is known, such as for a company intranet, the calculation can start from the total population figure.

13.4.2 Demographic Data on Individual Differences

Audience size estimates are improved with demographic data, but where can you find demographic data? Data in Table 13.3 came from the U.S. Census, and a variety of reports are available on U.S. demographics (for disability data in particular see Census 1997; Census 2000; Census 2001). Other data must be obtained from relevant sources based on the audiences you are interested in, or you can conduct your own surveys of your market. Data on browsers and computing platforms are available through a variety of sources, with richer data available at higher cost. For example, Table 13.5 shows browser market share and percentages of Web users with different screen resolutions from W3Schools (W3Schools). When obtaining this information, consider using multiple sources to get a feel for sampling biases in each. Various sources measuring hits will tend to be biased toward particular computing platforms or particular income or education levels. These data can also be obtained to some extent from the hit logs of your Web site, but should be interpreted with caution: are people with 640 × 480 screens uncommon, or do they avoid your site because your site doesn't work well at that screen size?

Notice that in Table 13.5 the Mac Safari browser is not represented. This is partly a result of the fact that its market share was relatively small in July 2003 (although it has grown quickly since). It also may be because of a sampling bias in this data source, thus emphasizing the importance of watching for biases and being prepared for rapid changes, especially in which technology is used.

13.4.3 Uncertainties in Estimation

There are a few cautionary notes in these market size calculations. At least three types of estimates are behind any calculation: 1) the size of any given market segment, 2) the number of people in each market segment who will benefit from any given accommodation, and 3) the overlap between two populations when estimating the impact of more than one type of accommodation. The level of uncertainty involved in these estimates is par for the course in making financial projections from market data, and a good understanding of the issues will enable a better sense of the range of possibilities, from a conservative to a liberal estimate.

Table 13.5 Percent of Web Users for Each Browser and Screen Resolution (W3Schools, 2003)

Browser Version	*July 2003 Share*	*Screen Resolution*	*July 2003 Share*
Internet Explorer 6.x	59%	1024 × 768 or greater	49%
Internet Explorer 5.x	34%	800 × 600	44%
Internet Explorer 4.x	1%	640 × 480	2%
Netscape 4.x	1%	Other or unknown	5%
Other Netscape compatible	1%		
Opera	1%		

Size of a Given Market Segment

Statistics like the ones provided in Table 13.5 on disabilities, browsers, and screen resolutions will vary based on the source, the quality of how data is measured and analyzed, and how the questions are asked. These statistics must be interpreted with regard to how broadly they apply.

A simple example of the ambiguity of statistics can be seen with the data on adults who have difficulty seeing. While the U.S. Census reports that 3.7% of adults have difficulty seeing, at least 61% of people in the U.S. wore some form of vision correction in 2001 (Access Media, citing Jobson Publishing LLC). It's fair to assume, as well, that some people without vision correction do not have 20-20 vision. Thus, creating Web pages so that font sizes can be adjusted easily is likely to benefit not just 3.7% of adults, but perhaps the vast majority.

The Number Who Benefit from an Accommodation

Solutions for visual accessibility issues are unlikely to be adequate for *everyone* who has difficulty seeing. A given accommodation, such as making font sizes scalable, will benefit some people with visual impairments but not others, such as those unable to see.

In addition, a given population may have a different rate of Web use than other populations. While people with mobility impairments will often use the Web (to minimize travel), those with visual impairments are often put off by the inaccessibility of so many Web sites. They may also have a different rate of purchasing products. They may purchase fewer products with accessibility concerns, such as unreadable documentation, and may purchase more of those products in which the brick-and-mortar purchase process is inconvenient. Thus, making your Web site accessible to those unable to see, 0.8% of the population, will likely result in either more or less than a 0.8% increase in usage depending on a range of usage patterns.

In a similar vein, even if you address every accessibility guideline, you shouldn't expect a 20% increase in market size just because 20% of adults have a disability. Many common disabilities are ones that don't interfere with Web use, such as difficulty walking, so they weren't being excluded from your Web site in the first place. Others are sufficiently severe to prevent extensive Web use, such as dementia. The guidelines you choose to follow will determine how much penetration you achieve within the new market segment you are addressing. The short list of 16 guidelines provided by Section 508 will not help as many people as more extensive lists. As such, even after applying all 16 of the Section 508 guidelines, we might estimate an increase in potential audience of perhaps 5% instead of 20% (which was the basis for using 5% in the calculation in Table 13.4).

Overlap of Populations

Even after addressing the needs of a given disability, such as visual impairments, some people are likely to have other disabilities or barriers, such as language difficulty or lack of computer experience. To compute the number of people affected when you address more than one barrier at a time, you need to figure out how much the populations overlap. If 3.2% of people have difficulty using their hands and 44% have a screen size of 800 × 600, you don't get an additional 47.2% benefit from addressing both. You have to subtract the people who are counted twice in those percentages because they fall in both populations (except in the rare circumstance when populations don't overlap at all). The actual population overlap depends on how much these factors co-vary.

Depending on your target market, disability rates may be much different. In particular, disability rates increase significantly with age. For example, 17% of Americans 45 or older report a moderate or severe visual impairment, and visual impairment rates vary significantly based on gender, race, income, education, overall health, education, and marital status (Lighthouse, 1994).

13.5 THE COSTS OF ACCESSIBLE DESIGN

The process of producing an accessible Web site has a number of variables that significantly impact cost. Some of these variables are as follows:

- Whether you plan accessible design from the beginning or make fixes after finishing the site
- Your personnel, their training and experience, and broad individual differences in their design skills, proofreading skills, and coding times
- Whether you employ automated tools to assist in the evaluation (Brinck and Hofer, 2002)

Automated tools can report on many types of guideline violations, much like a spell checker. Automated evaluation can reduce the time for reviewing your site, but doesn't eliminate manual checking; some manual evaluation and repair time is always necessary. An automated tool is unable, for example, to determine if the sequence of spoken text makes sense or if an image has a correct and meaningful text label.

13.5.1 Retroactive Process to Make Your Site More Accessible

As an example for this section, the following is a baseline process for testing your Web site and making fixes after it is completed. Whether you designed your site for accessibility from the outset or not, a retroactive review is appropriate as a quality check. The following is a reasonable set of steps for reviewing your site, with each step going deeper in analysis. The number of steps taken should be based on a cost analysis of the time taken in each kind of review, and the consideration of benefits in Table 13.1.

1. **Perform a Quick, Preliminary Review of the Accessibility of Your Site.** Do a quick review and ask yourself how someone would use your site if that person were blind, deaf, or had a motor impairment that limited them to the keyboard or limited the accuracy of their pointing. Simulating the use of your site by trying it with your eyes closed or your hands bound, can give you some initial idea of the extent of the problems.
2. **Define Your Audience and Your Accessibility Goals.** Define the audience for your site. What types of disabilities might they have? Does your target audience include older people or children? Does it include international users or people with limited literacy? Does it include people who might have temporary or situational disabilities? The answer to most of these is certainly yes, but to what extent? Decide the strategy you want to adopt for accommodating these groups—a design that works for everyone, a design that is merely adequate for people with disabilities, or a design optimized for specific disabilities. What range of disabilities will you target? How convenient does it need to be for people with disabilities? What inconveniences or design limitations are acceptable for the portion of the target audience with no serious disabilities?
3. **Review Your Site Without Images, Audio, or a Mouse.** Review your site with images turned off and make sure it makes sense. Then review it in a text browser, such as lynx (lynx.browser.org). If your site has any audio, review it again with the sound turned off. Try browsing your site without using the mouse and typing with only one finger.
4. **Meet Section 508 Standards.** Review and modify your site to comply with Section 508 rules. One way to save time is to use an automated tool that checks these standards.
5. **Meet W3C Web Content Accessibility Guidelines (WCAG).** Review and modify your site to meet WCAG guidelines. Most people find that some of

these guidelines are too strict for their site (especially those defined in the guidelines as Level 3), but a rationale should be given for each exception.

6. **Test Your Site With Assistive Technologies.** Review your site with a screen reader and screen magnifier. Does it make sense? Is it efficient?
7. **Review More Detailed Guidelines.** Review your site with even more thorough accessibility guidelines. For some other guidelines, see Theofanos and Redish (2003) and Coyne and Nielsen (2001).
8. **User Test and Improve Your Site.** Do user testing with people with disabilities. Find out what causes difficulty for them and fix it. Find out what is most useful for them and focus on that.

13.5.2 Estimating Costs

As a starting point, a simple way to estimate the time necessary to review accessibility problems on your Web site and repair them is to assume that each page on your site requires a fixed average amount of time to review and repair. Table 13.6 works out a cost estimate based on this approach. Reviews should typically be integrated with your standard quality assurance (QA) process, which reviews the site to make sure it is functioning properly and to proofread content. When quality assurance is divided among several people, such as separate content editors and technical reviewers, the accessibility review will need to be similarly distributed but managed centrally to ensure no piece is overlooked. In Table 13.5, we estimate that each page requires 0.2 hours average (12 minutes) for the standard QA process, including fixing bugs (this time is highly dependent on each organization and its type of product—it involves ensuring, for example, that links are not broken, that text reads correctly, and that the site works correctly under standard configurations). The calculation of Table 13.5 estimates that checking each page against Section 508 guidelines and making fixes requires an additional 0.2 hours, and that achieving additional browser compatibility (beyond your standard process) requires an additional 0.1 hours per page. These time estimates are reasonable but will vary considerably depending on the experience of the personnel and the complexity of the Web site. In this example, assuming a Web site of 300 pages and an hourly cost of $100, the total cost of quality assurance is $15,000, and the portion used to achieve additional accessibility is $9,000.

The testing itself is usually completed by a QA team, and the repairs are completed by the development team. A typical example is that a QA specialist finds an image without an ALT tag, noting whether the problem occurs on multiple pages. The developer would then enter the code and add the ALT tag, verifying that the fix is made on all pages that are affected. The system then goes back to the QA specialist, who then re-tests (called *regression testing*) and verifies not only that the fix is made but that no additional errors have been introduced (for example, a coding error may have resulted in the image failing to display). In a more complex but common example, the QA specialist may find that in a certain browser and version, the page does not display correctly because of a formatting problem with a table element misaligned. This occurs frequently because of small incompatibilities in the interpretation of HTML. The QA specialist then carefully documents the circumstances of the formatting problem, including the browsers, versions, and operating systems that are affected. The developer then must explore the cause of the formatting problem and may have to test various solutions. Once the problem is resolved, it again goes to the QA specialist for regression testing. Since the developer may not have been able to test the variety of platforms available to the QA specialist, additional problems may be found. While this testing and repair can be quite time-consuming, many problems are either rare or occur throughout many pages of the Web site, so they can be fixed all as a batch. Thus, the average time per page may be kept at a manageable level.

The calculation of Table 13.6 makes sense for small sites that are hand crafted, but almost all larger Web sites are database-driven, so that a small set of HTML templates are used to construct each page, and review time can be saved by checking the templates for accessibility. Regardless, individual pages need to be checked because the content will vary on each page and the content may have accessibility problems (e.g., images without text labels). When templates are

Table 13.6 A Simple Estimate of the Cost of Evaluating and Repairing Accessibility Problems in a Web Site

	Number of Pages	*Test and Repair Hours per Page*	*Hourly Cost*	*Overall Cost*
Standard quality assurance	300	0.2	$100	$6,000
Section 508	300	0.2	$100	$6,000
Additional browser compatibility	300	0.1	$100	$3,000
		Total quality assurance cost		$15,000

reviewed ahead of time and standards are followed in content design, the review of individual content pages can be much faster, but once again, the time taken depends on the complexity of the page. As mentioned earlier, automated tools can check some items to save time, but some manual review remains necessary. With large sites of thousands of pages, some prioritization makes sense—put more time into reviewing mission-critical pages than others, so that critical paths, like shopping cart check-out, are made especially accessible. Table 13.7 shows an estimate similar to Table 13.6 that breaks out templates, mission-critical pages, and low-priority pages.

Cost Variables

Table 13.6 shows the simple model. There are a variety of other cost variables and alternate approaches. Obvious variables have to do with skill level, degree of advance planning, and degree of use of automated tools. The cost and benefit calculations described here have all been in the context of Web design, but all the general principles apply to any product design.

A more detailed model of evaluating a site against guidelines can be constructed by evaluating the cost of applying each guideline. This will vary by the choice of guidelines and your chosen policies about how best to fulfill each guideline. For example, if you use media that require inaccessible plug-ins, you

Table 13.7 A More Detailed Estimate of Evaluating and Repairing Accessibility Problems in a Web Site, with Different Categories of Pages Requiring Different Levels of Effort

	Number of "Templates"	*Test and Repair Hours per Template*	*Mission-Critical Pages*	*Hours per Critical Page*	*Low-Priority Pages*	*Hours per Low-Priority Page*	*Hourly Cost*	*Overall Cost (Hours × Hourly Cost)*
Standard quality assurance	5	6	200	0.25	800	0.1	$100	$16,000
Section 508	5	1	100	0.4	900	0.1	$100	$13,500
Additional browser compatibility	5	1	50	0.25	950	0.1	$100	$11,250
						Total quality assurance cost		$40,750

must choose between policies such as: 1) stop using that media type, 2) create text equivalents every time you use that media, 3) document the use of that media and make it optional for the user, or 4) find an equivalent accessible plug-in. Then, to evaluate the time cost of your guidelines, you must evaluate the following for each rule:

- Frequency: How often does a page contain code that might not comply?
- Detection time: What is the average time it would take to review a page to spot all violations of this rule?
- Correction time: What is the average time it would take to fix a page that had violations?

For each rule, Rule cost (in time) per page = frequency × detection time × correction time. Finally, add the cost of each rule for the total time cost per page.

There are also metacosts that surround the activity of preparing an accessible Web site but that are not part of the design and coding itself. A key cost is determining what it means to prepare an accessible Web site. You need to educate yourself and train your team on what guidelines to apply and what process to use to create and improve your site. Unfortunately, *everyone* is still learning how to make Web sites accessible, and best practices are continually evolving, so doing a good job will require periodic research on best practices, which are not gathered in one source. Those who don't care to be on the leading edge can minimize this research cost by choosing the prevailing standard. Currently, I suggest using the Section 508 standards, as they are the lowest cost and most clearly defined. However, even the Section 508 standards require time to interpret, learn, and integrate into the production process of a Web design team.

Other metacosts of accessibility are goal definition, planning, audience definition, user-needs analysis, documentation, and auditing (auditing in this context means verifying for legal or other purposes that the team has in fact followed a process that achieves accessibility and has created the appropriate documentation). In small projects, these may be done informally and inexpensively integrated into the standard process. On larger projects, these may be somewhat separate activities, but they are fixed costs on a project—they don't tend to grow much with the size of the project.

In addition to evaluating guidelines, you may choose to do *access device testing* and *user testing*. Access device testing is testing against the technologies your users would use. For people unable to see, you ought to test with a screen reader, and for those with limited vision, a screen magnifier. In user testing, you bring in people with disabilities (or whatever population you are interested in including) and apply usual usability testing methods. User testing ought to be done *in addition to* using guidelines; it is never practical to skip the guidelines by testing the full range of disabilities for which the guidelines are designed. User testing for people with disabilities will involve some higher costs. Recruiting is typically more difficult because there is a smaller pool of people to recruit from. Because of high population variance among people with disabilities, your testing procedure will be less systematic than usual and may require a larger sample size. It may be necessary to meet people in their homes or other preferred locations, both to minimize their travel needs and to have access to the equipment they typically use to browse the Web.

One Hundred Percent Accessible is Infinite Cost

The goal of *universal design* is to create designs that work for everyone, but designing for absolutely everyone is impossible. Designing a highly usable Web site for an increasing market segment can involve serious compromises and extremely high costs. For example, some Web browsers may not display graphics (such as text browsers in Unix), may only display a few lines of text at a time (such as cell phones), and may have bugs that aren't compatible with standard HTML (such as Internet Explorer). For most browser differences, following W3C HTML standards will enable basic compatibility, but because standards aren't always supported correctly by browsers, testing a variety of browsers becomes necessary to ensure compatibility, and the more obscure the browser, the higher the cost of obtaining and testing it. Thus, while the cost is usually reasonable to go from 80% market penetration to 90%, the cost of going from 98% to 99% can be enormous.

In the domain of disabilities, designing a Web site for hearing limitations is usually easy. In most cases you can avoid using audio. Designing for visual deficits takes more effort. But designing for many types of cognitive disabilities, such as an inability to read, may require a fundamental change in what a Web site does and may require substantially different design skills. In turn, such modifications may create a site that does not practically serve the needs of the majority market. In such cases, alternate Web sites can be created for each market, but this can mean substantially higher cost.

Interface Design Strategies for Audience Diversity

Having considered both the benefits and costs of accessibility, what approach should be taken in designing for accessibility, and what are the tradeoffs with each approach? There are two aspects to consider: defining the target market and how different interface designs optimize for market segments (in this section), and deciding exactly how much effort to put into accessibility to achieve different levels of accessibility (next section).

The basic question in deciding the user interface approach is how optimized the site should be for each market segment. Do you want to put more emphasis on one market over another? Are you willing to have greater optimization for one market segment produce degradation for another? If you'd like it to work well for everyone, are you able to spend the additional cost for universal design? While improving the design for one group often improves the design for other groups, that is certainly not always the case, so these alternative design approaches must be considered:

- **Optimized design**. Often the lowest cost approach is to pick a narrow target audience and optimize the design for them, neglecting other audiences. This enables a potentially better user experience for that narrow audience, but as we have argued already, an overly-narrow target audience offers few benefits, and the apparent homogeneity of a targeted audience is often an illusion.
- **Universal design**. Providing a design that is accessible to nearly everyone can be affordably achieved by simplifying the design, staying focused on the minimal core features, and strictly following all standards, making no compromises that would impair accessibility. This creates a design that is able to be accessed by the largest population but is often optimal for no one.
- **Equivalent facilitation**. Separate audiences can be served by providing them equivalent but separate user interfaces, such as allowing people to branch off to a separate text-only or audio-only page. This approach is criticized when, for instance, a text-only page is not supported as well as the main site and so is not truly equivalent. To work well, both versions need to be driven from the same database and supported equally, which often involves a higher design and maintenance cost than a single user interface.
- **Multimodal design**. Another way to provide options for different audiences is to combine those options into a single interface, as when closed captions are displayed on video, or by providing text labels of images. Another relatively new technique is *audio descriptions*, which are essentially audio captions for video, where a narrator provides a voice-over to a movie with a description of what is appearing visually. When planned from the outset, a multimodal design approach provides a good cost/benefit balance, as media too difficult to translate into multiple modalities can be avoided. However, trans-

lating archival material (videos, animations, PDF files, complex interactions in Flash) can sometimes be an expensive proposition.

- **Customizable design**. Designs that give users some choice in how they appear enable them to fit the design to their needs. A simple example is to avoid setting the font size to an unmodifiable value so the user can adjust the font size. Some sites display buttons at the top of the screen that allow someone to easily select a font size. Customizability has great potential benefit but can add greater up-front development cost and won't work for all accessibility problems. Also, the user interface for customization must be very clear or the customization may itself be inaccessible.
- **Graceful degradation**. A gracefully degrading interface is one that is optimized for some, and while it won't work *well* for everyone, it is still accessible at a basic level to nearly everyone. An example is displaying a map or chart that is clear at a glance to those who can see it, but providing a more cumbersome text explanation for those who can't.

The following are criteria for choosing among these approaches:

- **Cost**. How much is involved in developing and maintaining each alternative? What is the cost to users who have to live with the design? Development costs can go up exponentially as you approach working for 100% of users (except possibly for very simple systems). Where is your cutoff? If you will ever plan to change an accessibility approach (or lack of an approach), consider the cost of retrofitting.
- **Market**: How large an audience can you reach? Who will be missed?
- **Optimization**: How good will the user experience be for each user type? Do you take advantage of specific knowledge about each population?
- **Confusion**: Will a generalized, simplified, or mixed design be harder to understand? For example, people using screen readers can get frustrated with having to listen to repetitive text on every page, especially since most Web sites include identical navigation links at the beginning of every page. A solution, required in Section 508, is to provide *Skip Navigation* links at the top of each page that provide a within-page link that jumps past the navigation links to the main content of the page. To comply, some sites have added text links that say "Skip Navigation" at the top of every page, but this link confuses the vast majority of users. The most common practice today that avoids this confusion is to put a small one-pixel image at the top of each

page that is linked. This tiny image appears as an ordinary link in a screen reader but avoids confusing sighted users. This is a good example in which design innovation was able to find a solution that works well for everyone.

- **Benefit**: How much value will a design approach have? What is the impact on each of the accessibility rationale: market size, market perceptions, legal compliance, public good, cost savings in service provision, and so on?

Levels of Accessibility

In the final analysis, exactly how much effort should be put into accessibility for any given project? I have provided the reasoning behind the benefits of accessibility and provided some techniques for estimating costs and financial returns. Many organizations will be guided by small budgets or may be swayed by legal requirements or altruism. Often overcoming organizational inertia can be one of the biggest roadblocks. Thus, Table 13.8 provides a sequence of levels of accessibility that might be achieved by applying progressively more detailed analyses and thus optimizing the design for greater and greater accessibility. Each level builds on the last.

The simplest level, called *Low-hanging fruit*, is to do good traditional design and catch some obvious accessibility problems. Some examples include avoiding pop-up windows, including ALT tags for images, avoiding audio and video, and using standard HTML markup, like the <h1> tag for headers, as opposed to specifying exact font sizes. All of these are either simple or are recommended standard practice with a number of practical advantages. This approach won't solve a lot of accessibility problems, but is feasible on an extremely small budget and when there is no imminent concern for accessibility.

The second set of levels, *Standards amd Guidelines compliance*, applies progressively more detailed guidelines. These are all relatively low cost, provide good coverage of many major issues, and will provide very useful improvements. Most such guidelines do not go far in helping make the designs substantially more usable, but merely make it possible to access information in a reasonable way. Section 508 is listed first because it is simple, well-defined, and relatively easy to learn and apply. The W3C WCAG (W3Ca, 1999) are related to the Section 508 guidelines so there is considerable overlap, but the WCAG contains additional guidelines. Using guidelines other than these will be more time-consuming because other guidelines are not as widely available nor have they been standardized. Some worth considering are the proposed WCAG 2.0 guidelines (in draft version [W3Cb, 2003]), and guidelines derived from user testing experiences (Theofanos and Redish, 2003; Coyne and Nielsen, 2001).

Table 13.8 Levels of Effort That Can Be Applied to Accessibility and What Types of Organizations and Projects Typically Are the Best Fit for Each Level of Effort

Level of Accessibility	*Who Should Target this Level*	*What Activities It Entails*
Low-hanging fruit	Small Web sites with limited budgets and no reason to expect people with disabilities	The straightforward basics: ALT tags; good visual contrast and scalable fonts; standard markup, e.g., <h1>. Things you need to have anyway: compatibility and standards; usability; robustness and technical correctness (QA).
Standards and guidelines compliance (primarily technical accessibility, making it *possible* to access the information)		
Section 508	A minimal legal standard, applies to U.S. federal government Web sites with no additional accessibility needs (some other governments, such as U.S. states, may have slightly different, but largely similar standards)	Apply the 16 Section 508 guidelines; use the guidelines at the design stage and in quality assurance.
W3C Web Content Accessibility Guidelines (WCAG)	Organizations intending to meet international industry-standard minimum requirements, e.g. Web design firms, software development firms, and professional societies	WCAG Level I and Level II should be fulfilled, and Level III should be considered. Document exceptions you've decided to make to the guidelines, with rationale.
+Further guidelines	Organizations wishing to create a somewhat better-than-standard level of	Apply additional guidelines, which involve automated or hand-evaluation. The biggest

Table 13.8 *Continued*

Level of Accessibility	*Who Should Target this Level*	*What Activities It Entails*
	accessibility without substantially higher costs	problem is in establishing the pertinent guidelines. See Theofanos and Redish, 2003; Coyne and Nielsen, 2001.
Universal accessibility (focus on ease of use)		
Access device compatibility and testing	Organizations able to establish a larger budget for accessibility or with a greater need to make sure their site is inclusive, such as educational institutions, public utilities, non-profits, government agencies, and healthcare organizations	Design for and test multiple access mechanisms, including extensive cross-platform testing (multiple platforms and browsers and older browser versions), text browsers, screen readers, screen magnification tools, keyboard-only input (with one-finger input), mouse-only input, and audio off.
+User testing	Any organization *truly* committed to making their Web site accessible, with adequate funds to achieve it	User testing with people with disabilities, exploring multiple types of disabilities.
Optimized accessibility	Web sites that need to work particularly well for people with disabilities, including many non-profits, government agencies, public utilities, and healthcare organizations; sites designed specifically for rehabilitation and to provide resources to people with disabilities	All of the above, plus user-needs analysis (interviews, focus groups), task analysis, and an ongoing mechanism for obtaining feedback from users with disabilities and updating the site to reflect that feedback.

The next two levels, described as *Universal accessibility*, require access device testing and user testing. Some access devices that should be tested are screen readers, screen magnifiers, and keyboard-only access. Because these often require additional training, resources, and costs, they are appropriate for those organizations that can set aside an explicit budget for accessibility endeavors. While they can be done without first applying the guidelines, it's best to apply the guidelines first to avoid learning about problems that could be found more easily and inexpensively and fixed using the guidelines. The guidelines also provide broader, though shallower, coverage than this more advanced testing.

Finally, the highest level, *optimized accessibility*, is appropriate when a budget is available to really make a design work in an excellent way for people with specific disabilities, as you would do in a health Web site targeted at a specific disease or injury category.

Achieving accessible designs will be more affordable over time as awareness increases and more designers and developers receive appropriate training. In addition, guidelines will evolve and tools for building sites will more effectively support those guidelines, enabling improved solutions that involve fewer compromises and take less time to implement.

The ROI outlook for accessibility is therefore quite promising. Today, the choice to achieve at least some level of accessibility is compelling, because there are certainly benefits at the level of low-hanging fruit, and there are levels of achievement that provide additional advantages at additional costs, enabling organizations to systematically increase their accessibility gains by aiming for the next level of accessibility. In the future, the hurdles for achieving additional levels will become more cost-effective, and the benefits will continue to increase as governments and the public recognize the continuing importance of accessibility.

13.6 ACKNOWLEDGEMENTS

I'd like to dedicate this chapter to Stephen Markel, my business partner and friend for more than 8 years, who passed away prematurely on September 18, 2004 at the age of 37 resulting from complications of head trauma in an earlier ski accident. Stephen was president of Diamond Bullet Design and was an ardent champion of both accessibility and an ROI focus in usability and design. Stephen was an inspiration in his commitment to this work. Stephen Markel, Seunghee Ha, Derren Hermann, and Htet Htet Aung helped to develop and refine the ideas in this chapter.

REFERENCES

Access Media Group. (2004). *Vision Statistics.* Retrieved from All About Vision on February 29, 2004. www.allaboutvision.com/resources/statistics.htm.

Architectural and Transportation Barriers Compliance Board. (2000). *Electronic and Information Technology Accessibility Standards,* Federal Register, 36 CFR Part 1194 [Docket No. 2000-01] RIN 3014-AA25. www.access-board.gov/sec508/508standards.htm.

Brinck, T., and Hofer, E. (2002). Automatically Evaluating the Usability of Web Sites, CHI 2002 Workshop, Conference on Human Factors in Computing Systems. www.usabilityfirst.com/auto-evaluation.

CNN.com. (2004). *Travel sites to be more accessible to the blind.* Retrieved from CNN on August 19, 2004. www.cnn.com/2004/TECH/08/19/Website.accessibility.ap/index.html.

Coyne, K. P., and Nielsen, J. (2001). *Beyond ALT Text: Making the Web Easy to Use for Users with Disabilities,* Nielsen Norman Group. www.Nngroup.com/reports/accessibility.

Lighthouse International. (1994). *The Lighthouse National Survey on Vision Loss: The Experience, Attitudes and Knowledge of Middle-Aged and Older Americans.* Retrieved on February 29, 2004. www.lighthouse.org/pubs_lhsurvey_purpose.htm.

Slatin, J. M., and Rush, S. (2003). *Maximum Accessibility: Making Your Web Site More Usable for Everyone.* Boston: Addison-Wesley.

Thatcher, J., Bohman, P., Burks, M., Henry, S. L., Regan, B., Swierenga, S., Urban, M. D., and Waddell, C. D. (2002). *Constructing Accessible Web Sites.* Birmingham, U.K.: glasshaus.

Theofanos, M. F., and Redish, J. (2003). Guidelines for Accessible—and Usable—Web Sites: Observing Users Who Work With Screenreaders. Interactions, Vol. X, No. 6, pp. 38–51.

U.S. Census Bureau. (1997). *Disabilities Affect One-Fifth of All Americans,* Census Brief, CENBR/97-5. www.census.gov/prod/3/97pubs/cenbr975.pdf.

U.S. Census Bureau. (2001). *Americans With Disabilities: Household Economic Studies,* Current Population Reports, pp. 70–73. www.census.gov/hhes/www/disable/sipp/disable97.html.

U.S. Census Bureau. (2003). *Disability Status: 2000,* Census 2000 Brief, C2KBR-17. www.census.gov/hhes/www/disable/disabstat2k.html.

U.S. General Services Administration (GSA). (1998). *Section 508.* Retricved on February 18, 2005. www.section508.gov/.

Vanderheiden, G. C. (1992). *Making Software More Accessible for People with Disabilities,* White Paper from University of Wisconsin-Madison Trace R&D Center. codi.buffalo.edu/archives/computing/.making/.

W3Ca. (1999). *Web Content Accessibility Guidelines 1.0,* W3C Recommendation. www.w3.org/TR/WCAG10/.

W3Cb. (2003). *Web Content Accessibility Guidelines 2.0,* W3C Working Draft. www.w3.org/TR/2003/WD-WCAG20-20030624/.

W3 Schools. (2003). *Browser Statistics.* Retrieved on December 17, 2003. www.w3schools.com/browsers/browsers_stats.asp.

14 CHAPTER Ethnography for Software Development

Anne Kirah MSN/Microsoft Corporation
Carolyn Fuson MSN/Microsoft Corporation
Jonathan Grudin Microsoft Research
Evan Feldman Microsoft Corporation

Ethnographers have contributed to industry as organizational consultants and in mediating labor disputes for the better part of a century. In addition, they have contributed to the design of commercial products since well before the software era. Companies and industries as varied as General Motors Corp (Kane, 1996), Kimberly-Clark (Feldman, 1999), Nokia (Lindholm *et al.*, 2003), and Motorola (Kupfer, 2000) all have long seen the value of ethnographic research in product design. Creating products that resonate with the buying public, that fill the wants and needs of consumers and that create loyal returning customers is the aspiration of commerce. It rarely happens serendipitously. Investing in research designed to identify the customer's core values, cares, and desires and to establish a deep understanding of how the product or service will be incorporated into the consumer's life is time and money well spent, indeed, it is invaluable. The business principle is a simple one: to create a product or service that consumers love, you must first understand those customers on the most fundamental level. Find or create a need and then offer the solution that touches on key fundamental values.

The first use of this principle for software was when Xerox Palo Alto Research Center (PARC) hired Eleanor Wynn in 1976. PARC was actively engaged in developing software for their leading-edge personal computing systems that foreshadowed the IBM PC and Apple Macintosh. Wynn's studies of secretarial work contributed to the design of early Xerox information systems, which in turn influenced Apple and eventually Microsoft Windows and Office products. Wynn hired Lucy Suchman, who went on to form an influential ethnography research group at PARC. Their early publications include Wynn (1979), Suchman (1983), Suchman and Wynn (1984).

Around 1980, the cognitive anthropologist Edwin Hutchins shifted his focus to computer-based training, and from there to software and hardware design. Working at the Naval Personnel Research and Development Center and University of California, San Diego, he contributed to the ground-breaking Artificial Intelligence STEAMER project (Hollan *et al.*, 1984) and to an influential analysis of direct manipulation interfaces (Hutchins *et al.*, 1986).

In the mid-1980s several computer companies hired ethnographers and other qualitative field researchers. The cultural anthropologist Constance Perin studied software use among other topics, working with faculty at MIT and consulting to industry (Perin, 1989). A research and development team at Digital Equipment Corporation (DEC) led by John Whiteside focused on integrating ethnographic techniques into design (Whiteside *et al.*, 1988).

The biennial ACM-sponsored Computer Supported Cooperative Work conferences (held since 1986), a European CSCW series (initiated in 1989), the Participatory Design Conferences (held since 1990), and the *CSCW* journal (first issued in 1992), have published scores of ethnographic studies (e.g., Bowers *et al.*, 1995). The challenges of integrating ethnographic research with design and development practice has also been a major topic, and the focus of at least a dozen papers (e.g., Hughes *et al.*, 1992). Some of these have focused on preserving and applying traditional ethnographic approaches. Others, such as the Contextual Design method that emerged from the initial work at DEC (Beyer and Holtzblatt, 1998; Holtzblatt and Jones, 1993) have adapted them.

Today, ethnographic studies of software use are by no means routine, but they are used by a growing number of software and design consulting companies. Articles centered on anthropologists in industry appear regularly in the popular media. Several consulting companies specialize in such work. Intel has a well-known ethnographic research group, and Microsoft has for several years hired ethnographers to work with a wide range of product groups.

The ethnographer's goal when working for a software company is to experience the world of technology from the people's perspective instead of the perspective of the software company. Ethnographers observe people in their own environment, where the activities the participants choose to do have meaning and have a direct impact on their daily lives. The translation of this experience and application of this learning into product design and development so that the products and features will be meaningful and appeal to "real people" is the key. In essence, ethnographers bring the voice of the customer into product development.

The goal of this chapter is to introduce how ethnography can be used in the product development cycles of software and internet services and to illustrate the tremendous value of doing such in-depth consumer research at the outset, and to a lesser extent periodically, throughout the software development cycle. Knowing what to design for, what core values, wants, and needs can be addressed by the product, is the heart of design. Understanding what consumer need or want the product will address and how the product will be used in the customer's personal environment from the beginning reduces the high cost risk of creating a product that does not resonate or succeed. This "needs based" knowledge and innovation is a vital approach to growing a business.

This deep research is an investment at the beginning of the process that can save both capital cost and time cost further down the development cycle. Ethnography can help focus the team initiatives, drive innovation, feature design and prioritization, and ultimately be a key foundation for customer response. In theory, this customer insight should help prevent working on the wrong innovations, and help focus on those that are commercially viable because they meet the customer need or want.

Ethnography, particularly if done in multiple countries needing localization work, can cost anywhere from a few thousand dollars and the ethnographer's time to more than $200,000 per study, but releasing a product that does not meet consumer needs can be fatal to the company. The return on investment (ROI) of ethnography will vary depending on the length and depth of the study and how successfully the findings are included in the product.

Some of the variables affecting cost of ethnographic study are:

- Length of study
- Number of markets
- Travel expenses for the research team
- Need for localization
- Vendor support for logistics
- Equipment rental or ownership (video/audio)
- Video editing needs

Make no mistake—software design and development is an expensive, time-intensive endeavor, and, as with all consumer products, how well it is embraced by the buying public can have enormous impact on company profits (or losses). Fiscally conscious companies should use whatever research can be done to offset these high stakes, and ethnography is a powerful resource.

This chapter gives a quick summary of what ethnography is, what distinguishes ethnography from site visits, how ethnography can be used within the product development cycle, best practices, and three case studies from within Microsoft. The first case study covers the entire product development cycle of

Windows XP and how the work of an ethnographer was successfully used to impact the design, development, and refinement of Windows XP, how listening to and observing the consumer contributed to creating a product that "real people" would want to purchase and use. Within this case study, specific guidelines are given to optimize an ethnographic project. The second case study highlights the field trials of the Windows Tablet PC and how the data from field research was used to inform its design and development. The third case study shows how ethnography was used to inform the strategy and vision of Microsoft's MSN 8 Internet software.

14.1 A BRIEF DESCRIPTION OF ETHNOGRAPHY

Ethnography is a form of qualitative research that is done in a natural setting. It contains methods developed and applied specifically for obtaining information about what people actually say and do, and is based on the traditional methods from the field of cultural anthropology: the study of humans, their values and beliefs. The terms ethnographer and anthropologist are often used interchangeably, however, whereas an anthropologist is trained as an ethnographer, an ethnographer is not necessarily trained as an anthropologist. Ethnography is a discipline and one method used by the cultural anthropologist to understand how humans interact with the world around them, with objects, beliefs, and values. A trained anthropologist has a wider knowledge base of human behavior and can employ many methods, including ethnography, in his or her research. Participant observation and interviews are core ethnographic methodology and are best described as participating in and observing as much as possible the daily lives of the individuals who are being observed.

A crucial part of the ethnographic data collection is learning the skill of understanding the "native" point of view without imposing one's own ideas, frame of reference, or conceptual framework on top of the participant's point of view. All observations and statements are taken from the participant's point of view. In software development, this equates to taking off the "hat" of the software company and putting oneself in the place of the people who will be using the company's products.

Fieldwork is both a descriptive and interpretive work. The ethnographer, however, walks a delicate balance between describing behavior and ascribing meaning to the behavior. It is up to the ethnographer to describe behavior and to find out what meaning the participants place in these behaviors. Then and only then must the ethnographer find a way to apply this information to the design and development cycle of a product or use it for the purpose of marketing and strategy. Bridging this gap between understanding human behavior and translating this knowledge into product development or marketing strategy and tactics is the ultimate goal.

Ethnographers not only focus on a particular subject of interest, they focus on the situational context of the study participant in time, location, space, emotion, and so on. This involves lengthy visits with participants and, when time permits, multiple visits to the same households to observe behaviors over a longer range of time. This accounts for the large quantities of data that emerge from ethnographic fieldwork, data that must be organized and analyzed in relation to the whole experience, and related back to the business purpose and question.

While focusing on the participant's perspective, the ethnographer is looking not only for patterns of behavior within the individual but also across individuals. The ethnographer is also looking for behaviors where patterns may not be readily apparent.

The information that is gathered comes from two domains of reality: reality formed by the notions or ideas people hold and reality formed by what people actually do (Holy and Stuchlik, 1983). Data are collected on what people say they do in the form of verbal statements obtained by asking questions, through survey work or by just listening as people describe what they are doing. Data are also collected on observable behavior by watching the participant in the course of a day, noting how the participant would term the day: either typical or unusual. Observable data consist of all nonverbal behaviors that come up during the sessions (Holy and Stuchlik, 1983).

The distinction between what individuals say they do and what they actually do is often the foundation for ethnographic work. When there is no difference found between verbal statements and observable behavior, less analysis is needed and a descriptive process of the observations, identified values, patterns, trends, and so on is easily written out. When there is a difference (which, over time, ethnographers have found to be in almost every case), the ethnographer's job is to find out why there is a difference and then to determine what takes precedence, the perception or the actual behavior, to what degree, and when.

Because ethnography entails data collection in a natural environment (be it work, home, or on the go), the ability to be flexible and adapt to new situations is a key to successful data collection. Ethnographers use a combination of tools depending on the context of the study. Some examples of tools include: note taking, tape recording, video recording, camera shots, artifacts taken from the participant, and collages created by the participants. In addition, a variety of questioning methods can occur—anywhere from structured interviews with both scripted and open questions to unstructured interviews that merely reflect observations or comments the participant makes.

The process of analyzing the data begins the moment the ethnographer finishes his or her first visit and continues through the draft of the final report and presentation of the data and conclusions. Ethnographers read and re-read copious amounts of notes. These notes, transcripts, artifacts, photographs and film are then analyzed and coded for content. Some ethnographers use qualitative software like Atlas/ti and Nudist to code and analyze their data; others do not. Within Microsoft, both methods have been used successfully and the ethnographer may choose the method he or she is most comfortable with.

14.2 DISTINCTION BETWEEN SITE VISITS AND ETHNOGRAPHY

A few years ago, there was a round of e-mail at Microsoft on the subject of ethnographers. The idea was put forth: Can't anyone do good site visits? Why do we need ethnographers? The answer to "Can't anyone do site visits?" is yes, *but* with what results? The answer to "Why do we need ethnographers?" is found in the specific area of their expertise. Ethnographers understand how to discern people's underlying motivations, belief, values, and behavioral triggers. One of our ethnographers responded with the following quip:

> A Microsoft researcher visits a man in his home. She wants to know how he uses his frying pan. The researcher asks the man a bunch of questions: "When did you buy this frying pan? Why did you buy a non-stick pan? Do you always spray your pan with PAM?" She asks for the man to tell her even more about the frying pan. First one hour, then two hours pass—the researcher is asking many, many questions about the frying pan. She totally missed the *really* interesting thing the man does with his blender.

The real question, however is when to use the skills of an ethnographer and when to use the skills of, for example, a usability engineer for site visits? The answer lies in an understanding of the differences between ethnography and site visits.

Ethnographic field research and site visits have many overlapping areas, and are, at times, indistinguishable. Both are observational in their nature; both strive to understand the user in a naturalistic setting, for example, in the context with how the user actually uses technology. It is arguable, however, that there are at least two important distinctions between them.

The first distinction is determined by whether the data collected is driven more by the end user (research methods primarily using participatory observation, longitudinal in nature and guided by the end user and not by the researcher) or by usually guided questions about specific products or features

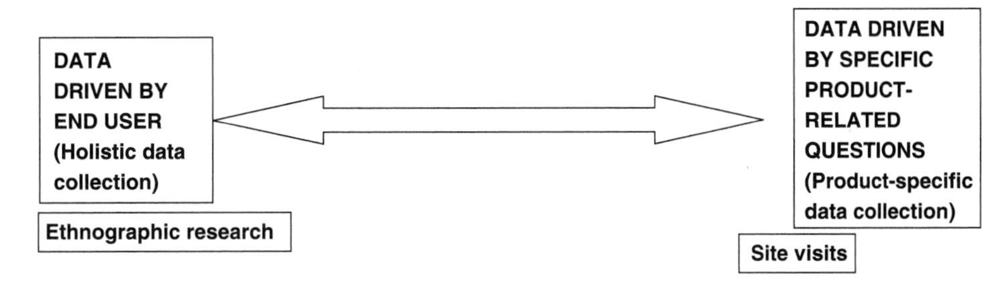

FIGURE 14.1 The continuum of research methods.

(site visit methods focusing on semistructured or structured interviews with an observational component).

Ethnographic research and site visits meet at the midpoint of the continuum of research methods. As ethnographic questions become more product driven and guided by the needs of the company, and as site visits become more about the user's experience and less about specific questions that need to be answered by the product/feature development team, the two become the same.

The second distinction comes from whether the focus of the research is to understand the holistic gestalt of the environment(s) in which software is used (and not used) or to change specific feature area. The latter can and should be done by anyone who has a strong knowledge of the feature area, whether or not that person is an ethnographer. The former should be done by an ethnographer trained in understanding the holistic nature of environments. In both cases, someone trained and skilled in interviewing should be leading this facet of the research.

When the user's point of view and the software company's point of view merge, product teams are more likely to create and design products that satisfy users and keep them loyal.

The goals of an ethnographer in the software industry are:

- To observe users from the user's point of view, not from the software company's
- To find out what users really want and need in the software products and not what the software company thinks they want and need

14.3 ETHNOGRAPHY AT DIFFERENT TIMES IN THE PRODUCT DEVELOPMENT CYCLE

As foundational in nature as ethnography is, it is tempting to relegate it solely to the outset of product innovation and development. To do so is to short change all parties. The big myth is that ethnography is only for product planning purposes. Although exploratory ethnographic research in the first phases of innovation and product development cycle is extremely important, this does not rule out the use of ethnography during the entire development process from product planning to public relations. Have we hit it right? How will the prototype be received? Has anything changed in the culture or environment that we need to adapt to? All of these questions need to be addressed throughout the product cycle of development and redevelopment.

14.3.1 Early in the Development Cycle: Near-Term Product Planning and Future Planning

At an early stage in the development cycle, exploratory research is an excellent way to capture the "here and now" of how technology influences our users' lives and what our users want and need. At this time, we can take the user point of view to create "observation-based maps" or scenarios of some realm of the world that is specifically interesting to our company. This could be something such as addressing the question: A user needs to write a term paper; what does this user need? Observing the steps and scenarios that a potential customer goes through to complete this task gives insight into product features, their relevance as seen through the users' eyes, and a prioritization of needs. Based on this insight, a solution can be built to meet and, ideally, surpass these needs. Exploratory research, including ethnography, can be used for:

- Strategy and vision
- Planning for next product or product release and future products and product releases
- Development and prioritization of scenarios and features
- Product evaluation

Questions that can be explored during the early stages of development are:

- How do people interact with technology?
- What are people doing?
- What might we build?
- What is going on with a specific phenomenon, such as online social networking?

- What is going on with a specific generation in the context of the business question?
- What are people doing with specific technologies?

During this phase, ethnography allows for innovation and new product ideas, answers about why features or products are successful or unsuccessful with any given scenario or culture. The key to success in this phase is the open-endedness of the inquiry, listening to and allowing the participants to express their usage patterns without any guidance from the research or product team.

14.3.2 Middle of the Product Development Cycle: Feature-Specific and Product-Specific Questions

Much of the work in the middle of the development cycle can be done by virtually any usability engineer, product planner, product manager, or developer working on specific features with a minimal amount of training, although some aptitude is needed. Of course the optimal situation would be for a team to go out with an ethnographer. Usability engineers and development teams often have more knowledge of how the features work and can therefore see successes and failures with a different perspective than the ethnographer, who generally has a broader understanding of the features (especially when talking about an operating system). However the ethnographer often has the skills to formulate the questions in a nonintrusive, nonleading way and will be skilled in understanding other factors, such as what core human value is the basis for a behavior, or what external variables are affecting the product usage and the research.

These visits are often shorter and focused on specific feature areas and need not be longitudinal in nature, although measurement of feature usage over time is valuable information and should not be discounted.

In the middle of a product development cycle, it is helpful to use ethnographic methods (unguided observation) combined with site visit methods (guided and unguided questions) to look at the following features:

- Implicit as well as explicit needs with relation to products and features. By doing *X* the user is showing a need of *Y*.
- Depth and understanding of quantitative data
- Systematic understanding of feature areas
- Scenario checks (refinement)

It is entirely possible to combine both ethnographic and usability methods during one visit.

14.3.3 Final Phases of the Product Development Cycle

During this part of the development cycle, field researchers can highlight specific features to participants or have a focus on a set of features. Again, this can be done by many different individuals within the product development team, not just ethnographers although is a task that is best done with at least one team member skilled in anthropological research techniques. Usability engineers and development teams often have more knowledge of how the features work and can therefore see successes and failures with a different perspective than the ethnographer who generally has a broader understanding of the feature or product. Bear in mind that because this is now near the final phase of the product development cycle, any changes that will make it into the soon-to-be-released product will be very minor ones. However, key areas on which to continue development, features that were left out by necessity of time and/or budget can all be noted. This is also the best time to run a close-to-complete product through a participant's life and environment, and an excellent time to note areas that will facilitate the marketing efforts for the product.

Beta testing (or field testing) is a unique opportunity to test the product in as close to a realistic setting as possible. While it takes a lot of planning, having a team run and observe a beta test in the field provides a deeper understanding of how a product will affect the lives of its users and may indicate whether the product will be successful. Real People, Real Data (Case Study 1) was the first program of its kind to study the Windows XP operating system in naturalistic settings during a beta test.

These beta visits are longitudinal in nature because it is important to see how individuals change their behavior over time when using the same (or an improved version of the) product or feature. Visits tend to be focused on specific feature areas with experimental designs built in to the visit. These designs may highlight a "next rev" feature or a close-to-final cosmetic look. It is important that the development teams have a chance to experience the real-world usage as the product is being tested so that they can see how it will be used outside the development buildings and by customers of varying technology skill sets and enthusiasms, patience levels, environments, and product needs.

An added benefit here is that a company's product support team can also train individuals on the new products and features in live settings. Although field visit teams should remain small in number for the obvious reasons of lessening the intrusion and forced artificiality of the situational use, having key product support personnel observe this field research can give them very good insight into what type of questions the support team is likely to encounter. This "heads up" will allow for training of the support staff and can be key in the writing of the help files of the product itself.

14.3.4 Both Marketing and Public Relations Are also Interested in Field Research

From a market research perspective, field research can uncover unexpected usage scenarios, and associated user benefits that had not been previously considered. Market researchers at Microsoft point out that some of these scenarios could have been missed if these qualitative methods had not been used. The outcome of the research has had direct impact on product positioning and messages in marketing materials.

Public relations (PR) benefits from this type of research as well. The result is having noncompany, "real" people talking about a product and positioning a product as it relates to their real needs in their real lives. These testimonials increase the credibility of a company's messages about its product, and often the participants will articulate a message in a way that is better received by the public than the enthusiasm of the company spokesperson. Microsoft uses these participant video clips and testimonials in internal product reviews, external industry conference presentations, and in its PR messaging as appropriate.

14.4 SOME BEST PRACTICES FOR FIELD RESEARCH

The key to the success of field research is not only interest from the teams, but involvement. Research has its greatest impact when it is embraced by all parties of the product team and the research managers are an involved part of the design process. When field research becomes a valued and trusted resource and is part of an ongoing collaborative process within the product team and any major players outside of the direct product team, it aids in the design of great products. Rather than make research a separate part of the design process, successful teams keep the research managers as core members of the entire building process. When time gets short and hard decisions must be made on which features to include and which to relegate to the next product revision, the field researcher can be one of the best advocates of the customer voice.

Including the product team in the data analysis is essential. Team ownership in the analysis makes it relatively easy to bridge the gap between field research and design. Including the entire team during this process allows many different perspectives that can influence the recommendations to be made from the analyses.

This data analysis is done through many different methods: from sticky notes with comments that need to be sorted, to video clips, to presentation of main observations and findings and a discussion about what they mean. The specific method of analysis is left to the discretion of the field researcher.

When bringing team members along on a site visit, it is important to prepare them ahead of time for what to expect from a day in the field. In addition, team members should be given specific roles and responsibilities when they are visiting users in the field. This not only helps train them in what ethnographers look for and need during the observation period; it keeps them involved in the process in a way that lets the participant be the most relaxed and natural. Preparing the team members and assigning roles allows the ethnographer to focus solely on his or her interactions with the participant instead of the team.

The ethnographer can assign people to:

- Take notes
- Help run the camera
- Take specific note of a particular environmental or social factor

For home visits, the rule of thumb that Microsoft uses is no more than four team members total, with only one member actually interacting with the participants; two to three is optimal. For office visits, the general rule of thumb is no more than two employees, as it can be very distracting to have too many people in a business setting.

Team members should follow a field visit protocol. The protocol describes how to collect as much information as possible with the least amount of contamination. Contamination occurs when company introduces its perspective or influences the participant in such a way that the participant no longer follows a natural path during the visit.

14.4.1 Protocols for Visits

Having a clear goal, plan, and protocol for the entire visit is crucial to the success of any one visit and to the overall success of the program. Following is a basic outline of a site visit protocol:

Previsit Planning

- Define the goal of the visit
- Decide who is going and why
- Prepare the team with a mini-class on field research and roles and expectations
- Choose a leader
- Define protocol for the visit
  - For instance, guard against the temptation to ask about a specific product or activity. If that product or activity is key to the participant's life, it will be mentioned in its appropriate time. Often what is *not* said is as important as what is.
- Prepare checklists for roles and responsibilities, equipment, protocol, what to look for, reminders of what to do or not to do.

Visits

- Plan and prebrief before each visit
  - Review general aims of the study and the governing principles of the approach (diary, visits, etc.)
- Follow protocol during the visit
- Check interpretations if a diary approach is a component of the research
  - The ethnographer may introduce selected incidents from the diary and ask the participant to talk through the event.
  - The ethnographer seeks to contextualize events noted in the diary and expand the details and understanding
- Plan a debriefing immediately after each visit that covers the major findings of the visit

Plan Meetings for Product Impact Before Going on Trip (Revise Afterward)

- Meetings to analyze the data
  - Debrief—the ideal when conducting multicountry research on the same product is to have all of the team finish the field work by all meeting in the same city (if teams have split up) and "camp out" in a place outside of the normal work environment. This can be a hotel or simply a different office building on the corporate campus, but the purpose is to have

everyone together to debrief and write up the research findings without other work distractions.

- ✦ Define key findings
- ✦ Patterns that trigger behavior
- ✦ Novel and/or unexpected triggers, needs, or wants
- ✦ Context surrounding behavior

- ✦ Define areas for further study or for possible next revision feature design
- ✦ Discuss implications
- ✦ Make strategic and tactical recommendations for product design based on the research findings.
- ✦ Presentations and reports based on the analysis
- ✦ Keep the researchers and ethnographers involved past the presentation phase—repeatedly include them in design, spec reviews, and so on. The research doesn't stop with one presentation—the ethnographer is a key resource available to the development team.

14.5 CASE STUDY 1: WINDOWS XP PROJECT: GUIDELINES FOR A SUCCESSFUL ETHNOGRAPHY—ANNE KIRAH

This first case study outlines the methods used for the first longitudinal field research project at Microsoft, Real People, Real Data. Real People, Real Data originated as a longitudinal field research study of Windows in the home both before and during the Beta testing phase of Windows XP. Forty families were visited in their homes from two to eight times to generate feedback that could not be obtained using traditional laboratory studies or other methods. During the study, these families virtually became a part of the product team and a subset were brought to Microsoft for participatory design studies as well. Development teams received timely feedback on Windows usability and functionality which was incorporated in subsequent beta tests of Windows XP and the final product. The feedback was also used for future Microsoft products.

The study was successful primarily because it was possible to study the same people over time using a combination of an ethnographic approach and a usability approach in the field (e.g., beta testing in the field). Because of its success, Microsoft continues to use ethnography throughout the development cycle to help bridge the gap between the target audience and the goals for the software.

The program was broken in to several phases. The following is an outline of the complete program including basic findings for each phase.

14.5.1 Phase 0: The Planning Phase

The goal of the planning phase was to create an infrastructure for the fieldwork. The final deliverable for this phase was a project plan with detailed milestones.

Defining the Team and Team Involvement

It was important during product planning to develop "buy in" from members of the development team, the product planning team, market research, the usability teams, product support, and public relations. Over the course of the project, different members from these teams were involved in the development of protocols for the visits, the actual visits, the analyzing of the data from the visits, and the generation of findings and recommendations from the visits.

The key to the success of this program was not only a developed interest from the teams just listed, but an empowered one. Through collaboration, it became a joint effort and succeeded in bringing product and research teams together to create user-centric end products. This inclusive teamwork resulted in a smoother integration of ethnographic findings, recommendations, and specific product features.

For success of initial planning, create a collaborative team with members from at least the following groups:

- Product planning
- Market research
- Usability
- Design
- Development
- Support
- QA

Project Plan

A project plan needed to be formalized with milestones for each phase. These milestones became place holders for later evaluation as planning and data changed the direction of the product.

Project plan phases included:

- Phase 0: planning phase
- Phase 1: recruitment of ethnographic participants
- Phase 2: exploratory research
- Phase 3: participatory design and conferences
- Phase 4: feature and topic specific research
- Phase 5: beta testing in the field

Feedback Model

It was important to integrate the feedback early on in the project. Because this was not done during the first phase, valuable information did not reach the right people in a timely manner. Key questions to answer when creating a feedback model are the following:

- Who needs to be informed about what, when, where, and how?
- Who are the primary stakeholders?
- How will the data be collected, edited and sent out to stakeholders?
- What is to be done with one-off data (data that may not be related to your target stakeholders but nonetheless is important information to send further)?
- What meetings should be held regularly?

Suggestions for Feedback

- Design a program status e-mail
- A status e-mail can contain anything from weekly or monthly project plans to actual reports on findings. At the very least, it should have appropriate links to reports. Often arranged by product feature, a status e-mail will give bulleted lists of the following:
  - Meetings that took place since the last status e-mail
  - Meetings planned in the immediate future
  - Schedule check or pulse
  - Feature development progress
  - Dependencies status checks
  - Red flags or issues needing immediate redress
  - Green flags noting issue resolution

  - Team or virtual team member changes
  - Links to internally posted reports and documents
- Design a newsletter that covers specific topics
  - A newsletter differs from a status e-mail in that its focus is to bring attention to a specific topic. It includes links to other information about the topic and links to the general project as well. Newsletters are distributed to a larger internal audience than the status e-mails, and are designed to keep the broader organization aware of the work in progress.
- Design an internal Web site with links to valuable information from the project
- Agree on report formats and where reports will be posted
- Have regular meetings with important stakeholders
- Create posters that are displayed with pictures and key topics discovered in the field
- Have miniconferences to evangelize the topics that are uncovered during field visits

Project Coordinator

With 40 families to keep track of over time with individual wants and needs, it became obvious that a project coordinator or project manager was needed who could be in charge of customer contacts, the infrastructure of the project, and milestones in the project plan. Do not underestimate the amount of work a project coordinator has; this quickly became a full time position for managing the 40 families and coordinating the visits.

Responsibilities include:

- Project tracking
- Customer contact (i.e., all issues that arise from individuals, scheduling individuals for visits, determining which Microsoft personnel will attend visits)
- Trip planning (e.g., organizing trips: where people will go, who will go, when individuals visits will be held, making sure checklists and protocols are in place)
- Maintaining the project Web site and tracking information as it comes and goes from the group
- Keeping track of legal documents like consent forms and nondisclosure forms.

Collateral Planning

Gifts to participants was a key source of goodwill during the many trying visits that were encountered (especially when technology in beta testing phases didn't work as expected). For a year-long study, it is reasonable to offer a computer, a printer or other peripheral devices depending on the particular needs of a family, company, or individual. Other possibilities are cash or gift certificates given at each visit. It is also wise to keep a stash of *tchotchkes* (small gifts in the form of balls, watches, cups, etc. that have the corporate logo on them). A little tchotchke goes a long way to promote goodwill.

14.5.2 Phase 1: Recruitment of Participants

Screeners

When working with marketing researchers, product planning teams, and usability engineers, it was important to create screeners that provided general data about the population of interest as well as specific data for selecting the people who participated in the study.

Key issues for screeners are as follows:

- Collect data with the screener that can be used not only to find the right participants but also to learn more about a sample of the population of users
- Ask pertinent questions with the screener that will determine the estimated gratuities the participant would need before telling the potential participant of the study, or the gratuities involved in the study. This allows you to avoid participants who want to participate in the study solely for the purpose of obtaining desirable equipment or gratuity.

Recruitment of Consumer or Home Users

The focus for the Windows XP project was on the family as a unit as well as individuals within a family. When planning this project, the kinds of participants chosen came directly from the targets provided by market researchers, product planners and members of the product teams and feature teams. The profile for the study reflected the home user as defined by the needs of the product and feature teams.

Some variables to think about during the selection process:

- Target populations for the product or feature
- Rural and metropolitan representation; there is a belief that differences in computer usage exist between rural and metropolitan computer users
- Generational, socioeconomic, and cultural representation reflecting census and demographic information for each targeted location and marketing data
- Willingness of the participant to fully engage in the study
- Level of expertise of in the subject area; for software, a balance of "tech savvy-ness" is needed

Double Screening: Telephone Calls and In-home Interviews. Using both telephone screeners as well as in-home interviews before the final choice for participants in a family study is recommended. This will allow for a more complete evaluation of how well the participant meets the selection criteria and what level of feedback can be anticipated. In-home interviews will let the ethnographer get an initial sense of the environment and the social interaction within the participant family.

Screening Calls for In-home Interviews. In the initial study, 60 families were screened through phone calls to participate in an in-home 1- to 2-hour interview. Independent vendors specializing in focus groups and market research were used to find these 60 families. Vendors were given the prewritten screener questions, which they asked their pool of participants. The outcome was mediocre. They discovered that a number of families were really "professional" focus group attendees.

During the second round of recruitment, the research team used www.houseandhome.msn.com to find zip codes in a variety of cities based on demographic information supplied by the Web site about the neighborhoods in those cities. Next, an independent vendor was hired to purchase telephone numbers affiliated with the zip codes and make cold calls to the families. While this was a tedious process and there was a limitation to the number of people who were willing to participate in a phone screening, it was successful in getting participants who were "more real" than the families from market research vendors. A limitation to using a Web site for neighborhood demographic data is that these data are deepest for the US market, although there may be similar sites for international locations.

Interviewing Participants for Selection: In-home Interviews

All 60 families were then interviewed by Microsoft representatives before they were accepted into the program. Because potential families were found through cold calling, it was important that the safety of the Microsoft employees was kept in mind. It is highly recommended that you call the local police department to ask about specific addresses and safety so that you can avoid mistakes like one made in this study: The researchers ended up with an address to a crack house and realized that the neighborhood wasn't safe. The research team went to a local police station and was welcomed by the police, who explained the dangers at the address (a well-known crack house), that the team had been in a very dangerous neighborhood and should not under any circumstances return there. It was at the police department's suggestion that the team implemented the step of calling ahead to hear about safety issues.

Step 1: Cold Calls to Families Who Are Asked to Respond to a Small Screener. If the potential participant answers correctly on targeted questions, ask if they are willing to participate in an in-home interview (they will be paid $100 for a 60 minute session). Call local police to verify the safety of the neighborhood.

Step 2: At the In-home Interview Ask a Series of Useful Questions. In this case we wanted to give the participant a computer in the study that would be equivalent to something they would actually buy. Many questions were asked about interest in buying a computer, how much he or she would pay for one, what peripherals they were interested in, and so on. By doing this before potential participants gained knowledge of the study, the research team was able to capture realistic data around the participant's purchasing behaviors and interests (e.g., digital photography). The importance of doing this before the potential participants gained knowledge of the study cannot be underestimated. If you ask a question about interest level in a home interview, they are more likely to answer truthfully because they will have fewer assumptions of why you are asking. If, however, you ask them of their interest in cameras after they know about the study, they might answer affirmatively just to get a free camera!

Step 3: At the end of the home interview, let them know about the study and ask if they would be interested in participating. Let them know that they would be one of many candidates for the study that will be chosen by a team as a "best fit."

Step 4: Create an interview document that contains crucial information about each potential participant. This document will be used to select participants based on the best fit for the project, but will also be used throughout the project to provide background on each chosen participant when needed throughout the course of the study and beyond.

Step 5: The team decides among the candidates as to who will most likely provide data using the following criteria:

- Are members articulate; can they express their views in a manner that the meaning is clear to all team members?
- Are they warm and friendly?
- Would you want to visit them again?
- Do they want to share with you, or is it like pulling teeth?

14.5.3 Phase 2: Exploratory Studies

The salient feature of exploratory studies is that the data must be formed primarily from the participants' perspectives. Rather than the researcher coming with questions that need to be asked, the participants guide the researcher. The participants were visited by a Microsoft ethnographer accompanied by interested members from the team. The ethnographer was the same person each time and spent considerable amounts of time shadowing (following) the participants in their daily activities. There was minimal intrusion, a relaxing environment which allowed the ethnographer to build rapport and collect as accurate and naturalistic a picture as possible of the day in the life of the participant combined with usage of PC and mobile phones. Building rapport is a crucial aspect of ethnographic work, something that short site visits can begin to do but that is best accomplished over longer or repeat visits.

During the exploratory phase:

- Participants were only questioned directly about their current (in the here and now) behavior and behaviors associated with it.
- Ethnographers did not answer questions that might influence the participant in some way (e.g., technical solutions, purchasing decisions)
- Participants were not given any surveys (to preserve the natural behavior of the individuals).

The goal for the exploratory phase is to gather clean ethnographic data to give a background for subsequent longitudinal data from a natural setting about the following:

- Features that are frequently used and why
- Features that are not used and why
- Features that are easy to use and why
- Features that are difficult or frustrating to use and why
- Expectations participants have about computer usage and the overall computer experience
- Needs and desires of the users that relate to Microsoft products and development.

The results of the exploratory phase allowed Microsoft to learn new things that were not necessarily the focus of any given product or feature team. Results from these studies have been useful not only for Microsoft, but for partners as well.

Examples of non-focus areas that were acted on are as follows:

- Learning how difficult it is to connect to the Internet for the average family
- Hardware–software integration issues that affect the initial experience
- The importance of personalization and the feeling of extension of self with communication tools
- The importance of having "my" space on the family computer

14.5.4 Phase 3: Participatory Design Sessions

In addition to visits in the field, it was useful to bring some of the participants to Microsoft for an event. This gave others within the company who were interested in the same consumer target, but who could not get out in the field, the opportunity to meet with "real life" people. During the event, different teams within Microsoft were able to obtain data to validate product planning, to test out scenarios and features, and overall to *listen and learn* from the users. Simultaneously, this turned out to be a way to reward participants for their long-term commitment to the project, and allowed them the opportunity to see how their ideas were implemented into the product development process.

Topics for such participatory design events are as follows:

- Specific feature discussion
- Specific concerns out in the field
- Collages

- "Building a feature together" activity
- Brainstorming

14.5.5 Phase 4: Feature Specific Site Visits and Usability in the Field

After the exploratory and participatory data were collected, participants were selected for the lab studies in the field. This was useful for studying specific features and was not done in the beginning of the study to avoid tainting the environment of our participants with knowledge of Microsoft's wants and needs.

An example of this type of study is the introduction of a specific feature or product with a longitudinal follow-up of usage after the initial introduction. This is also the time to focus on specific features and probe about usage. These studies do not need to be run by ethnographers. In fact, it is arguable that they *should not* be run by ethnographers but by usability engineers and the product managers who are focusing on specific features. Feature specific visits allow for an intensive study of a feature in a more naturalistic setting than in the laboratory. It also saves on costs because these families are already recruited and have "bought in" to being questioned and probed by researchers.

During the Real People, Real Data study for Windows XP, participants were given both hardware products such as printers and cameras to use, and specific features were introduced that either related to products they were given or related to some task they aspired to learn but didn't know existed in the software they originally had.

14.5.6 Phase 5: Beta Testing in the Field

Beta testing in the field allowed the product development team to try out their Windows XP system with "real" people. These real people were able to provide unique feedback that was different and more realistic than our regular beta testers, who often are more like the developers of our products than the consumers we target. Long before release, the product development teams learned valuable information that helped them improve the product and the accompanying user experience.

The participants were observed using their computers with a beta version of software and questioned about their usage with structured interviews and other methods (the following list talks about several methods in addition to interviews):

- All participants were asked to perform an abridged version of tasks in the presence of Microsoft representatives (usability engineers and product managers assigned to the feature area).
- Problems arising from their Windows usage were addressed at the participants' homes or by e-mail via contact with the Beta Technical Support Team. This team included members of Microsoft's public technical support team and the actual developers who were building the Windows XP system. The team was able to follow up directly and address usability issues and bugs in the software as they were discovered. Watching a real user struggle with a product can be a humbling experience for developers who believe they have created an easy feature. This learning went far in helping our development team become advocates for the consumer, not only for Windows XP, but for future products as well.
- Participants were sent surveys frequently concerning usability issues, and experience with reported bugs for both pre- and postbeta installations.
- In addition to one set of repeated measures on each survey, a set of questions related to current issues from the development teams, usability teams, and beta teams were asked.
- Through random sampling, participants were selected to send one-week diaries. This allowed us to obtain an accurate portrayal of usage issues while limiting the amount of time each family needed to spend filling out paperwork.
- All of the participants were given technology relevant to key product features based on the screener (about current and planned usage and aspirations around computer usage).
- Instrumentation was a requirement to join the study.

The key to a successful Beta study in the field is the communication infrastructure.

For the participants, prepare simple instructions to follow when crashes, bugs, and design changes need to be relayed back to the company.

- Have a card with red, yellow, and green ways of contacting technical support (based on the critical nature of the issue)
  - *Red* is only for complete failures
  - *Yellow* is for failures that are annoying but not critical
  - *Green* is for issues that aren't critical but can be brought to the product team's attention
- Remind participants that no issue is too small or uninteresting to Microsoft

For the team, weekly meetings and daily contact with technical support are crucial. Keep communication open about participants, issues, and rules for closure.

- Keep communication open with the participants:
  - Daily or weekly calls about issues are important (sometimes participants are afraid to let you know about some issue, or they think the issue isn't really important; a phone call is of great help)
  - Around-the-clock help for essential products (to the participant)
  - E-mail icon (shortcut) on desktop for unimportant issues (in the eyes of the participants) that come to mind but aren't worth a call to support (instead, participants send an automatic e-mail)
- Feedback process must be in place ahead of time with clear roles:
  - Who is responsible for filing bugs?
  - Who is responsible for follow up with the participants?
  - Who is responsible for follow up with the feature and product teams?
  - Who is keeping track of all issues? This is a great indicator of success of the program.

Real People, Real Data allowed for the goals of the product or feature to surface both in general and in terms of specific new features and real scenarios based on real people. This was in large part a result of the efforts of a team working together across groups, all of whom had the opportunity to be out in the field with potential customers at different times during the product or feature cycle.

14.6 CASE STUDY 2: TABLET PC BETA TESTING IN THE FIELD—EVAN FELDMAN

Most products are submitted to a version of trial by fire in that they are constructed and tested before release, but it's ultimately the real usage in the marketplace that determines whether the product actually is useful, usable, satisfying to use, and meets the needs of the user. When the Tablet PC team was formed at Microsoft, the main role of the user research team was to make sure the company built a product that people could use in their work environment in a way that provided real value to the user; the goal was not to build another Apple Newton or Windows for Pen Computing—products designed around pen input but not widely embraced by users. From the beginning the approach was to build prototypes of both the hardware and the software that could be given to a user and to study how that device fit into the user's environment. When the first version of the Tablet PC was shipped, the user research team had completed three field trials of the Tablet and was in the midst of a fourth trial.

Testing new products with users can be difficult, particularly when the product's value is a longer-term proposition. The product team had multiple questions about the ideal interface and hardware needs for a tablet computer and many of these questions couldn't be answered well in traditional laboratory testing, observational interviews, or contextual inquiry techniques. This isn't to say that these other methodologies weren't being used, but rather that the team needed more hands-on data from the end users to really help understand what was needed.

Before the first field trial, the team was not sure if the Tablet PC should be the user's primary computer, a secondary computer, or a companion computer (e.g., a Pocket PC type of device) nor was there consensus on whether the interface should be Windows or a new interface completely. Many of these ideas were tested in the usability labs and during a series of contextual interviews with mixed results. The results from these studies appeared to indicate a preference for a light-weight, dedicated note-taking device that was completely pen centric. Users wanted a device they could take notes on and then later use with their main PC as a companion computer. However, the current economics of the market didn't appear to make this type of device practical in that it would be extremely expensive to buy given the desired functionality. Yet during this process the team did get hints that the users didn't necessarily understand how to "integrate" this device into their environment, and as such their expressed desires and needs might not be representative of their real desires and needs.

14.6.1 Field Trial 1

The first trial was not an originally scheduled study. In fact the second field trial was really the main trial that the team was geared up for, but an opportunity came along that proved too good to resist. The Tablet PC team acquired a product early in its development. This product was a Windows CE based tablet that had a custom interface built on top of the operating system. This tablet had a very limited set of abilities. It could operate as a personal information manager (PIM), an e-mail client, or note-taking system. This product fit what users had expressed as the ideal device and thus gave the team the opportunity to have users try the device for an extended period of time.

For this trial Microsoft had 19 users try the device for a period of 1 to 2 weeks so that it could be determined whether this product actually fit into the users' environment. The results from this field trial were dramatic in that the longer-term usage in this case provided users with greater insights into what they really wanted out of the device, which was in sharp contrast to the earlier findings.

Users indicated that they wanted a device for note taking, yet when having such a device for more than a week, many of the users indicated that because they had the device with them they realized that they wanted to do *X* task with *X* being a different task for every user; thus, users really didn't want a dedicated note-taking device, but rather a general computing device with an interface skewed towards note taking. The team also learned a lot about the note-taking behavior with a computer and other particulars about the usage model that had not been captured previously. The level of insight into the usage was something that simply wasn't uncovered by any other testing we had performed. This trial also provided some great practice at performing a large trial and building the methodology and logistics support.

14.6.2 Field Trial 2

When the second field trial was over, Alex Loeb, the vice president for Tablet PC, likened this exercise to shipping version 1 of the product but not releasing it. The product team spent almost a year getting ready for this trial by building its own hardware prototypes and building nearly shipping-quality software.

This trial had 21 users who were going to use the Tablet PC as their main PC for 4 to 5 weeks. During this time, the research team would be actively following participants and interviewing them to get a sense of how the device fit into their environment. During the course of the week, researchers spent at least 2 to 3 hours with every user. For each observation of each user we had 2 or 3 team members watching the user use the computer.

This field trial was originally meant to be the "proof of concept," in that the team wanted to know if Microsoft could build the right software to really enable the user to have a pen-based computing interface that was usable, useful, and needed in the work environment. Given the scope of this trial, it was a very expensive endeavor for Microsoft, but the team and executive leadership agreed that we wanted to make sure that we got it right this time. The investment at this point in the process was large, but the cost of a failure to deliver a product accepted and embraced by the target retail market would be even higher.

By the time the fourth week of the field trial rolled around, it was becoming clear to the product team that, although the concept was a good one, the right product hadn't been built. In fact the team had missed the mark in several key areas that had not been uncovered in other testing that had been performed—note taking in our application was too laborious, the note-taking features were not appropriate, and we had deviated too far from some common Windows conventions. Had we not performed this trial, the Tablet would most likely have been released more than a year earlier than it finally was, but with a user model that would not have been accepted in the long term.

14.6.3 Field Trial 3

In the previous studies, the team learned a lot about the individual differences in usage and got good insights into the main areas of the product. However, the goal was to construct a different type of trial the third time around, because this time the focus was whether the changes made really got the team closer to the right product.

For this trial there were only seven participants who had used a Tablet PC as their main PC for 6 weeks. This did indeed give the confirmation that was looked for in that the changes made the product more usable, useful, and needed. However, it did uncover yet another hurdle: The learning curve for understanding how to use a tablet efficiently was still quite steep. The problem of an extended learning curve was again something that we had seen in the laboratory and in smaller field studies, but it was not clearly identified as a major issue until we performed this trial.

14.6.4 Field Trial 4

The fourth trial occurred when the product was mostly complete and was in the process of being finalized to ship to Microsoft partners. This trial was again different in that near-production-level hardware and software now existed and the team could now look at the longer-term effects of using a Tablet PC.

Here the team wanted to understand what processes would change or issues would come about with users who had months of time using a tablet. For this study eight users were given Tablet PCs that they could simply keep (Microsoft would never ask for them back) and in return agreed to be interviewed periodically over the next year. Members of the Tablet team spent about 9 months with these users watching and observing their behavior with the tablet and fed their results into the next version of the Tablet PC operating system. Again some of the things that were observed here were things that had not been captured in other studies and thus helped the understanding of how to build software to meet experience that may not be uncovered in short-term studies using other methods.

The value of field trials has been paramount to the Tablet PC team. The trials allowed the team to reset their plans and redo some of the key elements of software prior to releasing it to the world. The investment here allowed the team to truly focus on the user's needs and ship a product that really met those needs rather attempting to iterate in the market. In the end analysis the team was extremely committed and connected to the users and reset plans to insure that those needs were met.

14.7 CASE STUDY 3: MSN:FORMING THE PILLARS FOR A RELEASE OF OUR INTERNET SERVICES THROUGH EXPLORATORY RESEARCH—ANNE KIRAH

Field research allows Microsoft to build products that resonate with a broad base of consumers spanning the continuum of technology expertise. It allows researchers to follow customers over time and to understand usage patterns from a situational and holistic perspective. Anthropologists are trained in avoiding ethnocentrism (Microsoft-centric behavior in this case) and focus on user-driven scenarios as opposed to feature- or product-driven scenarios. During an exploratory phase of field research, the data collected for MSN were used to form the main pillars that features were built on for a recent release of its Internet service (MSN 8). The findings fell into four key areas: communication, safety and security, managing lives, and personalization.

14.7.1 Communication

While communication is an important need, researchers found that people wanted not only to be connected with others but also to have an emotional connection with those they cared about. In addition, we found that the Internet had become a transgeographical virtual neighborhood, playground, and community for all ages.

14.7.2 Safety and Security

We also found that people's need to make the experience safe mirrored the same security values that exist in the "real" (nonvirtual) world, consisting of wanting to protect families from intrusion, theft, harassment, pornography, inappropriate behaviors, information overload in the form of junk mail and SPAM, and other unpleasantries.

14.7.3 Managing Lives

Managing the lives of the members of a household was also important; people want to schedule, keep track of, find, and re-find important information sources.

14.7.4 Personalization

Additionally, it was found Web users wanted the experience to be about themselves, that they felt that the internet was truly an extension of self. They wanted their personality mapped through a sense of design, ease of use, and ability to change things based on mood. Being able to change the color of the screen, have a personal photo on the desktop, have a messenger icon that could be easily changed to reflect current mood were expressions that resonated with users. People consistently wanted to make their own choices and not have the software make the choices for them.

14.7.5 Development of Main Pillars

The field research data just listed combined with market research data yielded the four pillars on which MSN was developed:

1. Better security
2. Better communication
3. Better browsing
4. Better service

These pillars were used to define and refine designs of features throughout the product development cycle and included teams of product planners, usability engineers, program managers, and developers.

An example of a feature that bridged the gap between field research and design was monitoring. Parents repeatedly talked about wanting to know where their children were both in the real world, and on the internet. Parents fear the worst, yet want to allow their children the ability to safely make mistakes and learn from their mistakes, or just be in a safe and known environment. The problem was that many parental control features were too limiting to the children.

It wasn't uncommon for parents to say things like: "I just wish I could see where they were going when they go to the mall," or "I just want to see where they are going on line to make sure it's OK." The idea of a monitoring tool came from this field research data and was tested out in focus groups and online surveys. Once the product development team developed the feature, the researchers were able to test the monitoring tool during a beta phase. The team found that it was not only useful to the parents, but also to the children. Both parents and children used the information obtained from the data to negotiate not only time spent on the Internet but also allowances, ability to go out with friends, and so forth.

SUMMARY

Ethnography has impacted product development by creating observation-based insights into the technology world from a user-centric perspective, it adds depth and focus to questions left unanswered by quantitative data; it provides an underpinning for strategy and vision, and makes possible the development, prioritization and refinement of scenarios and features.

Ethnographic research is an investment that may seem expensive, but that can ultimately save both capital cost and time cost further down the development cycle. It can help focus the team initiatives, drive innovation, feature design and prioritization. In theory, this customer insight should help prevent a team from working on the wrong innovations, and help focus on those that will create commercially viable products that meet the customer need or want.

Ethnographers are, however, only one part of a larger team of developers, designers, usability engineers, and the others in the product development cycle. Meaning, for the development process of our software, comes when we combine our knowledge and data to create the best possible software for our users.

REFERENCES

Beyer, H., and Holtzblatt, K. (1998). *Contextual Design.* San Francisco: Morgan Kaufmann.

Bowers, J., Button, G., and Sharrock, W. (1995). Workflow from within and without: Technology and cooperative work on the print industry shop floor. *Proc. ECSCW'95*, 51–66.

Feldman (1999). Contextual Analysis: Panacea or Problem? *Pdma Visions Magazine.*

Hollan, J., Hutchins, E., and Weitzman, L. (1984). STEAMER: An interactive inspectable simulation-based training system. *AI Magazine,* 2 (5), 15–27.

Holtzblatt, K., and Jones, S. (1993). Contextual Inquiry: A participatory technique for system design. In A. Namioka and D. Schuler (Eds.), *Participatory Design: Principles and practice* (pp. 177–210). Erlbaum.

Holy, S., and Stuchlik, M. (1983). Actions, Norms and Representations: Foundations of anthropological inquiry. Cambridge: Cambridge University Press.

Hughes, J. A., Randall, D., and Shapiro, D. (1992). From ethnographic record to system design: Some experiences from the field. *CSCW,* 3 (1), 123–141.

Hutchins, E. L., Hollan, J. D., and Norman, D. A. (1986). Direct manipulation interfaces. In D. A. Norman and S. W. Draper (Eds.), *User centered system design.* Hillsdale: Erlbaum.

Kane, K. (1996). Get me Margaret Mead! The biggest names in business—GM, Intel, Nynex—enlist anthropologists to decode the rituals of corporate life. *Fast Company,* 5, 60.

Kupfer, P. (2000). Designing Products Based on Real Life: High-tech firms seek clues in anthropology. *San Francisco Chronicle* January 31, 2000.

Lindholm, C., Keinonen, T., and Kiljander, H. (2003). *Mobile Usability: How Nokia Changed the Face of the Mobile Phone.* McGraw-Hill.

Perin, C. (1989). Electronic social fields in bureaucracies. Presentation and paper for American Anthropological Association session on Egalitarian ideologies and class contradictions in American Society. Revised version in *Communications of the ACM,* 12 (34), 74–82.

Suchman, L. (1983). Office procedures as practical action: Models of work and system design. *ACM transactions on office information systems,* 1, 320–328.

Suchman, L., and Wynn, E. (1984). Procedures and problems in the office. *Office: Technology and People,* 2 (2), 133–154.

Whiteside, J., Bennett, J., and Holtzblatt, K. (1988). Usability engineering: Our experience and evolution. In M. Helander (Ed.), *Handbook of human-computer interaction,* 791–817. North-Holland.

Wynn, E. (1979). Office conversation as an information medium. Ph.D. dissertation. University of California, Berkeley.

15 CHAPTER

Out of the Box: Approaches to Good Initial Interface Designs

Douglas J. Gillan Department of Psychology, New Mexico State University
Merrill V. Sapp Department of Psychology, New Mexico State University

User interface (UI) design combines elements of cognitive science (e.g., Card *et al.*, 1983), usability engineering (e.g., Rosson and Carroll, 2002), software development (e.g., Spolsky, 2001), and artistic design (e.g., Ambach and Repenning, 1996). In recent years, the importance for UI designers to understand business principles has also become apparent (e.g., Bias and Mayhew, 1994; Donoghue, 2002). Although any given interface designer or design team will have a mix of strengths and weaknesses in these various disciplines, UI design is enriched by knowledge in multiple disciplines. At the minimum, all the designers on a team should be able to communicate with their teammates and be able to understand the others' perspectives. (See the discussion on social return on investment [ROI] in Wilson and Rosenbaum's Chapter 8.) On the other hand, no one designer is likely to be expert in all the disciplines. UI designers should recognize their own strengths and try to maximize their contribution by playing to those strengths. For example, cognitive psychologists who go to work in industry should maintain their strengths in understanding the users' cognitive abilities and limitations rather than adopting a strictly-business approach to interface design. For example, when a cognitive psychologist working as a usability engineer focuses *only* on ROI, the user loses a critical advocate in the design process. However, usability engineers who can communicate the impact that the application of cognitive principles in UI design has both for users *and* for the bottom line will be the most effective advocate.

We begin this chapter with the acknowledgment that our strengths lie in the areas of human perception and cognition and the application of that knowledge to the interactions between humans and technology. Our approach to UI design (and to this chapter) is unflinchingly in support of the user; when software development requirements or business needs necessarily trump the user in the design process we may understand and acquiesce, but not without first having worked as hard as possible for our beliefs that computing systems, whether application software, operating systems, or the Web, ultimately will be used by humans and must be easy for humans to use. So, for this chapter, we discuss cost effectiveness, but our focus will be on how to produce the best UI and Web site for the various types of users.

15.1 GETTING IT RIGHT OUT OF THE BOX

Human factors practitioners commonly prescribe an iterative design process (e.g., Boehm, 1989; Gould and Lewis, 1985). Related to that prescription is the belief that "You can't get a perfect interface right out of the box." However great the need for iterative design, designers face limits on the time and money available for iterations. Accordingly, the better the initial design, the better the final product is likely to be. In addition, the cost of the design process is typically based on labor costs for the interface designers and developers. Accordingly, the fewer the iterations needed to reach a final design, the lower the cost.

So how can a designer improve the chances of producing a good interface with the first design? In this chapter, we suggest four different approaches to the initial design, any of which may improve the likelihood of a good design. Note that the use of any of these approaches is predicated on the prior development of a set of well-defined requirements. Also, keep in mind that we are not limiting the interface design process in any way—the design concept may be implemented in any form, from a paper prototype to fully functional software. The four approaches on which we focus this chapter are 1) creative design, 2) design by analogy, 3) rational design, and 4) design based on knowledge of perceptual and cognitive principles. Although this list probably doesn't exhaust the possible approaches to producing a good initial design, it includes all of the approaches that we have observed designers use to produce an initial design.

As psychologists, in addition to trying to understand and advocate for the user, we also, in this chapter, try to understand the psychology of the designer. Thus, our discussions of creative design, design by analogy, and rational design address the psychological processes of creativity, analogical thinking, and rational thinking, respectively, to identify ways in which to improve the process of designing the initial interface. Similarly, our discussion of the role of knowledge of perceptual and cognitive principles in UI design does not attempt to enumerate a list of those principles; a chapter does not provide sufficient space for anything approaching a complete list. We leave such lists to sets of design guidelines (e.g., Mayhew, 1992; National Cancer Institute, 2004). Rather, we discuss important issues that any designer should consider when applying these principles to the problem of initial design.

15.2 CREATIVE DESIGN

We are aware that the term "designer" may evoke some degree of controversy about who really counts as a designer. Rather than diverting from the focus of this chapter by joining that controversy, we will simply define a designer as anyone centrally involved in any aspect of UI design—from a human factors expert drawing paper-and-pencil prototypes, to a graphic designer creating high-quality graphics, to a software developer making decisions about the implementation of the widgets. Many designers enter their discipline because they enjoy the creative aspect of design. Creativity is commonly divided into two categories: artistic creativity (i.e., novelty for purely aesthetic concern) and problem solving (i.e., novelty as a solution for a specific problem). The process of interface design seems to couple these two types of creativity. Designers are motivated by the opportunity to create something unique, but must solve the problem of making usable products for human interaction within technological, political, and cost constraints.

Cropley (2001) proposed that a major part of what motivates creative people is a challenge to overcome or a problem to solve. Similarly, the Gestalt psychologists suggested that creative people can be highly motivated by an incomplete or deficient gestalt. A flawed gestalt has a "dynamic gap" and a creative person can restructure the stimulus environment to develop more complete or "whole" gestalts, which can result in novel solutions to problems. Many people have had the experience of working on a problem (the problem in its unsolved state would have the dynamic gap); then, with a sudden flash of insight, the solution may reveal itself (underlying the sudden insight is the restructuring of the stimulus environment).

Guilford (1959) proposed four stages of the creative process:

1. Recognition that a problem exists
2. Production of a variety of relevant ideas
3. Evaluation of the various possibilities produced
4. Drawing appropriate conclusions that lead to the solution of the problem

Guilford's four stages resemble the Gould and Lewis (1985) approach to interface design, with its stages of analysis, design with diagnostic testing, and evaluative testing.

Guilford's description of the four stages of the creative process can be applied more specifically to the design process. First, an interface designer or design team becomes aware of an existing problem, for example, a client presents a problem or a particular project requires a design. The designer then produces a variety of possible solutions to the problem (i.e., designs). Typically, reaching the design solutions involves many arbitrary decisions from a vast array of choices. For example, where a particular icon should be placed on the display to be most noticeable by the majority of users. The arbitrariness could be reduced by Gould and Lewis' iterative design process, but for the initial design, the designer must evaluate those decisions. Typically, the more principled the evaluation (e.g., based on knowledge of human cognition and perception—e.g., American users tend to read displays in a left-to-right, top-to-bottom fashion), the greater the reduction in the arbitrariness of the decision process. The ability to make a principled evaluation should vastly improve interface success rates. Finally, drawing the right conclusions from the evaluation should also involve a systematic, principled approach. For example, a designer might trade off the estimated costs of a decision (e.g., placing an icon in a certain position prevents the designer from placing a different, possibly more critical object in that position) against its benefits to the user. In most cases, designers will need to acquire their principles of evaluation and conclusion through explicit training.

In addition to the four stages of creativity that Guilford proposed, the design process may include a stage in which the designer has to sell the design. Designers must be able to communicate the quality of their design to other design and product team members, to management, and, ultimately, to customers and users. Accordingly, in addition to design skills and knowledge of relevant principles, designers should acquire communications skills.

15.2.1 Preparation and Information

As the previous discussion of the stages of creative design suggests, one of the necessary preconditions for creativity is knowledge of interface design and the specific domain area of interest. In terms of the Gestalt approach to problem solving (e.g., Köhler, 1925), extensive knowledge of a field is necessary to recognize incomplete gestalts and to fill the gaps. The trick for the designer is to have that knowledge but not be trapped by it. In other words, for creative design, the designer must have the knowledge, but be able to restructure it.

The ability to restructure knowledge can be difficult if a designer has been indoctrinated into a way of thinking or approaching problems. According to Gestalt psychologists (Mayer, 1983), problem solving depends on a structural understanding of the problem. Each part of a problem has a relation to the other parts and to the problem as a whole. Structural understanding is the cognitive representation of the parts and how they might be organized in the solution. The key is to be able to reorganize the elements of the problem. If designers can avoid being mentally stuck by the current structure of a problem and can reorganize the parts in a variety of ways, they are more likely to find a good solution. A classic example of "functional fixedness," the inability to devise a solution because of established ways of structuring problems, is Duncker's (1945) box problem. Duncker gave one group of participants a book of matches, a box of thumbtacks, and a candle. The second group received the same items, except that participants in this group were given the box and the thumbtacks as separate items. The participants' task was to mount a candle on a wall to be used as a lamp. For some participants, the matchboxes were filled with other items, whereas for other participants, the boxes were empty.

The solution to the problem was to secure the candle on the end of one box by using the match to light the candle, dripping wax onto one end of the box, and then placing the candle in the wax. The candle could be mounted on the wall using the tacks. Participants who viewed the boxes as merely containers for the tacks were much slower to find the correct solution than were participants who saw the box as separate from the tacks. Duncker's explanation was that the participants who received the empty box were better able to restructure their understanding of the relations between items and use the box for another function. In contrast, the participants who received the box filled with tacks developed a structure that they found difficult to discard. Similarly, designers whose ideas were "fixed" by their current understanding of a problem might have a difficult time moving to a better solution.

15.2.2 Working Within the Rules

Part of mastering existing information is to know the "rules" of the game. According to Cropley (2001), creative individuals must possess certain personality characteristics, including 1) independence and nonconformity and 2) knowledge of the social rules and willingness to operate within them. Although these characteristics seem to be polar opposites, the point is that creative designers have to be able to think for themselves, but live in a world with others. This is especially true for interface designers, because the limitations can be less from the designer's imagination than from constraints imposed by others, including the various shareholders in the design. The ability to work successfully within those constraints is a hallmark of a creative designer.

Creative design often requires an understanding of others' good ideas and the ability to extend those ideas in the service of the creation of novel ideas. This contrasts with design by analogy, which is described later. In addition, creative design must occur within the context of domain knowledge. The rules of this game all relate to what the designer can expect from users. Without this knowledge, successful novel or unique solutions to usability problems are impossible.

15.2.3 Creativity as a Skill

An important issue in the study of creativity is the degree to which it is 1) a relatively persistent personality trait or 2) a temporary state that is influenced by the current environment of the creative person. Creativity can be enhanced by the design environment, as well as by the goals and motivation of the designer. Creativity can be considered to be a skill that can be developed with practice (Feldhusen and Goh, 1995). Weber (1992) has identified heuristics that describe how creative designers might generate novel solutions. For example, in applying the fine-tuning heuristic, a creative interface designer might rearrange or tweak the elements of a design after identifying those elements to produce better user performance. The goal of the fine-tuning heuristic is to get the elements in the right configuration. Weber (1992) also suggests that the application of the heuristics of creativity require that the designer be competent in evaluating the effects of applying the heuristic.

Weber's heuristics might be considered as a set of principles for creative design that can be acquired with practice. The inventive heuristics point out viable methods to modify existing structures in an abstract way. Design heuristics can help with both the production and evaluation of ideas. Using heuristics does not mean that the process ceases to be creative; rather, it can highlight opportunities for creativity. With guidance (and especially motivation to practice), individual creativity can be cultivated. The policies and the environment in the workplace can have a powerful effect on encouraging or discouraging the development of creativity. This can be seen in the following example (S. Pazuchanics, personal communication, August 4, 2004). A variation of Holtzblatt's (2003) affinity diagram strategy has been used successfully by software design teams to devise user archetypes. The team begins by putting user data summaries onto notecards then sorting them into meaningful categories during a meeting. The team then mounts the posters along the hallways surrounding their offices. For the next few days, the team walks through the information repeatedly. By organizing the data in this way and then posting them on the wall, it gives them a chance to think about and casually discuss the information. The goal is to organize the data, but the method used provides the advantage of making it easier to identify areas that may need further explorations (i.e., Gestalt psychology's gaps in the data) or to stimulate thoughts about technological answers to existing problems. The team reconvenes later and reevaluates the information in order to move to the next decision step. The idea cultivation atmosphere, however implemented, is a key to initiating creative design.

15.3 DESIGN BY ANALOGY

One of the main ways that humans transfer knowledge from one domain to another is through analogical thinking—recognizing similarities between aspects of two entities, then assuming that other aspects of the entities also are alike (Sternberg, 1977). In fact, analogical thinking may be one of the characteristic ways in which all primates think (see Gillan *et al.*, 1981). Given that creative thought involves a lot of analogical thinking (Weber, 1992), design by analogy may be a subset of creative design, but a sufficiently important subset to deserve its own examination.

The basic idea underlying design by analogy is to make use of relevant elements of interfaces that have been shown to be successful. Accordingly, an alternative name for the approach might be "design by appropriating good ideas." After all, appropriating *good* ideas is a better approach than appropriating *bad* ones. The phrase "design by appropriating good ideas" may be facetious, but it highlights one of the keys to analogical UI design—being able to discriminate good interfaces from bad interfaces. Thus, to use this approach, a designer has to apply some criterion of quality. One approach is to assume that any interface or Web design that has made it to the public arena is good; in other words, anything that is out there to be looked at is fair game. However, one might want to be more discriminating in selecting an analogical model, especially with Web sites, many of which may have been created without much concern for the user. Accordingly, a further criterion might be success in the marketplace—that is, a designer might want to examine only those interfaces or Web sites that have met with popular success. However, focusing on market share ignores the contribution of functionality and marketing in market success. That is, users may ignore a poor interface design to gain access to functions or products that only a specific Web site has. Thus, a designer may have to personally evaluate the usability of a Web site. This may seem like an anathema to one trained in human factors—given that one human factors truism is that designers aren't the only users, so they shouldn't design for themselves. However, in initial design, a designer doesn't have access to a large group of representative users to help in selecting a good analogical model (although they can use metaphor brainstorming and participatory design methods to enhance their ability to develop a good initial metaphorical model). To overcome idiosyncratic choice, a good designer might investigate the target user group's characteristics and needs and, as a consequence, might better be able to consider the user's point of view. Bailey (1993) found that initial interface designs by people with training in human factors or psychology were more usable than those by people with software development backgrounds—designs by people with human factors training produced few errors by users and the tasks times on their Web sites were about 60% faster than designs by programmers. Perhaps training in psychology and human factors provided Bailey's participants with a greater ability to adopt the user's point of view and subsequently represent that point of view in their designs.

A second key to identifying the analogical model is to determine the similarity between an existing interface and the one under design. In other words, you can't just select the most usable Web sites to appropriate from. Those usable Web sites must also share features with the Web site being created. For example, in designing a Web site that will be used to search large data spaces (e.g., for a library), you may want to restrict consideration of model Web sites to those that involve a search of some kind. Thus, early in initial design, good designers should understand the functional characteristics of the interface that they are designing to select good analogical models. One way to identify the functional characteristics is by identifying the requirements prior to beginning the design of the UI.

Although we have argued for criterial and functional selectivity in identifying an analogical model, a designer should consider both real world (i.e., noncomputer-based artifacts) and computer-based analogical models. The Google Web site (in 2004) provides a commercially successful and seemingly usable example of this approach to design. Both for simple and advanced searches, the Google interface takes advantage of people's experience in filling out forms. Users simply fill in boxes with words or phrases. Contrast this with the New Mexico State University library Web site (see Figure 15.1), which requires users to 1) go to multiple screens to input the request to search for an item, and 2) understand how to perform a Boolean search to find certain items.

Analogies with real-world functions may be inappropriate or unhelpful if they fail to take advantage of the unique capabilities of computers and of the Web (see also Mayhew, 1992). So, for example, a site that sells books does not

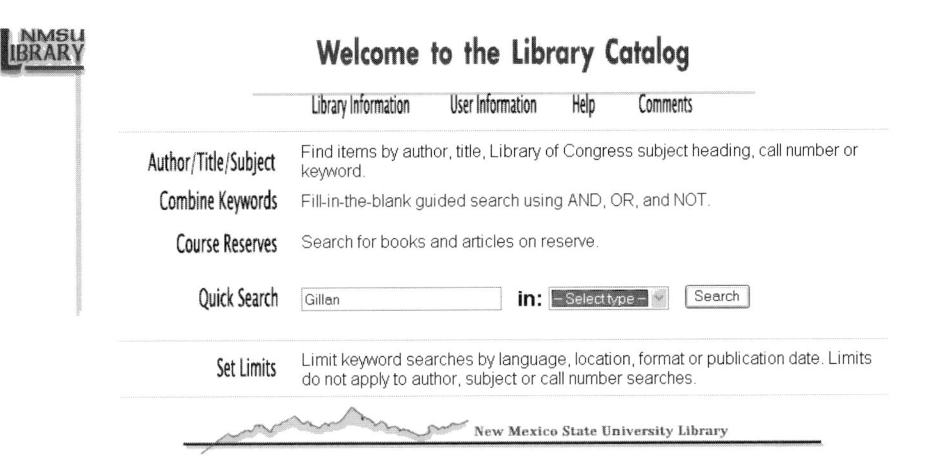

FIGURE 15.1 The entry screen to the search function on the New Mexico State University library Web site (in 2004). To reach the second screen for search by author, title, subject, or keyword combinations, the user must click on one of the bold words.

have to mimic the physical layout of a Barnes and Noble bookstore. The procedures that book buyers apply in the real world are not so necessary to the purchase experience that they overcome the advantages of a Web site, (e.g., the ability to quickly search for a specific item). This caveat highlights a more general concern in analogical design—how much of the analogical model should you use? Thinking by analogy requires both the transfer of relevant knowledge from the model and the recognition that the model differs from the new thing that is being considered. Taken in the design context, this means that the designer needs to decide which elements of the analogical model to transfer in the design, which to extend, and which to discard. One criterion to use for this decision is to consider whether the elements that you want to use or extend contribute to the success of the interface. A good initial designer should be able to identify both a good (i.e., usable) analogical model and the elements from the model that contribute to its usability.

15.4 RATIONAL DESIGN

Rational design involves three steps: 1) characterizing the objects and their properties in a situation, 2) identifying the rules for those objects in that situation, and 3) applying the rules and logically determining the likely outcomes

(Winograd and Flores, 1986). In the case of UIs and Web sites, rational design focuses on the properties and rules of visual objects such as frames, forms, tables, graphics, and maps, and interactive objects like buttons, selectable text, and clickable regions in graphics. A rational analysis of an interface requires the designer to run a mental simulation of a design to determine how users might interact with it (e.g., Polson *et al.*, 1992).

An obvious problem with rational design is that humans in general are not rational thinkers. Rather than making rational decisions (that is, decisions made simply on the basis of a logical analysis of all of the data), humans typically apply heuristics (e.g., Kahneman and Tversky, 1984). For example, we don't take into account all the information when making a decision, but more heavily weigh information that is most salient (i.e., most noticeable) or most easily accessed from memory. In addition, a designer is likely to be even less rational in analyzing a design in which he or she has a vested interest. A general bias in decision making is the confirmation bias—people tend to look for data that will support their ideas rather than the more logically compelling data that disputes those ideas. In analyzing their own designs mentally, designers find it especially difficult to overcome the confirmation bias. A second problem with rational design of the UI or a Web site is that the designer can rarely identify all the requirements and related assumptions before beginning to design the interface (this is a general problem in all software design, e.g., see Parnas and Clements, 1993).

Under what conditions can humans simulate rational thought? Training seems to be help people learn to apply a rational decision-making process. For example, logicians and economists can, in their work, identify the path to a rational decision. However, we know of no evidence to suggest that logicians or economists are more rational in their daily lives or anywhere outside of the narrow domain in which they have been trained. In addition, Parnas and Clements (1993) suggest that a well-structured rational design process can help produce a more rational design. They also promote documenting the design process, including a description of design alternatives that were rejected and explanations of why they were rejected.

15.5 DESIGN BASED ON KNOWLEDGE OF PERCEPTUAL AND COGNITIVE PRINCIPLES

An argument that seems obvious but is rarely stated and maybe even more rarely believed goes as follows:

> Most software and, specifically, Web sites, are intended to be used by humans. Those human users come to the human-computer interaction with certain perceptual and cognitive abilities. We have substantial knowledge concerning perceptual and cognitive psychology from experimental psychology.

If we abstract relevant principles of cognition and perception from that research, we can create interface designs that will be easier to learn and use. There is even evidence in support of this argument, particularly for the initial design of interfaces: Bailey's (1993) finding that people with training in human factors (and, thus, knowledge concerning perception and cognition) design interfaces that are more usable than do people without human factors training might provide support for the previous argument (see also Hewett, 1998).

So why is this argument partly, if not wholly, in disrepute? We believe that, in the past, designers have often either applied the wrong information from perceptual and cognitive psychology or applied that information incorrectly. Perceptual and cognitive researchers do not typically generate knowledge for applied purposes; rather, they are interested in testing theories and understanding the mechanisms that underlie human performance in certain constrained situations. To test theories, researchers tend to focus more on the reliability of their research than on its external validity. That is to say, a well-controlled experiment in an artificial laboratory setting may be better for the researcher's purpose than a more poorly controlled observational study in a real-world setting (although see Klein, 1998, for a contrasting view). As a consequence, psychological researchers may be discovering the perceptual and cognitive mechanisms that underlie performance only in these laboratory-based tasks. Another way to put this is that, as scientists, psychologists are typically more concerned with competence models (models of what the cognitive system *can* do) than performance models (models of what the cognitive system usually does) of human cognition.

Much evidence suggests that perception and especially cognition are sensitive to the context in which they occur, including the environmental, task, and even cultural contexts. As you change the nature of the context, the perceptual or cognitive mechanisms may not apply in the same way. An example relevant to human-computer interaction comes from an analysis of the application of Fitts's Law to word processing (Gillan *et al.*, 1992). Fitts's Law is concerned with human motor performance and proposes that the time to move from Point A to Point B is a linear function of the log (base 2) of the ratio of the distance between Points A and B and the size of the target that is the end goal of the movement. Theoretically, the accounts of Fitts's Law have focused on the information transmitted in a movement (e.g., Fitts, 1954) and the tradeoff between speed and accuracy as related to ballistic and homing submovements (Meyer *et al.*, 1988).

Fitts's Law is often applied to interface design (cf., Card *et al.*, 1983). However, Gillan *et al.*, (1992) found that how one applies Fitts's Law to UI design depends on what the user identifies as the target of his or her movement. For example, if a user is pointing the cursor at a block of text to be selected with a single click, the time to move the cursor is a function of both the distance and the size of the text block. In other words, the target is the text block, and the appropriate Fitts's Law model is $MT = a + b(\log_2[D/TS])$, where D is the movement distance and TS is the entire text block. In contrast, if the situation changes slightly and the user points the cursor at a block of text to be selected by dragging the cursor across that block, then the time to move the cursor to the text is unrelated to the size of the block of text. In this latter case, the target is the closest letter in the text block (usually the left edge of the leftmost letter in the text block), so the appropriate Fitts's Law model is $MT = a + b(\log_2[D/TS])$, where D is the movement distance and TS is the leftmost point of the text block. Because the size of the left edge of the leftmost letter doesn't change in size as the text block changes, the Fitts's Law model can be reduced to $MT = a + b(\log_2[D])$. Figure 15.2 provides graphic examples of the fit of Fitts's Law to these various situations.

Another important aspect of the context that affects the application of Fitts's Law is the size of the target relative to the distance of the movement. Dual submovement models (e.g., Meyer *et al.*, 1988) suggest that the slower, more accurate, homing movement is only necessary if the target can't be reached by the initial, faster, but less accurate, ballistic movement. Thus, in interfaces in which users need only move a small distance to reach a large target, they typically can forego the homing movement. As a consequence, in those situations, movement time is related only to the movement distance (e.g., Gan and Hoffman, 1988). This is especially important for UI designs that make use of impenetrable borders—for example, in moving the pointer to a menu bar—the user can't overshoot the target, effectively creating an infinitely large target (e.g., Walker and Smelcer, 1990). These various examples suggest that Fitts's Law can be applied to UI design, but that any such application requires an analysis of the specifics of the interface and the user's task to have that application be successful.

Given that many of the perceptual and cognitive effects that might be applied to UI designs are influenced by the context, are there effects that might be cross situational? One place to look for these cross-situational effects is at the initial level of information processing, that is to say, early in perception. For example, color vision is based on three types of photosensitive receptors in the fovea (i.e., the central part of the retina that light from an object falls on when

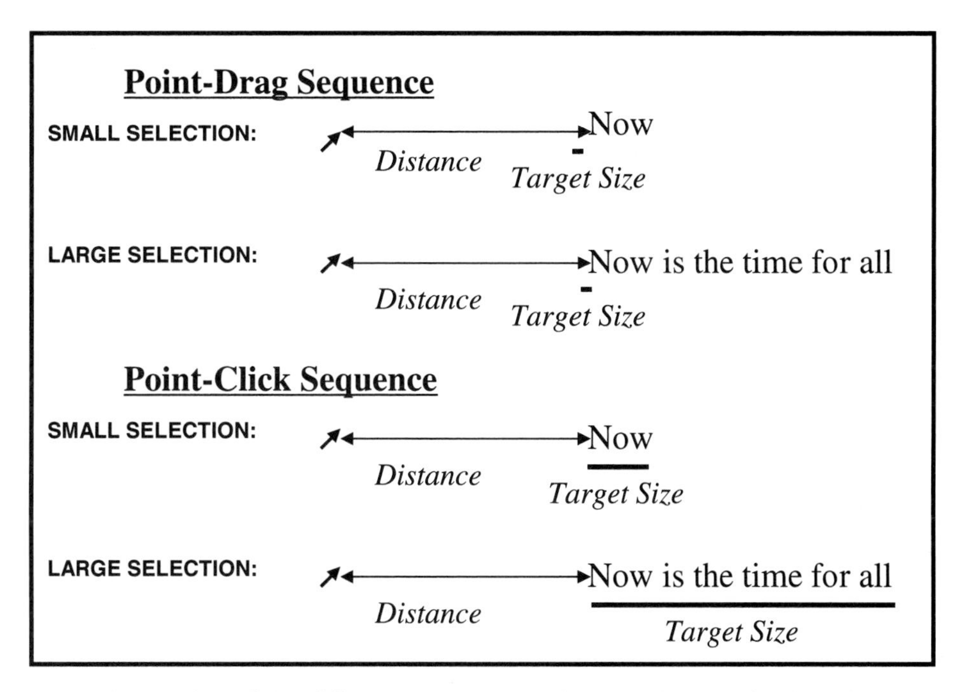

FIGURE 15.2 An illustration of the difference between the functional targets in the point-drag and point-click sequences. The pointing target (the underlined area) does not change with the size of the text to be selected in the point-drag sequence because the text is selected by dragging across it. In contrast, the pointing target changes with the size of the text to be selected in the point-click sequence because the text is selected by clicking anywhere in the text.

a human looks at that object). Those receptors, called cones, are sensitive to bands of light frequencies centering approximately around red, green, and blue (i.e., long, medium, and short wavelengths, respectively). However, the cones are not equally distributed in the fovea: most of the cones are sensitive to red light or to green light, with relatively few sensitive to blue light. Furthermore, the blue sensitive cones are absent from the very center of the fovea. Rather, they are found beginning about one degree outside of the center of the fovea. As a consequence, seeing small details that are colored blue can be difficult. So reading small light blue print is typically more difficult than reading small print in other colors (when contrast and light intensity are equivalent). In fact, reading appears to be best on displays that maximize the contrast between the text and background—typically black text and very light background (e.g., Dillon, 1992).

Despite the occasional broadly applicable perceptual effect, even early perception may involve a combination of bottom-up (or data-driven) processing that is cross-situational, and top-down (or knowledge-driven) processing that is contextual. For example, comprehension of phonemes, the basic unit of spoken and heard language, is influenced by contextual factors. For example, a phoneme is easier to identify if it is part of a word, that is, a set of meaningful phonemes, than if it is presented in isolation (Pollack and Pickett, 1964). The presence of visual information from a speaker's lips also provides a context that can influence a hearer's interpretation of a phoneme (McGurk and MacDonald, 1976).

Rather than attempting to find specific research findings or effects that would be broadly applicable, we might look for more general themes from perceptual and cognitive psychology that could be applied across a wide number of diverse contexts in initial UI design. One theme from attention and working memory that is critical for designers to keep in mind is that the user's ability to process information will be limited at times. In other words, don't ask the user to keep too much information in mind at one time. But what do we mean by "too much information?" One of the most cited and misused bits of knowledge from psychology is the "magic number seven plus or minus two" (Miller, 1956). UI design guidelines have gone so far as to suggest that displays should only provide five to nine pieces of information on a screen at a time because people can only handle that amount of information at once. In fact, limitations of screen real estate provide a good reason to limit the amount of information visually available at one time, but such restrictions are not a result of any magic numbers. A good interface design could make many more than nine pieces of information available at once if the design organized the information on the screen so that the user was guided to the related information (e.g., by contiguous placement of information that will be used in temporal contiguity) and didn't ask the user to attend to all of the information at once.

The danger in specifying a number of items or pieces of information is that, in certain cases (e.g., if a user could chunk semantically-related pieces of information), a user's processing ability might exceed that number, whereas, in other cases (e.g., if the information were complex, like phrases), the number might exceed the user's capacity. Alternatively, one might propose a guideline to avoid producing too great a cognitive load for the user, but specifying the amount of that load again requires interpretation by the designer.

Attempts to develop a set of principles from a science base in perception and cognition that could be followed by any designer are not likely ever to be successful, given that, as the previous discussion suggests, application of most effects in perception and cognition depend on an understanding of the context or an interpretation of the principle for the conditions of the interface and its use. Rather, application of the science base of perception and cognition is likely to require extensive experience observing perception and cognition under a wide variety of circumstances. One important way to gain that experience may be by specific training that involves testing people in psychology experiments. Indeed, it may be that the advantage that Bailey (1993) observed in interface design for people with human factors training was a function of their training in perception and cognition.

15.6 CONCLUSION

It seems likely that devotion to any single approach to initial UI design would not produce as good designs as would the use of multiple approaches. Combining knowledge of perceptual and cognitive psychology with creative design would probably lead to better design than either approach individually. So how can companies increase the likelihood of good initial designs?

The discussion in this chapter indicates that training in creativity, analogical reasoning, rational thinking, and/or the principles of perception and cognition can increase skill in these areas. Accordingly, companies might invest in training to improve on the design of initial interfaces.

Alternatively, companies could hire people who already had acquired the appropriate skills for initial design. To identify those people, companies would still have to invest in some means of identifying people with the appropriate skills. Current interview techniques at many companies (e.g., asking prospective employees to design an interface on the spot, with little opportunity for reflection, analogical reasoning, or rational thought) may not be effective in identifying the best designers. Developing valid instruments for selecting employees who have skills in creative design, analogical design, rational design, and/or the application of cognitive and perceptual principles will also require a substantial investment. The difference in cost between the training approach and the selection approach may be that developing and implementing an effective training system will have both a high initial cost (developing the system) and a relatively high continuing cost (the cost of paying instructors), whereas the bulk of the cost for selection would be upfront for the development of the selection instruments. The cost of any instruction that might give the employees the requisite skills would be borne elsewhere (e.g., by the employee as a student or the tax payers of the state that funds the university that trains the students). The application of the instruments might require some degree of skill, but that cost would be less than that of instruction.

A third approach is to continue with the current system in which most training is done at universities before a designer is hired, and selection is based on factors that may or may not correlate with design skill. Of the three approaches, this approach has a low cost but also has the lowest chance of producing good initial designs. Those interested in initial design will have to consider whether the opportunities for improvement afforded by training or selection instruments outweigh the additional cost. Because the training systems and selection instruments have not yet been developed, the costs would have to be estimated, as would the benefits of initial interface design.

ACKNOWLEDGEMENTS

The authors thank Chauncey Wilson for his extensive comments on an earlier version of the chapter. We also thank Skye Pazuchanics for her helpful discussions concerning the role of cognition in interface design. Finally, thanks to the editors of this book for their patience and the belief that not all analyses related to cost effectiveness have to be based on ROI.

REFERENCES

Ambach, J., and Repenning, A. (1996). Puppeteers and directors: Supporting artistic design by combining direct-manipulation and delegation. *Proceedings of the Second International Symposium on Creativity and Cognition,* (pp. 67–76), Loughborough, UK: LUTCHI Research Center.

Bailey, G. (1993). Iterative methodology and designer training in human-computer interface design. In *Proceedings of INTERCHI,* Amsterdam, The Netherlands, April 24–April 29 (pp. 198–205). New York: ACM.

Bias, R. G., and Mayhew, D. (1994). *Cost-justifying usability.* Boston: Academic Press.

Boehm, B. W. (1989). A spiral model of software development and enhancement. In B. W. Boehm (Ed.), *Software Risk Management* (pp. 26–37), Piscataway, NJ: IEEE Press.

Card, S. K., Moran, T. P., and Newell, A. (1983). *The psychology of human-computer interaction.* Hillsdale, NJ: L. Erlbaum and Associates.

Cropley, A. J. (2001). *Creativity in education and learning: A guide for teachers and educators.* London: Kogan Page Limited.

Dillon, A. (1992). Reading from paper versus screens: A critical review of the empirical literature. *Ergonomics,* 35, 1297–1326.

Donoghue, K. (2002). *Built for use: Driving profitability through the user experience.* New York: McGraw-Hill.

Duncker, K. (1945). On problem solving. *Psychological Monographs*, 58, Whole No. 270.

Feldhusen, J. F., and Goh, B. E. (1995). Assessing and accessing creativity: An integrative review of theory, research, and development. *Creativity Research Journal*, 8, 231–247.

Fitts, P. M. (1954). The information capacity of the human motor system in controlling the amplitude of movement. *Journal of Experimental Psychology*, 47, 381–391.

Gan, K. C., and Hoffman, E. R. (1988). Geometric conditions for ballistic and visually-controlled movements. *Ergonomics*, 31, 829–839.

Gillan, D. J., Holden, K. L., Adam, S., Rudisill, M., and Magee, L. (1992). How should Fitts' Law be applied to human-computer interaction? *Interacting with Computers*, 4, 289–313.

Gillan, D. J., Premack, K., and Woodruff, G. (1981). Reasoning in the chimpanzee: I. Analogical reasoning. *Journal of Experimental Psychology: Animal Behavior Processes*, 7, 1–17.

Gould, J. D., and Lewis, C. (1985). Designing for usability: Key principles and what designers think. *Communications of the ACM*, 28, 300–311.

Guilford, J. P. (1959). Traits of creativity, in H. H. Anderson (Ed.), *Creativity and its cultivation* (pp 142–161). New York: Harper.

Hewett, T. T. (1998). Cognitive factors in design: Basic phenomena in human memory and problem solving. *Tutorial at CHI 1998, Los Angeles, CA.*

Holtzblatt, K. (2003). Contextual design. In J. A. Jacko and A. S. Sears (Eds.), *The handbook of human-computer interaction: Fundamentals, evolving technologies, and emerging applications* (pp. 941–963). Mahwah, NJ: L. Elrbaum Associates.

Kahneman, D., and Tversky, A. (1984). Choices, values, and frames. *American Psychologist*, 39, 341–350.

Klein, G. (1998). *Sources of Power: How People Make Decisions.* Cambridge, MA: MIT Press.

Köhler, W. (1925). *Mentality of apes* (E. Winter, Trans.). London: Routledge and Kegan Paul. (Original work published 1917).

Mayer, R. E. (1983). *Thinking, problem solving, cognition* (2nd ed.). New York: W. H. Freeman and Co.

Mayhew, D. (1992). *Principles and guidelines in software user interface design.* Englewood Cliffs, NJ: Prentice-Hall.

McGurk, H., and MacDonald, J. (1976). Hearing lips and seeing voices. *Nature*, 264, 746–748.

Meyer, D. E., Abrams, R. A., Kornblum, S., Wright, C. E., and Smith, J. E. K. (1988). Optimality in human motor performance: Ideal control of rapid aimed movements. *Psychological Review*, 95, 340–370.

Miller, G. A. (1956). The magical number seven plus or minus two: Some limitations on our capacity for processing information. *Psychological Review*, 63, 81–97.

National Cancer Institute. (2004). *Research-based web design and usability guidelines.* http://usability.gov/guidelines/

Parnas, D. L., and Clements, P. C. (1993). A rational design process: How and why to fake it. *IEEE Transactions on Software Engineering*, 19, 251–257.

Pollack, I., and Pickett, J. M. (1964). Intelligibility of excerpts from fluent speech: Auditory vs. structural context. *Journal of Verbal Learning and Verbal Behavior*, 3, 79–84.

Polson, P. G., Lewis, C., Rieman, J., and Wharton, C. (1992). Cognitive walkthroughs: A method for theory-based evaluation of user interfaces. *International Journal of Man-Machine Studies*, 36, 741–773.

Rosson, M., and Carroll, J. (2002). *Usability engineering*. San Francisco: Morgan Kaufmann.

Simon, H. A. (1979). *Models of thought*. New Haven, CT: Yale University Press.

Spolsky, J. (2001). *User interface design for programmers*. Berkeley, CA: APress.

Sternberg, R. J. (1977). *Intelligence, information processing, and analogical reasoning*. Hillsdale, NJ: L. Erlbaum.

Walker, N., and Smelcer, J. B. (1990). A comparison of selection times from walking and pull-down menus. *Proceedings of the CHI' 90 Conference on Human Factors in Computing Systems* (pp. 221–225). New York: ACM.

Weber, R. J. (1992). *Forks, phonographs, and hot air balloons*. New York: Oxford University Press.

Winograd, T., and Flores, F. (1986). *Understanding computers and cognition: A new foundation for design*. Norwood, NJ: Ablex.

16 CHAPTER Keystroke Level Modeling as a Cost Justification Tool

Deborah J. Mayhew Deborah J. Mayhew & Associates

16.1 INTRODUCTION

Keystroke level modeling (KLM) is one of a variety of cognitive modeling techniques that have been reported in the literature over the last two decades. It is the original modeling technique described by Card, Moran, and Newell in their classic book *The Psychology of Human-Computer Interaction* (1983). It is the simplest and most basic of a family of modeling techniques (e.g., GOMS, which stands for goals, operations, methods, and selection rules) that have evolved from it over the past two decades (John, 1990, 1995; John and Kieras, 1996; Atwood *et al.*, 1996). Another book provides a good summary of past work in cognitive modeling, including the KLM technique (Carroll, 2003).

KLM, simply put, involves identifying and counting all of the discrete human operations—physical (e.g., mouse click, keystroke, or moving the hand from the mouse to the keyboard), cognitive (e.g., read a syllable of text or make a mental comparison) and perceptual (e.g., locate something on screen)—that a user must execute to most efficiently accomplish a specific task on a specific user interface design. System response time operators are then added to the model where appropriate and when there are data to estimate them accurately. Time parameters per operator (available in the literature for human operators) (Table 16.1) are then plugged into a task model to predict a total task time. The total task times generated by such models predict the fastest time (on average) that highly trained and experienced users will be able to perform a given task on a given user interface with a given set of system response times, assuming they perform the task with no user errors and no interruptions.

Because you can model a user interface that is nothing more than a specification on paper, modeling allows you to predict the relative *ease of use* of alternative user interface designs for a particular set of tasks. Comparing alternative designs in this way allows more informed decisions such as which of a set of alternative interface designs to build, whether you should build a new system (with a new user interface) to replace an existing one, or which of a set of competing products to buy, when ease of use is the critical usability goal.

Like formal usability testing, modeling is a method for assessing the expected usability of a particular user interface design. Formal usability testing is an excellent technique to use during product development when the main usability goal is *ease of learning* because it is relatively easy to find potential users who are untrained and inexperienced with a proposed design to participate in your study. Testing with such users allows you to discover what aspects of your design are unintuitive and difficult to learn for users who haven't had the benefit of training, user manuals, and frequent practice and experience. Public Web sites such as e-commerce sites tend to have casual and infrequent users and are an example of a type of software application in which ease of learning for novices is a much more important usability goal than efficiency of well-trained, high-frequency, and highly practiced experts. Thus formal usability testing (as well as other techniques such as heuristic evaluations or walkthroughs) is an excellent technique to use early in the development of such sites.

However, in many cases the key usability goal of a Web-enabled application (e.g., a traditional business application for use by internal users that just happens to be Web-enabled, as opposed to a public Web site)—as well as of many applications built on other platforms—is not ease of learning for new or casual users, but *ease-of-use*, that is, productivity, for well-trained, highly experienced, and proficient, high-frequency users. This aspect of usability is much harder to measure by formal usability testing early in the design process for a new application, because you never have well-trained, highly experienced, and proficient users of a system that nobody has used yet (precisely because it is still in the early design stage). It is very hard and perhaps impossible to accurately simulate peak proficiency usage when a design is only in the prototype stage, let alone when it is only on paper. Assessment of ease of use with alternative evaluation techniques such as heuristic evaluations and walkthroughs is also difficult. However, before a design is built, it would be extremely useful to know whether it will in fact meet a set of productivity (e.g., ease-of-use) goals, and in particular, if the design does *not* meet such goals, to know why it does not. If these things are known, the design could be iteratively developed and assessed until it can be demonstrated that the application would meet productivity goals if implemented as designed.

Thus cognitive modeling provides a useful and practical technique for assessment of a proposed user interface design against ease-of-use goals (i.e., goals for average error-free task time for trained, high-frequency, expert users). It can easily be applied long before implementation or even prototyping, at a point in the development process at which formal usability testing is just not practical, but at which early and reliable assessment of a design against productivity goals could support a relatively quick and inexpensive iterative evolution of a design to meet business goals.

Note that modeling is only relevant to predicting performance under the following circumstances:

- Expert users (high frequency, highly practiced)
- One very specific variation of a general task
- Specific variations of tasks that are quite linear and structured, not allowing a great deal of variation or creativity in how they are approached
- No user errors
- No interruptions or distractions

Conversely, modeling is *not* an appropriate technique for predicting performance under the following circumstances:

- Novice or casual users (low frequency, still in learning mode)
- Average performance over many variations of a general task
- Even specific variations of tasks that are highly unstructured and allow many different approaches
- Performance includes making and recovering from user errors
- Performance is typically interrupted or distracted

In addition, it is not a technique that can be used to predict user interface qualities such as time to learn, typical user error rates, and effectiveness in high-interrupt environments. It *only* predicts *productivity (i.e., task time) under ideal circumstances.* Nevertheless, when productivity in low-interrupt environments is the issue, modeling can be a powerful evaluation tool.

Mayhew and Tremaine (Chapter 3) provide a general framework for the cost justification of usability engineering methods and processes; their focus was on applying a generic cost-justification technique to make a business case for including a usability engineering effort on a software development project. In this situation, the cost-justification analysis compares assumed performance on two *hypothetical* user interface designs—one that would result without applying usability engineering and one that would result from adding a usability engineering effort. The benefits and costs of the latter are compared to those of the former to predict whether the latter would in fact pay off in the long run. The relative ease of use or ease of learning of one design versus another is predicted not on the particulars of the alternative designs, but on the general belief that usability engineering methods will result in a design that is easier to use or easier to learn than a design developed without the benefit of such methods and on the empirical findings.

In this chapter, in contrast, I describe the use of the specific usability technique of KLM to cost-justify the implementation of one very concrete and specific user interface design (as opposed to a hypothetical one) rather than another design. That is, instead of making general predictions about expected user performance on a hypothetical design to cost-justify incorporating a usability engineering process into a development project to achieve that hypothetical design, I discuss using a usability technique to support an objective choice between competing particular and detailed user interface designs. The point in this case is to cost-justify a particular user interface design *before* implementing it (or buying it) to help ensure a return on investment (ROI) on the development project as a whole.

A few scenarios come to mind in which cost-justifying one specific, concrete user interface design in comparison with another could be a productive exercise. First, many development projects are intended to replace older applications rather than to automate work that has not previously been automated. In this case, before the replacement application is built, it would be invaluable to be able to predict with some level of confidence that a proposed application user interface design will actually result in greater user productivity levels than the existing application to the extent that the new development effort will in fact pay off over a given period of time. Thus the assessment of the level of improvement in overall user productivity that will result from a proposed application design (which modeling can provide) can be used to cost-justify the whole development effort. Within this scenario, it is also true that if modeling of the initial proposed design does *not* predict a level of improvement in productivity relative to the existing applications that will in fact cost-justify the development and implementation of the new application, the models will also provide insights into *why* the hoped for improvement is not predicted to be realized. These insights in turn can drive redesign ideas for the new application. Modeling can then be applied iteratively to improve the proposed design until it in fact *does* achieve predicted improvements in productivity large enough to cost-justify the development effort. All of this can potentially be accomplished before development begins in earnest.

Second, when an organization is comparison shopping to purchase a commercial software package, being able to predict with confidence which of the competing packages would support the highest level of user productivity once users are trained and experienced would be invaluable, as this would probably result in the greatest return on investment for the purchase. Thus modeling can be used to cost-justify the purchase of one competing package over others.

And third, during the design and development of a new application, competing user interface design ideas inevitably arise. Modeling can be applied when user interface design is still just on paper to predict which of any number of competing design ideas will result in the greatest user productivity, thus cost-justifying one of the competitive design proposals relative to all others. Again, this can help ensure a particular desired ROI for the overall development effort by helping to optimize the design to achieve the business goals of the development project (see John, 1995, for a list of references including case studies of the application of KLM and other modeling techniques, and also Carroll, 2003).

The rest of this chapter presents a case study describing the use of modeling to cost-justify the development of a new Web-enabled application intended to replace a set of old mainframe-based applications. Although the case study has been heavily disguised, it is based closely on an actual recent project, and the results reported here are in important respects very real. For example, the organization in which the project was carried out is described as a credit card company, but in fact the actual organization the case study is based on was in another line of business altogether. The user tasks described in the case study are different from the tasks done in the real organization, but comparable in general type and complexity. Numbers are provided for predicted and actual task times which, although different from the times in the actual project, are representative of actual times obtained on the project.

16.2 CASE STUDY

The mail processing department of a credit card company was planning to replace a collection of 14 disparate mainframe-based applications, all with different user interfaces, that mail processing clerks ("clerks") currently used to process incoming requests from account holders. The replacement was to be a single integrated Web-enabled application with a consistent and improved user interface. The premise for the business case for this replacement project was an expected increase in clerk productivity that over time would cancel out the cost of the development effort and then continue to accrue cost savings for the company.

The new application was intended to support clerks who processed incoming mail from account holders. These clerks performed a variety of tasks, including the following:

- Accept and process monthly payments
- Accept and process requests for balance transfers from other credit cards
- Accept and process requests to add new credit cards to an account holder's account
- Accept and process requests to close out an account
- Accept and process requests to refuse a fraudulent charge

In all tasks, paper request forms and checks received from account holders by mail were scanned into the system by other staff members, and the images of these forms and checks appeared as tasks in a work queue on the workstations of the clerks. The clerks would bring up the images for a given task on their screen and then open up and work in a variety of other applications to get the requests (which were documented in the images) recorded in the appropriate databases.

The business manager overseeing the replacement development project worked out a business case to justify the project cost. Based on knowledge of the current yearly volume of transactions for each mail processing task (e.g., accept monthly payments, add a new card) and on identified opportunities for increasing productivity in the proposed Web-enabled application, she set a goal for each task that would have to be met for the overall cost justification of the new application to become a reality. These goals were expressed as a required productivity gain for each task on the proposed application relative to the existing applications. For some tasks, this goal was an 18% gain, for some it was a 10% gain, and for others it was a 0% gain (i.e., at least no loss). In other words, the goals were for trained and highly practiced mail processing clerks who used the system at a high rate of frequency to be able on average to perform specific tasks anywhere from 0% to 18% faster on the new Web-enabled application than they could on the multiple mainframe-based applications they were using. Unless these specific task goals were met, the cost of the overall development effort ultimately would not be cost justified.

Having set these goals, the business manager next began to wonder whether and how she could be assured that these task productivity gain goals would in

Table 16.1 Initial Model Operators with Time Values

Operator	*Description*	*Original Time Value (seconds)*	*Source*	*Revised Time Value (seconds)*
K	Single keystroke (average skilled typist, 0.20 second = 55 words/min)	0.20	Card, Moran, and Newell (1983)	0.12
C	Single mouse click (finger down, finger up)	0.12	Card, Moran, and Newell (1983)	0.12
P	Point with mouse (hand already on mouse, move mouse to move pointer across screen to a given target)	1.10	Card, Moran, and Newell (1983)	1.10
H	"Home"—move hand from one input device (e.g., keyboard) to another (e.g., mouse)	0.40	Card, Moran, and Newell (1983)	0.40
M	Mental operator—used only to capture time taken to move eyes across windows	1.35	Card, Moran, and Newell (1983)	0.00
W1Existing	A system response time ("Wait") for, e.g., field or window to take focus in response to mouse click	0.25	Internal measures of existing applications	0.25
W2Existing	A system response time ("Wait") for, e.g., display of a new page with no lookup	0.50	Internal measures of existing applications	0.50
W3Existing	A system response time ("Wait") for, e.g., display of a new page with simple lookup or carrying data from previous page	1.50	Internal measures of existing applications	1.50
W4Existing	A system response time ("Wait") for, e.g., display of a new page with database search	3.00	Internal measures of existing applications	3.00
SP	Speak/read a syllable in a highly practiced sentence	0.13	John (1990)	0.13
SU	Speak/read a syllable in an unpracticed sentence	0.17	John (1990)	0.17
Skim	Skim a page of text	5.00	Personal estimate	1.00

Table 16.2 Excerpt from the Model of the Accept Monthly Payment Task on the Existing Applications

	A	*B*	*C*	*D*	*E*	*F*	*G*
1				TASK	APPLICATION		
2				Accept Monthly Payment	Existing		
3	STEP #	STEP	SUBSTEP #	SUBSTEP	INTERACTION	OPERATORS	ASSUMPTIONS
4							
5	1	Get Task Description	1	Select and open Task Description	Eyes to Task List window; double click on Task	CC	Task List window is in focus from end of previous task; start task from double click, skip MHP
6					Wait for Task Description window to open	(W3Existing)	
7	2	Get Account	1	Search by credit card #	Eyes to Account window; click on search field once to give window and field focus	**MPC**	One click gives both window and field focus

8					Wait for window/ field to take focus	**(W1Existing)**	
9					Eyes to Task Description window, find credit card #	**M**	Visible on first page
10					Read and type credit card number	**H(16K)**	No hyphens/spaces to type in credit card # field; "1234567890123456"
11					Hit Enter key	**K**	Instead of moving hand to mouse
12					Wait for account holder data to populate in Account window	**(W4Existing)**	
13					Eyes to Account window, confirm that data populated	**MM**	Second M for compare to knowledge in head

Table 16.3 Excerpt from the Model of the Accept Monthly Payment Task on the Existing Applications (Continued)

	A	*B*	*C*	*D*	*E*	*F*	*G*
1				TASK	APPLICATION		
2				Accept Monthly Payment	Existing		
3	STEP #	STEP	SUBSTEP #	SUBSTEP	INTERACTION	OPERATORS	ASSUMPTIONS
14	3	Verify Account Holder Data	1	Verify that have correct account holder record	Find and read account holder name/address	(19SU)	Time to read = time to say; data are on first page
15					Eyes to Task Description window, find, read, and compare	M(19SU)	Account holder data are correct; "Ralph Koler, 321 Morrow Street, Orange Hill, NJ, 12345"; user does *not* check birth date or phone
16			2	Verify that have valid task	Eyes back to Account window; Check Task Type field in Account window to see if valid task	MM	Second M for compare to knowledge in head

17	4	Review Task Description	1	Skim Task Description form	Eyes to Task Description window	**M**	
18					Skim page 1	**Skim**	There is only one page, shows check and payment form, all visible at once, no scrolling
19	5	Perform Task	1	Get to Payment window	Eyes to Payment window	**M**	Payment window is open (user leaves it open all day)
20					Click on Load Account button	**HPC**	One click gives window focus and executes button click
21					Wait for account data to populate in Payment window	**(W3Existing)**	User does not check to see if correct account loaded—assumes it will

Table 16.4 Excerpt from the Model of the Accept Monthly Payment Task on the Existing Applications (Continued)

	A	*B*	*C*	*D*	*E*	*F*	*G*
1				TASK	APPLICATION		
2				Accept Monthly Payment	Existing		
3	STEP #	STEP	SUBSTEP #	SUBSTEP	INTERACTION	OPERATORS	ASSUMPTIONS
22			2	Select credit card to apply payment to	Find the only credit card number, click on its radio button	**PC**	This account holder only has one credit card on file
23					Click on Payment button	**PC**	
24					Wait for Payment screen in Payment window to come up	**(W2Existing)**	
25			3	Enter payment data	Eyes to Task Description window	**M**	Eyes stay on Task Description window; user knows how to tab through screen on Payment window

26				Find, read, and type payment amount	H(7K)	Cursor defaults to payment amount field; “3293.42”; must type decimal; time to type = time to read
27				Tab to Checking Account Number field	KK	Use keyboard rather than mouse, 2 tabs to field—user knows how many tabs, no need to look at Payment window
28				Find, read, and type checking account number	8K	Time to type = time to read; “34401776”; as soon as last digit entered, search for city/state happens automatically
29				Tab to Bank Number field	K	Use keyboard rather than mouse, 1 tab to field
30				Find, read, and type bank number	9K	Time to type = time to read; “344017769”

Table 16.5 Excerpt from the Model of the Accept Monthly Payment Task on the Existing Applications (Continued)

	A	*B*	*C*	*D*	*E*	*F*	*G*
1				TASK	APPLICATION		
2				Accept Monthly Payment	Existing		
3	STEP #	STEP	SUBSTEP #	SUBSTEP	INTERACTION	OPERATORS	ASSUMPTIONS
31					Tab	K	Tab automatically searches for and brings up Bank Name and Address field data; cursor left in last address field
32					Wait for Bank name and address to populate	(W3Existing)	
33					Read Bank Name in Task Description window; eyes to Payment window, find and compare Bank Name	(4SU)M(4SU)	"Wells Fargo Bank"; it is the same; time to compare = time to say

34				Tab to Check Number field	K	
35				Eyes back to Task Description window	M	
36				Find, read, and type check number	(4K)	4 digits; time to type = time to read; “3248”
37				Get cursor in Payment Date field	K	User uses tab
38				Find, read, and type payment date	(6K)	6 digits; no need to enter slashes, must enter leading zeroes; time to type = time to read; “103103”
39				Hit Enter key	K	Accepts Payment screen
40				Wait for Confirmation Screen	(W3Existing)	

Table 16.6 Excerpt from the Model of the Accept Monthly Payment Task on the Existing Applications (Continued)

	F	*G*	*H*	*I*	*J*	*K*	*L*	*M*	*N*	*O*	*P*	*Q*	*R*	*S*
1														
2														
3	OPERATORS	ASSUMPTIONS	C	K	P	H	M	W1Existing	W2Existing	W3Existing	W4Existing	Skim	SP	SU
19	M	Payment window is open (user leaves it open all day)					1							
20	HPC	One click gives window focus and executes button click	1		1	1								
21	(W3Existing)	User does not check to see if correct account loaded—assumes it will								1				

Table 16.7 Summary Calculations for the Accept Monthly Payment Task in the Existing and Proposed Applications models with Original Operator Time Values

Existing Applications

	G	*H*	*I*	*J*	*K*	*L*	*M*	*N*	*O*	*P*	*Q*	*R*	*S*
1													
2													
3	**ASSUMPTIONS**	C	K	P	H	M	W1Existing	W2Existing	W3Existing	W4Existing	Skim	SP	SU
55	**Operator count**	8	54	10	6	20	2	2	4	1	1	0	139
56													
57	**Operator times(s)**	0.96	10.80	11.00	2.40	27.00	0.50	1.00	6.00	3.00	5.00	0.00	23.63
58													
59	**Total task time(s) =**	91.29											

Proposed Application

	G	*H*	*I*	*J*	*K*	*L*	*M*	*N*	*O*	*P*	*Q*	*R*	*S*
1													
2													
3	**Assumptions**	C	K	P	H	M	W1Proposed	W2Proposed	W3Proposed	W4Proposed	Skim	SP	SU
67	**Operator count**	9	57	11	9	23	2	2	7	1	1	0	139
68	**Relative to existing**	**1**	**3**	**1**	**3**	**3**	**0**	**0**	**3**	**0**	**0**	**0**	**0**
69	**Operator times(s)**	1.08	11.40	12.10	3.60	31.05	0.50	1.00	10.50	3.00	5.00	0.00	23.63
70	**Relative to existing**	**0.12**	**0.60**	**1.10**	**1.20**	**4.05**	**0.00**	**0.00**	**4.50**	**0.00**	**0.00**	**0.00**	**0.00**
71													
72	**Total task time(s) =**	102.86											
73													
74	**Relative to existing**	**11.57**											
75	**As % difference**	**12.67%**											

Table 16.8 Summary Calculations for the Accept Monthly Payment Task in the Existing and Proposed Applications Models with *Final* Operator Time Values

Existing Applications

	G	*H*	*I*	*J*	*K*	*L*	*M*	*N*	*O*	*P*	*Q*	*R*	*S*
1													
2													
3	**Assumptions**	C	K	P	H	M	W1Existing	W2Existing	W3Existing	W4Existing	Skim	SP	SU
55	**Operator count**	8	54	10	6	20	2	2	4	1	1	0	139
56													
57	**Operator times(s)**	0.96	6.48	11.00	2.40	0.00	0.50	1.00	6.00	3.00	1.00	0.00	23.63
58													
59	**Total task time(s) =**	55.97											

Proposed Applications

	G	*H*	*I*	*J*	*K*	*L*	*M*	*N*	*O*	*P*	*Q*	*R*	*S*
1													
2													
3	**Assumptions**	C	K	P	H	M	W1Proposed	W2Proposed	W3Proposed	W4Proposed	Skim	SP	SU
67	**Operator count**	9	57	11	9	23	2	2	7	1	1	0	139
68	**Relative to existing**	**1**	**3**	**1**	**3**	**3**	**0**	**0**	**3**	**0**	**0**	**0**	**0**
69	**Operator times(s)**	1.08	6.84	12.10	3.60	0.00	0.50	1.00	10.50	3.00	1.00	0.00	23.63
70	**Relative to existing**	**0.12**	**0.36**	**1.10**	**1.20**	**0.00**	**0.00**	**0.00**	**4.50**	**0.00**	**0.00**	**0.00**	**0.00**
71													
72	**Total task time(s) =**	63.25											
73													
74	**Relative to existing**	**7.28**											
75	**As % difference**	**13.01%**											

Table 16.9 Summary of Modeling Results and Productivity Tests across Eight Tasks—after *Existing* Applications Productivity Test

	A	*B*	*C*	*D*	*E*	*F*	*G*	*H*	*I*
1		1	2	3	4	5	6	7	8
2		Accept Monthly Payment	Balance Transfer	Add Card to Account	Close Account	Change Address	Cash Advance	Refuse a Charge	Credit Line Increase
3									
4	**Existing MODELED S**	55.97	140.33	170.23	127.58	45.61	102.35	132.11	132.37
5	**Existing ACTUAL S**	59.45	138.72	143.98	124.33	54.53	96.65	114.98	161.89
6	**Existing ACTUAL N (t)**	8 (22)	8 (19)	5 (14)	6 (18)	8 (21)	7 (21)	8 (23)	7 (18)
7	**Existing ERROR RATE**	–5.85%	1.16%	18.23%	2.61%	–16.36%	5.90%	14.90%	–18.23%
8									
9	**Proposed MODELED S**	63.25	106.68	126.49	129.88	46.61	93.26	89.27	130.17
10	**Proposed ACTUAL S**								
11	**Proposed ACTUAL N (t)**								
12	**Proposed ERROR RATE**								
13									
14	**GOAL % DIFFERENCE**	–10%	–18%	–18%	0%	0%	0%	–18%	0%
15	**MODELED % DIFFERENCE**	13%	**–24%**	**–26%**	2%	2%	**–9%**	**–32%**	**–2%**
16	**ACTUAL % DIFFERENCE**								

Table 16.10 Excerpt from the Analysis of Differences Between Existing and Proposed Application Accept Monthly Payment Task Models

	A	*B*	*C*
89		**Payment screen in Payment window:**	**4.92 seconds MORE in PROPOSED Application:**
90			***No default cursor placement in two places: when screen first comes up, and after Bank Number search requires 2 extra point (P) and click (C) operations (2.44 seconds) and 2 additional W1Proposed operations to wait for above fields to take focus (0.50 second)**
91			
92			
93			***3 additional Ks (in existing applications it is 2 tabs from payment amount to checking account number, 1 from checking account number to bank number,**
94			**1 from bank number to bank zip code; in proposed application it is 1 tab from payment amount to checking account number, 1 from checking account number to bank number, 5 from bank number to bank zip code) (0.36 second)**
95			***Proposed application has extra HPC after entry of bank number to do search for bank name/address; in existing applications, search happens automatically when tab out of bank number field (1.62 seconds)**
96			

Table 16.11 Summary of Modeling Results and Productivity Tests across Eight Tasks—after *Proposed* Application Productivity Test

	A	*B*	*C*	*D*	*E*	*F*	*G*	*H*	*I*
1		1	2	3	4	5	6	7	8
2		Accept Monthly Payment	Balance Transfer	Add Card to Account	Close Account	Change Address	Cash Advance	Refuse a Charge	Credit Line Increase
3									
4	**Existing MODELED S**	55.97	140.33	170.23	127.58	45.61	102.35	132.11	132.37
5	**Existing ACTUAL S**	59.45	138.72	143.98	124.33	54.53	96.65	114.98	161.89
6	**Existing ACTUAL N (t)**	8 (22)	8 (19)	5 (14)	6 (18)	8 (21)	7 (21)	8 (23)	7 (18)
7	**Existing ERROR RATE**	–5.85%	1.16%	18.23%	2.61%	–16.36%	5.90%	14.90%	–18.23%
8									
9	**Proposed MODELED S**	63.25	106.68	126.49	129.88	46.61	93.26	89.27	130.17
10	**Proposed ACTUAL S**	45.55	109.09	120.23	109.23	45.21	86.22	79.32	153.49
11	**Proposed ACTUAL N (t)**	8 (23)	7 (15)	8 (20)	5 (12)	6 (18)	8 (20)	5 (13)	8 (24)
12	**Proposed ERROR RATE**	38.86%	–2.21%	5.21%	18.91%	3.10%	8.17%	12.54%	–15.19%
13									
14	**GOAL % DIFFERENCE**	–10%	–18%	–18%	0%	0%	0%	–8%	0%
15	**MODELED % DIFFERENCE**	13%	**–24%**	**–26%**	2%	2%	**–9%**	**–32%**	**–2%**
16	**ACTUAL % DIFFERENCE**	–23%	–21%	–16%	–12%	–17%	–11%	–31%	–5%

fact be achieved by the new application *before* it was built and launched. Having this information early would of course be preferable to waiting until after launch and possibly finding out that the goals had *not* been met and that significant additional time and cost would be required to redesign and rebuild the new application to meet goals, further eating into the ROI. She turned to her internal usability engineering staff, whose time was already limited in supporting this and various other internal development projects. Together the business manager and the usability engineering manager decided to look for an outside consultant to help them assess whether or not the proposed design for the new Web-enabled application would in fact result in an application that would meet the identified productivity goals. This is where I came in.

Although I was generally aware of the cognitive modeling techniques presented and discussed in the literature over the past 2 decades, I had never encountered a consulting project for which they seemed like the right tool. Here was a case, however, in which they seemed to potentially be appropriate and useful tools. The primary usability issue was ease of use (i.e., productivity of high-frequency, expert users) rather than ease of learning. The project was in a stage that made traditional usability testing unrealistic: a subset of functionality for the new application was partly in the specification stage and partly in the prototype stage—even the prototyped parts, however, were not stable and realistic enough to even try to simulate ease-of-use testing. On the other hand, the documented design was specific and detailed enough to allow KLM. I proposed that we try to apply modeling, and the business manager and usability engineering manager accepted my proposal based on my high-level description of the KLM technique, my review of validation of the technique in the literature, and my clarification of what conditions must be met for modeling to be an appropriate and effective tool.

We carried out the project very collaboratively in the following steps:

1. *Model tasks.* For each of a set of eight high-priority user tasks, we modeled the task both on the existing and on the proposed application user interfaces.
2. *Run a productivity test on existing applications and use the results to refine the modeling technique.* We carried out productivity tests on the existing applications with highly-trained, experienced, and proficient users, and compared these actual task times to the task times predicted by the models of the existing applications. We then used these data to refine our modeling technique so that it minimized the error rate of the task times *predicted* by the modeling technique relative to the *actual* task times on the existing applications (note that here I am referring to the error rate of the modeling technique as

distinct from user errors during performance of tasks). Finally, we applied the refined modeling technique to the models of the proposed application.

3. *Compare modeling results to goals.* Next, we compared the predicted productivity gains (or losses) of the proposed application relative to the existing applications for each task with the productivity gain goals for each task.

4. *Address unmet goals.* For tasks in which the models predicted that productivity gain goals would *not* be met, we then analyzed the models in detail to determine why and to help drive suggested redesign directions.

5. *Run a productivity test on the proposed application.* After redesign and implementation of a beta version of the proposed application, we ran productivity tests to measure actual productivity on the proposed application as a final validation of the productivity gain goals as well as of the model predictions.

I describe each of these project steps in more detail in the following sections.

16.2.1 Model Tasks

For each of eight user tasks identified as high priority by the business manager, we modeled the task both on the existing and on the proposed applications. This modeling was accomplished in a highly collaborative process in which I worked closely with business and technical experts within the credit card company. The steps in the modeling process were as follows:

1. *Identify task variations.* Business management decided on a particular variation of each task to model. This was important because the modeling technique captures an exact sequence of user interactions and predicts an overall task time for that specific sequence only. General tasks have many possible variations. For example, in the balance transfer task, sometimes account holders want to transfer balances from a single other credit card, whereas other times they want to transfer balances from multiple other credit cards. Similarly, in the address change task, sometimes account holders have multiple addresses (e.g., primary home, vacation home, and business) and sometimes they have only one. Sometimes they want to change just one whereas sometimes they want to change more than one. Thus each general task had multiple points within it where variations such as these might occur. However, a given model can only capture one of each of these potential variations at each point where they occur. Thus business management first had to decide for each general task exactly which very specific variation of it they

wanted to model. This decision was based primarily on choosing the variation of each task that was most common, that is, the one that had the highest yearly volume of transactions. However, sometimes we deliberately chose less frequently occurring variations of a task because there was a specific expectation that the proposed application user interface would provide a particular advantage in that variation of the task, but perhaps not provide a significant advantage in a simpler variation of the task. Perhaps, for example, the proposed application was expected to provide faster processing when multiple addresses needed to be changed, but not when only one address needed to be changed. Even though changing a single address was a higher-volume transaction, we might model the multiple address change variation of the task for this reason.

Related to these sorts of variations were variations at an even lower level of detail. Additional decisions had to be made about variations such as complexity of addresses (for example, corporate addresses tend to have more lines and characters than residential addresses) and other free-form text entries (because modeling counts keystrokes), use of mouse versus keyboard entry techniques when either was possible, and whether users would cross check and validate their entries at any point during the task. We tried to base these decisions on what was known about actual usage and in fact in many cases interviewed and observed users to try to discover which practices were most common.

2. *Generate task scenarios.* Business staff then generated "task scenarios" that documented for me how the selected variation of a task would play out on both the existing applications and on the proposed application. This was necessary because I worked primarily from my own office rather than at the offices of the credit card company, and I did not have direct access to either their existing applications or their prototyped proposed application. We worked out a documentation format that consisted of a sequence of screen shots pasted into Microsoft PowerPoint. Each slide presented a screen shot in the sequence, with notes attached to it describing generally how the user would interact with that screen at that point in the task. (We have since adopted a relatively inexpensive software package by TechSmith called Camtasia Studio that records user interface interactions as a video clip as a way of documenting tasks on the existing applications, and this is much more efficient and accurate, but we did not have access to this program at the time we did the modeling described herein.)

3. *Prepare draft models.* Based on the task scenarios, I then generated two draft models for a given task—one on the existing applications, and one on the proposed application. These models are described in more detail later.

4. *Conduct model walkthroughs.* The task scenarios were rarely complete and detailed enough for me to generate complete and accurate models, but they were more than adequate for generating draft models. After drafting models, I would talk on the phone with the business experts who had generated the task scenarios for the tasks. We would walk through my draft models together, and they would clarify and correct as necessary. For example, the task scenario might indicate that the user would type a two-letter state abbreviation in one field and then a phone number in the next field, but I might have to follow-up in a walkthrough to learn whether the application auto-tabbed from one field to another after input in the first or whether a keystroke (tab) or mouse click was required to move the cursor to the next field. Usually we did this sort of walkthrough once, then I made additions and corrections, and then we did it again, just to be sure the final models were truly accurate and complete and also that they reflected task variations that were of interest to the project stakeholders.

5. *Report model results.* Once the models for a given task were validated as accurately reflecting the desired variation of that task, we reported the results of the modeling to the business manager. The results were in two forms. The first was numerical: We reported the total task time predicted on the existing applications, the total task time predicted on the proposed system, and the difference between them as a percentage (e.g., "the models predict a 24% productivity *gain* on the proposed application relative to the existing applications" or "the models predict a 13% productivity *loss* on the proposed application relative to the existing applications").

   The second reported result was more qualitative: I compared the two models for each task in great detail and analyzed exactly where and how the proposed user interface design and the existing design differed and how this difference resulted in the reported productivity gain (or loss). For example, if the proposed application was predicted to provide a faster task time than the existing system, it might be because a function could be accomplished within a single window (proposed application) rather than requiring opening and moving back and forth between multiple windows (existing applications). Or, if the proposed application was predicted to provide a *slower* task time relative to the existing applications, the reasons for the slower time might be that on certain displays (proposed application) users had to place the cursor in a field before typing because there was no appropriate default cursor position, whereas appropriate default cursor positions were provided on the current applications, and that more tabbing (keystrokes) was required to enter data in all fields on the proposed application relative to the existing applications. For the tasks for which goals were not met, this analysis helped the designers to strategize redesigns that might overcome these failures to meet goals. An example of these results is presented later.

I created the models in Microsoft Excel. Each task was captured in a separate Excel file. There were three worksheets in each file with the following purposes:

1. Capturing all parameter values to be plugged into the models
2. Documenting the model of the task on the existing applications
3. Documenting the model of the task on the proposed application

The models included the set of operators presented in Table 16.1 (see color insert).

The first five operators listed in Table 16.1 were drawn directly from Card *et al.* (1983) who included these operators in the KLM technique they originated and who reported average values from the literature for these operations. We initially chose an average typing speed (K = 0.20 second) based on input from the business staff. Later, when refining the modeling technique after comparing predicted and actual times for tasks on the existing systems, we changed this parameter to reflect a higher average typing speed (K = 0.12 second, see later discussion). Whereas Card *et al.* recommended inserting "M" (mental operators) in models to represent a variety of operations such as recall, simple skimming, and planning motor movements, we initially chose to include them to represent only two things: the need to move the eyes from one window to another and get oriented on the display currently in that window, and the need to compare something on screen to something in the head (for example, to recognize from the value in a particular field whether to route a task to another department). The rationale for this strategy was the expectation that these users were so practiced and proficient that it was likely that they would be able to conduct most M operators *in parallel with* the physical operators (i.e., H, K, P, and C) so that the M operators would not really need to be factored into the model to get accurate predictions of task time (the KLM technique, unlike some modeling techniques developed later such as cognitive perceptual motor(CPM)-GOMS, is unable to account for such parallel processing, which is known to occur). In fact it turned out that eliminating even the few M operators included in the initial models made the model predictions more accurate (see later discussion).

The next four operators listed in Table 16.1 represent four different system response time variables. We assigned time values to these operators based on data provided by the credit card company technical staff and on what we knew about the tasks being performed. We actually created two separate operators for each of these system response time variables—one for the existing applications (as listed in Table 16.1) and one for the proposed application. Initially we set them to the same values, but creating two distinct operators and incorporating one into the existing applications models and the other into the proposed application models allowed the possibility of easily "what-iffing" at some later time to see what impact different system response times on the proposed system might have on user productivity and on the hoped-for productivity gains.

Next in the list in Table 16.1 are two operators used to take into account the fact that, in these tasks, users were reading textual data on the screen—some in the images of paper forms and checks and some in application windows (for example, credit card numbers, names, addresses, phone numbers, payment amounts, etc.). Although the original KLM technique of Card *et al.* did not address tasks that included reading or speaking, later modeling techniques did, and I found empirically obtained average time values in the literature for speaking, one for speaking familiar, highly practiced syllables (SP), and another for speaking spontaneous, unpracticed syllables (SU) (John, 1995). I did not find comparable time values for reading, so I chose to simply use the values for spoken speech in my models, based on the rationale that reading is to some extent simply silent speaking.

Finally, one thing that was common across most of the tasks being modeled in this project was the need to skim an online image of a paper form to get an overview of a task before beginning it. These forms could be as long as 10 pages, including prose text and fill-in forms, and users typically skimmed every page of a form before beginning a task. They did not carefully read these forms word by word or page by page, however, so the reading parameters referred to above did not seem appropriate. Instead, we added an operator to our models to represent skimming a full page of a paper form. Because I could not find time values in the literature for this operator, we initially, somewhat arbitrarily (by having me informally test myself), assigned a time value of 5 seconds to this operator ("Skim" in Table 16.1). Later, when refining the modeling technique after comparing predicted and actual times for tasks on the existing applications, we changed this parameter to 1 second (see later discussion).

By assigning values to each of these model operators in a worksheet separate from the models themselves and then referring to them in the model worksheets, we created an easy way to "what-if." That is, once the models were constructed, we could simply go to the operator worksheet and change operator time values and see what effect this had on the models. We used this what-if strategy later to refine the modeling technique (see later discussion). (Although we did not do so on this project, note that another kind of what-iffing that can be done with these models is to explore small changes in design. For example, you could replace a small set of steps involving selecting an entry from a pull-down menu with steps involving filling in a text entry field to see what effect this would have on the overall task time.)

Tables 16.2 through 16.5 (see color insert) show excerpts from one of the models for a particular task on the existing applications. The first 7 columns and the first 40 rows in the worksheet contain part of the Accept Monthly Payment task model on the existing applications. The task is described conceptually in columns A to E in a hierarchical breakdown. Columns A and B identify the major steps in the task, for example, bring up a task description (i.e., select and open a task description from a list of tasks) on screen (row 5, column B in Table 16.2), search for the corresponding account record in the database (row 7, column B in Table 16.2), verify the account holder data currently in the record against the data provided by the account holder on the request form (row 14, column B in Table 16.3), review the task request form and check images (row 17, column B in Table 16.3), and perform the task (row 19, column B in Table 16.3).

Columns C and D then break each major step into conceptual substeps. For example, the step of performing the Accept Monthly Payment task involves first bringing the Payment application window into focus and loading the account holder data into it (row 19, column D in Table 16.3), then selecting the credit card to apply the payment to (account holders often have more than one credit card associated with an account) (row 22, column D in Table 16.4), then entering the payment data (row 25, column D in Table 16.4), and so on.

Column E then describes each task substep in terms of a sequence of actual interactions, given the user interface the model represents. For example, giving focus to the Payment application involves first moving the eyes to the appropriate window (row 19, column E in Table 16.3), then clicking on a button labeled "Load Account" (which both gives the window focus and invokes the button, which in turn brings the account holder data that was previously searched for in another application into the payment application) (row 20, column E in Table 16.3), and then waiting for the system to populate the account holder data in the Payment window (row 21, column E in Table 16.3).

Column F then breaks the interactions described in column E into the actual operators of the modeling technique. For example, moving the eyes to the payment window is captured as an M (row 19, column F in Table 16.3), clicking on the load account button involves first moving the hand from the keyboard to the mouse (H), then moving the mouse until the cursor points to the button (P), and then clicking the mouse button (C) (row 20, column F in Table 16.3). Waiting for the account data to populate in the Payment window is assigned the operator W3Existing, which represents the time on the existing system that it takes to carry already searched for data from one application to another (row 21, column F in Table 16.3).

Column G provided a place to capture assumptions being made in column F. The purpose of this column was threefold: it was partly to support the walk-through process described earlier to ensure that the validity of the draft models was verified; it was partly to document key assumptions made so that anyone reading the models in the future would be able to easily understand exactly what system behavior was being modeled; and it was partly to highlight aspects of a design that might help in the later comparison of existing and proposed models to analyze and diagnose an unmet goal.

For example, in row 7 (Table 16.2) in the Assumptions column (column G), note that when the account window does not currently have focus, a single click in a visible field in that window gives both the window and the field focus. That is, it is not necessary to click once to bring the window to the forefront and a second time to place the cursor in a particular field. Clearly this makes a difference in how many P and C operators are counted at this point in the model, and it was important to be very specific at this level of detail to get accurate predictions from the models. And as was found, in some cases in which a single click did the job on the existing applications, two clicks were required to serve the same purpose on the proposed application. This difference was discovered by comparing models and the proposed application was redesigned in response. The observations in the assumptions column helped highlight these differences.

Columns H through S (a sample is shown in Table 16.6) each represent a model operator (see Table 16.1). A number representing how many of a specific operator are involved in executing an interaction (as indicated in column F) is entered into the row and column representing that interaction. For example, because the interaction described for moving the eyes to the payment window entails a single M operator, a "1" is entered in the M column for the row containing that eye movement interaction (row 19 in both Table 16.3 and Table 16.6 in the color insert). Similarly, because the interaction describing clicking on a button in that window entails an H, a P, and a C operator, the columns for each of those operators contain a 1 in the row corresponding to that operation (row 20 in both Table 16.3 and Table 16.6).

Table 16.7 (see color insert) shows excerpts from the end of the model worksheets where the overall calculations are performed. The chart at the top of the table shows the overall calculations for the model on the *existing* applications. Columns H through S are the same as the columns shown in Table 16.6, representing each operator in the model. Here though, we see in row 55 the total count for each operator across the whole model. For example, the model for this task on the existing applications includes a total of 8 mouse clicks (row 55, column H), 54 keystrokes (row 55, column I), and so on. Row 57 calculates the total time spent during the whole task on each operator type. This is calculated by multiplying the count in row 55 by the time value of the parameter the column represents. For example, there are 8 mouse clicks (row 55, column H), and each mouse click is assigned a time value of 0.12 seconds (see Table 16.1), so the total time spent on mouse clicks during the whole task is 0.96 seconds (row 57, column H). The total task time is calculated by adding up all the time values across operator types in row 57. This value is calculated and displayed in column H of row 59.

In the chart at the bottom of Table 16.7, comparable results are given for the model for the same task on the *proposed* application. In addition, highlighted in maroon, the differences between the two models are displayed. In rows 68 and 70, the counts and times, respectively, for each operator are given for the proposed system *relative to the existing systems.* Positive numbers in these two rows indicate greater operator counts and times on the proposed application relative to the existing applications, negative numbers indicate fewer counts and times. Thus it can be seen, for example (row 68, column H), that across the whole Accept Monthly Payment task, there is one additional mouse click on the proposed application relative to the existing applications. In row 70 it can be seen that this adds 0.12 second to the overall time for this task on the proposed system relative to the existing applications.

In row 74, column H, the total difference in time is calculated. Here it can be seen that the whole task is predicted to take 11.57 seconds longer on the proposed application than on the existing applications. This represents a predicted 12.67% productivity *loss* on the proposed system for this task (row 75, column H), calculated as:

$$\frac{(\textit{proposed}\text{ modeled task time}) - (\textit{existing}\text{ modeled task time})}{(\textit{existing}\text{ modeled task time})}$$

16.2.2 Run Existing Applications Productivity Test and Use Results to Refine Modeling Technique

Next—and actually somewhat in parallel with our modeling effort—we carried out a productivity (ease-of-use) test on the existing applications. Our purpose here was twofold. We planned to eventually compare *actual* task times on *existing* applications to *actual* task times on the proposed application as a final validation of the proposed application, so the data collected now on the existing system would be used in that comparison later.

However, even though it was as long as a year later before we could run a productivity test on the implemented proposed application, we ran the existing applications productivity test earlier rather than later because we planned to use the data from it to refine and validate our modeling technique to increase our confidence in our comparison of models of tasks on the two applications. That is, if we at least knew that our models of the existing applications were relatively accurate, we would have more confidence that our models of the proposed application—and thus the predicted productivity gains (or losses) of the proposed relative to the existing applications—were also relatively accurate.

In the productivity tests we had eight representative (i.e., expert) users do each of the eight tasks on the existing applications three times. The three trials each user performed on a given task varied in their actual content (e.g., credit card number or payment amount) but were identical in the sequence of required steps and substeps and the number of keystrokes, mouse clicks, and other operations required to complete the task. Recall that these were already highly-trained and experienced users of the existing applications—people who had been using those applications daily for years—so no training was required before testing. Test tasks were presented on screen just as in real life, and a stopwatch was used to time completion of the tasks. In the end we had data points from a potential 24 trials (8 users doing 3 trials each) from which to compute an average actual total task time for each task on the existing applications.

After testing, we reviewed videotapes as necessary to determine whether users were in fact carrying out the tasks in the specific way that they were modeled. Recall that the modeling technique predicts the total task time for a very particular sequence of interactions. If the data from a productivity test that a modeled result is being compared to is based on a sequence of interactions that is different in some significant way from that which was modeled, it is not a valid comparison. In our productivity tests, users did not always do a task in the expected way. For example, although users were expected to always make sure they had pulled up the record for the correct account holder before proceeding with the task, and the models included this step, users did not always do this. When we found such discrepancies, we did one of two things. If a small number of users on a small number of trials departed from the way a given task was modeled, then we simply threw out those trials from our sample before computing the across-trial-across-user average time for that task. If most users on most trials departed significantly from the way a task was modeled, then the model was modified to reflect what most users did on most trials. In either case, only trials in which the way the user carried out the task matched how the task was ultimately modeled were included in the set of data points used to calculate the average total task time for a given task.

We then used the average total task times computed as described earlier to refine our modeling technique so that it minimized error in the modeled times relative to the actual task times on the *existing* applications. For example, the original model for the Accept Monthly Payment task on the existing applications, with all operators set to the *original* values given in Table 16.1, predicted a task time of 91.29 seconds (see Table 16.7). The actual time for this task on the productivity test, however, was 59.45 seconds. This represents an error rate of the model of 53.56%, which is very high. The error rate is calculated as

$$\text{(existing } \textit{modeled} \text{ task time)} - \frac{\text{(existing } \textit{actual} \text{ task time)}}{\text{(existing } \textit{actual} \text{ task time)}}$$

In fact, across the eight modeled tasks, initially about half of them had error rates exceeding the error rates of less than 20% typically reported in the modeling literature (Card *et al.*, 1983).

A variety of what-if scenarios were played out with the models, varying the time values assigned to the operators K (typing speed), M (mental operations), and Skim (time to visually skim a page of online text). The time values for these three operators that yielded an optimized set of existing applications model error rates across all eight tasks were K set to 0.12 second (which equals 90 words per minute—it had originally been set to 0.20 second or 55 words per minute), M set to 0 seconds (it had originally been set to 1.35 seconds), and skim set to 1.0 second (it had originally been set to 5.0 seconds). With these operator time values, the Accept Monthly Payment model for the existing applications now predicted a total task time of 55.97 seconds (Table 16.8 in the color insert). Again, the productivity test yielded an *actual* time for this task on the existing applications of 59.45 seconds. This represents an error rate in the modeled time of –5.85%, computed as in the previous equation.

This is a much lower and more acceptable error rate, more in keeping with error rates typically reported in the modeling literature. In fact, assigning these new time values to these three operators resulted in bringing the error rates of the models of all eight tasks on the existing applications into an acceptable range.

Another way of stating this is as follows: If we assumed that users were faster typists and faster skimmers than was assumed in the original models, and if we assumed that these users could in effect "parallel process" any very simple mental operations with the physical ones (thus making it unnecessary to include the time of any simple mental operators), then the existing applications models predicted actual times much more accurately (that is, with lower error rates) across all eight tasks. Thus, in the end, we used the time values for operators mentioned earlier, rather than the original time values given in Table 16.1, to generate our predictions for task times on both the existing and the proposed applications.

Note that in refining our modeling strategy we altered only the time values of those operators not firmly established in the literature. The K operator, which represents typing speed, will by definition vary across user populations. The Skim operator is one we made up and thus is not reported in the literature at all. For the M operator, we did not change the well established time value of 1.35 seconds—rather we simply did not include the operator in our models at all in the end, under the assumption that these very experienced users would always be able to parallel process these simple mental operators with the physical ones. For all other operators we retained time values well established in the literature.

The spreadsheet in Table 16.9 (see color insert) shows the following for each of the eight tasks:

- The *modeled* total task time on the existing applications (row 4)
- The *actual* total task time on the existing applications (row 5)
- The number of test users (N) and trials [(t)] in the productivity test on which the actual total task time was based (remember that eight users performed three trials each, but data from some users and some trials were thrown out for the reasons described earlier) (row 6)
- The error rate of the existing applications model, calculated as (row 7):

$$\frac{(\text{existing } \textit{modeled} \text{ task time}) - (\text{existing } \textit{actual} \text{ task time})}{(\text{existing } \textit{actual} \text{ task time})}$$

Here it can be seen that across the eight tasks, the error rates of the final models of the existing applications are all less than 20%, and some are quite low indeed. The average error rate is around ± 10%.

16.2.3 Compare Modeling Results to Goals

Having refined and validated our modeling technique by comparing modeled times to actual times on the existing applications, we next compared the final *modeled productivity gains* (or losses) of the proposed application relative to the existing applications for each task to the *productivity gain goals* for each task.

The spreadsheet in Table 16.9 presents the modeled total task times for each of the eight tasks on both the existing (row 4) and proposed (row 9) applications and the percent difference between them (row 15):

$$\frac{(\textit{proposed}\text{ modeled task time}) - (\textit{existing}\text{ modeled task time})}{(\textit{existing}\text{ modeled task time})}$$

These percent differences between models were compared to the *goal* percent differences for each task, also presented in the spreadsheet (row 14). The modeled percent differences are presented as red (and positive) numbers when the proposed application was predicted to be *slower* (that is, take more time) than the existing applications, green (and negative) when the proposed application was predicted to be *faster* (that is, take less time) than the existing applications, and orange (and positive) when the proposed application was predicted to be *slower* than the existing applications *but not by much.*

An inspection of these numbers in the spreadsheet shows that on tasks 2, 3, 6, 7, and 8, the models predicted that the productivity goals for the proposed application, given the user interface that was modeled, would be met and in fact exceeded, in some cases dramatically. This of course was good news to the business manager.

On the other hand, task 1 was predicted to fail to meet the productivity goal assigned to it rather dramatically, and tasks 4 and 5 were also predicted to fail to meet goals, although only minimally. We next focused our attention on those tasks that were predicted to fail to meet productivity goals.

16.2.4 Address Unmet Goals

For tasks for which the models predicted that productivity gain goals would not be met, we analyzed the models in detail to determine why and to suggest redesign directions.

The one task that stood out dramatically in the spreadsheet in Table 16.9 was task 1, the Accept Monthly Payment task. The goal for this task was that users would be *10% more productive* on the proposed application relative to the existing applications, but the models predicted that, given the proposed user interface design for this task, in fact users would be *13% less productive.*

We analyzed the models for this task to determine exactly what part of the proposed user interface was accountable for this predicted productivity loss. Our method for this analysis is illustrated in Tables 16.2 through 16.5, which represent an excerpt from the model of the Accept Monthly Payment task on the *existing* applications.

In Table 16.2, look at the Operators column, column F, in the spreadsheet. Notice that the text in the cells of this column, which represent model operators, appear in different colors. Each color bounds a coherent task step or substep. Thus for example, red represents all the operators in the get-task-description step, blue denotes all the operators in the get-account step, and so on. Also note that in Tables 16.2, 16.3, and the top of 16.4 the cells in this column have a light blue background with no pattern, but in Tables 16.4 and 16.5 the cells in this column have a pattern added to the background blue. No pattern indicates cells in this model of this task (*existing* Accept Monthly Payment) for which operators *are identical to* those in the corresponding cells in the model being compared (*proposed* Accept Monthly Payment). A pattern indicates cells in which operators in the corresponding cells for the two models do in fact differ. Thus patterned cells highlight differences between the two models, and the color of the operator symbols indicates the particular step or substep in the task for which the models differ.

The overall time difference between the two models for this task was thus broken down into details. Table 16.10 (see color insert) shows an excerpt from this analysis of the two Accept Monthly Payment models. Here we see that 4.92 seconds of the difference between the two models (which totaled 7.28 seconds; see Table 16.8) occurs because on a single page more operators were required on the proposed application to place the cursor in fields, tab between fields, and execute a bank number search. Other differences, similarly analyzed and documented, accounted for the remaining difference of 2.36 seconds, and these were documented in a similar fashion. The designers then revisited the design of the Accept Monthly Payment task on the proposed application and looked for ways to eliminate the identified additional operators.

This sort of analysis was also carried out for tasks 4 and 5, which were also predicted by the models to fail to meet productivity goals. Design changes were made to these proposed task user interfaces as well.

In theory, we could have then remodeled these tasks on the redesigned proposed application to predict whether the redesigns would in fact result in achieving the increased productivity goals for these tasks. Instead, however, the business manager decided to simply go ahead and make changes driven in this way by the model analyses, and then wait until it was possible to perform a productivity test on the proposed application beta version to determine whether the goals had been met by the redesigns.

Thus the proposed user interface for three of the eight tasks (tasks 1, 4, and 5 in Table 16.9) were redesigned in response to the modeling results. Note that they were *not* redesigned in ways that introduced inconsistencies into the overall conceptual model and page design conventions of the proposed application as a whole. Care was taken to eliminate operators and streamline interactions without violating the overall user interface architecture of the proposed application.

16.2.5 Run Proposed Application Productivity Test

After redesign and implementation of a beta version of the proposed application, we ran a productivity test to determine actual productivity on the proposed application, as a final validation of the goals themselves, as well as of the model predictions.

We had to conduct this productivity test slightly differently from the one on the existing applications, because there were in fact no highly-trained and experienced users of the proposed application yet; we ran the test as soon as the beta version was stabilized, as always, hoping to get information on productivity gains (or losses) sooner rather than later, when changes would be cheaper and easier to make. Thus we had to simulate expert usage. We did this by providing test users with training (several hours long) and an opportunity to practice (for several additional hours) before running our test. Other than that, the testing was run just as described earlier for the productivity test of the existing applications. We were not entirely sure we would be able to get users to potential expert peak performance in a brief training session, but in fact it appeared that we did. This was good both from the point of view of simulating potential productivity and as a sign of how easy the proposed application would be to learn.

Table 16.11 (see color insert) shows the same spreadsheet as Table 16.9, with the data from the productivity test on the *proposed* application now filled in. The last row (row 16) in the spreadsheet in Table 16.11 shows the percent difference between the *actual* average total task times on each task on the existing and proposed applications:

$$\frac{(\textit{proposed}\text{ actual task time}) - (\textit{existing}\text{ actual task time})}{(\textit{existing}\text{ actual task time})}$$

Note that these are *actual* performance differences—no *modeled* results are involved here. Note also that for all eight tasks, the proposed application shows a productivity gain (a negative percent difference)—despite the fact that the training and practice sessions were relatively brief.

Remember that task 1, Accept Monthly Payment, had shown a productivity loss of 13% when modeled (row 15, column B). Also remember that this productivity loss was analyzed in detail through the models (see 16.2.4 Address Unmet Goals) and that the user interface to the proposed application was then *redesigned* to address this productivity loss. In the productivity test on the redesigned proposed application, a 23% productivity *gain* was seen relative to the existing applications (row 16, column B)—surpassing the business goal of a

10% gain (row 14, column B). Clearly the redesign succeeded in solving the problem the modeling had revealed.

Also note that the error rate of the model for the proposed application relative to the actual productivity data on this task (row 12, column B) is quite high (38.86%). This in part reflects the fact that the user interface *modeled* and the user interface *tested* for the proposed application were *not the same*—the one *tested* had in fact been *redesigned* in response to the modeling results. We did not bother, as stated earlier, to go back and remodel after redesigning (as we might have if a beta version was not going to be available for a long time). We simply went next to getting actual productivity data. Note that the high error rate is what you would expect in this situation. Because the user interface modeled and the user interface tested were not the same, you would expect a high error rate between them. In addition, you would expect the error rate to be positive, rather than negative, because this indicates that the actual (redesigned) proposed user interface allowed *faster* task times relative to the modeled user interface. Another way of expressing this finding is that the *model* overpredicted task times, and this is what you would expect because the *actual* user interface was redesigned specifically to improve upon the original, modeled user interface. The fact that the models show logical results in such situations provides further confidence in the modeling technique.

Tasks 4 and 5 were the other two tasks—like task 1—that were revealed by the modeling to fall short of business goals and that were redesigned in response to the analyses the models facilitated. Table 16.11 (row 16, columns E and F) shows that for these two tasks, the model results showed a small productivity *loss* for the proposed application as modeled, but the productivity tests showed a significant productivity *gain* for the proposed application *as redesigned in response to modeling results and analyses.* These findings again highlight a major benefit of the modeling technique—the ability to detect and eliminate design flaws interfering with productivity improvement goals *before* implementation of a proposed application design.

Thus the productivity test on the proposed application in its beta form—even though test users were minimally trained and not highly experienced, and the beta application was not 100% stable—validated that both the original proposed application design on a number of tasks and a redesign of other tasks would succeed in meeting the business goals that in turn would cost-justify the whole development effort.

Note also that on tasks for which the design did not change from the design modeled (i.e., tasks 2, 3, 6, 7, and 8), error rates for the proposed application models are reasonable, and *modeled* percent differences are fairly consistent with *actual* percent differences. That is, the refined modeling technique seemed to do a good job in predicting actual productivity gains.

Finally, note that on task 3, the final comparison of productivity tests on the existing and proposed applications did show a productivity gain for the proposed application, but the gain (16%) did not actually meet the goal gain (18%). Although this result was noted, the business manager decided that because the goal gains were exceeded on a number of other tasks, this was acceptable—the overall productivity gain *across all tasks* would be enough to cost-justify the overall development effort.

16.3 CONCLUSIONS

Modeling proved to be a valid and very useful cost-justification technique for this project. It allowed the business manager to predict fairly accurately—long before the proposed application was developed and launched—whether or not productivity gains required to cost-justify the new development effort would in fact be achieved. In addition, it enabled an analysis of those tasks predicted to fail to meet business goals, which in turn drove redesign to achieve those goals—again, long before launch, when it was much cheaper and easier to make design changes.

Running productivity tests on the existing applications early in the whole process allowed us to refine and validate the modeling technique, making us more confident in our interim conclusions from modeling alone well before we were able to collect productivity data on the proposed application. The productivity data from the proposed application—when we were finally able to collect it—in turn validated the model results we had generated earlier. In addition, we demonstrated that goal productivity would be achieved quite quickly; it was in fact achieved on most tasks after only a brief training program and several hours of practice time.

In the end, both the client organization and I felt that a valuable tool had been refined and validated and that this tool could be used in a similar way in the future within this organization to minimize risk and ensure ROI in software development projects. At the time of this writing, we are continuing to model new tasks for a new application now under development. This time we are also focusing on a transfer of the modeling skills from me to staff within the client's organization.

REFERENCES

Atwood, M. E., Gray, W. D., and John, B. E. (1996). Project Ernestine: Analytic and empirical methods applied to a real-world CHI problem. In M. Rudisill, C. Lewis, P. B. Polson, and T. D. McKay (Eds.), *Human-Computer Interface Design* (pp. 101–121). San Francisco: Morgan Kaufmann.

Card, S. K., Moran, T. P., and Newell, A. (1983). *The Psychology of Human-Computer Interaction.* Hillsdale, NJ: Lawrence Erlbaum.

Carroll, J. M. (2003). *HCI Models, Theories and Frameworks.* San Francisco: Morgan Kaufmann.

John, B. E. (1990). Extensions of GOMS analysis to expert performance requiring perceptions of dynamic visual and auditory information. In *CHI '90 Conference Proceedings* (pp. 107–115).

John, B. E. (1995). Why GOMS? *Interactions,* 2 (4), 80–89.

John, B. E., and Kieras, D. E. (1996). Using GOMS for user interface design and evaluation: Which technique? *ACM Transactions on Computer-Human Interaction,* 3, 287–319.

17 CHAPTER The Rapid Iterative Test and Evaluation Method: Better Products in Less Time

Michael C. Medlock Microsoft Human Resources
Dennis Wixon Microsoft Games Studios
Mick McGee Oracle Corporation
Dan Welsh Microsoft MSN Division

17.1 INTRODUCTION

Pretend you are running a business. It is a high-risk business, and you need to succeed. Now imagine two people come to your office:

- The first person says, "I've identified all problems we might possibility have."
- The second person says, "I've identified the most likely problems and have fixed many of them. The system is measurably better than it was."

Which one would you reward? Which one would you want on your next project?

In our experience, businesses are far more interested in getting solutions than in uncovering issues. In addition, businesses want solutions sooner rather than later and want assurance that the solutions provided are good ones. We feel that these fundamental truths have often been forgotten in the literature on usability methods (Wixon, 2003). The rapid iterative testing and evaluation (RITE) method is a usability method designed with these business truths firmly in mind. Its focus is on quickly determining the sufficiency of solutions rather than solely identifying problems.

In this chapter we will do the following:

- Provide a definition for RITE
- State the business case for RITE
- Outline the pitfalls of using RITE
- State the conditions for using RITE effectively
- Provide successful case studies of RITE
- Draw a set of conclusions about RITE

Like the rest of this book, this chapter is about helping usability engineers convince their businesses that usability activities are a good return on investment (ROI). Thus we are purposefully not recounting many arguments used to convince other audiences more familiar with traditional usability methodologies, such as academics or other researchers, about the validity of the RITE method. For this audience, reliability and validity are mechanisms for establishing truth. A business audience is less concerned about ultimate or scientific truth. Business is usually concerned about effectiveness and efficiency and the costs related to both. Consequently validity in a business context is established through observation, anecdotes, examples, clarity of presentation, the credibility of the messenger, and most importantly the immediate perception of the relevance of the information to the business goals. This chapter is written with that business perspective in mind. For further, and more academically oriented, discussions on the validity of the RITE method, see Medlock *et al.* (2002).

17.2 DEFINITION FOR RITE

In essence, the RITE method is a discount usability test conducted in a *fast* and *highly collaborative* manner. As a general rule it can often be used in situations as an alternative to a discount usability test, for example, for medium-size and medium-complexity products or large and complex products whose components can be broken into medium-sized or less complex chunks. A RITE test shares the same initial four basic principles that usability tests share as outlined by Dumas and Redish (1993):

1. The primary goal is to improve the usability of the product. For each test you also have more specific goals and concerns that you articulate when planning the test.

2. The participants are real users.
3. The participants do real tasks.
4. You observe and record what participants do and say.

However, RITE differs in the following ways:

1. Key *decision-makers must observe the participants* with the usability engineer. In this case a "decision-maker" is someone who has the *authority* to authorize a change to the user interface. Identifying the minimal set of these individuals is critical to the success of RITE.
2. The data are analyzed after *each participant* or at least after each day of testing.
3. Changes to the user interface can be made *as soon as a problem is identified and a potential solution is clear.* In situations in which the issue and solution are "obvious," the change may occur after one participant. Later in the chapter we will delve into this validity issue in more detail.
4. *Resources are available to make changes* to the user interface during the course of testing. In this case a "resource" is someone who has the *ability* to make a change to the user interface.
5. *The changed interface is tested with subsequent users* to see if the changes solved the issues previously uncovered without introducing new problems.

The time savings from a RITE style test comes from the fact that the most of the data analysis, presentation, discussion of results, determination of changes, and testing of changes made are done during testing rather than after testing. Therefore the elapsed time from the start of the test to an improved product is shorter. A visual representation of RITE compared with a "standard" discount usability test is shown in Figure 17.1.

17.3 BUSINESS CASE FOR RITE

When used correctly in the appropriate context, RITE will create a measurably better product in a shorter period of time than standard usability testing by finding and fixing more issues in that shorter time. Therefore the key business ROI impact is on faster verifiable product quality. In addition, RITE has the benefit of leading to better team dynamics by transforming the usability lab into

Standard Usability Testing

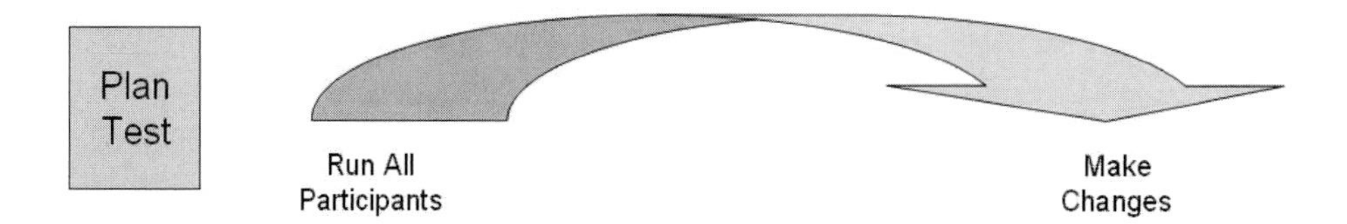

Rapid Iteration and Evaluation testing (RITE)

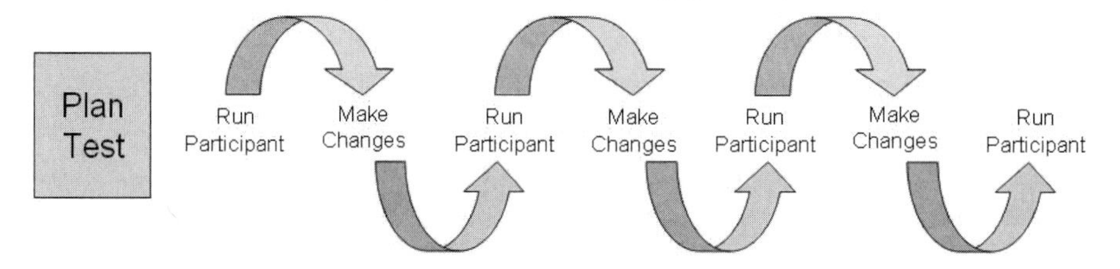

FIGURE 17.1 RITE testing versus "standard" usability testing.

a design environment as well as a testing environment. The focus of RITE is on using participants to determine whether a change fixed an issue, in addition to uncovering new issues. Ultimately it is more important to get the team to fix problems and to determine the likelihood that a "fix" has solved a problem than to agonize over whether every problem has been found. As such the RITE method shares the goals, philosophy, and approach of quality assurance in software.

Traditionally, the literature on sample sizes in usability studies has focused on the likelihood that a problem will be found (Lewis, 1990, 1991, 1993; Nielsen and Landauer, 1994; Nielsen *et al.*, 2002; Spool and Schroeder, 2001; Virzi, 1992; Woolrych and Cockton, 2001). This research has often *not* focused on what we as practitioners and what many business people view as the primary goal of usability in a commercial setting: shipping an improved user interface as rapidly and cheaply as possible (Wixon, 2003). The quickest and most effective way to achieve an effective user interface is to fix problems as soon as the development team has identified them, agreed upon the probable cause, and proposed a plausible solution. These solutions can then be tested with subsequent participants to establish confidence that they would work in practice. "Solutions" that do not work or that cause other problems can be removed or revised.

17.3.1 RITE Method Results in More Issues Fixed

Uncovering issues in software products is rarely difficult—what is difficult is getting them fixed. The following four reasons explain why usability issues that are discovered do not get fixed:

1. Usability issues are discounted. The decision-makers believe that the issues uncovered will not occur when the product is shipped or will not occur for most users.
2. Usability feedback arrives late. Feedback that is available when decisions are being made is far more likely to be taken into account than feedback that arrives after the decisions have already been made. The delay between when a feature is implemented and when usability feedback is delivered to the team is a barrier to using those recommendations.
3. Teams are uncertain whether a proposed solution will successfully fix the problem. The team doesn't want to undertake a potentially difficult or time-consuming fix if they are not convinced that the changes will not fix the problem. The lack of verification that the fix will solve the problem forces the team to implement the usability recommendations on faith, rather than on a demonstrated history of accuracy. Furthermore, if subsequent tests show that some fixes haven't worked, the feedback usually arrives too late to be useful (i.e., see item 2).
4. Fixing problems takes time and resources, and development and design resources are often scarce.

The RITE method effectively addresses all of these issues:

1. The usability issues observed during testing are "believed," because the decision-makers are in the room observing when the issues arise. The usability engineer in conjunction with decision-makers usually predefines what tasks participants should be able to accomplish, increasing buy in. In addition, through their constant involvement and observation, the decision-makers also buy into issues for which there are no a priori pass/fail criteria.
2. The usability feedback is delivered to team members immediately, right after the issues occur or as they are occurring in "real time."
3. The team has measurable assurance that the solutions are successfully fixing the problems because the fixes are tested by subsequent participants. In

addition, testing the fix so quickly allows the team to verify good fixes as working and catch poor fixes and correct them. If desired the team can calculate the likelihood that an issue has been fixed by calculating binomial confidence intervals. The appropriate equations for binomial confidence intervals can be found on the Web by searching for the words "binomial confidence intervals" or in articles on the subject specifically for usability engineering situations (Lewis, 1996). For example, if you have made a change and subsequently run five participants without seeing an issue, you can use binomial confidence intervals to calculate that you are now 90% confident that between 100% and 55% of the population would not fail the task. Note that this is much better than having no confidence if you run no participants.

4. Fixing the discovered issues is planned for and agreed upon before testing. Of course, as fixes are identified, a prioritization of the things to fix is needed. This is particularly true if the number of issues exceeds the amount of resources available, as is often the case.

17.3.2 RITE Method Results in More Issues Found

When issues are fixed, they often allow users to travel much deeper into an application—meaning they can uncover new issues in areas where previous participants could not venture. In addition, once a particular portion of an interface has been looked at by a certain number of participants, it becomes much harder to find "new" issues with the area. Much of this can be explained by analogy to theories of optimal foraging strategies in the biological sciences. Optimal foraging strategies are based on the idea that an organism will choose to gather resources in such a way as to invest the smallest amount of time/energy possible for the maximum gain. During a panel discussion at the 2002 Usability Professionals Association conference Nielsen gave an example of a fox hunting for rabbits (Nielsen *et al.*, 2002). Initially the fox stays in the best hunting grounds and catches rabbits there, but after some time, rabbits get more scarce, and the fox moves to another part of the forest. The key insight is that the fox will move *before* having eaten the last rabbit in the first hunting area. It's not worth the energy to capture all the rabbits. Similarly, in usability, it is better to abandon testing for a given iteration *before* you have tested it thoroughly, because it would take too many resources to track down the last remaining usability problems. The difference between the foraging metaphor and actual usability engineering is that in usability, any "rabbit" that survived in the old hunting area is likely to follow the "fox" over to the new hunting area. That is, any usability problem you didn't find in a particular iteration is still present in subsequent iterations; thus you get another chance to find it.

This analogy can be seen in practice in the case studies later in this chapter and in other RITE style tests we have done. It seems that this could be a fruitful area for further research into the efficacy of the RITE method.

17.3.3 RITE Method Results in Better Team Dynamics

In our experience the RITE method has been more enthusiastically embraced by development teams that have used it than discount usability testing has been. All decision-makers are more likely to become an integrated and excited part of the design process without becoming an interference.

- At Oracle team members have given rave reviews of the RITE method. For example, one team member wrote:

  > There's no question our user-centered design process has been enhanced by adopting the RITE methodology. To me, the biggest change is the removal of the implicit adversarial relationship between the usability specialist and developers that is an unwanted byproduct of usability testing. A typical usability test focuses on the problems identified. The RITE methodology is inherently about the problems solved. . . . A typical usability test presentation says to the development team, "your products sucks," right before saying, "here's our opinion on how to make it better." A difficult communication situation at best—It should be no surprise that there are many usability specialists complaining that developers do not listen. A typical RITE test presentation says to the development team, "here's all the problems we fixed already" and "here's how much better the product is than before we fixed the problems" all in one presentation. . . . Commence the high-fives between developers and usability specialists.

- Mark Terrano, the lead designer for Age of Empires II, was highly satisfied with the method and wrote about it in a designer diary on the Web (Terrano, 1999). He also coauthored the original paper on the method (Medlock *et al.*, 2002).
- Most teams who have worked with the Microsoft Game Studios User testing team and have experienced use of the RITE method now want more future usability tests be run in a similar fashion.

- Microsoft Games Studios upper management has advertised the use of the RITE method as a *selling point* for potential partners. The games industry is incredibly competitive, and the ability for publishers to sign game companies to make games for them is crucial to business strategy.

17.3.4 RITE Method Degrades Gracefully

Like all usability methods things can and will go awry when RITE style tests are run. Fortunately, in our experience, when a RITE test does go awry, it almost always gracefully degrades into a basic discount usability test in which all the participants scheduled are run before any changes are made. The reason is that in our experience if a RITE test has fallen on its face it most often revolves around the participation of the "decision-makers" or the "resources" used to make the changes to the interface during testing. For example,

- The true decision makers have not accurately been identified. In this instance the people involved do not have the authority to make the changes to the interface, which means that the issues seen and the recommendations to make changes will have to be shown to this person first before changes can be made. Often this person first has to be identified and found. This often means that the course of participants has to be run through just like a standard discount usability test before changes are made.
- The decision-makers do not show up. As in the previous item, you have to wait until you can show the decision-makers the issues and recommendations before moving forward. Once again this lengthens the time until a fix can be made. However, this issue is usually less serious than the former issue because the usability engineer can potentially set up set times to go over the results and recommendations during the course of testing if the particular decision-maker cannot make it to the tests.
- The resources for making the changes are not identified or unavailable to make changes during testing. If this happens the usability engineer is virtually guaranteed that the best they will be able to do is run a standard discount usability test through to the end before changes are made.

A good example of this graceful degradation is exemplified in the original article on the RITE method during the documented RITE testing of MechWarrior IV (Medlock *et al.*, 2002). The RITE test of MechWarrior IV did not go according to plan. Although the problems were analyzed at the end of each session, it was impossible to implement a solution until four participants were run because the resources to make the changes had to be used for something else. Fortunately we were able to secure the resources in time to make changes after the fourth participant and make changes. The subsequent four participants already scheduled were run, and the "fixed" issues were not seen again. In that sense the MechWarrior IV test in question was very close to a traditional discount usability test with a set of changes made once in the middle.

17.3.5 The Pitfalls of Using the RITE Method

RITE is not a panacea; it is a tool for a particular context of use. And like all tools it has disadvantages and can be potentially misused. For starters, RITE is not the correct method for every situation because its goals are not appropriate for every situation. For example, RITE should not be used in situations in which

- The team needs to uncover actual work practice and real tasks (Squires and Bryne, 2002).
- The team needs to decide who the user really is (Pruitt and Grudin, 2003).
- The system is almost entirely built and there is the ability to only make one or two large expensive changes (Whiteside *et al.*, 1988).
- The engineer is trying to perform a test to use for comparison against other similarly run tests (e.g., a benchmark study).
- The decisions-makers are not committed to spend the time to go through the RITE process.
- There are too many decision-makers, making it infeasible to get all of them to participate in the sessions as they are in progress.
- Resources for making changes to the interface cannot be secured for the RITE style test.

In a previous section we listed ways in which a RITE test can go awry. In addition to these there are also dangers to using the RITE method. Some of them are the following:

- Making changes when the issue and/or the solution to the issue is unclear. Poorly solved issues can "break" other parts of the user interface or user experience.

- Making too many changes at once. If one of these changes degrades the user experience, it may be difficult to assess which of the changes is causing the problem.
- Not following up at the end of the test with enough participants to assess the effectiveness of the changes made. Without this follow-up there is no guarantee that the changes made were any more successful than the previously-tested user interface.
- Missing meta issues that are hard to detect when looking at things in the detail that RITE style testing does. Sometimes fixing things one at a time either creates, obfuscates, or does not fix a serious meta problem.
- Missing less frequently occurring, yet important, usability issues. By using very small samples between iterations it is possible that less frequently seen issues will slip through unnoticed. This is particularly true if the task is broad and the domain is known to have many such issues (Spool and Schroeder, 2001) or if the topic is one in which the practitioner cannot afford to miss an error. In either case the solution to this issue is to run more participants overall and between iterations. How many more depends on a number of things that are beyond the scope of this chapter. That topic by itself could be a whole article. Many researchers have tackled this topic, and we have found the work of Lewis to be excellent (Lewis 1990, 1991, 1993; Nielsen *et al.*, 2002). He has dealt extensively with this issue and provides well-substantiated formulas for calculating how many participants are needed depending on the situation.

Additional risks in applying RITE method or any other method that uses participants are the following:

- The tasks are not representative of typical user work or tasks that make the product attractive to new users or users of competitive products.
- The users tested are not representative of the target users or customers.

17.4 USING RITE EFFECTIVELY

Users of RITE should adapt the method to their needs and context. In practice the RITE method constitutes a methodology, that is, a collection of methods reflecting a philosophy and approach to applied product testing. Having said this, the RITE method can fail if certain conditions are not met. The major preconditions for RITE regardless of the way a team chooses to use it are the following:

1. The "real" decision-makers must be identified for inclusion. If the people you have with you do not have the authority to make changes, then the enterprise is destined for failure. Note that this condition is generally true regardless of the method used.
2. The decision-makers of the development team must participate to address the issues raised quickly. Note that this is more than a commitment to attend and review. This is a commitment to actually make decisions and changes during the time the test is being run.
3. The usability engineer must have experience both in the domain and in the problems that users typically experience in that domain. Without this experience determining whether an issue is "reasonably likely" to be a problem is difficult. For someone who is inexperienced with a domain, a more traditional usability test is more appropriate.
4. The design team and usability engineer must be able to interpret the results rapidly to make quick and effective decisions regarding changes to the prototype or whatever is being tested.
5. The ability to make changes to the system or prototype very rapidly must be present (e.g., in less than 2 hours to test with the next participant or before the next day of testing depending on context). This means that development time and resources must be scheduled in advance to address the issues raised during testing, and the development environment must allow for those rapid changes to be made.
6. Enough time must be available between participants to make changes that have been decided upon. How long this time is depends on how quickly teams can make decisions and implement them. For example, text on Web-based interfaces can be changed very quickly (minutes), whereas changes to the behavior of an underlying database can take days.

It should be noted that often the set-up for a RITE style test takes more time and energy up front and pays off in all the items listed in previous sections. For an idea of the revised timeline for a RITE style test see Figure 17.2. It should be noted that this schedule seems correct for the context in which the authors have done their work, but many factors may influence the reader's context.

A number of things should be done during the planning phases to make everyone's tasks much easier during a RITE test. Some of these include the following:

Monday	Tuesday	Wednesday	Thursday	Friday	Sat/Sun
1 Plan	2 Plan	3 Plan & Schedule Participants	4 Plan & Schedule Participants	5 Plan & Schedule Participants	6 / 7
8 Plan & Schedule Participants	9 Run Participants & Analyze Data	10 Fix Issues, Analyze Data	11 Run Participants, Analyze Data Fix Issues	12 Fix Issues, Analyze Data	13 / 14
15 Run Participants, Analyze Data & Fix Issues	16 Fix Issues, Analyze Data	17 Run Participants, Analyze Data & Fix Issues	18 Fix Issues	19	20 / 21

FIGURE 17.2 An example of a timeline for RITE testing.

- Have someone on the team keep a record of the changes made and when they were made. It is very difficult for a usability engineer to accomplish these tasks during testing. Make sure that this record is easily accessible so that all team members can see it.
- Do not schedule too many participants per day and allow for plenty of time between participants. We have often found that even three participants per day can push everyone's boundaries when changes need to be made between participants. In other domains, three participants or more per day may be fine—or even one participant every 2 days. The number of participants depends entirely on the engineer's situation.
- Set up a dedicated place for the people making the prototype or coding changes to work close to the observation site, ideally, a place where they can physically see the usability participants.
- Consider saving full versions of the application between changes so that previous versions can be examined or reverted to if problems are encountered with fixes.
- Set expectations for the decision-makers about what a RITE test is and what their roles will be. A good way to do this is to give the decision-makers a script to show what a RITE test is, and what their roles in it will be.
- Set up a format for recording issues that can instantly be translated into a report format to assist turnaround time for the final results. We have provided a recording template that some of us have used along with an example

of how we have used it. Table 17.1 is a fictitious example of tracking the effectiveness of a RITE study. The columns represent participants and the rows represent issues. The participants are numbered in the order they were tested. To make this into a blank template simply erase all the shading and text.

In the fictitious example in Table 17.1 we have run 12 participants. This is an idealized example and illustrates possible outcomes of a RITE test. However, in practice its outcome is not far from the actual outcome of a RITE test:

- For issue 1 the problem was very clear and very severe. The team felt they understood it well enough to make a fix after just one participant. They made the fix and the next 11 participants did not experience the problem.
- For issue 2 the problem did not get fixed during this test either because the team did not feel it was that serious or it was too difficult to attempt a fix in the given time frame of the test.
- For issue 3 the problem was seen by the first three participants, but the team could not agree on a fix or thought that the problem might not be seen again. They waited until after participant 6 and then fixed the problem and it was not experienced by users again.
- For issue 4 a fix was tried after the first three participants, and it was not successful (i.e., the problem recurred). A new fix was tried after participant 6 which appears to have been successful.
- Issue 5 did not surface until participant 7 but was noted for three participants. It showed up because fixing issues 1 to 4 allowed users to complete more tasks.
- Issue 6 caused failures and was fixed successfully after three participants.
- Issue 7 showed up with participants 4 through 6 and caused failures. The fix reduced its severity but did not eliminate it. A new fix was tried after participant 9 and eliminated the problem entirely.
- Issue 8 required three fixes to eliminate.

17.5 CASE STUDIES

The RITE method is not new—practitioners do this and other activities like it all the time. Usability engineers at many different companies have used RITE,

Table 17.1 Sample RITE Table: Fictitious Issues and Fixes History

Issue/Participant	*P1*	*P2*	*P3*	*P4*	*P5*	*P6*	*P7*	*P8*	*P9*	*P10*	*P11*	*P12*	*1st Change*	*2nd Change*	*3rd Change*
Issue 1: brief description	X												Note fix		
Issue 2				X		X				X					
Issue 3	X	X	X			X							Note fix		
Issue 4	X	X		X	X								Note fix	Note new fix	
Issue 5							X	X	X				Note fix		
Issue 6	Fail	Fail	X										Note fix		
Issue 7				Fail	Fail	Fail	X	X	X				Note fix	Note new fix	
Issue 8	X	X		X		X		X	X				Note fix	Note new fix	Note new fix

or some method that is very similar to it, for some time, but there is surprisingly little written about methods like it in the literature. A quick internal survey of usability engineers and practitioners at Microsoft and on a private usability discussion group found that 33 of 39 respondents had used a similar method of very rapid iterations and fixes at least once.

What follows are three case studies in which the RITE method has been used successfully in different business situations by different businesses. Two are from Microsoft, and one is from Oracle. The first is on the use of RITE to improve the tutorial of the game *Age of Empires II*. The second is on the use of RITE to improve interactive voice-response e-mail. The final case study is on Microsoft's use of RITE to improve the Digital Imaging Library for storing and displaying photos.

17.5.1 Case Study 1: Microsoft Age of Empires II Tutorial

Age of Empires II is a real-time strategy game created by Ensemble Studios and published by Microsoft Game Studios. The overall goal in *Age of Empires II* is to build a civilization and conquer the other civilizations that oppose you. You gather resources that you then use to create military or technological advances to achieve your goals. It is a complex game that requires users to understand new rules and master new skills to be effective. The original *Age of Empires* was a very successful game, but the game designers wanted to expand their demographic, and this meant going after people who didn't play this type of game or didn't normally play games. To achieve this goal the game designers could have chosen to change the complex rules by which the game is played. However, after much fiddling they determined that the rules appeared to be the very thing that made the game fun—so the challenge became teaching new users how to play the game in a fun way. A tutorial was deemed to be the solution.

Before testing, the usability engineer and the team developed a list of tasks and concepts that participants should be able to do and/or understand after using the *Age of Empires II* tutorial. Issues were identified for which there would be zero error tolerance, for example, learning how to move units or gather resources. Once testing ensued at least one decision-maker on the development team, who was authorized to make changes, was present at every session (e.g., the program manager, game designer, or development lead). After each participant, the team would quickly meet with the usability engineer, go over issues seen, and do one of the following things:

1. Attempt a fix, and then use the new prototype with the remaining participants.

2. Start to work on a fix, and use the new prototype with the remaining participants as soon as it was available.
3. Collect more data (e.g., more participants run) before any fix is attempted.

The results are summarized in Figure 17.3, which is a record of failures and errors over time (as represented by the participants) on the *Age of Empires II* tutorial. In addition, the graph shows the points at which the tutorial was revised. Every vertical line on the graph indicates a point at which a different version of the tutorial was used, that is, when a change was made. Changes were implemented between participants 1, 2, 5, 8, 9, and 10 (6 iterations).

You can see in Figure 17.3 that the prototype was changed after the first participant. It is instructive to examine an issue that caused the team to make a fix after having observed so few instances of an issue. In the second part of the tutorial participants are supposed to gather resources with their villagers. One of the resources that they are instructed to gather is wood by chopping trees. However, initially there were no trees on screen and as a result the first participant spent a long time being confused as to what to do. The problem was clear, and so was the solution—place some trees within view and teach users how to explore to find trees off-screen. Both of these were done, and the issue did not appear again with the next 15 participants.

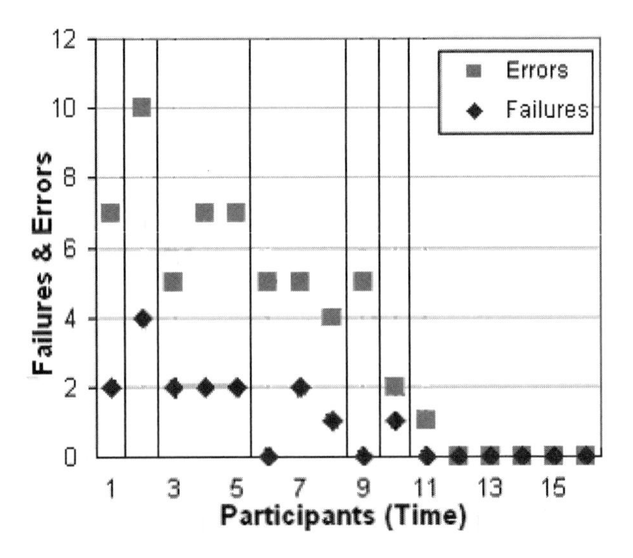

FIGURE 17.3 A record of errors over time as changes to the Age of Empires II tutorial were made using the RITE method.

It's also clear in Figure 17.3 that the number of failures and other errors generally decrease over time as iterations occur and eventually go to one error after the final iteration. Sometimes "spikes" occurred after a revision. This appeared to be caused by participants being able to interact with more of the game by removing previous blocking issues. For example, once the wood-chopping issue was resolved after the first participant, subsequent participants were able to move farther into the tutorial, thus encountering other issues along the way that otherwise may not have been found.

In addition, we could calculate the probability that the fixes we made actually fixed the issues we uncovered. After the last participant there were six additional participants run with no detection of any previously fixed issue reoccurring (the one issue that turned up never received a fix). This means that for these issues we are 90% confident that between 100% and 61% of our users would not have these issues. The issues that were fixed earlier had an ever greater degree of verification; for example, the wood-chopping issue fix was verified by 15 participants making us 90% confident that between 100% and 82% of the users would not have these issues (Lewis, 1996).

As stated in the Introduction, finding problems in a user interface is only half the battle for a practitioner; the other half is getting them fixed. One of the weaknesses of traditional usability tests is that there is often not enough time or resources to discover which fixes were not effective. In contrast, the RITE method allows designers to uncover fixes that need refixing. Here is an example of a refix. Participants had problems with terminology in *Age of Empires II* that required repeated fixes. Specifically, participants thought that they needed to convert their villager units into swordsmen units, when in fact the two types of units were completely unrelated. When the tutorial directed the user to "Train five swordsmen," participants would move their villagers over to the barracks (a building that creates swordsmen), assuming that the villagers could be *converted* into swordsmen. In fact, swordsmen had to be *created* at the barracks using other resources (e.g., food or gold). When this issue was discovered, there were inconsistencies in the terminology used in the product. In some places the user interface said "train units" and sometimes it said "create units." After watching six of eight participants fail to produce new units, the team decided that the text inconsistency was the problem and chose to use the term "train units" throughout the user interface. When the next participant was run, the problem reoccurred in the exact same way. As a result the team changed the terminology to "create unit" throughout the rest of the user interface. Once this problem was refixed, it was not seen again after seven additional participants.

The RITE method can clearly eliminate user interface problems during a user test, but assessing its overall effectiveness for practitioners requires looking at additional information outside of the lab setting such as awards, reviews, and sales. For example,

- The *Age of Empires II* tutorial received an "Excellence" award from the Society for Technical Communication in 1999.
- The game play and the tutorial of *Age of Empires II* received critical acclaim from the gaming press. In many cases the tutorial was singled out as excellent. In addition many RTS games have now used tutorials similar to *Age of Empires II.* Aggregated reviews of *Age of Empires II* can be seen online at Game Rankings.
- It is practically impossible to relate sales of a product to any single change or feature or to any particular method or technique. However, it is worthwhile noting that *Age of Empires II* was very successful and that it clearly reached a broader audience than the previous product. Since its release in August of 1999 until the end of October 2000 *Age of Empires II* never left the top 10 in games sales according to PC Data. In addition, it continued to break back into the top 10 in 2001 from time to time almost 2 years after its release. The original *Age of Empires* was also very successful, selling 456,779 units between October 1997 and December 1998. *Age of Empires II* sold almost double the number of units as the original *Age of Empires* in a similar time frame—916,812 copies between October 1999 and December 2000. The sustained sales of *Age of Empires II* over the original demonstrates that it reached broader markets, many members of which may not have had experience with games like it before (the segment at which the tutorial was aimed).

17.5.2 Case Study 2: Oracle Interactive Voice Response E-Mail

Oracle has utilized the RITE methodology for emerging software technologies for which there are more problems to be solved than traditional usability methods can handle (e.g., interactive voice response [IVR], mobile hand-held units, and Linux-based e-business applications). The standard summative usability tests (i.e., 8 to 12 users, one consistent set of representative tasks, and a focus on performance metrics) take considerable time to set up and prepare (including political and corporate concerns about product comparisons beyond simply setting up the usability activity), have strict constraints that limit flexibility, and evaluate one version of an application. In the end, they can measure usability quite well, but answer few formative questions on specific tasks and issues. This summative process works well for a completed product but not for products in formative development, particularly products in rapid development cycles.

The first case study Oracle conducted using the RITE methodology was for IVR e-mail (McGee, 2004). Mobile, IVR, and wireless technologies have advanced considerably over the past several years. Generally, mobile end-user applications are easier to create than their desktop counterparts because their size is limited by the display, memory, and/or interaction. Rapid development cycles are a de facto standard for these products. The RITE methodology seemed a good fit for the mobile domain. At the time, the product to be tested, IVR e-mail, was undergoing "rapid-prototyping" with the specific intention of trying to rapidly improve the product; however, formal usability testing was not going to be part of the process because of the limited time. Beyond the general fit with the mobile domain, the RITE method seemed tailor-made for the IVR e-mail rapid-prototyping exercise.

In studying the RITE methodology's applicability to the IVR situation, we realized that RITE is essentially optimization usability. It attempts to rapidly improve a product's usability to an optimal state by implementing changes suggested by user feedback (through errors, think aloud, or some other usability metric or process). These user-driven changes limit the theoretical large space of all changes that might be made. This is similar to algorithmic methods common in industrial engineering or computer science that reduce massive search spaces of possible solutions to an optimal solution. Optimization algorithms do not guarantee a perfect solution, but for "optimal" usability and interface design, many solutions for product design are acceptable.

With the IVR prototype we had previously identified many interactions and designs that we wanted to evaluate (e.g., we were interested in system responses to unrecognized user commands). Furthermore, we had already created a variety of possible design alternatives for several of these interactions. We outlined a potential experimental design in which the interactions of interest were listed as variables and the design alternatives as variable levels. We also knew that additional design variables and levels would probably be identified through RITE testing. The potential design solution space was quite large, far too large for any known traditional experimental methodology. In fact, at the conclusion of the study, we had manipulated or observed 21 design variables of interest, which, when factorially combined across all the variable levels, resulted in 1,271,981,408,256 unique combinations! At the onset of the study, these design variables needed to be addressed one way or another, so we used the RITE methodology and algorithmic optimization concepts in an attempt to quickly reach a consensus.

We used the RITE methodology with a master usability scale (MUS) as the discriminating usability metric instead of errors (McGee, 2004). Errors were not useful for differentiating our IVR voice e-mail tasks because the tasks were generally all completed without any user errors (similar situations occurred for all performance-based usability metrics). The MUS is based on the psychophysical method of magnitude estimation in which users provide a ratio estimate of usability for each task. Reference comparisons are used to combine different tasks and designs between users onto a single master scale of usability.

MUS is a measurement method for developing a usability continuum across multiple studies (McGee, 2004). It is based on holistic perceptions (i.e., usability) of physical environments (i.e., user interfaces [UIs] and tasks), also known as psychophysics. Its fundamental strength lies in a resultant ratio measurement scale (an underlying assumption for many statistical tests). In practice, users rate the usability of study-specific tasks and then generic reference tasks. Using the same reference tasks between studies allows usability comparisons to be made between different UIs or products. These universal usability comparisons make MUS a flexible, powerful measurement and analysis tool for usability practitioners.

With RITE as the experimental framework and MUS as the discriminating usability metric, we were able to emulate algorithmic optimization for our very large IVR design space. For example, given a task of "delete an e-mail," completion of the task would require the user navigate to an e-mail and then say the appropriate command to delete the e-mail. However, we could induce a misrecognition error that initiated a specific system error response with a specific voice gender. At the completion of the task, the user might assign the task a usability rating of 50 in comparison to all other tasks. The specific system error response and gender would be assigned the 50 rating (along with all other design levels present in the combination tested for this task). These variable levels would then be compared to all other system error responses and genders rated during other tasks and ranked and adjusted before design combinations for the next user's tasks were selected. As an exemplar, one system error response may consistently be ranked as the best error-handling strategy and therefore be subsequently more likely to be included in future tests and the ultimate optimal design.

Despite the astoundingly large number of possible design solutions, in actuality, we learned that RITE testing quickly distilled the alternatives down to clearly preferred solutions within 10 users. Users were quickly able to choose, through MUS ratings, among various design alternatives presented to them. Even 21 design variables with many design levels were, for the most part, easily discriminated after 3 to 4 users. Many problems discovered in designs were fixed immediately within the context of the study. "Winning" design alternatives were adopted into the product. Other issues for which discrimination was not decisive with usability were slated for redesign or selected based on business or technology reasons. In addition, although most design alternatives were choices among best candidates, some designs were known to have a poor UI; however, certain stakeholders remained beholden to them for various reasons. However, on rapid and objective user review of poor UI design, these options were eliminated or modified, averting potential design arguments and ultimately poor design in the final product. Overall, a higher-quality product was the result.

The RITE method alone would have advanced the product design to a near-optimal solution. The optimization techniques provided direction and more control over the design solutions exposed to each successive user. The MUS allowed all the tested combinations to be quantitatively compared along the same usability scale. The flexibility of the RITE framework afforded the combination of these disparate advanced techniques that yielded a powerful overall usability evaluation. We achieved far more cost-effective formative improvements to product design than with traditional usability and experimental methods. In comparison to reducing more than 1 trillion design combinations that our RITE test effectively accomplished, a typical formal usability test is constrained to a single design solution, a priori assumed to be an optimal or near-optimal solution. We have repeatedly learned that such genius in design is a myth. A typical analysis of variance experimental philosophy would require thousands of experiments with millions of subjects to statistically conclude differences among so many conditions. Clearly, both typical methods are inappropriate choices for formative evaluation in which many issues need to be resolved.

17.5.3 Case Study 3: Microsoft Digital Image Library

The Digital Imaging Group at Microsoft produces the Picture It! brand digital image–editing products. The digital imaging software market is currently experiencing rapid growth and change because of the mass market adoption of digital cameras. Historically, consumer products such as digital imaging software have been released on an annual cycle. These annual release cycles, which coincide with seasonal purchasing behavior, are short, and they impact the amount of time usability engineers have to evaluate and effectively implement design changes. The RITE method is ideally suited to this environment because it allows usability engineers the opportunity to evaluate more features or products during an individual release cycle.

The Digital Imaging Group recently introduced a new product, Digital Image Suite, that was designed to provide a complete end-to-end digital imaging experience for amateur photographers and digital camera owners. The product contained two separate applications that were designed to work together to enable complete digital imaging scenarios from image acquisition from digital cameras to photo printing. One of the applications, the Digital Image Library, was designed to acquire, organize, and manage digital images stored on the computer. This application is also a single launch point for editing and creative activity using the Picture It! photo editor. The Digital Image Library is a relational database of digital image files and associated metadata (some of which comes from the camera as EXIF data and other of which is assigned by the user). The metadata can be used to create so called "logical views" of the photos, which are ways of displaying and sorting the images by metadata type. The result is a more intuitive method of digital image organization and management.

Creating this new product was a significant investment in a rapidly changing and competitive software market. Product planning research had indicated that digital camera owners with all levels of computer sophistication experienced usability problems both in finding and in managing digital images on their computers. The RITE method offered an opportunity to evaluate the Digital Image Library thoroughly within the time constraints of the product development cycle.

Although the usability engineer was familiar with and had confidence in the RITE method, the product team was not familiar with the method. In addition, a small group on the product team had become invested in the early UI prototypes of the Digital Image Library. To familiarize the product team and gain acceptance of the RITE method, the usability engineer gave a presentation on the RITE method to the team. The presentation highlighted the effectiveness of the method but also uncovered some of the "typical" reservations found in introducing this method to product teams:

1. Concerns about the "validity" of usability issues because of the limited number of observations.
2. Concerns about time commitments from team members not traditionally directly involved in usability lab studies.
3. Concerns about the ability to identify the "correct" usability recommendation or fix based on limited observations.

These concerns were addressed by the following reassurances:

1. The team's understanding of the product and the users provides the basis for determining the validity of the problems, and if the validity of an

observed problem is in question, then the team can simply wait and see if the problem is replicated with the next users(s).

2. The resource commitment is not different from that for a more traditional approach. The time is concentrated during the test period, rather than being spread out through testing, analysis, presentation, discussion of results, and determination of changes. The elapsed time from start of test to changed product is shorter because changes occur during the testing process.
3. Fixes occur more quickly and can be tested in subsequent trials. Iteration can occur until the product reaches a sufficient level of performance as determined by a usability specification or explicit goal.

Before testing, the usability engineer and the team developed a list of tasks and concepts that the participants should be able to do and/or understand after using the Digital Image Library. Issues were identified for which there would be zero error tolerance (e.g., learning to scroll in and manage the database of images to identify specific photos). All members of the development team (e.g., the program manager, product designer, and developer) were present at each session. After each participant completed the test, the team met with the usability engineer to review usability issues observed and do one of the following things:

1. Attempt a fix, and then use the new prototype with the remaining participants.
2. Start to work on a fix, and use the new prototype with the remaining participants as soon as it is available.
3. Collect more data (e.g., more participants run) before any fix is attempted.

The results are summarized in Figure 17.4, which displays screenshots of the product as it was designed before the first session and the product as it was designed after the last session recommendations had been implemented.

The final design is shown in Figure 17.4. Some of the final design reflects both changes made during the RITE study and changes intended to address problems uncovered during the RITE study but addressed later in development. An unanticipated benefit of RITE was that the deep involvement of the development team produced a lasting motivation to fix issues. The study determined that although the controls and navigation elements available were useful in certain scenarios, only the basic controls were essential to finding specific images. The exposure of all navigation controls by default overwhelmed and

Original Design

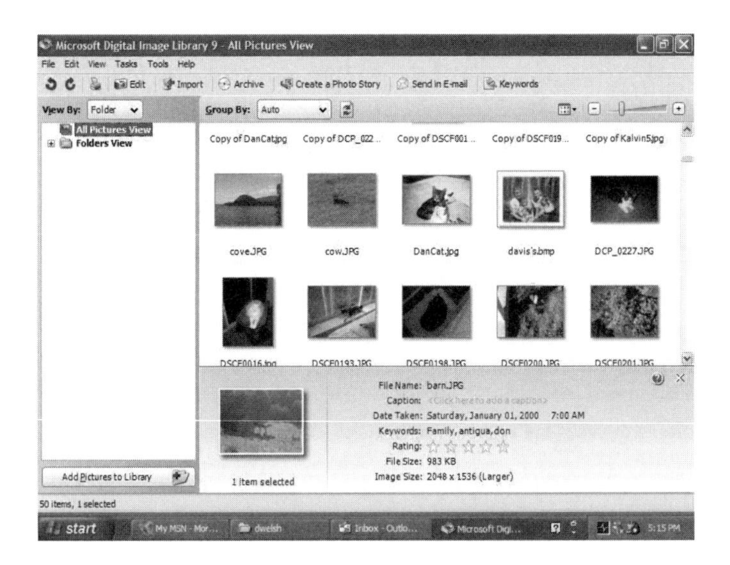

Current Design

FIGURE 17.4 A visual record of design changes to the user interface for the Digital Image Library during testing using the RITE method.

interfered with the participant's ability to navigate successfully. These essential controls included the "Group by" control which changed the way the digital images were sorted (e.g., folder they were in: date taken, event image taken at, or keyword) and the scroll bar. This study and subsequent studies determined that when digital images are sorted by known criteria, individual digital images are easily found, even in large collections of digital images. The RITE method was able to help the team eliminate unneeded complexity from the photo sharing library and retain desired functionality.

Lessons learned from this study are the following:

1. The method can be used to effectively communicate results and design changes without the burden of creating large and detailed study reports. The limited time imposed by the short product cycle made timely communication of results and recommendations to the team important. Because the product team was involved in the process, they simply used the final tested and validated interaction design and visual design to finalize the product specification needed to begin development. To ensure appropriate communication, the usability engineer prepared a verbal presentation for the wider product team that used screenshots of the UI progression and highlighted the data collected that supported the changes.

2. As indicated earlier, the daily triage meetings provided the usability engineer with an opportunity to review the data with the team and determine changes to be made to the prototype. An unintended but beneficial result is that these interactions helped foster confidence in the design and galvanized the product team's efforts behind creating a positive user experience. This strengthening of efforts behind the design was also bolstered by the opportunity for the team to observe the validation of changes by subsequent participants.

3. The team needs to agree on criteria that will be used to determine whether a change will be made. In this study, the team had not anticipated a specific problem, and this unexpected occurrence resulted in an inability to reach consensus on the problem and the appropriate solution. For example, in this study we observed user confusion with a search control that was included in the original designs. However, we could not agree on a clear solution to the problem. As a result, we decided not to make changes to this control during this study. However, during subsequent unrelated studies we monitored this issue. As a result of these extra observations, we determined that the metadata likely to be used by our target user was a short list of items (date, event, and subject) and that if we enabled the user to sort the digital images by these metadata using the "Group by" control, the Search control became unnecessary.

4. The RITE method provided the engineer and the product team with the ability to focus on several key questions and rapidly iterate and finalize

designs for that feature or product. This allows the usability engineer, and more importantly the product team, to focus efforts on improving other features in the product. The application of the RITE method several times during this product cycle enabled a single usability engineer to provide effective usability coverage for a brand new product offering as well as continued product efforts.

The Digital Image Suite received positive attention from the press, and sales increased significantly. Clearly many factors contribute to the success of a product. However, it is equally clear that a difficult-to-use product will be unlikely to succeed unless it offers some unique benefits. It is also clear that an easy-to-use interface lowers the barriers users experience to perceived product benefits; that is, the user's tolerance for figuring out a confusing interface to experience a promised but not yet experienced benefit can be quite low for consumer products. Thus, an easy-to-use interface can hasten adoption as shown by the following:

- ZDNet review of Digital Image Suite—"My new favorite way to organize digital photos" (Coursey, 2003).
- Digital Image Suite was granted the Good Housekeeping Seal of approval for 2004. This is the first digital photography software package to be granted the seal.
- Financially, sales for Digital Image Suite are ahead of projections for the product. Year to date (for July 2003 through September 2003), the retail sales quota was 126% of projections, which translates into a 153% year over year growth in revenue, and an increase of 5.4 market share points.

17.6 CONCLUSIONS

The goals of the RITE method are to identify and fix as many issues as possible and to verify the effectiveness of these fixes in the shortest possible time. These goals support the business reason for usability testing, that is, improving the final product as quickly and efficiently as possible. The case studies presented in this chapter suggest that in these contexts the RITE method was successful in achieving its goals. The goals for the RITE method are very similar to those for other iterative methods (e.g., cognitive walkthroughs and heuristic reviews, GOMS [goals, operations, methods, and selection rules] analysis (Gray *et al.*, 1993) and usability tests in general). It differs from other methods in the degree of collaboration it engenders and the rapidity with which fixes occur and are verified.

We are not proposing the elimination of summative usability methods or other study designs. We are proposing that the RITE method be used to solve as many formative problems as possible when rapid iteration is practical and appropriate. We expect the trend of rapid programming cycles to accelerate in the future. Desktop applications and other development teams are adopting "extreme programming" and other techniques to decrease the development cycle as much as possible while raising the quality of work (Rudd and Isensee, 1994; Rudd *et al.*, 1996). Rapid formative usability techniques will be needed to meet these development demands. The RITE method is an ideal candidate to meet this need.

It is important to reiterate the things that make the RITE method succeed. They are as follows.

Before the test:

1. Success metrics for each task are identified
2. Time and resources are set aside to be used during testing time
3. All key decision-makers plan to attend the test

During the test:

4. Product decision-makers attend the tests (e.g., developers, usability engineer, testers, and program manager).
5. Issues and potential solutions can be rapidly identified.
6. A usability engineer is present who has a great deal of experience watching participants in similar situations in the past and is well versed in the product itself.
7. Team members who have a strong knowledge of design and a deep understanding of the system architecture are present.
8. Observers have the opportunity to brainstorm different fixes as testing occurs.
9. Decision-makers agree on changes to be made.
10. Applications are available that have a powerful and flexible architecture, which allows for rapid changes to be made to the product.

As demonstrated by the case studies in this chapter and the correspondence we have received from other usability engineers, the RITE method has an established track record of success on a variety of commercial products. As such, it follows in the tradition of other methods that have been shown effective in real world contexts, such as GOMS (Gray *et al.*, 1993). This chapter embodies a "case study" approach to examining usability methods (John, 1998; Wixon and Jones, 1996). We hope that others will choose to publish case studies of RITE and other methods. We look forward to lively discussions on instances of success and failure with a focus on commercial success. We believe that this is the most effective way to advance the practice of usability.

REFERENCES

Coursey, D. (2003, August). *My New Favorite Way to Organize Digital Photos.* ZDNet. Retrieved January 6, 2004, from http://reviews-zdnet.com.com/4520-7298_16-4208095.html.

Dumas, J., and Redish J. C. (1993). *A Practical Guide to Usability Testing.* Norwood, NJ: Ablex.

Game Rankings Review Scores for Age of Empires II (n.d.). Retrieved January 6, 2004, from http://www.gamerankings.com/htmlpages2/63605.asp.

Gray, W. D., John, B. E., and Atwood, M. E. (1993). Project Ernestine: Validating a GOMS analysis for predicting and explaining real-world task performance. *Human Computer Interaction,* 8 (3), 237–309.

John, B. E. (1998). A case for cases. *Human Computer Interaction,* 13 (3), 279–280.

Lewis, J. R. (1990). *Sample Sizes of Observational Usability Studies: Tables Based on the Binomial Probability Formula.* Technical Report 54.571. Boca Raton, FL: International Business Machines.

Lewis, J. R. (1991). *Legitimate Use of Small Samples in Usability Studies: Three Examples.* Technical Report 54.594. Boca Raton, FL: International Business Machines.

Lewis, J. R. (1993). *Sample Sizes for Usability Studies: Additional Considerations.* Technical Report 54.711. Boca Raton, FL: International Business Machines.

Lewis, J. R. (1996, May). Binomial confidence intervals for small sample usability studies. In *Proceedings of the 1st International Conference on Applied Ergonomics.* Istanbul, Turkey.

McGee, M. (2004). Master usability scaling: Magnitude estimation and master scaling applied to usability measurement. In *CHI Proceedings 2004,* Vienna.

Medlock, M. C., Wixon, D., Terrano, M., Romero, R., and Fulton B. (2002, July). Using the RITE method to improve products: A definition and a case study. Presented at the *Usability Professionals Association (UPA2002),* Orlando, FL. Also available at http://www.microsoft.com/usability/publications.htm.

Nielsen, J., Lewis, J. R., and Turner, C. W. (2002, July). Current issues in the determination of usability test sample size: How many users is enough? Presented at the *Usability Professionals Association (UPA2002)*, Orlando, FL.

Nielsen, J., and Landauer, T. K. (1994, April). A mathematical model of the finding of usability problems. In *Proceedings of ACM INTERCHI'93 Conference* (pp. 206–213). Amsterdam.

Pruitt, J., and Grudin, J. (2003). Personas: Practice and theory. In *Proceedings of DUX 2003.* Retrieved January 6, 2004, from http://research.microsoft.com/research/coet/Grudin/Personas/Pruitt-Grudin.pdf.

Rudd, J., and Isensee, S. (1994). Twenty-two tips for a happier, healthier prototype. *Interactions,* 1 (1), 35–40.

Rudd, J., Stern, K., and Isensee, S. (1996). Low vs. high-fidelity prototyping debate. *Interactions,* 3 (1), 76–85.

Squires, S., and Bryne, B. (2002). *Creating Breakthrough Ideas: The Collaboration of Anthropologists and Designers in the Product Development Industry.* Westport, CT: Berwin and Garvey.

Spool, J., and Schroeder, W. (2001, March). Testing websites: Five users is nowhere near enough. In *Proceedings of CHI 2001, Extended Abstracts* (pp. 285–286), Association for Computing Machinery. Seattle.

Terrano, M. (1999, October). *Age of Empires II: Designer Diary.* Retrieved January 6, 2004, from http://firingsquad.gamers.com/games/aoe2diary/.

Virzi, R. A. (1992). Refining the test phase of usability evaluation: How many subjects is enough? *Human Factors,* 34, 457–468.

Whiteside, J., Bennett, J., and Holtzblatt, K. (1988). Usability engineering: Our experience and evolution. In *Handbook of Human Computer Interaction* (pp. 791–816). New York: Elsevier.

Wixon, D. (2003). Evaluating usability methods: Why the current literature fails the practitioner. *Interactions,* 10 (4), 29–34.

Wixon, D. R., and Jones, S. (1996). Usability for fun and profit: A case study of the design of DEC Rally Version 2. In. M. Rudisill, C. Lewis, P. Polson, and T. McKay (Eds.), *Human Computer Interface Design: Success Stories, Emerging Methods, and Real-World Context* (pp. 3–35). New York: Morgan Kaufman.

Woolrych, A., and Cockton, G. (2001) Why and when five test users aren't enough. In J. Vanderdonckt, A. Blandford, and A. Derycke (Eds.), *Proceedings of IHM-HCI 2001 Conference* (Vol. 2, pp. 105–108). Toulouse: Cépadèus Éditions.

18 CHAPTER

Summative Usability Testing: Measurement and Sample Size

Jurek Kirakowski University of College Cork, Ireland

18.1 INTRODUCTION

Usability testing can take many forms, for instance, expert inspections, employment of "user representatives" for their opinions, and various forms of automated testing. Nevertheless, in a survey published by the Usability Professionals' Association in 2001, "usability evaluation with real users" was the most popular activity cited by the respondents (Usability Professionals Association, 2001). Usability *evaluation* (sometimes called *testing*) refers to two different activities. It is customary, following terms used in educational psychology (Scriven, 1967), to call these *formative* and *summative* modes of test.

Formative modes of test take place during the activity of developing the software or Web site, and the goal of these activities is to improve the product by providing specific feedback to the design team of which aspects of the software or Web site are usable and which aspects need improvement.

Summative modes of test take place when the product has reached a certain stage of development, and the primary goal of these latter activities is to show how much progress has been made in the development of a usable product. *Summative testing* is a term widely used in usability, although it has not yet found general acceptance. Other terms used have included *usability validation, user validation* and *user acceptance test.* In some quarters, the term *test* is depreciated in favor of the less-intimidating words *study* or *evaluation.*

This chapter focuses on summative testing. In contrast to sample sizes usually employed in formative testing (rarely more than five users per test), summative testing sample sizes are generally larger (between 12 and 20 users per test, sometimes many more). The biggest source of expenditure in any kind of usability testing is the time taken to identify and recruit users, to bring them to a place of testing, and to perform the tests with them. The more we know about how many users we need, the more cost effective our testing will be.

My primary reason for writing this chapter is to point out that we have to bring together two strands of knowledge in the service of cost-justifying usability testing: The first strand is about what we can measure to do summative testing and the second strand is about the statistics of estimating the smallest effective sample sizes for such tests. The purpose of this chapter is not didactic. Because I need perforce to bring up points on issues that may not be familiar to all, I have tried to balance exposition with argument and have made reference to some of the most famous resources when appropriate.

In Section 18.2 I outline the difference between summative and formative testing. Summative testing is assumed in the standard issued by the International Organization for Standardization (ISO) as standard 9241 part 11 (ISO, 1998). This part of the standard gives a definition of usability that can be operationalized in terms of how usability can be measured. Although other definitions have been proposed, the 9241 part 11 definition, by virtue of the process of how these standards are put together, is the widest consensual definition to date.

In Section 18.3 various base metrics that can be used to test usability following the ISO definition are discussed. The general topic of measurement in usability testing most probably dates from Chapanis' book *Research Techniques in Human Engineering* published in 1959 in which the author attempted to describe "some of the methods available to the human engineer for collecting trustworthy data on men and machines and the relationship between them." In a recent contribution to the area, Constantine and Lockwood (2000) present some interesting methods but still little empirical validation is shown. Questionnaire methods testing for the quality of user *satisfaction* are in fact the best developed from a measurement point of view (Kirakowski, 1998a).

In Section 18.4 I discuss statistical issues relating to sample size and the size of the effect that it is required to demonstrate as a result of testing. Although there is a good deal of debate and argument over practically any issue among statisticians, they seem to be extraordinarily unanimous on one point, and it is this: Never use the results of statistical testing as an oracle that must be obeyed; use the results as a guide to decision making. As has been said more than once, usability testing on its own will never improve anything; it is what you do as a result of usability testing that makes the difference.

Section 18.5 contains brief conclusions and recommendations: In the end, a usability test is only as good as its initial assumptions about who the target users are, what they will do with the product, and where they will carry out their interactions.

18.2 TYPES OF USABILITY TESTS

Enough arguments have been made for the advantage bestowed on a development process by incorporating *formative* testing within the cycle, especially a "lean" formative testing process that includes small, regular evaluations feeding back into development (see, for example, the Chapter 17 in this volume by Medlock *et al.*). But is there any point to a *summative* evaluation? What advantages does such a mode of evaluation give? The usual answer is that if the context in which the product is tested (so-called *context of test*) is sufficiently similar to the context in which the product will eventually be used (so-called *context of use*), then a summative evaluation is a *prediction* of what will happen when the product or Web site is launched on the market.

There are actually three stages in a product lifecycle at which such a prediction may be made, which are the following:

1. When the product has reached sufficient development, so that users can interact with it, as a benchmark or indicator of progress in achieving usability at that point.
2. Before a product is launched, to warn the company launching it of what kind of reception their product is likely to achieve, to help position the product on the market, and to prepare documentation and help desk support.
3. After a product has been launched, for the company to better understand how their product is actually being evaluated by the target market, for potential clients to understand the benefits to them of acquiring the product, or for the development team to know what is good about the product and what needs improving.

Summative evaluation at all three stages may also provide information that will cost-justify work done on usability.

Although one may consider these two modes of usability testing as a dichotomy, it can be seen throughout this chapter that gathering *summative* metrics can also tell us a lot about why the user interface is a problem, although admittedly performing *formative* evaluation may not adapt itself so readily to quantification leading to summative metrics.

Formative usability testing has as its objective the identification of usability bugs or errors in the system being evaluated. End users may or may not be involved in such tests. Although this is the predominant mode of usability testing, little agreement among exponents about methodology or indeed about what constitutes a reportable "usability bug" exists. The Comparative Usability Evaluation (CUE) trials headed by Molich (Molich *et al.*, 1998, 2004) show that when different centers of excellence carry out usability tests on the same software or Web site, there can be a difference of at least a factor of 10 between them as to how many usability bugs are reported and little consensus as to what to report. What emerged from the first CUE trials in 1998 is that assessment teams who carry out this kind of usability testing see themselves as being very closely connected to the development teams of the systems being evaluated. In general formative usability testers like to position themselves as part of an ongoing development effort and therefore recommend frequent testing with small samples of users throughout the development cycle (see, for instance, Larry Constantine's review of the session on Feature Testing at the CHI 2003 conference in which he notes that "rapid iteration struck a responsive chord within the audience"; Constantine, 2003).

Summative testing has as its objective the production of a report that explains to interested parties how usable the system being evaluated is. Of course, how you test will depend on how you define *usability*. Different definitions of usability have been proposed: each of these sets a slightly different testing agenda and meets needs for testing in different environments. A significant step forward was made in the ISO standard 9241 part 11, in which usability was defined as the "extent to which a product can be used by specified users to achieve specified goals with effectiveness, efficiency and satisfaction in a specified context of use" (ISO, 1998).

Effectiveness refers to the ability of the user to achieve the intended results; *efficiency* refers to the use of resources, such as time or effort; and *satisfaction* refers to users' appraisal of the quality of their interaction with the system. "Specified users . . . goals . . . context of use" provides the conditional clauses within which measurement of usability is considered meaningful.

Usability tests should be documented using standard formats. A widespread consensual agreement on what a usability report should contain and how it should be written will facilitate the dissemination of best practice: standards, methodologies, and content. This means that one can, in principle, estimate how much a given report will cost in terms of effort, that reports on the same system do not have to be replicated by different centers and that centers may interchange reports, and that general criteria for excellence may be developed.

Standardized reporting with an accepted methodology behind it leads to cost effectiveness and development of the professional status of the discipline. Several standards have been proposed for how to report the results of a summative analysis: from the BASELINE report format (Kirakowski, 1998b), which drew on earlier proposals generated by projects partly funded by the European Commission, to the Common Industry Format (CIF) sponsored by the National Institute of Standards and Technology in the United States (INCITS, 2001). Together with ISO 9241 part 11, this latter format represents a strong consensual agreement between those involved in carrying out such usability testing.

At present no agreed upon standards of best practice for carrying out formative usability testing or agreed upon report formats for such testing exist. Although Rubin (1994) and Dumas and Redish (1993) give good advice what to include in such reports, this advice is not always followed (see the comments in Molich *et al.*, 2004, for instance). An important issue is the severity rating of a usability bug. Severity rating is not often carried out, so when controversy arises over the minimum number of users required to locate some desirable percentage of usability bugs, it is not clear whether the disagreements arise because of differences in the methodologies being adopted, the complexity of the software being evaluated, or a definition as to what constitutes a usability bug in the first place (e.g., is a poor choice of background color in fact a usability bug and, if so, of what severity given the context in which it was found?)

18.3 SUMMATIVE TESTING: METRICS

Unfortunately, at the time of writing, summative metrics that can be used to assess efficiency, effectiveness, and satisfaction are not all developed to the same standard of quality.

It is customary to distinguish between *base* and *synthetic* metrics. Base metrics are raw measures that can be gained from direct inspection: counts of events or objects, measurement of time elapsed (some physiological measures can also count as base events—see later discussion—but these are rarely encountered in practice). Synthetic metrics are combinations of different base metrics. Thus the amount of time taken to complete a task and the number of errors made are two base metrics. However, the number of errors per hour is a synthetic metric made up of two base metrics.

In the remainder of this section I deal mainly with the base metrics that could be used to operationalize the ISO 9241 part 11 definition, paying particular attention to metrics for evaluating the usability of Web-based systems. Synthetic metrics can be multiplied *ad infinitum*; they will have to await another review.

A discussion of the measurement of the three components of usability (effectiveness, efficiency, and satisfaction) gives rise in each case to characteristic methodological issues. Thus, each of the three components will be reviewed in turn, trying to bring out the essential character of the problem with each.

18.3.1 Effectiveness, Dimensionless, and Embodied Metrics

Effectiveness is best measured in terms of how closely a user achieves results that conform to what is required of the software system. Very often, in testing, the intended results are specified in behavioral terms, assuming a known universe. For instance, in evaluating a Web site for booking flights, if the user is instructed to "find the cheapest airfare for a flight from Cork to Boston in the week of the . . . ," the true result is known and the ability of the user to discover this can be precisely measured, in this case, usually on a pass/fail basis. One would normally pose a series of such tasks for users to carry out.

Such effectiveness tests yield *counts* of events. As such, these are relatively uninteresting and are not easily generalizable from one testing situation to another. When the counts of different kinds of events are divided by the total number of events (for instance, number of users who succeeded in a particular task divided by the total number of users who carried out the task), we get *probability* or *proportional* data, which have interesting properties and for which there are well known groups of statistical tests—tests that can tell us, for instance, whether the observed proportion of users succeeding in task *x* and the observed proportion succeeding in task *y* is different enough for us to conclude that the tasks are not equally supported by the software.

This discussion of effectiveness measurement raises an important issue that recurs throughout this chapter: the difference between a *dimensionless* and an *embodied* metric. A dimensionless metric is one that stands independently of the characteristics of the trials and that is therefore comparable between trials, between trials of different systems, and even between trials of different kinds of systems. The probability of task success is an instance of such a dimensionless metric. To be sure, the value of the metric will depend on situational factors such as the amount of user expertise and the difficulty or length of the tasks, but there are known upper and lower bounds to the metric (the result will lie somewhere between zero and one), and one does not have to control for the situational factors when citing the metric.

The raw count of how many users succeeded in carrying out each task is an instance of an embodied metric. Although there is a lower bound to this metric (no users may have succeeded), there is no upper bound and there is no understanding of how data will be distributed between zero and the upper bound. So there is no meaningful way of comparing the relative standing of two sets of trials on this metric unless it is assumed that the situational factors between trials are identical.

Going back to probability of task success, in practice, a trial is either successful or unsuccessful for one of the three following reasons:

1. The user gives up before the time limit is passed.
2. The time limit for the trial is passed, and the user is asked to stop.
3. The user finishes within the time limit but does not complete the task to the criterion (whether or not the user realizes that he or she has not completed it).

In each case, it may be possible to engage in assessment of how close the user's result is to the criterion result. This is often not easy in practice and a fourfold assessment may be used, employing verbal anchors such as "nowhere near," "honest attempt," "close," and "perfect." Figuring out what numbers to use to represent each of these anchors is difficult because how far apart they are from each other is not clear. For instance, is the difference between "nowhere near" and "honest attempt" the same as that between "close" and "perfect"? Therefore, responses to such anchors ought to be treated as loosely ordinal (i.e., don't make them into a scale of one to four and don't compute averages).

The same remarks apply to the situation in which the amount of progress made by the users is estimated. Let us suppose we can enumerate the way-points from the start of the task to its successful conclusion. Again, even if our counting of the way-points achieved by the user is very carefully done, we have no obvious way of assigning numbers to them. Suppose one user completes five way-points out of seven and a second completes six. Is the measurement value of the sixth way-point the same as that of the previous five? Unless one can prove that each way-point is of equal difficulty, the data obtained here are also loosely ordinal and the same strictures apply (i.e., don't make a scale and don't compute averages). The best one can do is compute the probability of each way-point being attained and then display this sequence of probabilities as a line graph.

18.3.2 Efficiency and the Underlying Form of Data

Efficiency is defined as the use of resources. The most commonly studied resource is the users' time—time to complete a task, for instance, or time spent recovering from an error. This is what is called *chronometric* data, and it has a long pedigree in psychological measurement going back to the nineteenth century. These days, mental effort and stress are also proposed as metrics, for which good measuring tools based on self-report are available (see, for instance, the NASA Task Load Index or TLX, Hart and Staveland, 1988 and the Delft Subjective Mental Effort Questionnaire, Zijlstra, 1993). Variation in pulse rate has also been proposed for use as an efficiency metric (Mulder, 1992), although in practice it is seldom used. Other physiological measures of effort are also proposed but in practice are used even less often (Mulder, for instance, also discusses respiration rate).

Related to chronometric measures are counting measures, for instance, the number of actions that a user executes to get a task done, the number of error conditions that a user raises, or the number of references to help messages or other documentation. Although these are discrete events and time is a continuous measure, forms of distribution of many such discrete events will usually follow the form of distribution of time on task.

With regard to time, one must remember that all chronometric measures display a tendency to peak at low values, with a large tail-off (so-called *positive skew*). Figure 18.1, for instance, shows the amount of time required to carry out a booking task in the evaluation of an airline reservation system. There are several points to note about these data that will strike a chord with anyone who has been involved in such an evaluation.

First, data are presented from all users; although two users took up to 10 minutes to complete the task, one of these users was not able to complete the task within the 10 minutes, and the other user completed the task after 9 minutes and 38 seconds. Second, tester intervention was allowed, and although any interventions were noted in the evaluation, they are not shown on this graph. Third, some users gave up on the task or insisted that they had completed it correctly

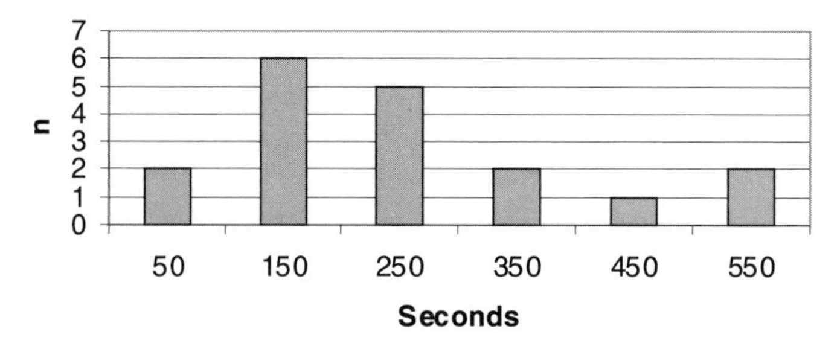

FIGURE 18.1 Histogram of seconds taken to complete booking task.

although they had not. All these problems can be overcome by partitioning the data and showing the data for each of these three categories of user separately; but such a display becomes complex and the required number of users to achieve any kind of smoothness to the three resulting curves becomes large.

Because of the tail-off toward high numbers, a suitable measure of the representative number of seconds required to complete the task is not given by a straightforward average. The arithmetic average of the data shown in the graph is 268 seconds whereas the median is 247 seconds. The median is a more appropriate statistic, because it is not affected by extreme data at the ends of the distribution, and it assists in ironing out the problems mentioned in the preceding paragraph.

Although computing *central tendencies* such as averages or medians is important as a summary of the direction toward which the data tend to go, looking at a measure of *dispersion* around the central tendency is also important to estimate the amount of error associated with the statement of central tendency. However, various methods for estimating dispersion are available. The choice of statistics for central tendency usually influences how we will compute dispersion. For dispersion around an average, one would use a standard deviation; for dispersion around a median, the most frequently used statistic is the *interquartile range*, which unfortunately is not as directly comparable to a standard deviation as the mean is to the median.

If you have problems with the *description* of such data, this is only a prelude to the problems that will be encountered when you attempt to make *inferences* about the underlying populations from which the data have come. The best statistical treatment for such data is statistics using ranks—an earlier generation of psychometricians would have transformed chronometric data using an arcsin transformation and then proceeded with normal distribution assumptions, but this is rarely done today. Instead I have frequently encountered treatment of chronometric data as normally distributed without regard for the actual underlying form of the obtained data. The justification is that tests based on normal distribution assumptions are "robust." Arguments about the robustness or otherwise of inferential statistics procedures have arisen in the past and no doubt will continue.

Suppose that a statistical procedure assumes that the population of data from which the sample data have been derived is in fact distributed according to the Gaussian normal pattern (and therefore that the data are well described in terms of two *parameters*: 1) *mean* and 2) *standard deviation*).

Are the various lumps and irregularities in the data because of accidents of sampling, that is, data gathering? In our example, we would consider whether, if we tested many hundreds of users, the proportion of users who took more than

10 minutes would become infinitesimally small. Those who hold that parametric procedures are not unduly affected by such bumps and irregularities are proponents of the "robustness" approach, whereas their opponents claim that using these "parametric" procedures on such data is folly and leads to unreliable results—just as in the case cited in Figure 18.1, we receive a different impression of the central tendency of the data, depending on whether we consider the mean or the median to be the better descriptor.

Aron and Aron (1994) gave a summary of the two approaches, but be warned that these authors favor the position that classic parametric testing procedures are quite robust and therefore can be used with greater impunity than a purist would dare. The robustness approach is partly, I suspect, fueled by the fact that much more statistical research has been done on elaborating parametric statistical techniques than nonparametric ones and that a disproportionate amount of time is spent on parametric methods in introductory statistics courses.

Raw measure of time on task is an embodied metric, as outlined earlier. We have no upper or lower bound for most chronometric measures, although it may be argued that the speed of the computer program responding to a previously generated script without human intervention can set a lower bound. This is not very helpful or easy to attain in practice.

A metric for efficiency that tends to the dimensionless was first publicly documented by Bevan and MacLeod (1994) and called the Relative User Efficiency (RUE) metric, although this metric seems to have been discovered independently by numerous researchers before this date. RUE involves measuring the time taken to carry out a task by an expert user, perhaps after some practice, and then dividing the time taken by an ordinary user by this minimal time:

$$\text{RUE} = \frac{\text{Ordinary user time}}{\text{Expert user time}} \times 100$$

RUE is normally cited as a percentage. Although there is no upper bound to this metric, and some users may provide artificial data if they are stopped after a certain number of minutes (e.g., 10 minutes), the lower bound is, by definition, 100%. If ordinary users take less time than experts, then this suggests that the expert time estimates were ill chosen, and this consideration should be a self-correcting part of the method. In essence, RUE standardizes the time on task metric by reference to an expert performance and thus goes quite far toward a dimensionless metric: It is possible, for instance, to compare RUE for different tasks and for similar tasks carried out with different software systems or Web sites.

None of the problems of descriptive or inferential statistics as discussed earlier is actually solved by using RUE, but at least by using suitably cautious statistical procedures, comparisons between different versions of the software or against baselines are possible. I am not aware of any studies indicating whether RUE manages to avoid the problem of positive skewness: Because it is known that the product of any two distributions (whatever their underlying form) tends to the normal, perhaps RUE does approximate a normal distribution. We will have to see data to test this hypothesis. Anecdotal evidence has been reported to indicate that for first-time users of desktop products an average RUE (using the mean) of approximately 400% may be expected. However, we do not know what a generally acceptable value of RUE is nor by how much it can vary.

Chronometric methods for measuring efficiency raise the problem of time-lag resulting from system response when you are measuring time on task in Internet applications. Very often, when studied in real-life conditions on the Internet, time on task is affected by circumstances totally outside the control of both user and developer. Thus chronometric RUE is not a satisfactory metric for Internet applications.

Counts of user actions or events are slightly more satisfactory if there is a problem of asynchronous response by the computer system, and such counts may also be cast as RUE (e.g., the number of actions an ordinary user takes to complete a task as a function of the minimum number of actions actually required by the functionality of the software). The use of action or event RUEs overcomes the problems of connections, unanticipated events, and system response speed to some extent. A similar approach is proposed by Constantine and Lockwood (2000), who call it a metric of "essential efficiency." Unfortunately, beyond proposing the metric with some interesting mathematics, these authors do not give any reliability or benchmark data.

Self-report measures such as the NASA TLX and the Subjective Mental Effort Questionnaire (SMEQ) are also well suited to Internet environments; SMEQ in particular is quick to administer, and under perfect conditions has a correlation with task time of approximately 0.80 when tasks are held constant. These self-report measures of effort and stress are already standardized, so benchmarking against expert performance as an RUE is not needed.

18.3.3 Satisfaction and the Reliability of Metrics

Although user satisfaction seems at first glance to be the least promising component of the ISO definition, the tools for measuring it are the best developed and standardized. The usual method of measuring user satisfaction is by questionnaire. Users who have had some experience with the software to be evaluated (either in laboratory or real-life conditions) are asked to rate the software on a "user satisfaction" scale. The development of such a scale is not to be undertaken by the faint hearted; Kline (1998) in a thoughtful discussion, criticizes contemporary psychometrics for "failing to meet the criteria of fundamental scientific measurement" after 90 pages of a thorough review of contemporary measurement techniques and their statistical background. In the face of such divine discontent, the journeyman must indeed tread warily. Two principles serve as a guiding light to develop a scale that will be of any use.

First, one must be prepared to carry out at least two or three iterations of the scale, testing the scale and analyzing the results to improve the scale's *validity*.

Second, one should follow well-defined and agreed-upon technical criteria that indicate how *reliable* such a scale is. These reliability estimates must not be done using data that were gathered while the scale was developed: In the terminology adopted in this chapter, reliability data must be summative. Kirakowski (2003) provides an informative Web site of questions and answers about the use and development of satisfaction scales in usability testing from which the following two definitions are taken:

- The *validity* of a scale is the degree to which the scale measures what it is intended to measure. Note that satisfaction surveys are not the only ones to have validity issues; factual questionnaires may have serious validity issues if, for instance, respondents interpret the questions differently.
- The *reliability* of a scale is the ability of the scale to give the same results when filled out by like-minded people in similar circumstances. Reliability is usually expressed from zero (very unreliable) to one (extremely reliable).

From the point of view of testing for usability, it is important to realize that the use of satisfaction scales for which there is no reliability estimate is at best an inefficient use of resources and at worst is potentially misleading. It is inefficient, because even supposing that the scale does measure some aspect of user satisfaction, it is impossible to estimate the inherent reliability of the scale before starting and therefore also impossible to estimate how much difference between two versions of software being tested or between a version and a baseline is "enough", and it is potentially misleading, because a poorly formed satisfaction scale may be dominated by one or two questions that refer only to some properties of the software being evaluated or that are capable of being misinterpreted in various ways by the respondents. In questionnaire design, Murphy's law may be stated in this form: If a respondent can misinterpret a question, a respondent will misinterpret it.

In medicine, professionals sometimes use a row of six faces that enable a patient to communicate to the physician the level of pain or discomfort that he or she is feeling (i.e., a pain rating scale that progresses from a smiling face to a crying face). The analogous testing of user satisfaction by asking a single simple question, "How satisfied are you with this software?" (usually on a five-point rating scale, sometimes illustrated by five faces to make the scale look user-friendly) is extremely unhelpful for several reasons. Presumably the analyst takes as his or her basic hope that satisfaction will emerge as the central tendency among a large amount of variation, but the following may occur:

1. Respondents may interpret "satisfaction" differently so that the answers will not generally be about the same thing.
2. There may be an inherent bias against reporting low levels of satisfaction so that the scale will in practice be only two or three points at the upper end.
3. Large numbers of respondents will be required, because the inherent reliability of the test may not generally be known and therefore must be assumed to be low. Note that many commercial usability tests use very small numbers of respondents!
4. Test item bias cannot be extracted, and therefore the inherent validity of the test will also be unknown.

Satisfaction scales that have good reliability estimates include the System Usability Scale (SUS) scale by Brooke (1996), the Computer Usability Satisfaction Questionnaire (CUSQ) scales by Lewis (1995), and the Software Usability Measurement Inventory (SUMI) scale by Kirakowski (Kirakowski and Corbett, 1994). Two scales have been reported for the purpose of measuring aspects of the ISO 9241 part 10 "dimensions" of usability: ISOMetrics by Gediga *et al.* (1999) and ISONORM by Prümper (1999). One scale with good psychometric properties has been reported for use with Internet and World Wide Web applications: the Web Site Analysis and MeasureMent Inventory (WAMMI) scale (Kirakowski *et al.* 1998; Kirakowski and Claridge, 2001). At present, SUS, CSUQ, and ISOMetrics are free and in the public domain; the ISONORM, SUMI, and WAMMI scales are sold commercially, although the developers of the latter two are known to offer their scales for use in selected educational programs at nominal rates.

The omission of the Questionnaire for User Interaction Satisfaction (QUIS) scale originally developed by Norman (Chin *et al.*, 1988) in the preceding list is not intended as a slight. In the original report of the scale, the authors expressed some reservations about reliability properties of some of the subscales. Since then, the scale has been extensively modified by Norman and his colleagues, but no reliability data about it have been published either by Norman or by Shneiderman (see, for instance, Shneiderman, 1997), who has championed it over the years. It is sold commercially, and although it may not be wise to use it as a summative metric, it raises useful questions for formative testing.

Whereas the SUS scale is a good example of a unidimensional scale (i.e., only one summative number per respondent emerges as the result), CSUQ, SUMI, WAMMI, and the ISO 9241 scales are all multidimensional; that is, they report user satisfaction as a profile of numbers for each respondent. A profile enables the tester to see with greater precision the usability strengths and weaknesses of the software or Web site being evaluated. Figure 18.2 is an example of a WAMMI profile (WAMMI is used for testing user satisfaction with Web sites), in which, as with the SUMI questionnaire, Satisfaction is broken down into Attractiveness, Controlability, Efficiency, Helpfulness, and Learnability. A Global Satisfaction score is also provided (note that other questionnaires define User Satisfaction in terms of subscales differing slightly from those presented from WAMMI).

In Figure 18.2, reference is made to a normative standard that defines an average as a score of 50. WAMMI refers to a statistical digest of a standardization base of more than 200 different Web sites analyzed with WAMMI and places the Web site being evaluated with reference to the digest on each of the scales. That is, WAMMI not only has reliabilities, but it also has *population parameters* for each of its scales. This is an important addition: We know how WAMMI performs in

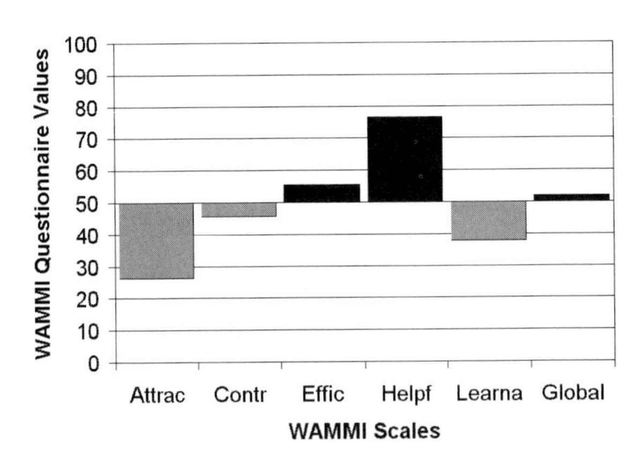

FIGURE 18.2 Summary of WAMMI results for anonymous Web site.

terms of what the level of "average" user satisfaction with a Web site is and also by how much this measure of satisfaction may vary.

If we ask a group of users to evaluate a Web site using WAMMI (for the sake of simplicity, just consider the Global Satisfaction scale), we will find an average figure for the site. We know from the standardization base what the population average is but, more importantly, what the dispersion around the population average is as well. Because of the fairly large numbers involved, and as a result of the standardizing process, we can assume that the population data are of Gaussian normal type, and therefore parametric statistics are applicable.

If we assume that the Global (G) score of the *i*th user is $G(i)$, we can now divide this score into two components: the Effect and the Error. Knowledge of the magnitude of the dispersion around the population average allows us to make estimates of the Error, giving us a more precise estimate of the likely Effect or rating of the Web site that the *i*th respondent actually wanted to give us:

$$G(i) = \text{Effect} + \text{Error}(i)$$

The Error component can be further subdivided into two parts: one part is the inherent error of measurement of the tool employed (in this case, the Global WAMMI subscale) and the other part is the amount by which individuals in the population from which the *i*th individual has been sampled vary between themselves as to how they are prepared to use a rating scale. This is sometimes called the "biological" source of error. That is,

$$\text{Error}(i) = \text{Measurement Error} + \text{Bio Error}(i)$$

Interestingly, if we know the population error, as estimated by the population standard deviation (SD), and we know the reliability r of the measure, we can estimate the Measurement Error quite simply:

$$\text{Measurement Error} = \text{Population SD} \times (1 - r)$$

and thence refine the original formula for the composition of the Global score:

$$G(i) = \text{Effect} + \text{Bio Error}(i) + \text{Measurement Error}$$

To summarize, the Global value is what the *i*th respondent provides; the Effect size is the amount of user satisfaction the *i*th respondent experiences; the Biological Error is the amount of variation to be expected between people from this population when we do this kind of measurement; and the Measurement Error is the amount of variation due to the inaccuracy of the questionnaire.

Because we have the standardization base for the Global subscale of WAMMI (as well as all the other subscales, of course) and because we know how reliable the subscale is, we can compute the previous quantities directly. This means we know how much Measurement Error is involved and how much variation to expect. These are important quantities to know: If we did not have the standardization base, we would have to *estimate* these quantities on the basis of the actual *sample* we obtained. (How else would we know?) Now the smaller the sample is, the wider we would have to leave the margin of error, and therefore the less precise our knowledge of what made up the *i*th respondent's Global score. If someone else completed the preparatory work for us, we can operate with smaller samples. To put it another way, if we don't know much when we start, then we have to run large samples.

Satisfaction metrics appear to be the only one of the three components of usability for which reliability of measurement has been presented, and within satisfaction questionnaires, the SUMI, WAMMI, and the ISONORM scales are the only ones on which population data have been reported. Note also that satisfaction metrics are quite neatly dimensionless—you could, if you wished, compare user satisfaction of one group of users with a word processor and another group of users with graphics software (if such a comparison made any sense to begin with).

Because this is a chapter in a book about cost justification, it is useful to include some cost information about the commercialized questionnaires. Although vendors will offer incentive schemes to promote their wares, a basic evaluation using this kind of technology cost approximately $700.00 in 2004, but this figure does not include the cost of recruiting respondents and handling the data. SUMI and WAMMI have standard report generators, and bulk purchase reduces the costs considerably.

18.3.4 Measuring Usability: In Summary

This review of basic usability measures has introduced us to the fundamental concepts of dimensionless versus embedded metrics; the effects of the underlying form of the numerical distribution (is it normal or not?); samples versus populations; and the notions of reliability, validity, and the need for standardization of metrics.

Effectiveness and Efficiency sound like "hard" metrics that are objective. They measure the behavior of users or the results of user behavior. They refer to publicly observable events and can be captured on videotape. Surely metrics such as these offer the best hopes for establishing the scientific credibility of usability evaluation? Unfortunately, these metrics have not been empirically researched in terms of their metric properties as thoroughly as the Satisfaction metrics have been. Not only do we not know how reliable such performance metrics are, but we also have no information about their population parameters (although they do have a lot of what is known as *face* or apparent validity). As we shall see later, this has serious consequences on the sample sizes needed to demonstrate an effect. Satisfaction metrics, on the other hand, are usually cost-effective and user-friendly to the analyst.

18.4 STATISTICAL ISSUES

We turn now to issues to do the analysis of the data we have collected. Introductory statistical textbooks for the behavioral sciences usually give a "cozy" version of statistical testing that does not correspond precisely with the ideas held by most statisticians who write about it. Two aspects of statistics that are under-taught are the concept of statistical power and the use of confidence intervals (of the mean or median). In addition, it would seem that many introductory texts treat the subject of null hypothesis testing in an unsatisfactory manner, and my intention in this section is to raise some of these important issues.

18.4.1 The Statistical Argument, Hypotheses, Rejection, and Power

As the ISO definition of usability implies, when we test samples of users, we do so with the intention of generalizing their behavior to the behavior of the populations from which we sampled them. The *statistical argument* is a method of framing a chain of reasoning in statistics that will lead us to a correct conclusion on the basis of the results obtained. Unfortunately for the user of statistics, statisticians are not clear about what a correct statistical argument is and numerous formulations have been presented. The entire issue is, from the point of view of statisticians and philosophers of science, a series of confusions to the uninitiated (see Denis, 2003, for an account of no fewer than five possible alternatives to null hypothesis statistical testing).

To start with, I will outline the ideas of Jerzy Neymann and Eugene Pearson (the so-called Neymann-Pearson formulation; Neymann and Pearson, 1933).

They are important because their ideas developed from the original ideas of R. A. Fisher, and they have been widely adopted by practicing scientists. Let's assume that we are conducting a statistical test to see whether two samples (e.g., the effectiveness scores for being able to purchase a certain book chosen at random on two rival book-selling Web sites) come from the same population of effectiveness scores (and therefore that the two sites are almost the same in terms of effectiveness) or whether they come from different populations of effectiveness scores (and therefore that one site leads to greater user effectiveness than the other).

We set up two hypotheses: a null hypothesis H(0), which states that the observed sample differences between sites X and Y are a result of chance or accident, and an alternative hypothesis H(1), which states that site X leads users to be more effective than site Y. We decide on an appropriate statistical test, and we look up a set of statistical tables associated with the test to find the rejection region, R, for the computed test statistic. R is usually from the tabled value to infinity so we want as large a value as possible for the computed test statistic.

We now carry out our observations, tabulate them, and compute the test statistic.

If the test statistic is within the rejection region R, then we can reject H(0) in favor of H(1). If the result is outside of R, we reject H(1) in favor of H(0). Readers familiar with conventional explanations in introductory textbooks of statistics may be startled to read that one can accept H(0), but this is explicitly stated repeatedly in the writings of Neymann and Pearson (see, for instance, Pearson, 1955). Neymann and Pearson saw statistical tests as "rules of behavior," so that if one behaved according to the rules, then one would not make a false conclusion about the two hypotheses more than a certain proportion of times, on average.

For instance, Neymann and Pearson argued that if one uses a 0.05 level of probability for determining the rejection region, in the long run, one is prepared to make a false conclusion that proportion of times—in fact, we are prepared to make a false conclusion not more than once every 20 analyses.

There are of course two sorts of false conclusions one can make. The particular false conclusion scientists are most concerned about is of admitting to the canon of scientific knowledge *something that is not true.* In other words, rejecting H(0) when it is true that there is no difference. This is known as a type 1 error, and the size of R (the rejection region) depends a lot on α, the proportion of times one is prepared to risk this kind of error (and on the sample size—more of this later). Conventional values of α are 0.05 and 0.01; that is, one is prepared to jump to a false conclusion of difference not more than 1 time in 20 (0.05) or 1 time in a 100 (0.01).

However, Pearson and Neymann also recognized that there was an equal danger in the opposite direction—that of rejecting H(1) when it is in fact true: accepting that no difference exists when there actually is a difference. This is called a type 2 error, and the proportion of times one is prepared to risk it is called β. The scientific community regards this as a much less serious error, because, given the pressure to publish scientific results, if one group of researchers fails to publish because they made a type 2 error, someone else is sure to come along later and remedy the situation. Thus for scientific work, type 2 errors are not given as much consideration. The power of a statistical test is given in terms of β, as:

$$\text{Power} = 1 - \beta$$

So that we do not to confuse power and β, we shall adopt the convention of noting power as a percentage in this chapter and β as a probability. Thus conventional values of power seem to be 80 and 90%, that is, 1 minus the β levels corresponding to the likelihood of making a type 2 error: 1 time in 5 (0.2) and 1 time in 10 (0.10).

Although in theory this would be adequate for scientific work, numerous analyses of published results in psychological journals have shown that although α is usually high (as it has to be to get accepted for publication), levels of β have actually been quite low—often as low as 0.65. However, for applied work, the detecting of no difference is as important as the detecting of a difference. Thus it would seem for applied work that the levels of α and β should in fact be about the same: We should not favor one over the other. That is, if you decide that you are prepared to risk rejecting the null hypothesis on false grounds less than 1 time in 20 ($\alpha = 0.05$), you should also set up your test symmetrically so you are prepared to risk rejecting the experimental hypothesis on false grounds in the same proportion ($\beta = 0.05$).

Interestingly, this information now supports the argument that both H(0) and H(1) are acceptable and rejectable. Whereas if one does not pay attention to β, H(1) is favored, because we are blinded by the possibility of a false acceptance of H(0). Cohen (1988) discusses implications for a researcher who does not wish to entertain the possibility that the null hypothesis may be rejectable.

As any reader of statistical tables will notice, R (the region where we reject H(0)) becomes wider as the sample size increases. So that in theory, if we show only a modest difference between site X and site Y, which is in the right direction but nevertheless not sufficient to reject H(0), all we have to do is to increase our sample size, and sooner or later we will get to the stage where the value of R is wide enough for us to be able to reject H(0).

For instance, suppose we have obtained measurements from 10 users apiece on site X and site Y, and the obtained value of the Student's t statistic is 1.70. The bigger this statistic is, the greater the difference between site X and site Y. The table of values of t (with df = 18, $p \leq 0.05$) tells us that the value of R ranges from 1.734 to infinity, so we cannot reject H(0). Now, we obtain more observations, until we have 50 observations at each site. Let's assume that the obtained value of the statistic stays about the same (i.e., the size of the difference effect does not change, although we have tested many more users) and now we see in the table (df = 98, $p \leq 0.05$) that R ranges from 1.676 to infinity. We seem to have rejected the H(0) with a considerable amount of power because the statistic obtained is now well in the region of rejection.

That is, as our sample size increased, the power of our procedure to reject the H(0) also increased. But note that we may have been able to reject the H(0) with fewer than 50 users.

In contrast, if we start with an enormous difference, even with a small sample size, we will be able to make it easily into R. Suppose the value of t had been 2.00 with only 10 users apiece. We would have easily been able to reject the H(0) at once, most probably with power to spare.

Thus to be able to predict the optimum sample size, we have first decided that α and β should be approximately the same and that poses some constraints on our solution to start with. Obviously, if we expect that the difference between our two applications will be large in terms of measured usability, then we would need few users to demonstrate this result. If the difference is not obvious, then we will need more users to demonstrate it. In other words, the larger the anticipated size of the effect (or effect size, usually abbreviated as ES), the smaller the sample size needed (usually called N). ES and N are inversely related. The precise forms of the relationships are complex and vary from test to test. Conventional statistical tables that only give values for α are not much help: We either refer to books such as Cohen's (1988) or compute N ourselves. Luckily, most large statistics packages now include power analysis as an option, which will take care of the computations for the applied statistician (see the final section of this chapter for a mention of some of these packages).

18.4.2 Different Kinds of Hypotheses

So far we have assumed that the kind of test we are carrying out is a comparative one (i.e., site X against site Y or site X1-before and site X2-after additional work). As we have seen in our review of usability metrics, very often, this is the easiest kind of design to set up, given that we have no population parameters to tell us what an average or *expected* usability value is. We do not have an absolute standard, but we have a comparison of the difference between two sites, perhaps one baseline site and one hopeful new or competitive version.

This of course doubles the cost of the usability evaluation, although the *X1-before compared to X2-after* type of design should be encouraged because it implies that usability records are kept in some ordered state by the organization.

However, because most evaluations are about one Web site or software package only, and comparisons take extra effort and therefore cost more, is there anything else can we use as a comparison value against which to differentiate the results?

The most obvious alternative depends on researchers compiling statistics of metrics as the SUMI and WAMMI teams have done. That is, a large collection of data that tells us that the average Web site gets such-and-such a score on this scale. We can now compare our Web site and see how far above or below the average it falls.

Collection of these data can credibly be carried out by an organization that does not have a particular business interest outside of the area of usability testing—a research group in a university or a panel of a professional society is ideal. Collections of metrics made within a development organization, reflecting the organization's practices, are usually misleading because there is no outside referent. Such internal collections of metrics can show that a particular Web site is better than the company average, but finding how the Web site compares to other similar Web sites from outside the company cannot be done with an internal database. A company can of course start their own database of competitor Web sites. However, in practice, to get reliable data on a site that is remotely complex enough (e-business sites, for instance, usually require registration and some kind of commitment from the users to allow them access to certain key parts) is a course of action beset with difficulties.

Occasionally, results are partially shared in discussion groups. In this way I learned that an average RUE for first-time users of a typical desktop software product is on the order of 400%, that is, novice users are on average four times as slow as experienced users (this finding also agrees roughly with our experience in our laboratory). However, such benchmarks are informal and should be used with caution.

There are two other approaches to developing a standard to use as a comparison. A panel of users with relevant experience is required for each of them. It is useful for the panel to meet face-to-face and to discuss and reach a consensual decision. End users with varied experience are preferred to usability experts (or, worse still, technical experts who are usually good with computers and tend to underestimate technical difficulties).

One approach is to establish by consensus what the general population parameters are and to use some statistical knowledge to convert these to data. Participants are encouraged to give anecdotal evidence and refer to specific instances in their experience.

For instance, how long, on average, does it take to book a flight on an airline Web site if you know precisely where and when to fly? The method of limits may be used: "How many around the table think it should take more than 10 minutes, on average?" (laughter) "How many think it should take less than 2?" The limits are successively narrowed until an acceptable figure is reached for the mean. Let's say that an acceptable mean is 2 minutes. Now, questions relating to dispersal are posed: "What is a good, fast task time?" and "What is a poor, slow task time?" Let's say the slowest acceptable time is 3 minutes and the fastest possible time (someone can usually be found to demonstrate this!) is 1 minute. Dividing the interval between the consensual fast time and consensual slow time by 4 will yield an estimate of the standard deviation (because in the Gaussian normal distribution, an interval of 2 standard deviations on either side of the mean includes approximately 99% of the observed cases). In our case, the standard deviation will be 30 seconds, 0.5 minute. We now have an idea of the population mean and standard deviation without having had to do any empirical testing!

The second approach is more related to the performance of a particular application and may feel uncomfortable to those whose statistics training has taught them that one can never "prove" a null hypothesis. What is the target acceptable time that a booking should take on a Web site? Statements are elicited in the form "if it takes longer than x minutes, it's unacceptable" or "the longest acceptable time is x." The various values of x are tallied and displayed, and a value is arrived at by discussion and consensus. Later, when the data are being analyzed, we hope to demonstrate that the H(0) is true and that our obtained value is indistinguishable from the value under H(0) (of course, we would be extremely glad if in fact we find that our obtained value is more advantageous—faster in this case—than the H(0) value, but then we may be guilty of overengineering or of a less than cost-effective evaluation, which only needed to show *comparability* with H(0)). That is, mindful of the fact that we are able to both accept and reject the H(0), especially if we have kept α and β the same, with our test we will expect to find that there is no difference between the obtained statistic and the target acceptable time as worked out following the stated procedure.

Alternatively a cost-benefit tradeoff may be negotiated within the team (for instance, so much time to book a local flight or so much for a transatlantic flight). We can now test the actual statistics obtained against these tradeoffs to derive a profile of how well the Web site serves different kinds of needs.

18.4.3 Do We Need Hypotheses?

A steady undercurrent claiming that we do not actually need the apparatus of null hypothesis formulation is present in statistical thinking. What we are most concerned with, this argument states, is not the blanket acceptance or rejection of a hypothesis, but an idea of where the true population parameters lie given what we know about the sample. This usually resolves into the estimation of what are known as the "confidence intervals of the mean (or median)." The 95% confidence intervals are the ones usually used. The confidence interval of the mean does not actually make any statements about where the real population mean may be found. In principle, the 95% confidence interval should be interpreted in this way: "If we take this size of sample 20 times and compute the confidence interval on the basis of the collected data, we will expect on average to find the population mean somewhere inside it 19 times." (Readers are warned that although there is a "law of large numbers" in statistics, there is no corresponding "law of small numbers," and although we may expect that *on average* we will find the population mean inside a 95% confidence interval 19 times out of 20, if we replicated these 20 trials many times, then we would often find that either the mean is *always* in the confidence interval or the confidence interval actually *excludes the mean more than once* in any batch of 20 trials.)

Confidence intervals around the mean are usually computed with the mean plus or minus a value we may call H_{CI}:

$$H_{CI} = \frac{\sigma}{\sqrt{n}} \times 1.96$$

where σ is the standard deviation and the value 1.96 is used when the sample size is sufficiently large (Snedecor and Cochran, 1980).

Usually, a sample size of 30 is considered sufficiently large, yet another rule of thumb with which the behavioral and biostatistical literature seems to be replete. However, the width of the confidence interval is affected by the degree of precision of measurement. Thus, in theory if you want your confidence interval to contain a certain number, then all you must do is use a small sample size and a sloppy measurement technique that produces a large standard deviation. This is something that an investigator will guard against if it is required to show that there is *little* chance that the confidence interval does actually contain a certain value, but if the requirement is to show that a certain target value *is actually included* in the confidence interval, then the sloppy approach seems like the perfect solution for pseudo-science and bad engineering.

A useful formula for working out the minimal sample size required to compute an acceptable confidence interval is available, but in addition to α and β, the formula also requires us to estimate the population standard deviation and a factor called Δ, the latter of which is usually thought of as the "acceptable margin" or a "difference worth detecting."

$$n = \left(\frac{(z_\alpha + z_\beta)\sigma}{\Delta} \right)^2$$

See Zap (1999) for the derivation. In this formula, z_α is the two-tailed normal score of the probability of a type I error, so if we were interested in working with $P = 0.05$, it would be 1.96. z_β is the one-tailed normal score value so if we want to retain β at $P = 0.05$ in symmetry with α, it would be 1.64. σ is our estimate of the population standard deviation and finally Δ is our difference worth detecting.

Suppose now we had a questionnaire with a population standard deviation of 10, population average of 50, and scores ranging from 15 to 80. We may expect that 5% of the questionnaire's range is certainly a difference worth detecting, and then applying the above formula to these values we come up with a size estimate that 12 users will give us a sufficient sample, with α and β balanced, on which to take an average. Suppose now that the database average is 50 (which we want to do better than), our goal is 60, and we produce an average of 57. We are able to claim (1) that we are better than the average and (2) that our confidence interval contains the desired target value of 60 at a probability level of 0.05.

If we do not have the population standard deviation (for instance, if we are using an unstandardized test or a metric such as RUE for which no standardization data exist), we must first determine the probable size of the population standard deviation by a pretest and then compute the range of values within which the true (population) standard deviation may be found. This procedure is outlined in Snedecor and Cochran (1980). Using the upper bound of the interval, we can then make a "worst case" guess for σ. This unfortunately has the effect of increasing the required sample size for the main trial.

For example, suppose we just have the basic questionnaire but no population information for either where the population mean lies or the size of the population standard deviation. Our group of expert users may not be of much help with such a questionnaire, so we have to do some pretesting. If we do a pretest with 10 users, the upper limit of the standard deviation rises to 18.25 and the prediction for an adequate sample size for the main trial rises to 40 users. If we do a pretrial with 20 users, the upper limit for the standard deviation is 14.61, and the number of users required for the main trial is 25.

It is difficult to make a better case for the cost-effectiveness of standardized metrics.

Incidentally, in relation to the time-honored probability values of 0.05 and 0.01 (and the 95% confidence interval) it should be pointed out that these have no actual weight other than that they are generally accepted within the behavioral scientific literature as standards. In an applied setting, although one should use the 0.05 probability value as a guide (do you really want to endorse findings that may turn out differently more often than 1 time in 20?) the more important criterion is to maintain a balance between the α and β levels, so as *not* to give unfair advantage to either the H(0) or the H(1).

18.4.4 How Do We Compute Sample Size and Effect Size?

The technology of computing statistical power was presented in the context of confidence intervals in the previous section. Generally statistical power analysis as conventionally presented assumes that the researcher wants to carry out some kind of inferential procedure using the statistical argument.

The value called Δ in the previous section was introduced as an example of a difference worth detecting. It is more generally known in power analysis as an ES. ES is the magnitude of difference expected between two or more groups of measurements or between a measurement and a benchmark. As we saw in the previous section, sample size estimates depend on our expectation of ES. Having some criteria as to what are acceptable values for ES is useful. Cohen (1988) suggested three "bands" of difference of ES that may be considered, which he calls *small, medium,* and *large.* Cohen's definitions are given in an arbitrary manner in his book, but his recommendations were verified by Lipsey (1990), who carried out meta-analyses of published studies in the behavioral sciences. To a surprising extent, Cohen's and Lipsey's values agree, and in the following account, the actual values are given—Lipsey's as a range, Cohen's as a nominal value.

Small (ES < 0.32, nominal 0.20). Examples of small effect sizes are the magnitude of difference between twins and nontwins, and the difference in mean height between 15- and 16-year-old girls. The small ES specification is recommended for use with variables producing differences that are not really visible to the naked eye. In usability testing terms, small ES may be expected in the comparison of software systems that differ in technical detail that does not readily make itself apparent to the user or in aspects of presentation that do not greatly affect the usability of the systems for the tasks studied in the evaluation. Examples could include a better technical file interface that makes an application more robust or a small realignment of menu items between versions.

Medium (ES = 0.33–0.55., nominal 0.50). Examples of medium effect sizes are the IQ difference between clerical and semiskilled workers and between professional and managerial groups and the difference in mean height between 14- and 18-year-old girls. The medium ES specification is recommended for use with independent variables causing effects that are large enough to be visible to the naked eye. In usability testing terms, one would expect that investment in usability engineering should pay off in terms of at least medium differences between software versions, and one may expect at least medium differences between successive major released versions of a piece of software or between a market leader and the rest.

Large (ES = 0.56–1.20, nominal 0.80). Examples of large effect sizes are the IQ difference between freshmen and individuals with PhDs and between college graduates and those with a 50-50 chance of passing an academic high school curriculum and the difference in mean height between 13- and 18-year-old girls. The large ES specification is recommended for use with independent variables causing effects that create gross differences. In usability testing terms this would be equivalent to the difference between a prototype or an interface that is designed from a software technical point of view and one that is designed as a result of task analysis.

For purposes of evaluation of software systems, quantitative evaluation goals may be stated as an ES statement specifically related to the software systems under consideration. The cost of evaluation and the cost of development interact. Thus if we suppose that it costs more to create a system that demonstrates a larger ES compared with its predecessor, other things being equal, a smaller ES means greater precision of measurement or larger sample size required for the detection of the difference. To detect a very small ES you need a lot of data gathered with an extremely reliable measurement technique. For a large ES, the evaluation constraints are more relaxed, although the cost of creating the system may be greater.

For example, if testing will yield means and standard deviations, then ES will be determined by the formula:

$$\text{ES} = \frac{\text{Av}(1) - \text{Av}(2)}{\text{Measurement Error}}$$

where Av(1) is the baseline or competitor average, Av(2) is the obtained or current system average. The Measurement Error is as computed above in Section

18.3.3. As the measurement error becomes smaller, the difference between the averages that is required to yield the expected ES also becomes smaller. In fact all four quantities can be played against each other: sometimes to estimate what ES should be, sometimes to estimate what the difference between the averages should be, and sometimes to obtain an estimate of the required measurement error to be able to achieve a particular ES.

Having obtained an ES, one can then enter tables such as those provided by Cohen (1988) for estimation of power and sample size to discover how large a sample will be required, given the preselected α and β, to demonstrate the ES with the measurement tools at hand. If the requirement is to reject H(0), then one should use a reasonably *large* ES (given the above discussion). If rejecting H(1) is required, then one should use a reasonably *small* ES (Cohen says 0.10 or thereabouts).

The concluding section of this chapter suggests computer programs that can be used for power analysis.

18.4.5 When Do We Compute Sample Size and Effect Size?

In a rather Aristotelian humor, one may contemplate doing these calculations at one of three moments: before an evaluation, during an evaluation, or after an evaluation. Let's look at these options in turn and consider the kinds of information we may gain at each.

Before an Evaluation. Before an evaluation, one either knows what the population parameters are or one doesn't. If the standard error is known, then the discussion trades ES against the required sample size.

If the population parameters are not known, then ES and sample sizes are traded against the probable measurement error. A pilot study may be carried out on a small sample to find the level of magnitude of the measurement error if this has not been estimated already using experienced users. There are two situations:

1. The pilot sample is more homogenous with respect to the behavior (or attitudes) of the end users than the real sample will be.
2. The pilot sample is less homogenous than the real sample will be.

In other words, the sample will either exhibit a broader or a narrower dispersion (it may or may not be part of the same problem that the sample will also be closer to the H(0) value or further away). A more homogeneous pilot sample will give the user a nasty surprise when the real sample is collected and found to be more heterogeneous; a less homogeneous sample will inflate the required sample size.

In practice, when pilot samples are used, a wise procedure is to compute the 95% confidence intervals of the sample variance and to base one's estimates of the population variance on the upper value of the confidence interval. The perils accompanying this procedure were depicted graphically in Section 18.4.3.

In general pilot samples and main-phase samples should be carefully selected according to what is known about the ability and attitudes of the intended user population. Context of use analysis (Kirakowski and Cierlik, 1998) is an important tool for this purpose. The worst possible case is when the data for a pilot sample (or a main sample) are taken from colleagues working in the vicinity and their friends without any care as to how closely this sample corresponds to the real user population. Unfortunately, this kind of thoughtlessness occurs far too often.

Characteristically, before an evaluation, one is in the business of making tradeoffs among the four major factors: sample size, effect size, α and β levels, and measurement error.

During an Evaluation. During an evaluation, this kind of analysis enters the realm of *sequential testing*. In sequential testing we gather some data and test it for statistical significance. If we do not reach the rejection region of either H(0) or H(1), then we go back and gather more data. See Wetherill (1966) for a good, overall introduction to this area, although significant advances continue to be made (see, for instance, a thoroughly technical discussion by Xiong, Tan, and Boycott, 2003).

The evaluator will be in roughly two situations, and two tests are appropriate. In the first situation, we know the standard deviation of the measurements with great accuracy beforehand. The *sequential probability ratio test* is appropriate for this situation. If the values of α and β are equivalent (and small; i.e., about 0.05) and the ES ratio is held constant, then the sample sizes required for making a decision with regard to H(0) and H(1) are about the same as if one carried out a preplanned study (i.e., as if we chose a sample size before the evaluation was started). Thus in this case, sample size estimates made before and during the study are equivalent. This corresponds to common sense: If we know what the measurement error is independently of the sample size, then it matters little when we compute the required sample sizes before, during, or after.

In the second situation, we do not know what the standard deviation of the measurements is. Our tests will take the standard deviations of the sample and attempt to predict the standard deviation of the population. Clearly, the closer we get to the population size, the better our estimate will become; however, because we are continually revising our statistic, it is important in this case to guard against the cardinal statistical sin of *capitalizing on chance.* Thus the ES needed for the statistic to be significant will be extremely large when the sample sizes are small. However, as the sample sizes get bigger, the rejection regions are "protected," so that if a hypothesis is rejected with a certain sample size, you may be confident that had the tests continued, it would also have been rejected with samples sizes larger than the one stopped at. In this case, sequential testing puts the evaluator at a significant disadvantage with respect to the required sample size. Needless to say, statisticians are working hard to improve this situation for us (see the article by Xiong et al., 2003, who develop a technique known as the *sequential conditional probability ratio test,* which they claim attains the smallest maximum sample size among group sequential tests). However, it appears that in sequential testing, as in many other forms of human activity, there is no such thing as a free lunch.

After an Evaluation. When an evaluation has been completed, averages and measurement error are, by definition, known. All that can be done at this stage is to model α and β levels against ESs, bearing in mind the recommendation that these rates, for usability testing, should be approximately equivalent. The result will actually be a graph of ES against α and β rates. Interpretation of the probability values may become important at this stage. Instead of talking about the probability of samples approximating population values, it is more in line with the applied nature of usability testing to speak of "decision rules." That is, a probability value of 0.05 corresponds to the decision rule that we are satisfied it is sufficient that if we were to do the same evaluation 20 times over, only once would we have to reverse our judgment.

18.4.6 Statistics and Reality

One of the issues that the forgoing discussion brings out is that a high premium should be paid for data which enable us to make confident assessments of the required size of samples for usability testing. A useful agenda for professional societies in usability is to encourage the collection of such databases for non-proprietary methods.

When one begins to research into the details of statistical issues such as the true nature of hypothesis testing and the benefits of sequential testing, one is struck by the vigor of the discussion on such issues between professional statisticians and the level of disagreement between them. This is, of course, only to be expected of a discipline in a phase of growth, but it does make consumers of such research anxious that what they have taken from the discussion is correct. In the end, the most consistent advice from all the statistical sources reviewed in the course of preparing this chapter has been to use common sense and to use the results of statistical computations as a *guide* to action rather than as a *set of oracles.*

In the end, the conclusions made from a usability test are as good as the assumptions made at the start: This involves finding an answer to the "conditionals" of the ISO usability definition. Who are the target users for the application? If functional differences between groups of target users exist, then the user groups ought to be examined separately. What will the users do with the application? Again, if the application is large and can support different activities, then these activities should be evaluated separately. This is especially true of large Web sites that may offer a variety of experiences for the users. In what environments will the application be used, or how will users access the Web site? It is fallacious to conclude that users who access a Web site using dial-up technology at 50 kilobytes per second many miles away can be adequately represented by users who access the same site in a laboratory close to the server operating at 20 megabytes per second. (Conditions may, of course, be set up to mimic the former kind of connection in a laboratory in the same way that a laboratory may be set up to mimic a "busy office," but is this ever done?)

"User panels" or user groups who can be recruited to be usability testing volunteers may represent a particular danger to the generalizability of evaluation results, because very often such users come to a Web site with a large critical apparatus on the basis of previous testing experiences and may not be involved enough with the application to want to use it seriously (even if they are given "tasks" to complete). In a recent evaluation of a Web site, we differentiated users on the basis of their responses to the statement that *"this Web site contains the kind of information I need for my day-to-day work."* Those users who answered this question positively gave the site much higher satisfaction ratings than those users who answered negatively. The results, using the WAMMI questionnaire, are shown in Figure 18.3.

Comments from these two categories of users were also informative. Users who considered that the Web site did not have the kind of information needed for their daily work gave only negative views on relatively superficial aspects of the site's implementation, such as the use of Flash or stylistic issues. Users who considered that the Web site did have the information they needed for their daily work gave informed comments about the kind of information being presented and how the organization of the site could facilitate their navigation for different kinds of required information. (Also, occasional references to the use of Flash, appearance, and colors were made.)

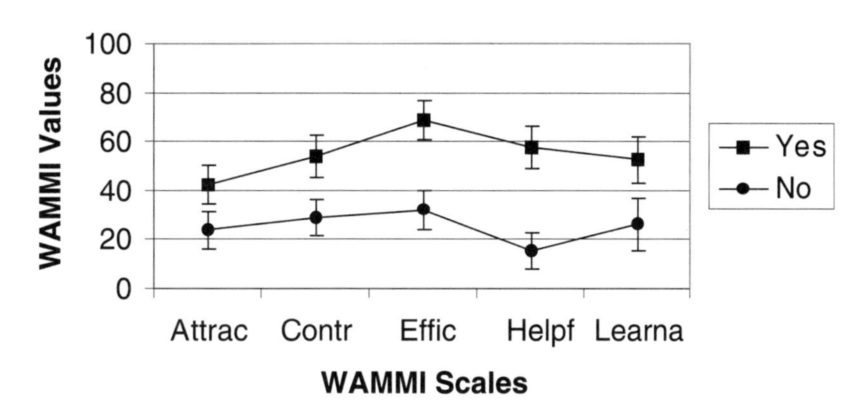

FIGURE 18.3 WAMMI results with 95% CI for two categories of user in response to the question: "This Web site contains the kind of information I need for my day-to-day work."

18.5 CONCLUSIONS

Having read all this, the reader may well reflect that summative user testing is not much more advanced than formative testing! However, it is surely worth the effort involved to know beforehand by appropriate testing how efficient, effective, and satisfying a product is. This information is certainly welcomed by managers, who put store on quantitative data. (Note that for a design team, the reverse is often true—qualitative data are more useful than quantitative.)

To carry out a summative test, the evaluator performs the following seven steps:

1. Plan the evaluation in terms of design, procedures, tasks, and environments.
2. Decide on a sampling frame—where the user sample(s) will be drawn from.
3. Decide on an adequate user sample size and a method of recruiting a random sample.
4. Contact, gather, and run the individual users in the sample through the procedure(s).
5. Analyze the data.
6. Consider the implications.
7. Report the data.

The good news is that step 1 usually takes least stress: Most usability testers will be working within a niche and will soon develop a set of tools and procedures that will satisfy the needs of their clients. Please note: This also means that data from past evaluations ought to be kept, where possible, in a professional casebook.

Step 4 usually takes the most time, and step 2 is usually the most critical step in terms of acceptance of the report. It follows that step 3 has to be as fine tuned as possible to offer the client value for money. Perhaps the most crucial step, step 6, is never done in costable time. Too often, in fact, testing runs quickly from step 5 to step 7, and yet work is most often judged on the outcome of step 6.

A theme in this chapter has been that things generally become easier if one lives in a known universe. Perhaps one of the reasons more attention is not paid to gathering information about efficiency and effectiveness metrics is because the cost of not knowing this information is not understood. The same is true, one may suppose, for home-grown satisfaction questionnaires. If you don't know, then you have to pay extra.

With open source tools such as RUE, it is perhaps the duty of professional bodies to institute and encourage subcommittees to obtain such data for standardization purposes and to encourage workshops on quantitative topics for the purpose of disseminating benchmarks and case studies. Not everybody will be fascinated by such topics to be sure.

A frequently asked question is, Where does one begin? I used to have the opinion that the answer to this question was highly contextual—different projects and different organizations will have different needs. Although this advice remains true, it has little value to the novice, because it sets no standards. The following list is therefore presented with the caveat that contextualization is always needed.

1. The usability tester should be familiar with a context of use analysis methodology: This is the foundation for summative testing, and it takes care of the conditionals in the ISO 9241 definition (making an appropriate selection of tasks, users, and environments).

2. After this, one methodology for each of the usability qualities, task completion and efficiency, and a good freeware product such as the SUS questionnaire for satisfaction are recommended.

3. Learning how to use a statistical package is important; spreadsheet statistics are still full of traps for the unwary, whereas statistics packages such as SPSS, SAS, MedCal, MiniTab, and SYSTAT have long served the needs of seminu-

merate researchers in the social and biosciences. (These commercial packages all contain power analysis options.) There are also many freeware packages specifically dedicated to power analysis, but such packages are more ephemeral (see Helberg, 2004, for a listing).

4. To understand how to report an evaluation, know the CIF standard (INCITS, 2001). In addition, use plenty of common sense to not get bogged down in formalities and details and to not be misled by quick fixes. Remember step 6.

The alternative, of course, is not to do any testing at all.

REFERENCES

Aron, A., and Aron, E. (1994). *Statistics for Psychology.* Englewood Cliffs, NJ: Prentice Hall.

Bevan, N., and MacLeod, M. (1994). Usability measurement in context. *Behaviour and Information Technology,* 13, 132–145.

Brooke, J. (1996). SUS: A "quick and dirty" usability scale. In P. Jordan, B. Thomas, B. Weerdmeester, and I. McClelland, *Usability Evaluation in Industry.* New York: Taylor & Francis. (pp. 189–194).

Chapanis, A. (1959). *Research Techniques in Human Engineering.* Baltimore: John Hopkins Press.

Chin, J., Diehl, V., and Norman, K. (1988). Development of an instrument measuring user satisfaction of the human computer interface. In *Proceedings of CHI '88* (pp. 213–218). Association for Computing Machinery. Proceedings of the SIGCHI Conference on Human Factors in Computing Systems, May, 1988, Washington, D.C. (pp. 213–218).

Cohen, J. (1988). *Statistical Power Analysis for the Behavioural Sciences* (2nd ed.). Hillsdale, NJ: Lawrence Erlbaum.

Constantine, L. (2003). *CHI 2003 Feature Testing . . . 1 2 3 4 5 . . . Testing. . . .* Usability News. Retrieved December 16, 2004, from www.usabilitynews.com/news/article1058.asp.

Constantine, L., and Lockwood, L. (2000). *Software for Use: A Practical Guide to the Models and Methods of Usage-Centered Design.* New York: ACM Press.

Denis, D. (2003). Alternatives to null hypothesis significance testing. *Theory and Science,* 4. Retrieved December 16, 2003, from theoryandscience.icaap.org/content/vol4.1/02_denis.html.

Dumas, J., and Redish, J. (1993). *A Practical Guide to Usability Testing.* Norwood, NJ: Ablex.

Gediga, G., Hamorg, K.-C., and Düntsch, I. (1999). The iso metrics usability inventory: An operationalisation of ISO 9241/10. *Behaviour and Information Technology,* 18, 151–164.

Hart, S., and Staveland, L. (1988). Development of a multi-dimensional workload rating scale: Results of empirical and theoretical research. In P. A. Hancock and N. Meshkati (Eds.), *Human Mental Workload.* New York: Elsevier.

Helberg, C. (2004). *Statistics on the Web.* Retrieved August 14, 2004, from my.execpc.com/~helberg/statframes.html.

INCITS. (2001). *Common Industry Format for Usability Test Reports.* International Committee for Information Technology Reports. Retrieved December 16, 2003, from www.techstreet.com/cgi-bin/detail?product_id=918375.

International Organization for Standardization (ISO). (1998). *Ergonomic Requirements for Office Work with Visual Display Terminals (VDTs). Part 11: Guidance on Usability.* ISO 9241.

Kirakowski, J. (1998a). *SUMI User Handbook* (2nd ed.). Cork, Ireland: Human Factors Research Group, University College Cork.

Kirakowski, J. (1998b). *Usability Validation Report Format.* Retrieved December 16, 2003, from www.ucc.ie/hfrg/baseline/filearchive.html.

Kirakowski, J. (2003). *Questionnaires in Usability Engineering—A List of Frequently-Asked Questions* (3rd ed.). Retrieved December 16, 2003, from www.ucc.ie/hfrg/resources/qfaq1.html.

Kirakowski, J., and Cierlik, B. (1998). *Context of Use Analysis—Background Notes.* Retrieved December 16, 2003, from hfrg.ucc.ie/baseline/filearchive.html#cou.

Kirakowski, J., and Claridge, N. (2001). A professional tool for evaluating Web sites. In J. Vanderdonckt, A. Blandford, and A. Derycke. *IHM-HCI 2001.* Toulouse: Cépaduès Editions.

Kirakowski, J., Claridge, N., and Whitehand, R. (1998). Human centered measures of success in Web site design. In *4th Conference on Human Factors and the Web, Basking Ridge, NJ.* AT&T. Retrieved December 16, 2003, from www.research.att.com/conf/hfweb/conferences/presentations98.zip.

Kirakowski, J., and Corbett, M. (1994). SUMI: The Software Usability Measurement Inventory. *British Journal of Educational Technology,* 24 (3), 210–213.

Kline, P. (1998). *The New Psychometrics.* New York: Routledge.

Lewis, J. (1995). IBM computer usability satisfaction questionnaires: Psychometric evaluation and instructions for use. *International Journal for Human-Computer Interaction,* 7, 57–78.

Lipsey, M. (1990). *Design Sensitivity.* Thousand Oaks, CA: Sage.

Molich, R., Bevan, N., Curson, I., Butler, S., Kindlund, E., Miller, D., and Kirakowski, J. (1998). Comparative evaluation of usability tests. In *Proceedings of the Usability Professionals Association, Washington, DC.* Usability Professionals Association.

Molich, J., Ede, M., Kaasgaard, K., and Karyukin, B. (2004). Comparative usability evaluation. *Behaviour and Information Technology.* 23 (1), 65–74.

Mulder, L. (1992). Measurement and analysis methods of heart rate and respiration for use in applied environments. *Biological Psychology,* 34, 205–236.

Neymann, J., and Pearson, E. (1933). On the problem of the most efficient tests of statistical hypotheses. *Philosophical Transactions of the Royal Society A*, 231, 289–337.

Pearson, E. (1955). Statistical concepts in their relation to reality. *Journal of the Royal Statistical Society B*, 17, 204–207.

Prümper, J. (1999). Test IT: ISONORM 9241/10. In H.-J. Bullinger and J. Ziegler (Eds.), *Human-Computer Interaction: Ergonomics and User Interfaces* (Vol. 1, pp. 1028–1032). Mahwah, NJ: Lawrence Erlbaum.

Rubin, J. (1994). *Handbook of Usability Testing.* New York: John Wiley.

Scriven, M. (1967). The methodology of evaluation. In R. Tyler, R. Gagne, and M. Scriven (Eds.), *Perspectives of Curriculum Evaluation* (pp. 39–83). Chicago: Rand-McNally.

Shneiderman, B. (1997). *Designing the User Interface* (3rd ed.). Reading, MA: Addison Wesley.

Snedecor, G. W., and Cochran, W. G. (1980). *Statistical Methods* (7th ed.). Ames: Iowa State University Press.

Usability Professionals Association (UPA). (2001). *2000 UPA Member Profile and Salary Survey.* Retrieved December 16, 2003, from www.upassoc.org/upa_publications/upa_voice/survey/2000_survey.html.

Wetherill, G. (1966). *Sequential Methods in Statistics.* London: Chapman and Hall.

Xiong, X., Tan, M., and Boycott, J. (2003). Sequential conditional probability ratio tests for normalized test statistic on information time. *Biometrica*, 59, 624–631.

Zap, J. H. (1999). *Biostatistical Analysis* (4th ed.). Upper Saddle River, NJ: Prentice Hall.

Zijlstra, F. (1993). *Efficiency in Work Behaviour: A Design Approach for Modern Tools.* Delft, Netherlands: Delft University Press.

19 CHAPTER Cost-Justifying Online Surveys

Scott Weiss Usable Products Company

NOTE TO READER

The content in this chapter was informed in part by literature searches, but the majority of the material was gathered through telephone- and e-mail–administered questionnaires. Respondents included independent research practitioners, research and usability managers in large corporations, and online survey vendors. Some of the contributors requested anonymity, whereas others permitted attribution of their statements.

19.1 ONLINE SURVEYS AND THEIR VALUE

Some readers of this chapter may be trying to decide between formal usability testing and online surveys to address their product usability goals. The purposes of this chapter are to explain the following:

1. Exactly what online surveys are and how they contribute to usability activities
2. Key differences between formal usability testing and online surveys
3. Types of online surveys, and the value of each type with respect to usability
4. Costs associated with online surveys and how those costs compare to lab testing
5. Details about online surveys that will help the reader deploy them effectively
6. Strategy recommendations for mixing online surveys with other usability methods

A *survey* is "a gathering of sample data or opinions considered to be representative of a whole" (Dictionary.com, 2003). Surveys can be used in research efforts such as concept testing, positioning, market forecasting, customer satisfaction, and pricing in addition to studies of usability opinions and/or performance. Online survey types can range from very simple to very sophisticated. Questions can be closed-end, open-end, conjoint (two or more associated concepts), or discrete choice and can even assist with best attribute scaling. Online survey questions can be text-based, pictorial, animated, and even highly interactive with drag-and-drop features. Sophisticated software exists for matching and ranking objects across rows and columns. The technology supports complex branching and skip patterns, which, if sent by direct mail, might be confusing to respondents.

Lumped in with online surveys are *online usability studies*, which are online surveys with additional technology that synchronizes them with a Web site, tracking both users' responses and users' on-screen activities. For the rest of this chapter, *online usability surveys* will be referred to as *synchronized surveys.* One additional survey type that focuses on usability is asynchronous, but tied to respondents' opinions about a Web site's ease of use. This type of survey will be described as an *asynchronous usability survey* (Table 19.1).

The Website Analysis and MeasureMent Inventory (WAMMI) is perhaps the most well-known of the asynchronous usability surveys. It features fixed-response and open-response questions. Fixed-response questions have a limited set of answers, via a dropdown menu or series of radio buttons, whereas open-response questions have type-in fields. The WAMMI questionnaire is short and targeted and gauges respondents' attitudes about ease of use. However, it is purely opinion-oriented, and although respondents' opinions about usability may

Table 19.1 Survey Types

Survey Type	*Description*
Online survey	Catch-all term, usually reserved for simple surveys
Synchronized survey	"Online usability" surveys that directly link survey questions and instructions with a live Web site in several states of use and also track the paths of users as they navigate through a Web site
Asynchronous usability survey	Opinion-oriented survey about usability, capturing respondents' attitudes

correlate to their actual performance, there is no guarantee that they will correlate. WAMMI results *do* correlate with heuristic/expert reviews of Web sites, according to Kirakowski and Cierlik (1998). The WAMMI questionnaire also maintains a database of results, making norm-based comparisons possible. The key value of WAMMI is its measure of *user satisfaction*, which *is* a key element of usability.

Like the WAMMI, most online surveys measure user satisfaction and other opinion metrics rather than user performance, which is the forte of in-person usability testing. Synchronized surveys also track the paths that respondents take during the tasks they are asked to complete. Path tracking is offered by several firms, some of which have built sophisticated tools and visualization techniques for analyzing the data. Path tracking analysis is an attempt to understand user performance, and these tools can be more effective than log file analysis for understanding user activity. Log file analysis is a cruder strategy, because the researcher must guess what the users' goals were, as opposed to synchronized usability surveys, in which the researcher determines the respondents' goals in advance through survey question design. Synchronized surveys also offer the ability to ask respondents open-ended questions, providing further insight into their intentions. However, there are drawbacks to synchronized usability surveys, which are detailed later.

Online surveys must be only a part of a comprehensive usability effort. When used alone, the weaknesses of online surveys become overwhelming. However, as part of a balanced approach, they are a valuable research tool.

19.1.1 Cost and Return on Investment

Online surveys can cost significantly less than surveys administered by telephone or surface mail for the simple reason that they are conducted via computer, which allows for a number of efficiencies: elimination of the human interviewer, no postage and/or telephone costs, and increased size of potential populations. These factors help to reduce the cost of additional survey completions compared with other survey methods.

A great many low-priced survey automation software packages are available, and most of them are adequate for opinion gathering (Table 19.2). However, the software cost is only a small part of the total, because recruiting and respondent incentives usually cost far more than the software itself.

Pricing for surveys ranges wildly from free, for very limited sample sizes and survey scope, to $1,000 for a small survey, to more than $200,000 for a "full-scale discrete choice" project. The following factors contribute to cost:

Table 19.2 Survey Cost Factors

Lowers Cost	*Increases Cost*
No postage or telecommunications costs	Software package purchase price or study administration fees
Lower per-respondent incentives	More respondents, thus potentially a higher incentive pool
No data entry costs	More data analysis required
No travel costs	

- ✦ Recruiting
- ✦ Incentives
- ✦ Study design
- ✦ Programming
- ✦ Data collection and analysis
- ✦ Report writing

D&B's Short Hills, NJ–based market research group bills $1000 to $4500 per survey project back to their internal clients. D&B (formerly known as Dunn & Bradstreet) does not generally pay incentives to survey respondents, and their software licenses cost approximately $20,000 per year. Online surveys are an extremely valuable component of D&B's market research strategy, comprising more than 20% of their annual research effort in 2003.

In most situations, synchronized usability surveys are not lower priced than traditional lab-based usability research. Some online survey companies do offer low-priced entry-level survey products that may be useful to some clients. Following is the text from the Web site of a vendor for a 50-person survey:

> Starting at $1500, Vividence eXpress offers a faster, more accurate, and more cost-effective evaluation of customer reactions and responses than conventional research methods such as surveys and usability tests.

Vividence eXpress gives you several price options, depending on the panelists you use for the evaluation and whether you want to reward them ($5 Amazon.com gift certificates) for their participation:

Panel Source	No Rewards	With Rewards
Customer intercept	$1,500	$2,000
Your e-mail list	$1,500	$2,000
Vividence research panel	n/a	$3,500

These low-cost quotes supply a very basic service set that includes only two tasks (and up to 60 questions) using either a panel provided by the client or, for additional cost, the vendor's pre-existing panel with a limited set of qualifying questions to screen respondents. However, custom recruits, larger sample sizes, and analysis of the results typically drive prices higher than those for in-person studies.

Table 19.3 contains pricing from four synchronized survey vendors who were asked to respond to a request for proposal (RFP) for an online survey *without analysis* (Anonymous, 2003a). The intent of the RFP was to identify a practical minimum number of survey respondents to match a similar in-lab study of between five and seven users. Most usability professionals consider a "small in-lab study" to consist of between five and seven respondents, based on simple math. A small in-lab study schedules all of the interviews in one calendar day. Each interview lasts between 45 and 90 minutes, with some time buffer included between interviews. Therefore, five to seven interviews is a reasonable number, considering that a typical work day is between 7 and 9 hours. The survey vendors were asked to propose an online study for interviews of a "practical minimum number of respondents."

A small in-lab study might cost as little as $12,000, working with an independent contractor who keeps a home office. A professional usability firm might charge $25,000 or more for a similar study, because of the requisite overhead for maintaining an office and support staff. Formal usability study costs vary significantly, and usability pricing varies considerably from vendor to vendor. From

Table 19.3 Synchronized Survey Pricing

No. of Respondents	50	75	100	200
Vendor A	21,794	24,878	28,349	32,963
Vendor B	32,750	34,750	36,750	
Vendor C	33,650	38,475	43,300	
Vendor D	38,000			

Table 19.3, you can see that synchronized surveys are often more expensive than comparable in-lab studies, even considering the variability in in-lab usability study pricing. Furthermore, the table indicates pricing *without analysis*, as per the RFP, so prices are actually deflated somewhat from those of a complete study.

Asynchronous usability surveys can be ordered individually, and the WAMMI survey is priced between €800 and €2,000 ($963 and $2,407 in September 2004) (Kirakowski and Claridge, 2004). One could sidestep the vendor and create an asynchronous usability survey with a standard online survey package. However, the benefits of working with an experienced usability survey vendor would then be lost. Quality experienced vendors will prevent their clients from making mistakes with recruiting criteria, survey construction, and survey administration. Without the guidance of an experienced survey vendor or consultant, the entire monetary investment—and precious time—could be lost. Worse yet, a bad survey could yield results that if followed could cost far more than the survey itself.

Actual return on investment (ROI) figures for online surveys are hard to derive, but the cost per interview for online surveys is definitely lower than the cost per interview for in-lab studies. One online retailer spends approximately $100,000 annually for development of custom online survey software for its proprietary use. They spend an additional $10,000 on off-the-shelf software used for online surveys. They provide dollars-off coupons to survey respondents, although the coupons require a minimum-dollar-amount purchase for the coupons to be valid—thereby guaranteeing a positive cash flow from the coupons. The research manager at this e-commerce venture estimated that online surveys have generated "millions of dollars of sales from these dollars-off coupons" (Anonymous, 2003b). This order-of-magnitude (10 times) ROI is impressive, to say the least.

19.1.2 A Case Study for Online Usability Research: Staples Inc.

Following is an excerpt from an interview in 2003 with Colin Hynes, Director of Usability for Staples Inc., the largest office product retailer worldwide:

> Several years ago, I had 136 categories for an in-house card sort and had two stacks of 136 sheets of paper and a big conference room table. I had people put the cards into the high level buckets, and I had a computer with Excel. After the first person, I'd type it in as the second person started on the task. It was like "Beat the Clock" and was maddening, but I wanted statistical significance in this exercise. After that frustrating exercise, it was clear that I

needed a better way to conduct taxonomy research to support our Web sites, catalogs, and store signage. I looked for survey companies and then I hired an employee with background in usability and statistics who also had a technical bent. I also looked for a vendor who could do these things. We spent time and worked on a proof of concept with a vendor but it was clear that there were no vendors at that time who did it the way we wanted. So we decided to build the tool in-house and wrote code ourselves. It wasn't very complicated at first. Our first online surveys consisted of showing pictures of products to customers with a list of categories in a pull down for each product. Then we asked them to choose the category where they most likely would expect to find the product on our Web site. Although it was not a technically or scientifically complex survey, it proved to be a powerful way to accomplish our goals.

Since those early days we have honed our methodology for taxonomy surveying. Our hierarchy of products is important to us. We take a three-step approach. First, we give customers a list of items—like white-out or self-adhesive notes or gel pens—and ask them to bucket the items into groups with similar products. We analyze the data with large sample sizes [and] come up with dominant groupings of those products. For the second phase, we use a different set of customers to name the groups of products. Respondents might say, "This group should be pens and correction supplies." In the final step, another group does a reverse card sort. In this survey we ask people, "If you're looking for this product, where would you expect to find it?" Since both the buckets and nomenclature are user centered, we generally see far greater perceived and actual performance after the final step in the process.

Staples' story is probably atypical. The company took a big risk in investing usability analysis talent in the online tool development arena, but for them it has paid off extremely well. The ROI in this case is that automated tools, once created, can be used repeatedly. Because they have proven to be of significant value to Staples, the ROI is clear.

19.2 MECHANICS OF ONLINE SURVEYS

Surveys require a *call to action*, the *survey form*, and *data presentation*. The call to action can be a Web address included within an e-mail message, a hyperlink on a Web site, or even a pop-up window. In all cases, the call to action invites the survey candidate to participate. Many surveys are presented without a respondent incentive, but many research efforts reward respondents with payment, gift certificates, coupons, or, at the very least, a sweepstakes entry. The survey process includes the following:

- Study design
- Recruiting
- Study administration
- Analysis

19.2.1 Study Design

The actual design of survey questions and strategy is presented best in *Mail and Internet Surveys* (Dillman, 2000). This book covers all the essentials and is recommended for those who are planning, designing, and implementing surveys. This section covers only the background appropriate to cost-justification and data gathered during practitioner and vendor interviews specifically for this chapter.

Completion Rates: Incentives, Duration, and Content Quantity

Respondent incentives play a contributing role in completion rates and overall project cost. In 2002, Bill MacElroy of Socratic Technologies cited studies indicating that the greater the incentive, the higher the completion rate. However, these benefits tend to flatten above a certain threshold. For fixed incentives, $5.00 values correlated to a 78% completion rate, and increasing incentives to more than $20 yielded diminishing returns. Sweepstakes incentives followed a similar pattern as prizes exceeded $1,000 in value.

MacElroy also reported the role of survey duration in completion rates, finding that the greater the number of screens (or questions), the greater the occurrence of midsurvey terminations. His best practices state that "surveys should consist of fewer than 30 questions or screens and no more than 55 clicks to complete the entire survey." Furthermore, online surveys should "last no longer than 17 to 18 minutes," but "ideally last 10–12 minutes."

Molly Langridge's experience at D&B has been different. She stated that a typical survey lasting between "5 and 7 minutes" translates to "20 or 30 questions." D&B does not conduct synchronized surveys, however, and respondents are more likely to terminate an asynchronous survey than a synchronized one. The WAMMI asynchronous usability survey recommends only six questions

(MacElroy, 2002), stating that longer questionnaires "tend to suffer significantly from reduced user response rate."

19.2.2 Recruiting

In lab-based usability tests, participant definition and recruitment require from 10 to 40% of the research project budget, with respect to time and resources expended (Anonymous, 2002a). This effort is extremely worthwhile, as readers who have ever witnessed a participant who did *not* meet the correct profile in a study will attest. Inappropriate study participants can skew results, and thus their data are typically discarded. It is common practice to over-recruit screened participants to avoid this type of misalignment, replacing inappropriate respondents as needed. For online surveys, large sample sizes are necessary to ensure an adequate number of respondents. For phone-based recruiting, sometimes 10 calls are needed to yield a single respondent. For online recruiting, the response rate can be 1 in 100 or worse. In some situations, fresh respondents are needed, such as when a concept to which respondents have already been exposed is being retested. In many situations, such as new card sort exercises, reusing an existing panel will be satisfactory. Six times a year is a practical maximum for online survey participation (Hynes, 2003).

Access to Hard-to-Reach Audiences

One research manager from a U.S. telecommunications carrier stated that online surveys enable her research to reach "hard-to-find populations such as information systems/information technology business decision makers and senior business process owners; i.e., CEO, CTO, Sales VP, Operations VP, etc. We are also able to get better geographic representation for large companies that are dispersed" (Anonymous, 2002b). Online surveys can be conducted anywhere and at any time. For this reason, they are an attractive option for research projects that require data points from audiences who are far apart or who have specific hard-to-find traits. Nevertheless, this research manager warned that "Samples are not representative of the general market. Online survey respondents skew to more education and higher income—which is fine if that is your target market." The reason is obvious: online survey respondents have regular access to a computer and are likely to own one. Computers with Internet access are still beyond the reach of many people, making those who have them likely to have higher incomes than those who do not.

Vendor Panel Recruits and Their Dangers

Online survey vendors build their panels continually. Their primary mechanism for expanding a panel is through advertisements on search engines such as Yahoo! and Google. A secondary mechanism for panel expansion is unsolicited e-mail campaigns. Targeting is based on keywords and/or self-reported registration data. Some researchers who use online surveys recruit from their own Web sites, which yields superior study participants, but most online survey users rely on vendor-supplied panels, which exhibit the aforementioned problems.

Without in-person validation, it is impossible to distinguish between a 14-year-old boy and a middle-aged woman. Small differences in age might not affect usability outcomes, but large age differences certainly do produce different results. For example, teenage girls are likely to have cosmetic color preferences different from those of 30-something women, just as 20-something men might have health concerns different from those of 40-something men. Not only can online survey respondents easily misrepresent themselves, but they can also create multiple personae, completing the same survey multiple times to receive multiple incentives. Although your target audience is likely to have no interest in such activities—incentives are typically low-value gift certificates—school-age children may find online survey-taking to be more lucrative than the available after-school jobs.

Spoofing (the act of pretending to be someone else) can be prevented through analysis of panelist responses to questions and validation of e-mail addresses. No doubt better vendors use a great deal of antispoofing heuristics. Antispoofing strategies include outlier elimination and trait-response comparison checks. Checking references on vendors of online research panels is always a good idea.

The costs for antispoofing efforts are minimal and, when effective, yield a more valid set of results. The least expensive strategy in terms of time and money is e-mail address validation. Simply e-mail participants and require a reply to ensure that their e-mail address is a valid account. This strategy prevents people from being "too anonymous," because they will at least require a unique e-mail address for each survey response. Outlier elimination requires additional human processing time, but is also effective. Eliminating outliers can be accomplished by invalidating panelists with unlikely combinations of traits, such as the following:

- ✦ High income and very low age
- ✦ Mismatch between completed higher education and age
- ✦ Mismatch between stated location and telephone area code or postal code

Of course, there are situations in which the aforementioned outlier characteristics would be honest. However, each researcher must determine the likelihood that such participants would represent their target audience. With the small populations interviewed, some culling of the outliers is prudent.

Additional outlier elimination can be done through analysis of actual survey responses. If 95% of survey respondents perform within a tight range to a particular question, the remaining 5% of responses are worthy of individual analysis. For example, if a particular task takes 95% of the respondents 90 seconds to complete, and 5% of the respondents 20 seconds to complete, take a look at the actual responses to see if they make sense. If all of the answers appear to be randomly selected, eliminate that respondent from your final calculations and include in your report the number of respondents whose data were stricken, and the impact of the eliminated responses on the final data. Wanton elimination is of course unwise, as some outliers will be honest and not spoofed.

Direct Recruiting and Panel Development

Recruiting can be done by telephone-based recruiting (often called *telephone intercept*"), e-mail, or advertising or from within the Web site to be studied. The best and most cost-effective recruits come from within the destination Web site, but this strategy requires a high number of visitors. Not all Web site visitors are likely to spend the time to respond to an online survey. Also, Web site visitors are less likely to complete a survey without an incentive, such as a coupon, gift certificate, or even a sweepstakes entry. Yields higher than 2% are rare when no incentives are offered, according to Molly Langridge, a research professional at D&B, in 2003.

Staples is fortunate to have a panel of 20,000 customers. According to Colin Hynes, Staples conducted an e-mail campaign in 2000, asking customers, "Would you like to be part of a community of change and get paid for your opinion?" They answered about 20 questions, including their gender, what products they buy, and the size of their businesses. The initial panel was developed without incentives. Staples found people who were interested in helping out the cause for dollars-off coupons on future purchases. Staples typically gets 35% or more completion rates on e-mail surveys. Not everyone is so lucky.

Spam, unwanted advertising e-mail, is an increasing problem, resulting in customer resentment and reduced reach because of spam-blocking software. Subject lines need to be designed carefully, for many spammers have already attempted "surveys" in their own, unwanted, e-mail.

Telephone intercept is perhaps the most successful strategy, but it is terribly time-consuming and expensive. Furthermore, residential do-not-call lists have made telephone recruiting problematic even in the business space, resulting in increasing resentment from call recipients. Some local governments have made any type of cold calling illegal, so check out the regulations before you attempt this strategy.

The best strategy is to recruit directly from the target Web site and offer incentives to increase participation—and to be patient, especially if a Web site does not have a high volume of traffic. E-mail is still perhaps the most prevalent means of online recruiting because of its low cost. However, it is best to use an opt-in list, as spam laws are toughening worldwide.

19.2.3 Study Administration

Online studies can go quickly or run for more than a week, but they typically do not require longer than a week. A pleasant side benefit of their online nature is that results can be observed during the administration period, and surveys can be "turned off" when the data trends appear solid, potentially cutting costs (Table 19.4).

Critical Comparison of Synchronized Surveys with Lab-Based Usability Testing

Synchronized surveys automate the interview process by presenting tasks to respondents in adjacent windows. The tools record click streams and typed comments and time respondents as they navigate through a Web site. To capture these data, synchronized survey tools require that software be installed on the participants' computers, although most providers can now install this software without user intervention.

One provider of online surveys directly questions the value of online study-only efforts. Axance, a Parisian usability vendor, states that online studies "[do]

Table 19.4 Time Frame for Surveys

No. of Respondents	*Vendor A*	*Vendor B*	*Vendor C*	*Vendor D*
50	21,794	32,750	33,650	38,000
75	24,878	34,750	38,475	
100	28,349	36,750	43,300	
200	32,963			

not allow detailed understanding of user behavior (because users must voluntarily type all comments)" and that "causes of complex usability problems cannot be reliably understood without [lab-based] usability testing" (Axance, 2003).

"Behavior data" from online surveys include task success and failure rates, task completion time, time spent on each Web page, and navigation paths taken. Although these data are valuable, they are meaningless without extensive analysis. Additionally, task completion times can be affected by setting effects, discussed later. All other data are collected from typed responses and answers to multiple choice and rating questions or must be gleaned from correlations between the respondents' traits and any combination of the previously mentioned metrics. Compared to the wealth of observations possible in in-person studies, behavior data represent a smaller data array from which to draw conclusions. On the other hand, online surveys provide a much greater number of data points, so conclusions from online surveys can provide more convincing arguments.

Uncontrolled Setting

Online surveys can be taken anywhere an Internet connection is available. Therefore, the same survey can be taken in a work setting, in a home, or in an Internet café. Phones ringing, televisions playing, and other distractions are unavoidable and will affect task performance and completion time. Other software applications, such as e-mail or instant messaging, can be running in the background. It is impossible to determine how online survey participants spend their idle time, whether they are looking at a newspaper, speaking on the phone, or genuinely puzzling over the user interface. Synchronized surveys utilize task times as a key indicator of usability, but task times can vary widely based on these distractions. A controlled setting and direct observation eliminate this problem, but neither is available in most synchronized survey studies.

Task Misunderstanding Effects

In lab-based testing, tasks are assigned to respondents to complete, typically in verbal form by the moderator. Conveyance of these goals to respondents with clarity is not always guaranteed. In in-person usability studies, moderators frequently must clarify task descriptions for respondents. Such clarifications are difficult or impossible in online surveys. The cost associated with clarifying questions to hundreds of respondents would be prohibitive.

No Observation Possible

Body language, facial expressions, sighs, laughs, grumbles, and jeers made by respondents during live usability studies together provide more information than task duration, success, and failure data. Online survey software is incapable of capturing human expression beyond what the respondent decides to type. One could argue that video could be captured, but the technological requirements, analysis time, and poor quality of online video make this argument moot today. Processing the video would be cost-prohibitive, making the ROI impossibly low.

Feedback Inconsistent

All qualitative feedback in online surveys must be typed in. Individuals' typing skills vary widely. Some respondents will provide extensive commentary, whereas others will be terse. The same is true in lab-based interviews, but it is easy for the moderator to compensate by asking additional questions, and observing facial expressions and body language. Live usability study interactions are more like a telephone conversation, whereas online survey interactions are more like e-mail exchanges. First reactions that spill out verbally are likely to be modulated by the time they are typed, and participants' first reactions produce some of the most prized data in a usability session. However, a direct relationship between first reactions and usability performance does not exist, as some respondents who like a user interface very much still perform poorly; the opposite is also true.

No Probing Opportunity

In live usability studies, moderators frequently ask follow-up questions, such as "What are you thinking?" Probing questions are useful to understand why respondents pause during a task or to clarify remarks. Terse comments are often the start of a useful dialog. Such probes are impossible in online surveys. Software can certainly pop up questions when survey respondents are idle, but pop-up alerts provide a radically different experience from a gentle, unbiased question from a live moderator.

Conclusions about Survey Administration

These critical points challenge synchronized usability surveys, but do not condemn them. Knowing the strengths and weaknesses of the technology will enable you to use it more effectively. As stated later, it is the mix of online with traditional usability testing that brings out the most value in online surveys.

19.2.4 Analysis

Data from online surveys are both quantitative and qualitative in nature. Quantitative data come from task completion times, from the number of steps taken by each user in response to a given survey question, and from user responses to discrete questions. Qualitative data come mostly in the form of open-ended (type-in) questions. Qualitative data take significantly more time to digest, despite increasingly sophisticated software tools offered by some vendors for analyzing the material generated by respondents.

Analysis is based on the trends that emerge and the data that are produced. Trends can indicate with some certainty that an issue does in fact exist. However, the deep understanding that is required to fix a usability issue requires something beyond click streams and task times. Just knowing a task takes a long time to complete does not indicate which aspects of the task are confusing. Direct observation of interaction with widgets, where users look for certain elements, and comments with follow-up probing questions are all necessary to identify usability problems. Without direct observation, many data are missed. Attempts to make up for missed data with increased numbers of online survey respondents exacerbates, rather than eliminates, this failing.

In some circumstances, online survey data may even be misleading. Because task successes can be accomplished in multiple ways, some successful task completions might even be recorded as failures. For example, one respondent may receive several phone calls during his online survey, thereby extending many of his task completion times. A single long pause can be addressed by the survey software, but frequent shorter pauses are harder to explain as anything but a usability problem.

Ultimately, analysis of online survey data will give the researcher indicators of areas to probe, rather than immediate answers to usability questions.

Reporting

Online survey vendors differ in their reporting technology. Some vendors present data in beautiful charts and graphs, whereas others keep reporting fairly simple. The automatically generated reports, however, tell only a small part of the story. Human analysis is the key to extracting the most value from survey results.

19.3 MIXING ONLINE WITH OTHER METHODS

The absolute best strategy for using online surveys of any type is to mix them with other types of usability and focus group research. Colin Hynes, Director of Usability at Staples Inc., stated that online survey research "enables us to get a wider number of respondents for our data gathering quickly. Our qualitative data are typically gathered in a usability lab or through field studies. For surveys, we don't just increase respondent quantities from the qualitative research, but focus on the nature of the data we need for statistical significance for large sample sizes. [Online surveys] don't replace usability testing or field studies, but complement it where appropriate." Staples conducts 15 to 20% of its usability research online, with the remainder conducted in-lab or in the field.

Log file analysis, or *analytics*, is another tool useful to researchers. According to Colin Hynes, "Analytics is the 'what,' while online surveys help target the 'why.' Analytics provides some direction on where to dig further through observing behavioral trends in the data. Surveys then enable us to target research to understand why people may be behaving in the way they are. After making improvements based on the survey data, we test to make sure that our sample has an improvement. Then we pump it to the site and then look at the analytics tool. If we did it right, we should see better traffic flows as a result."

19.4 SURVEY VENDORS

Full-service online survey firms will take your survey idea and do everything from survey design to final presentation, for a price. Prices vary but are typically tens of thousands of dollars. The key benefit is the ease of administering a survey in this manner, and the key drawback is the cost.

The two easiest methods for finding online survey firms are obvious. First, search on the Web. Second, ask around. Both methods will yield results, but be sure to check references, because quality varies significantly from vendor to vendor.

A survey vendor will want to conduct your research project as a consulting engagement rather than as a simple service. Most full-service synchronized survey vendors have analysts on staff who will support the research process throughout an engagement. This support enables a more robust study, with more insightful analysis. The better the analysis is, the better the value you will receive.

19.4.1 Requests for Proposals

Write an RFP before engaging a specific firm. The RFP will require you to focus on your needs and to shape the project in the vendors' terms. The RFP is unlikely to reduce your overall cost, but it will enable you to compare vendors on an equal footing. One researcher stated that she did not use RFPs but another stated that she always solicited at least three bids per project.

Usable Products Company, a boutique usability consultancy, engaged five synchronized usability survey vendors in early 2003 with an RFP. One vendor, after some back-and-forth e-mail, refused to respond to the RFP. The remaining four vendors responded, with a range of prices and response formats. Prices ranged from $25,000 to $40,000 for essentially the same services. However, the approaches all varied somewhat, and the level of presales service varied significantly.

Following are sections appropriate for an RFP:

- *Background.* Describe the research need and how you see the online survey component filling that need.
- *Objectives.* What do you hope to gain from this research? What sorts of metrics do you need?
- *Recruiting.* How will recruiting be accomplished? Do you want the survey vendor to provide participants from their panel, or will you create the call to action on your target Web site? What are the characteristics of the audience? What are the quality control processes and what is the refresh process for the vendor's panel?
- *Study guidelines.* What do you want the vendor to do? Who will produce the test script? Will custom programming be required? What is the time frame, and how will results be presented?
- *Scope.* How many participants will be used? How many questions will the survey include? What sort of branching will be required? Who will do the analysis—you or the vendor?
- *Specific issues important to you.* What parameters are important to you? You might be interested in the analytics that are available. You might inquire as to the technology, the documentation produced, client references, access to demonstrations, and so on.
- *Contact information.* Be sure to provide your full name, title, e-mail address, phone number, and street address.

Expect a lot of questions from vendors, and be prepared for a variety of approaches to your problem. Most vendors want to provide a complete solution, and most of them want big projects far in excess of $25,000. Although this amount may seem high, remember that the experience a vendor has with other, similar projects will provide significant insights into the problems discovered in your Web site. However, nondisclosure requirements restrict reuse of knowledge significantly, so do not expect vendors to divulge your competitors' secrets.

19.4.2 References

Request telephone contact information from three or more customer references. Complaints about large online survey vendors have included poor project management and problems with back-end reporting quality. Be sure to check references by telephone, because people are unlikely to provide any sort of negative feedback via e-mail.

19.4.3 Do It Yourself: Survey Creation Tools

There are dozens of do-it-yourself survey creation tools with a broad range of prices. Perhaps the most popular is *Zoomerang*, but one research manager cautioned that it is a "very basic tool, with [the possibility of] garbage in/garbage out. One still must know how to write questions correctly and *Zoomerang* gives one the illusion anyone that can write a survey." It is true that *Zoomerang* and similar tools are very easy to use—that is a plus. However, creating a survey easily is quite different from creating a great survey. Reading up on how to write effective survey questions is strongly recommended.

19.5 CONCLUSIONS

Online surveys, be they synchronous, asynchronous, targeted at usability, or at other customer perspectives, are an extremely valuable tool for the researcher. Utilizing them effectively poses significant challenges, because misusing them can produce large volumes of meaningless data. Mixing online surveys with traditional methods is the key to maximizing value, because online alone is inadequate. Online surveys, be they asynchronous or synchronous, bring tremendous value and high ROI when combined with traditional usability testing.

ACKNOWLEDGMENTS

The following people contributed to this chapter: Richard Martin, Analyst for Usable Products Company, Sean Angerman, Usability Analyst for Staples, Colin Hynes, Director of Usability for Staples, Jurek Kirakowski, Human Factors Research Group of University College Cork, Molly Langridge, D&B, Dr. Bill MacElroy, President of Socratic Technologies, Whitney Quesenberry of Whitney Interaction Design, LLC, Tim Semen, Usability Project Manager for Staples, and Chauncey Wilson of WilDesign.

REFERENCES

Anonymous. (2002a, September). Multiple e-mail responses gathered from a professional, private Internet usability discussion group.

Anonymous. (2002b, December). E-mailed survey response from research manager from U.S. telecommunications carrier.

Anonymous. (2003a, March). Quotes provided to Usable Products Company as part of research proposals from four online survey vendors.

Anonymous. (2003b, December). Telephone interview conducted with a research manager at a large e-commerce venture.

Axance. (2003). Retrieved November 15, 2003, from www.axance.com/08english/08english_03meth_04online.htm.

Dictionary.com. (2003). Definition of "survey." Retrieved November 15, 2003, from dictionary.reference.com/search?q=survey.

Dillman, D. A. (2000). Mail and Internet Surveys: The Tailored Design Method. New York: John Wiley.

Hynes, C. (2003, December). Telephone interview.

Kirakowski, J., and Cierlik, B. (1998). *Measuring the Usability of Web Sites.* Human Factors Research Group. University College Cork, Ireland. Retrieved October 15, 2004, from ftp://ftp.ucc.ie/hfrg/wammi/hfes98Q.rtf.

Kirakowski, J., and Claridge, N. (2004). Retrieved September 15, 2004, from www.wammi.com.

Langridge, M. (2003, November). Telephone interview.

MacElroy, B. (2002, October). New Tools and Techniques: Taking Online Research to the Next Level. In *Proceedings, Web-Based Surveys and Usability Testing.* San Francisco.

Vividence. (2004). Vividence eXpress. Retrieved April 14, 2004, from www.vividence.com/public/products+and+services/research+tools/vividence+express/vividence+express.htm.

20 CHAPTER Cost-Benefit Framework and Case Studies

Nigel Bevan Serco Usability Services

This chapter summarizes the benefits that can be obtained from taking a user-centered approach to design and discusses how to select appropriate methods and justify their cost benefits. It includes two case studies of the cost benefits of employing a usability maturity model to improve the usability capability of an organization.

20.1 POTENTIAL BENEFITS OF USABILITY

The objective of introducing user-centered methods is to ensure that Web sites and products can be used by real people (not just designers) to achieve their tasks in the real world. This requires not only easy-to-use interfaces, but also the appropriate functionality and support for real business activities and work flows. According to IBM (1999), developing easy-to-use products "makes business effective. It makes business efficient. It makes business sense."

User-centered design can reduce development and support costs, increase sales, and reduce staff costs for employers. The checklist in Table 20.1 can be used to identify the potential benefits, and indicates which chapters give examples of these benefits. Chapters 3 and 4 also include examples of how to calculate benefits.

20.2 ESTIMATING COSTS

Chapter 3 illustrates the cost of a relatively sophisticated usability engineering plan that takes into account the time of usability engineers, developers, mangers,

Table 20.1 Potential Benefits of User-Centered Design

Potential Benefits	*Chapters*
A. Development costs can be reduced by:	
1. Producing a product that has only relevant functionality	
2. Detecting and fixing usability problems early in the development process	2, 7, 10, 17
3. Reducing the cost of future redesign or radical change of the architecture to make future versions of the product more usable	10
4. Minimizing or eliminating the need for documentation	
5. Redesigning Web sites to increase revenue, not only to change the image	3
6. Reducing the risk of product failure	2, 7, 9, 10, 11
B. E-commerce sales can be improved by increasing the number of Web site customers who will:	
1. Be able to find products they want	2, 3, 4, 10
2. Find supplementary information easily (e.g., delivery, return, and warranty information)	
3. Be satisfied with the Web site and make repeat purchases	2, 4, 10
4. Trust the Web site (with personal information and to operate correctly)	2, 4, 10
5. Not require any support, or use the Web site for support rather than calling the support center	10
6. Recommend the site to others	
7. Support and increase sales by other channels	4, 11
C. Product sales can be increased as a result of the usability of the product	
1. Improving the competitive edge by marketing the product or service as easy to use	2, 3, 7
2. Increasing the number of customers satisfied with the product who will make repeat purchases and recommend the product to others	7
3. Obtaining higher ratings for usability in product reviews	3, 7
D. Employers can benefit from easier to use systems in the following ways:	
1. Faster learning and better retention of information	10
2. Reducing task time and increased productivity	2, 4, 7, 10, 16
3. Reducing employee errors that have to be corrected later	4, 7

Table 20.1 *Continued*

Potential Benefits	*Chapters*
4. Reducing employee errors that impact on the quality of service	4
5. Reducing staff turnover as a result of higher satisfaction and motivation	4
6. Reducing time spent by other staff providing assistance when users encounter difficulties	7
E. Suppliers and/or employers can benefit from reduced support and maintenance costs in the following ways:	
1. Reducing support and help line costs	2, 3, 7
2. Reducing costs of training	2, 3, 4, 7
3. Reducing maintenance costs	2, 4, 7

and users. This chapter shows how to calculate costs when only a limited number of usability methods are to be used, and the time of the usability engineers is the main cost. It also shows how to use a usability maturity model to assess what additional user-centered methods an organization should use.

20.2.1 User-Centered Design Methods

The essential activities required to implement user-centered design are described in ISO 13407 (User-centred design process for interactive systems, 1999) under the following headings:

1. Plan and manage the human-centered design process
2. Understand and specify the context of use
3. Specify the stakeholder and organizational requirements
4. Produce design solutions
5. Evaluate designs against requirements

The EC INUSE project developed a structured and formalized definition of the human-centered processes described in ISO 13407 (Earthy, 1998). An improved version has subsequently been published as ISO TR 18529 (2000).

The usability maturity model in ISO TR 18529 describes seven processes, each of which contains a set of base practices (Table 20.2). The base practices

Table 20.2 Human-Centered Design (HCD) Processes and Their Base Practices

1	*Ensure HCD Content in System Strategy*
1.1	Represent stakeholders
1.2	Collect market intelligence
1.3	Define and plan system strategy
1.4	Collect market feedback
1.5	Analyze trends in users
2	*Plan and Manage the HCD Process*
2.1	Consult stakeholders
2.2	Identify and plan user involvement
2.3	Select human-centered methods and techniques
2.4	Ensure a human-centered approach within the team
2.5	Plan human-centered design activities
2.6	Manage human-centered activities
2.7	Champion human-centered approach
2.8	Provide support for human-centered design
3	*Specify the stakeholder and organizational requirements*
3.1	Clarify and document system goals
3.2	Analyse stakeholders
3.3	Assess risk to stakeholders
3.4	Define the use of the system
3.5	Generate the stakeholder and organizational requirements
3.6	Set quality in use objectives
4	*Understand and Specify the Context of Use*
4.1	Identify and document user's tasks
4.2	Identify and document significant user attributes
4.3	Identify and document organizational environment
4.4	Identify and document technical environment
4.5	Identify and document physical environment
5	*Produce Design Solutions*
5.1	Allocate functions
5.2	Produce composite task model
5.3	Explore system design

Table 20.2 *Continued*

5.4	Use existing knowledge to develop design solutions
5.5	Specify system and use
5.6	Develop prototypes
5.7	Develop user training
5.8	Develop user support
6	*Evaluate Designs Against Requirements*
6.1	Specify and validate context of evaluation
6.2	Evaluate early prototypes in order to define the requirements for the system
6.3	Evaluate prototypes in order to improve the design
6.4	Evaluate the system to check that the stakeholder and organizational requirements have been met
6.5	Evaluate the system in order to check that the required practice has been followed
6.6	Evaluate the system in use in order to ensure that it continues to meet organizational and user needs
7	*Introduce and Operate the System*
7.1	Management of change
7.2	Determine impact on organization and stakeholders
7.3	Customization and local design
7.4	Deliver user training
7.5	Support users in planned activities
7.6	Ensure conformance to workplace ergonomic legislation

describe what needs to be done in order to represent and include the users of a system during the product lifecycle.

A full implementation of user-centered design would conform to ISO 13407 and use all the relevant base practices in the usability maturity model.

Our experience has been that the core set of methods illustrated in Figure 20.1 (see also Table 20.4) provide the essential activities necessary to achieve user-centered design in a wide range of projects (Bevan, 2000a). The exact nature of the activity should be customized to the needs of each organization and project. Not all activities may be needed for every project, and for some projects additional activities may be required (e.g., field studies to gather information from existing users).

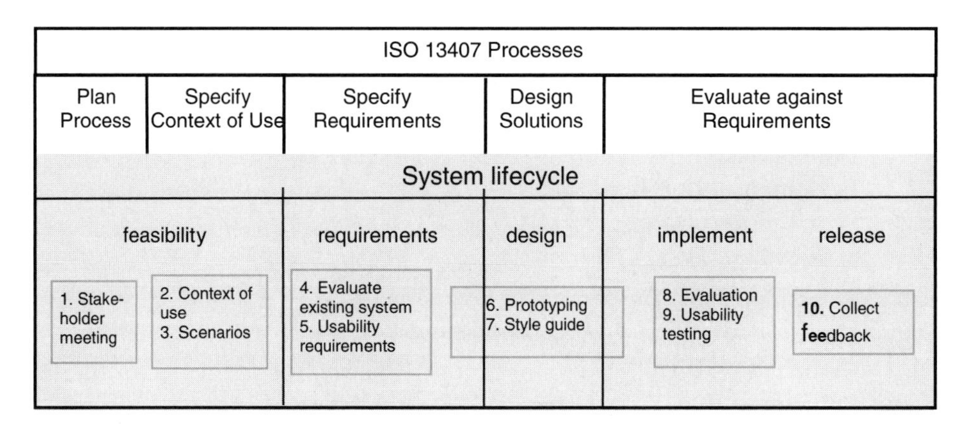

FIGURE 20.1 TRUMP methods for user-centered design.

20.2.2 Costs of User-Centered Design

Most user-centered design techniques are relatively simple to apply. The major cost is the time of the people who apply the methods. The methods chosen will depend not only on the overall budget but also on the available skills and experience, as well as practical constraints such as project deadlines and the availability of users.

Table 20.3 shows the typical range of effort and people required for each of the methods, providing a total that ranges from 26 to 80 person days.

For some projects, even the lower end of this scale may be too ambitious or beyond the available budget. In that case an essential subset of the activities could be used (see, for example, Bevan, 2000b).

The minimum figures estimate the effort required by experienced facilitators (sometimes working alone) to obtain basic results when there are no complications. The maximum figures could be exceeded in some cases, particularly for a larger project in which the activities are repeated for different parts of the system.

20.3 MAKING THE COST-BENEFIT CASE

Having made an estimate of the financial benefits and knowing the effort required for each method, cost benefits can be calculated for the intended set

Table 20.3 Number of People and Person-days Required

	Plan/ Report (Person-days)		*Execute (Person-days)*		*Total (Person-days)*		*Usability Experts (number)*		*Managers/ Developers (number)*		*Users (Person days)*	
	Min	Max	Min	Max	Min	Max	Min	Max	Min	Max	Min	Max
Maturity assessment	2	10	1	15	3	25	1	2	2	10		
1. Stakeholder meeting	1.5	3	.5	2	2	5	1	2	2	10	.5	2
2. Context of use	2	3	.5	2	2.5	5	1[a]	2	2	6	.5	2
3. Scenarios of use	1	2	.5	1	1.5	3	1	2	1	4	.5	2
4. Baseline existing system	2	4	.5	4	2.5	8	2	2	0	4	.5	2
5. Usability requirements	1	2	.5	1	1.5	3	1	2	1	4	.5	2
6. Paper prototyping	3	6	1	6	4	12	2	2	0	6	.5	3
7. Style guide	1	10[b]	.5	3	1.5	13	1	1				
8. Evaluate machine prototype	2	6	.5	6	2.5	12	1[a]	2	0	6	.5	3
9. Test against requirements	2	7	2	6	4	13	2	2	0	6	.5	3
10. Feedback from use	3	5	1	5	4	10	1	1			*[c]	*[c]
Total (excluding Maturity)	18.5	48	7.5	36	26	84						

[a]Possible with one person, but two people recommended.

[b]More effort would be required for a style guide that covers multiple products or platforms.

[c]Variable.

of user-centered activities. There are several situations in which a cost-benefit analysis can be useful:

- A financial case can be made for the budget required to carry out intended user-centered methods.
- The choice of user-centered methods within a limited budget can be prioritized and justified.
- Cost benefits can be calculated at the end of a project to provide a case study for future use.

Increasingly, the question will become not "usability, yes or no?" but "usability, which methods when?"

The steps to follow are:

1. Decide which user-centered design methods are intended to be used. This will depend on the nature of the project, the anticipated benefits, and any budget or time constraints.
2. For each method, add the person-days required to perform it.

$$(\text{Preparation time} \times \text{people}) + (\text{Application time} \times \text{people}) + (\text{Reporting time} \times \text{people})$$

3. Multiply the person-days by the appropriate day rate(s) to give a total labor cost, and add any other costs (such as laboratory hire or participant recruitment).

$$\text{Total cost} = \Sigma(\text{Person days} \times \text{Day rate}) + \Sigma(\text{Other chargeable costs})$$

4. Decide which of the benefits listed in Table 20.1 the methods will contribute to. For example, does the method contribute to lower development costs, increased sales, improved productivity, and/or reduced support?
5. Estimate the financial benefits that come from using the methods.

$$\text{Total financial benefit} = \Sigma(\text{Financial benefits for relevant items in Table 20.1})$$

6. Calculate the cost-benefit ratio.

$$(\text{Total financial benefit}) \div (\text{Total cost})$$

7. If there is a need to prioritize the use of methods within a limited budget, or to justify the use of a specific method then you should:
   a. Decide how each method will contribute to the overall benefits identified in step 4. (Group into one composite method any component methods that are used together, such as a usability test followed by an interview and a questionnaire.)
   b. Because later methods typically depend on the results of methods carried out earlier (e.g., scenarios depend on the context of use), the overall cost benefits cannot be partitioned between the individual methods. Instead the potential value of a particular method can be assessed by recalculating the estimated cost benefits when that method is excluded. For example, it might be concluded that the additional benefits obtained by carrying out a late usability test did not justify the additional cost. By comparison an evaluation to baseline the usability of an existing system might provide a much greater benefit for a similar cost.

For organizations already committed to user-centered design, a cost-benefit analysis is not essential, but it can provide valuable input when formulating a usability plan. Cost benefits could be recalculated as a development project progresses to reassess the importance of various activities.

20.4 CASE STUDIES

The objective of the EU-funded TRUMP project (Bevan, 2000a) was to improve the usability capability of the development processes in two organizations—Inland Revenue/EDS (IR/EDS) in the UK and Israel Aircraft Industries (IAI) in Israel—and to demonstrate the cost benefits of applying user-centered methods. The steps taken over a period of 2 years were to:

1. Identify needs for usability process improvement by using the usability maturity model in ISO TR 18529 to assess the current capability of each organization (Bevan and Earthy, 2001).
2. Make the identified improvements to the software development processes by introducing simple user-based methods implementing ISO 13407 (Bevan *et al.*, 2001).
3. Identify the cost benefits of the improvements and integrate the methods into the documented processes.

The usability maturity model in ISO TR 18529 was used to assess the usability capability at IR and IAI, and to identify any gaps in their ability to apply user-centered design. Each organization was free to decide which of the 44 usability maturity model base practices was within the scope for potential process improvement.

In the assessments, each base practice was rated as one of the following:

- Not performed
- Partly performed
- Largely performed
- Fully performed

20.4.1 Israel Aircraft Industries

The LAHAV division of IAI has a group of about 100 people developing aircraft avionics. IAI uses a well-established development methodology, but their process for specifying operational requirements was not supported by any specific methods or techniques.

A 1-day workshop provided the basis for agreeing on the scope for process improvement at IAI. The activities in the usability maturity model were used as a best-practice checklist. The author rated the extent to which each activity was currently performed based on a short discussion with one or two developers or managers who were most knowledgeable in each area. Although some ratings may not have been completely representative, they were sufficient to provide the basis for an agreed program of improvement.

LAHAV selected the development of a new mission planning center (MPC), using the Windows NT interface as a trial project. An MPC enables a pilot to plan an airborne mission that is then loaded onto a data cartridge and taken by the pilot to the aircraft.

The user-centered design methods used and the IAI comments are shown in Table 20.4.

A second workshop to assess the improvements was held 16 months later. IAI commented (Bevan *et al.*, 2000):

> The one-day assessment format was appropriate for LAHAV since it is a) a relatively small organization, and b) it has a lasting culture, commitment and infrastructure for process improvement. The first assessment revealed many

Text continued on p. 592.

Table 20.4 IAI and IR/EDS Experience with TRUMP Methods for User-Centered Design

Method	*Description*	*IAI comments*	*IR comments*
1. Stakeholder meeting	A half-day meeting to identify and agree on the role of usability, broadly identifying the intended context of use and usability goals, and how these relate to the business objectives and success criteria for the system.	Conducting a stakeholder meeting allowed IAI to identify previously unforeseen users and stakeholders, better understand the project scope and objectives, define the success factors, and identify some different interpretations for follow-up discussions and resolution. Involvement of senior managers and marketing personnel contributed to identification of some strategic issues.	No stakeholder meeting held, as usability activities were already planned as part of the normal development process.
2. Context of use	A half-day workshop to collect and agree on detailed information about the intended users, their tasks, and the technical and environmental constraints.	We never used this method before. The facilitator guided us through a long checklist covering many aspects of the user's skills, tasks and the MPC working environment. Most of the data captured was not new to the participants due to their existing familiarity with the users' environment. Some valuable information was captured, other parts did not seem to be relevant to the MPC. We concluded that the checklist	The user's skills, tasks and the working environment were defined. The value of documenting this corporate knowledge should not be underestimated. Our IT supplier does not have staff with an intimate knowledge of our core processes or organizational culture. There was a feeling that we had this knowledge "in our bones" and could pass this on to the IT supplier when

Table 20.4 *Continued*

Method	*Description*	*IAI comments*	*IR comments*
		of issues should be tailored to IAI's needs.	requested. Context analysis proved there was a better way to spread that knowledge around and the document has been used time and again by all involved to act as a reminder of what and whom we were trying to design for.
3. Scenarios of use	A half-day workshop to document examples of how users are expected carry out key tasks in specified contexts, to provide input to design and a basis for subsequent usability testing.	This method contribution for the MPC system was low because the operational scenarios required for the MPC are obvious to pilots. [In most development environments, this is a valuable means of transferring information about user tasks to the development team. As the people collecting requirements at IAI were themselves pilots, the main beneficiary was the usability engineer who needed this information for later parts of the usability process.]	We all took this technique to our hearts. It was relatively simple to pick up as it involved the end users documenting what they did on a daily basis back in the office. This knowledge could then be captured before every function design workshop and not only used to focus what the IT was being developed for but used in conjunction with other techniques such as task analysis and paper prototyping to verify the emerging design was meeting the needs of users

			and then used again to validate the final IT prototype was correct.
4. Baseline existing system	Evaluate an earlier version or competitor system to identify usability problems and obtain baseline measures of usability as an input to usability requirements.	Four users evaluated the existing system. Each user was given brief (15 minutes) training on the system. The user was given a mission to prepare and commented as he went along. Comments were captured by the facilitators, generating a detailed list of about fifty problems. The problems were reviewed by the pilots defining the new system to find ways to avoid them in the design of the new system. The users filled out SUMI questionnaires after the evaluation to give a baseline for satisfaction. The technique was very productive even though applied in a semiformal way.	A usability analyst and seven users evaluated the existing system out in the local office network. Each user was given a brief introduction and then observed using the system to do the same key tasks. Comments were captured by a usability analyst who generated a problem list and a report was produced that was fed into the development team before design of the new system began. We should also have used the opportunity to gain effectiveness, efficiency, and satisfaction figures for later use.
5. Usability requirements	A half-day workshop to establish usability requirements for effectiveness, efficiency, and satisfaction with the user groups and tasks identified in the context of use analysis and in the scenarios.	Goals for task time and satisfaction were agreed upon, and a list of potential user errors were identified. We realize the need for the technique and its potential but more work is needed to better define it.	The various estimates for effectiveness and efficiency had to be agreed upon and then verified in local offices on the existing system. We made the mistake of not growing and refining the requirements sufficiently as our understanding of the

Table 20.4 *Continued*

Method	*Description*	*IAI comments*	*IR comments*
			system matured. . . . In hindsight it is clear, however, that the advantages do outweigh the time spent. All parts of the project team had a clear, common understanding of what is an acceptable standard for the usability of the system and we were able to evaluate if that benchmark was being met, so helping improve and control the quality of the system. The skills necessary to set a requirement were easily transferred from the facilitator to the business and are already being applied on other projects.
6. Paper prototyping and affinity diagramming	Evaluation by users of quick low fidelity prototypes (using paper or other materials) and	IAI had not used this method before and had doubts about its value, mainly because it is very easy to create computer-based	We used affinity diagramming to construct a model of key functions and then to logically group

	construction of affinity diagrams to clarify requirements and enable draft interaction designs and screen designs to be rapidly simulated and tested.	UI prototypes. In practice the potential users and developers liked the method and its contribution to MPC usability. Mockups of screens were posted on the wall and provided the "Big Picture." Each screen was subsequently displayed using an overhead projector, resulting in very fruitful and productive discussions by potential users. A detailed list of usability comments was created.	them. After they were grouped the structural hierarchy was developed and verified by use of task scenarios. It proved a simple technique to use and helped us resolve a problem that would have had a major impact on the usability. Paper prototyping was already widely used on other projects but was formalized for the trial project and linked to the preparation activities before the workshop and the use of task scenarios during it. As a technique it was easily picked up by the analysts and end users.
7. Style guide	Identify, document and adhere to industry, corporate or project conventions for screen and page design.	Off-the-shelf style guides were provided to the developer. It turned out that these style guides are very detailed and difficult to use. Given intuitive visual development tools, developers prefer to learn by click-and-see rather than by	Our usual practice had been to leave Graphical User Interface standards to individual projects, which meant applications were delivered to the business with a different look and feel. A corporate style guide

Table 20.4 *Continued*

Method	*Description*	*IAI comments*	*IR comments*
		reading lengthy manuals. IAI realizes the need for a style guide, but currently doesn't have one with an appropriate level of detail.	and an overview of the chosen user interface style were provided to the development team. Developers involved in previous projects commented that as a result much less pointless discussion was spent on names and placement of controls.
8. Evaluation of machine prototypes	Informal usability testing with three to five representative users carrying out key tasks to provide rapid feedback on the usability of prototypes.	Software developers were present and observed the evaluation. In general the developers were very receptive and cooperative. A summary meeting was held at the end of the evaluation. Comments were listed and prioritized, and it was decided to fix 93 of the 97 problems. The problems were points of detail and not major issues showing that earlier design was sound.	The system was only partially developed. Nevertheless the major usability and window design issues could be verified. The developers were present, assisted in the evaluation, and could not have been more cooperative and supportive. Analysis identified 32 problems, including 3 major usability issues, all of which were formally logged on the IR

			problem management process and prioritized for fix.
9. Usability testing	Formal usability testing with eight representatives of a user group carrying out key tasks to identify any remaining usability problems and evaluate whether usability objectives have been achieved.	The system was tested against timing and satisfaction requirements. First, two hours MPC training was given to the pilots followed by individual hands-on practice for another two hours. Each pilot then received written instructions regarding the mission he had to plan and modify, and worked without assistance. He also could write down comments on printed versions of the screens. The facilitators and developers observed the work and documented their observations. All pilots were happy with the MPC, which was confirmed in the SUMI results, which were well above the industry average. The overall duration of the tasks was within requirements.	We have been using summative usability testing for a number of years, but this time introduced scoring against a detailed usability requirement for all the main business tasks. The use of a detailed requirement that formed part of the wider business requirement and that had the buy-in of the whole project meant the results of the exercise carried much more credibility and empowered the usability analysts in their discussions about resolution of the problems that have been discovered.

areas that needed improvement including some organizational issues. These were used to select UCD methods for trial. The second assessment purpose was to evaluate the improvements made. The detailed results are very valuable and will be used in further dissemination activities in LAHAV and other IAI divisions.

Cost Benefits

After the work was complete, IAI estimated what it would cost to carry out the methods again, and what it would have cost to make the same fixes and changes at a later stage (Table 20.5). The estimated time in hours assume that the work is carried out by a member of the development team experienced in the methods, and that no formal documentation is required.

Reduced Development Costs

IAI estimated that all the methods used (except Style Guide and Scenarios of Use) resulted in savings in development costs of between $5,000 and $70,000 for each method, with a total saving of $330,000. The cost of using the methods was only $22,000, giving a cost-benefit ratio of 1:15.

Sales Benefits of Increased Usability

IAI markets the MPC independently of other avionics. Increased sales were estimated to be $400,000.

Support

Reduced costs of developing and providing training and support were estimated at $50,000.

Overall Cost Benefits

The overall costs of the maturity assessments and use of methods was $27,000. The total estimated savings and increased sales is $780,000, giving a cost-benefit ratio of 1:29.

Conclusions

The IAI concluded that the techniques were both low cost and very cost effective. IAI knew from previous experience that introducing changes into an

Table 20.5 IAI Estimated Costs and Benefits

Technique	*Duration (Hrs)*	*No. People Participating*	*Preparation and Documentation (Hrs)*	*Total No. HF Person-Days*	*Equivalent Later Costs ($)*	*How Cost Effective Was the Technique?*
Stakeholder meeting	3	6	2	2.5	50K	High
Context of use analysis	3	4	6	2.3	30K	Medium
Affinity diagramming	6	5	5	4.4	40K	High
Scenarios of use	2	5		1.3	—	Low
Baseline existing system	2	4	5	2.1	30K	High
Usability requirements	3	5	3	2.3	5K	Medium
Paper prototyping	5	5	10	4.4	70K	High
Use style guide	2	1	0	0.3	—	Low
Evaluate computer prototype	2	5	6	2.0	40K	High
Test usability against requirements	4 + 1.5	10	5	4.0	20K	High

organization can be a lengthy, costly, and complicated process. It requires convincing many people to invest time and money and then demonstrating that the benefits outweigh the costs. In recent years it has become even more difficult because of staff shortages and the requirement to reduce the time to market.

TRUMP was the exception because of its low cost and obvious benefits. When the developers have to invest only a few days applying the methods and see the results on the spot, convincing the managers is very simple, and performing cost-benefit analysis is not needed. Because these techniques significantly improved the quality of the system, but required relatively little time and effort, they are being incorporated into LAHAV's development process.

20.4.2 Inland Revenue/EDS

The IR in the UK provides data processing support for 60,000 employees in more than 600 local offices. At the time of the trial, the IR employed a well-defined joint application design (JAD) and rapid application design (RAD) methodology with its IT partner EDS.

At IR, usability capability was assessed using a conventional software process assessment procedure based on Process Professional Assessment (Compita, 1997). This lasted one week and was carried out by two trained assessors, assisted by two usability specialists who identified opportunities for process improvement. A total of 13 stakeholders associated with the trial project at different levels in IR and EDS were interviewed in 12 3-hour sessions, resulting in a detailed profile and comprehensive information about where improvements would be beneficial.

The main conclusions were as follows:

- User-centered information exists, but not always at the right time or in the right place.
- Usability requirements are either not documented or are documented much too late.
- Building usability into the development process is only partially documented and managed.

A feedback meeting provided the basis for an agreed set of improvement activities.

1. Extend and integrate the user-centered design methods employed early in the lifecycle.

a. Define a range of people who will use the system and what tasks they will undertake.
b. Produce task scenarios to cover all the main tasks.
c. Set usability requirements for the success rate, accuracy, task time, and satisfaction for these tasks.

2. Employ more user- and task-based methods in the JAD workshops.
   a. Focus on real-life task scenarios.
   b. Use different prototyping approaches to design windows.
   c. Adhere to corporate and industry guidelines.
   d. Test the paper mock-ups from a user perspective using the task scenarios.
   e. Produce a preparation pack for each function that collates the context analysis, task scenarios, IT requirements, and design thoughts so that the organization shares a common view of what they need to deliver from the JAD.
3. Methods used after JADs were to:
   a. Evaluate the usability of an IT functional prototype to validate the emerging design.
   b. Test the business system against the usability requirement.

IR commented (Bevan *et al.*, 2000):

> It was however a wary project team that was brought together for the first maturity assessment, uncertain what they had let themselves in for. The maturity assessment however opened everyone's eyes to:
>
> ✦ The different ways users could and should be involved throughout the lifecycle.
>
> ✦ The benefits that could accrue to both the project and IR/EDS.
>
> ✦ Professional support available from usability engineers.
>
> Output from the assessment was not only a clear eyed assessment of the level of maturity in this area but it provided a straightforward model for raising that level aimed at the heart of the development lifecycle, the facilitated workshops which are the engine of design and development stages.

IR/EDS experience using the methods is shown in Table 20.4.

When the improvements had been made 12 months later, a second similar assessment was carried out to see whether the agreed-on improvements had been achieved. Significant progress had been made (Table 20.6). When the results

Table 20.6 Comparison of Assessments

	IR/EDS	*IAI*
Number of design and development staff	>200	40
Use a fully documented process?	Yes	No
Importance of end user needs	High	High
Experience with usability	Moderate	None
Attitude to process improvement	Committed	Committed
Number of stakeholders interviewed	13	8
Number of usability maturity model base practices judged relevant and assessed	39	33
Initial number of activities partially or not performed	19	24
Final number of activities partially or not performed	3	2

were presented to a meeting of senior stakeholders, the benefits were sufficient for the meeting to authorize incorporation of most of the methods into the standard IR/EDS documented processes. The meeting also suggested that regular usability capability assessments should be arranged to monitor improvement in the user-centered design process.

Cost Benefits

After the work was complete, IR estimated the resources that would be required to carry out the techniques again (Table 20.7).

Reduced Development Costs

The methods used to improve JADs were: context of use analysis, set usability requirements, task analysis, task scenarios, preparation pack, paper prototyping, managing issues, using smaller teams, a project glossary, and style guides.

These saved staff time by bringing a degree of engineering to the workshops that hadn't previously existed, and provided a framework for the users to make an effective contribution.

For a system with 20 functions the value of the total savings in staff time was estimated to be £231,000 ($390,000). The cost of using these methods in JADs was estimated to be £88,500 ($150,000), giving a cost benefit ratio of 1:2.6 for these methods.

Table 20.7 IR estimated benefits

Technique	*Duration (Hrs)*	*People Involved*	*Preparation and/or Analysis (Hrs)*	*People Involved*	*Total Resource in Person Days*	*Number of Skills Transfer Sessions Needed*	*Contribution to a Better System*
Usability maturity assessment	4	20	4	20	20		
Context of use analysis	2	4	4	1	1.5	None	Medium
Scenarios of use	2	4	0	0	1 per function	1	High
Usability requirements	4	8	32	1	3	1	Medium
Baseline existing system	1	7	32	1	5 per event	None	High
Affinity diagramming	2	6	0	0	2 per event	1	Medium
JAD: preparation pack	1	8	8	1	2 per function	1	High
JAD: paper prototyping	2	8	0	0	2 per function	1	High
JAD: manage issues	2	8	0	0	2 per function	1	High
JAD: smaller teams	2	8	0	0	2	None	Medium
JAD: project glossary	1	8	0	0	1	None	Medium
Style guides	1	8	160	2	21	None	Low
Evaluate usability of prototype	2	8	104	3	15 per event	1	High

Use

Evaluation of the existing system, several prototypes, and live running cost a total of £51,500 ($88,000). Evaluating an existing system clarified requirements, and the evaluations ensured that the requirements were met and enabled additional improvements to be made, which will lead to benefits in use.

It was not possible to estimate the use and support benefits, but usability testing verified that employees could complete tasks quickly and to acceptable quality standards on their first day of using the online system. The user-centered methods employed during development that ensured the system was designed to meet real work scenarios played an important part in achieving these results.

Overall Cost Benefits

The overall cost of the maturity assessments, development, and evaluation methods was £152,000 ($260,000). The cost benefits of using all these methods, based only on estimated savings in development costs, was 1:1.5. The potential benefits of savings in use were not estimated, but are likely to be substantial, with 30,000 users whose time costs more than $1 per minute. Thus, the actual cost benefits of using these techniques were almost certainly much higher.

Conclusions

The value of the methods to the business was so clear at the time of the second maturity assessment that IR formally adopted the methods into their developoment process without waiting for the results of a formal cost-benefit analysis. The cost-benefit results confirm the value of that decision.

The IR was subsequently awarded UK Central Government Beacon status for their work on user-centered design and usability.

20.4.3 Comparison

The usability maturity model was a valuable tool for identifying needs for process improvement in both organizations (see Table 20.6). IR valued the detailed information obtained from a summative assessment requiring three person-weeks of effort, while for the smaller development group at IAI many of the benefits were gained from a simpler, formative, 1-day assessment.

Particular user-centered design methods were not of equal value to both organizations. For example, IAI staff were more familiar with the usage environment, so that the context of use and scenarios were of less benefit than at IR, where this information was important in establishing a common understanding. At IR the use of an in-house style guide was an important factor in maintaining consistency in a large organization, while at IAI graphical user interfaces were not developed frequently enough to justify development of a style guide.

The base practices in ISO TR 18529 are generic, but the methods used to implement them need to be selected and tailored to meet the needs of the project, development environment, timescales, and budget.

20.4.4 Taking up the Methods

Both organizations found the results so beneficial that they have adopted the methods as a normal part of their development processes. At IR the methods are applied by usability specialists, whereas IAI found them sufficiently intuitive that they plan to train existing members of the development team to use them (calling on expert assistance when required).

Does this provide a model for how to introduce user-centered design in other organizations? Jokela and Iivari (2001) found assessments based directly on ISO TR 18529 less successful, which led them to develop a new assessment process based on the intended outcome of each process, rather than the specific practices. Although an ISO TR 18529 assessment can also be based on outcomes (which are listed for each process), most of the TRUMP assessment centered on the base practices, which were easy to interpret. The comparative success may have resulted from the assessor's high degree of familiarity with the ISO TR 18529 model, and the common goals of IAI and IR to:

- Provide systems that meet user needs
- Improve their processes

This management commitment to improvement and change may have been lacking in some of Jokela and Iivari's assessments.

But even IR/EDS and IAI had initial difficulty in understanding the assessment model and potential benefits of the user-centered methods, which differ in nature from other software engineering activities. Jokela and Iivari report similar difficulties in conveying the meaning of the model in advance of their assessments. It is still not clear how best to present the proposed user-centered design activities in a way that can be understood and appreciated by designers and developers.

We nevertheless believe that the assessment processes used in TRUMP are an effective way to successfully implement user-centered design, and that the cost benefits obtained by IAI and IR could be replicated in similar organizations.

ACKNOWLEDGEMENTS

The IR case study was carried out in conjunction with Nick Ryan, who provided the IR cost-benefit information. The IAI case study was carried out in conjunction with Itzhak Bogomolni, who provided the IAI cost-benefit information. The usability maturity model work was led by Jonathan Earthy of Lloyds Register of Shipping.

The TRUMP project was supported by the European Union.

REFERENCES

Bevan, N. (2000a). Getting started with user centred design. <*www.usabilitynet.org/trump*>

Bevan, N. (2000b). Basic methods for user centred design. <*www.usabilitynet.org/trump/methods/basic/*>

Bevan, N., Bogomolni, I., and Ryan, N. (2000). TRUMP case studies. <*www.usabilitynet.org/trump/case_studies/*>

Bevan, N., Bogomolni, I., and Ryan, N. (2001). Incorporating usability in the development process at Inland Revenue and Israel Aircraft Industries. In M. Hirose (Ed.), *Human-Computer Interaction—INTERACT'01* (pp. 862–867). Amsterdam: IOS Press.

Bevan, N., and Earthy, J. (2001). Usability process improvement and capability assessment. In J. Vanderdonckt, A. Blandford, and A. Derycke (Eds.), *Proceedings of Joint AFIHM-BCS Conference on Human-Computer Interaction IHM-HCI'2001,* Volume 2., Toulouse: Cépaduès-Editions.

Compita. (1997). Process Professional Assessment. <*www.processprof.com*>

Earthy, J. (1998). Usability Maturity Model: Processes. INUSE deliverable D5.1.4p, see <*www.usabilitynet.org/trump/methods/integration/assurance.htm*>

IBM. (1999). Cost Justifying Ease of Use. <*www-3.ibm.com/ibm/easy/eou_ext.nsf/Publish/23*>

ISO 13407. (1999). User-centred design process for interactive systems. Geneva: ISO.

ISO TR 18529. (2000). Human-centred life cycle process descriptions. Geneva: ISO.

Jokela, T., and Iivari, N. (2001). Usability capability assessments—experimenting and developing usability maturity models. In J. Vanderdonckt, A. Blandford, and A. Derycke (Eds.), *Proceedings of Joint AFIHM-BCS Conference on Human-Computer Interaction IHM-HCI'2001,* Volume 2. Toulouse: Cépaduès-Editions.

21 CHAPTER At Sprint, Understanding the Language of Business Gives Usability a Positive Net Present Value

Clyde C. Heppner Sprint Corporation
Jesse Kates Sprint Corporation
Jefferey S. Lynch Sprint Corporation
Robert R. Moritz Sprint Corporation

21.1 INTRODUCTION

21.1.1 It Is Time for a Mindset Upgrade

Every business, from sole proprietorships to multinational corporations, is faced with the same fundamental problem: how best to invest constrained resources—money, people, and time—to create long-term, sustainable value for business owners and shareholders. Note that this problem definition does not explicitly mention customers. Creating sustainable value for the owners of the business is the *end*, while creating things like highly satisfying, revenue-producing, or cost-saving customer experiences is *a means to the end*.

Most user experience professionals strive primarily to create exceptional customer experiences. For such professionals, the quality of the customer experience is the *end*, the predominant measure of success or failure. This perspective produces two primary effects; one positive, one negative. On the positive side, a quality-oriented focus can lead to quality designs. Negatively, user experience professionals may lose sight of the primary concern of the business, creating financial value. Such professionals are prone to either overemphasize the impact of user-experience blemishes or underestimate the impact of competing business concerns. Although *tactically* valuable, a myopic user-experience focus obscures other perspectives critical to *strageic* business decision-making efforts.

In the mind of a user experience professional, a flawed user experience may equate to a failed product. However, an organization may opt to launch products or services that contain significant known flaws for valid business reasons—to capture a first-mover advantage, for instance. Yet businesses also launch flawed products or services that should have been fixed, resulting in reduced usage that may negate expected business benefits. In extreme cases, launching a product that fails to meet customers' usability expectations may lead to failure of the business case.

Without the contribution of user experience professionals, businesses may fail to make the "right call" when faced with tradeoffs between user experience and business concerns (e.g., usability vs. time to market). Thus, user experience professionals must contribute to business decision making. To accomplish this, user experience professionals must 1) adopt the business decision-making mindset, and 2) establish and demonstrate the linkage between user experience concerns and shareholder return (see Donoghue, 2002).

21.2 ADOPTING THE BUSINESS DECISION-MAKING MINDSET

21.2.1 The User Is Not You

As user experience practictioners, we are accustomed to adopting the mindset of others. In most cases, we adopt the mindset of target customer groups. But to work successfully within the corporate world, we must adopt the mindset of the CEO, the archetypal business decision maker. In user-experience terms we must uncover the mental model of the key decision makers within the company so we can learn to articulate and demonstrate our unique skills and perspectives within the business decision-making framework.

21.2.2 The Business Decision-Making Mindset

As mentioned before, the most important constituency for any company is its shareholders—period. Not customers, shareholders. While this statement may seem counterintuitive, shareholders—whether there are one or one million—are the true owners of any business. A company is expected to make daily decisions that generate a return equal to or exceeding shareholder expectations.

In the case of corporations, shareholders do not manage the business day-to-day, so they elect a Board of Directors to be representatives of the shareholders. The Board is not engaged in running the business day-to-day either, which is why it appoints a CEO, who, in turn, creates a management structure to run the day-to-day business. This entire structure is created and managed with the sole intention of ensuring the shareholders' best interests are served and value is created in the form of earnings and stock price appreciation. Thus, the mindset of a business leader is framed on one item first and foremost—creating value for shareholders.

How, exactly, do companies and executives go about systematically analyzing the wide variety of investment possibilities and resource allocations that present themselves every day? The short answer is "not easily," but fortunately there are a number of methodologies available to ease the work. One in particular, net present value (NPV), is widely applied and understood, and is viewed by many business leaders as the most effective measurement tool. Brealey and Myers (2000) highlight NPV as the top idea in "The Seven Most Important Ideas in Finance" section of their seminal work, *Principles of Corporate Finance.*

Without going into detail, NPV represents the present value of all future cash flows associated with a project or investment, less the cost of the investment. The calculation is based on the principles of the time value of money—that is, a dollar today is worth more than a dollar tomorrow, because the money can be invested to receive a return. Because NPV leverages the shareholders' expected return as a baseline, projects with positive NPVs create shareholder value and projects with negative NPVs destroy shareholder value. Karat (Chapter 4) goes into more detail on NPV.

Given that NPV incorporates a wide variety of variables—cost, revenue, risk, uncertainty, and expected returns—it is a comprehensive, powerful tool capable of valuing disparate opportunities. Invest in a network upgrade or redesign the website? Choose the project with the highest NPV and rest assured that the right decision for both shareholders and the business has been made. Because of this power and simplicity, adherence to the principles of NPV is second nature to executives of virtually any company. NPV is the fundamental concept behind most business decision making, and serves as a numerical embodiment of the business decision-making mindset. Despite the concept's obvious importance, however, most user experience professionals do not "speak" NPV.

21.2.3 Ignored and Undervalued?

Because user experience professionals don't think or speak the way other business professionals do, many user experience professionals feel ignored. Such professionals may perceive a fundamental gulf between their mindset and the archetypal values of the CEO. In most cases, however, there is no such gulf. Once you understand how business decision makers see the world, you will quickly recognize that user experience simply has not been quantified in financial terms. If a functional group cannot demonstrate the linkage between their expertise and financial results, that functional group will be either ignored or their output will be considered interesting but non-actionable. All business decisions boil down to dollars and cents. Groups that cannot prove they affect revenue are consistently overridden by groups who can.

Recall that the role of most finance organizations is to run the numbers executives use to make decisions. The finance department is where we began our quest to understand how user experience fits in the fundamental business model of our company. We began by examining the business metrics that our company reports to Wall Street as key performance indicators. The yearly bonuses of Sprint's employees are tied to these metrics.

- Gross adds—Number of new customers we added during a specific period of time.
- ARPU (Average Revenue Per User)—Revenue (in dollars) that each customer creates per month.
- Churn—An indicator of how many of our customers leave our service during a specific period of time.

Most user experience professionals would say that customer experience impacts these three critical business metrics. For example, a poor customer experience could result in "word of mouth" that reduces sales, causing customers to find an alternate product for their wireless communications needs. However, if this supposition is not validated empirically, and if user experience practicitioners regularly raise an alarm without providing evidence as to why an alarm should be raised, business owners will either disregard user experience input as "crying wolf" or, at best, will heed only the warnings they personally agree with. When discussing user experience issues in a business context, user experience professionals must speak in financial terms.

Professionals who cannot speak in financial terms may feel undervalued. Many will experience a recurring pressure to justify their involvement. Needless to say, this situation produces stress and distracts the user experience professional, reducing performance. We agree that the contributions of user experience professionals are often undervalued. However, we propose that the root cause of this phenomenon is the inability of the user experience professional to isolate his or her contribution to the bottom line. For lack of better indicators, user experience professionals may point to global "Customer Satisfaction" measures or sales figures. Yet these measures are influenced by numerous organizations, including marketing, sales, and customer service. User experience professionals cannot leverage global measures to demonstrate their unique effectiveness. Instead they must identify and validate a unique measure that tracks their specific contribution to shareholder value.

21.3 LINKING FINANCIAL RESULTS TO USABILITY

21.3.1 The Price of Admission

Any functional area that wishes to participate meaningfully in a business decision-making process must be able to contribute quantifiable inputs. These inputs must have demonstrable validity (i.e., the input must be psychometrically identified as measuring the targeted phenomenon) and must be verifiably linked to financial impact and thus the NPV of specific projects.

The first challenge for user experience professionals is that the measures typically collected in a usability lab (e.g., frequency counts, completion rates, error rates) or during a field study are not expressed in financial terms.

The second challenge is that user experience research is often project oriented rather than program oriented. In a project-oriented approach, each study is optimized to measure specific aspects of a single product or service. As a result, data from one study (with one product) is not necessarily comparable with data from a second or third study (with other products). Incompatibility between data sets prevents standardization and hinders translation into financial terms.

A program-oriented approach provides the only reasonable means to isolate the financial impact related to usability. In a program-oriented approach, all studies are designed within a common framework. Within each study, the researchers administer the same collection of inventories—psychometrically designed questionnaires, such as Brooke's (1996) System Usability Scale, or SUMI (developed by the Human Factors Research Group [Porteous *et al.*, 1993]), segmentation questions, and demographic questions. By applying a consistent set of inventories across all products, systematic (inter-product) comparisons can be made.

The final challenge is that most usability data is collected within a lab setting. Lab data can guide design, but may not map exactly to the customer experience in the real world. As a result, it can be difficult to link lab data to post-launch financial results. Usability practitioners who wish to participate in business decision making must collect data from actual consumers who have purchased or experienced the actual product or service.

In summary, to collect data that maps to financial results, user experience professionals must identify or create a valid assessment tool, adopt a program-oriented approach, and conduct research with actual customers. Once the data are acquired, the link between the data and financial results must be identified. In the following paragraphs, we describe the execution of this strategy at Sprint.

21.3.2 Creating a Valid Assessment

Sprint's assessment instrument consists of three inventories (e.g., usability, product satisfaction, and user needs), a variety of attitude and use questions that are product specific, two sets of segmenting questions (e.g., marketing acquisition and technology adoption), and a set of demographic questions. The following list provides definition of the measures used within the assessment.

- Usability Inventory—A measure of how easy a product or service is to use. The output of this inventory is norm-referenced and classifies the product into one of four customer expectation categories: exceeds expectations, meets expectations, is below expectations, or is significantly below expectations.
- Product Satisfaction Inventory—A measure of how well a product or service meets the user's needs (i.e., "Did the product provide the right set of functions and features?") The output of this inventory is norm-referenced and classifies the product into one of four customer expectation categories: exceeds expectations, meets expectations, is below expectations, or is significantly below expectations.
- User Needs Inventory—Measures the magnitude of the user's needs.
  - Safety and security—The need to react or obtain assistance in an emergency.
  - Convenience—The need to simplify or make daily activities easier.
  - Personal Communication—The need to keep in touch with family and friends.
  - Business Communication—The need to facilitate or enhance business relationships.
  - Social Image—The need to raise the level of social influence and self-respect.
  - Cool Product—The need to have new and fun technology.

- Attitude and Use Questions—Product-specific questions that identify usage patterns and barriers across the stages of the product lifecycle.
- Segmentations and Demographics—Profiles the customer base and classifies each customer within the segment defined by the business.

Alone these measures provide insight. Combining these measures, however, provides an output that exceeds their individual additive value. For example, Figure 21.1 displays the interactive effect between usability and product satisfaction. Sprint leverages our understanding of this effect to identify whether a product's flaws are functional or user-experience related.

21.3.3 Sprint's Program-Oriented Approach

Assessments are conducted after launch with thousands of actual customers. Customer cohorts are formed based on equal levels of exposure to the product or service (e.g., a group of 2000 customers is identified who, at the current point in time, have all used the product or service for 2 months.) Each cohort

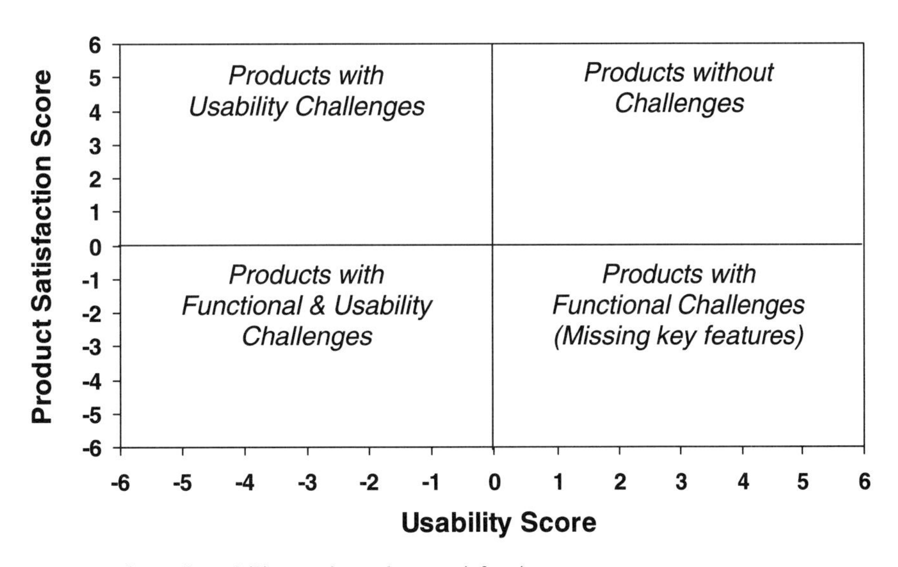

FIGURE 21.1 Interaction of usability and product satisfaction.

completes the standardized assessment described previously after 2, 4, and 10 months of use. This approach reveals how usage and behavior change for each customer across the product lifecycle. Sprint isolates the influence of usability on usage and adoption in a real world context with actual customers.

Sprint's program-oriented approach is executed by two full-time and one part-time research professionals, four lab interns, and one market research manager. Product Develop provides a budget allocation for the usability lab and Marketing provides a budget allocation to conduct the field research. This team completes more than 75 studies each year, equally split between lab and field. Data from these studies are readily and routinely linked to draw systemic conclusions.

21.3.4 Identifying the Link Between Usability Data and Financial Results

The usability assessment program identifies the influence of usability on usage and adoption. Usage and adoption are established business inputs and can be used with existing financial valuation methods. Thus, once the usability assessment program is in place, it becomes relatively straightforward to leverage financial expertise to determine the specific connection between usability and changes in value to the business (and thus NPV).

Once the link between usability and value is understood, it becomes possible to identify which usability transitions cause the largest changes in financial value (Fig. 21.2). For example, we found that for a set of Sprint's products the largest usage increase (i.e., number of times products are used) occured when the products transitioned from "meets expectations" to "exceeds expectations" with regards to usability. A smaller usage increase occured when the products moved from "below expectations" to "meets expectations," yet this shift represents the greatest financial gain for the company because of Sprint's business model. This example illustrates that the intuitive aim of the user experience practitioner ("Let's maximize usability!") is not necessarily aligned with the business goal of maximizing shareholder value. To reiterate, an intuitive understanding of the linkage between usability and financial return is insufficient to power business decision making. User experience professionals must replace their intuitive knowledge with data-driven knowledge.

Once the critical transitions for a given product set are identified, usability practitioners can leverage usability data to prioritize their efforts to produce the largest possible financial return for the business. For example, at Sprint it would not make sense to expend a great deal of energy on a product that is already

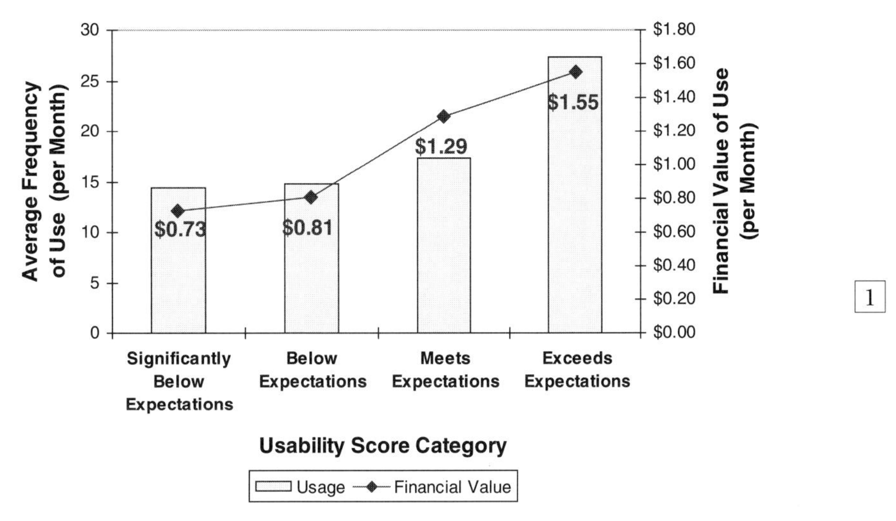

FIGURE 21.2 Frequency of usage and financial value as a function of usability.

meeting usability expectations if opportunities exist to improve products that are below expectations. Knowledge of the link between usability and usage can also be used to weigh the costs and benefits of proposed usability improvements and to present a compelling argument for those improvements that generate a positive NPV.

21.4 CONCLUSION

21.4.1 Brave New World

By isolating the impact of usability on your business's financial results and tying usability concerns to NPV, you can pave the way for greater recognition, influence, and effectiveness. Your contribution to shareholder value will no longer be subsumed within global customer satisfaction and sales figures. You will be able to point to a single number and say, "We did that. This is how we add value." When you raise a user-experience issue, business leaders will know that the resolution will directly affect revenue, and user experience will be taken more seriously. The intuitive belief that "ease of use matters" will be replaced by actual knowledge that translates to financial terms. User experience improvements will compete with other service upgrades on equal ground. You will gain credibility, with increased social return on investment as described by Wilson and Rosenbaum (Chapter 8).

Once you establish the link between usability and revenue, the institutionalization of usability within your corporate culture will be inexorable. The CEO may begin to measure the success or failure of ease-of-use initiatives, and management and executives will be expected to meet usability benchmarks. It may not happen overnight, but at Sprint, after 3 years of concerted effort, a portion of executives have their incentive compensation tied to the percent of products meeting customers' usability expectations over time. In other words, usability has become a key business performance indicator like gross adds, ARPU, and churn. Figure 21.3 is an example of how Sprint plots the percent of products meeting customers' usability expectations year after year.

21.4.2 You Are Aligned

By adopting the business decision-making mindset and isolating the impact of usability on revenue, you will gain the knowledge and insight necessary to align

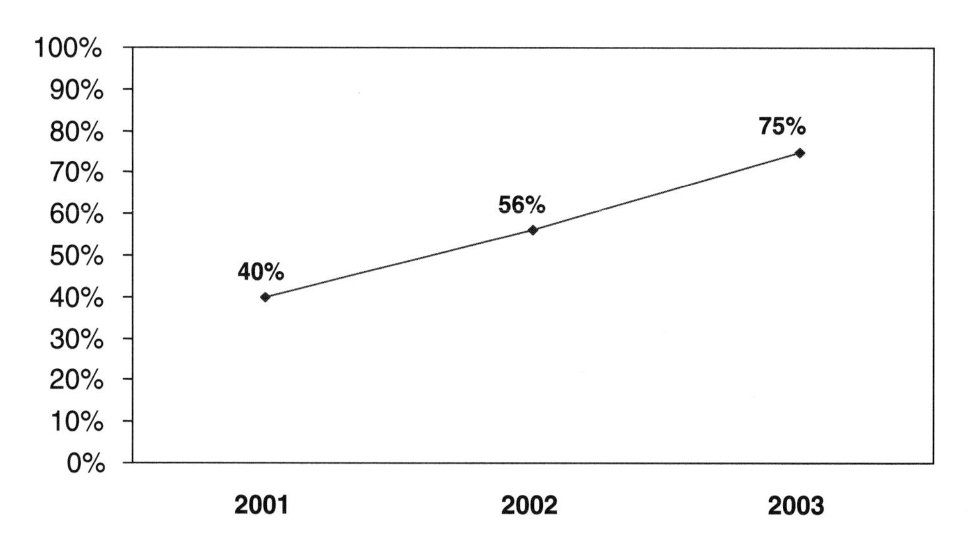

FIGURE 21.3 Percent of products meeting customers' usability expecations over time.

your constrained resources—money, people, and time—with the corporate mandate: Provide the greatest financial return for your shareholders. Nothing is lost. You will continue to drive quality into the products and services that you design, and you will continue to make the lives of your customers easier. You gain the ability to document and present your work in a way that the rest of the company will recognize, understand, and appreciate. Most importantly, you can rest assured that you are applying your efforts and energy where they are needed most.

REFERENCES

Brealey, R. A., and Myers, S. C. (2000). *Principles of Corporate Finance Sixth Edition.* New York: McGraw-Hill.

Brooke, J. (1996). SUS: A quick and dirty usability scale. In P. W. Jordan, B. Thomas, B. A. Weerdmeester (Eds.), *Usability Evaluation in Industry.* London: Taylor and Francis.

Donoghue, K., and Schrage, M. (2002). *Built for Use: Driving Profitability Through the User Experience.* New York: McGraw-Hill.

Porteous, M., Kirakowski, J., and Corbett, M. (1993). *Software Usability Measurement Inventory (SUMI).* Human Factors Research Group, University of Cork, Ireland.

22 CHAPTER Cost-Justifying Usability: The View from the Other Side of the Table

Randolph G. Bias School of Information, The University of Texas at Austin

22.1 INTRODUCTION

Consider, if you will, any "proposal for the funding of usability support" to be an artifact. The "users" of such artifacts would be the software development executives or other executives who hold the purse strings, and who decide whether funding such usability work will be a wise investment for the organization's money, time, and other resources. If we usability professionals believe what we say about "participatory, iterative design," we would be smart to gather some user data on the design excellence of our proposals.

This book has devoted 21 chapters to usability professionals telling other usability professionals about their experiences with cost justifying usability efforts. It is now time to listen to some of the folks on the other side of the table. What do those software executives who hold the purse strings think? What sorts of arguments in favor of usability engineering have they "bought," and to which have they said "no"?

What would they *like* to hear, to help them make good decisions, from those of us who believe usability efforts will be "worth it"?

With the help of some other contributors to this book, I identified four software executives who met the following three criteria:

- Individually, they had experience considering proposals to fund or otherwise support usability projects

- Collectively, their companies represented a broad expanse of the software or Web site landscape
- They were willing to participate in an e-mail- or phone-based interview for this chapter.

22.2 THE EXECUTIVES

22.2.1 Joyce Durst, Infraworks Corporation

Joyce Durst is the CEO of Infraworks Corporation, a 6-year-old, privately-held company that develops information security software solutions. Joyce has broad experience in all aspects of software development.

22.2.2 Sara Garrison, Inovant

Sara Garrison is the Senior Vice President of Network & Open Systems for Inovant—A Visa Solutions company. Sara is responsible for the development of business-to-business applications for Visa Members (financial institutions, merchants, and processors), including browser-based applications at Inovant, Visa's wholly-owned IT subsidiary.

22.2.3 Bill Mitchell, Microsoft

Dr. Bill Mitchell has led start-ups within Microsoft, including the Windows CE group, the Pocket PC group, the Smartphone effort, and Smart Personal Objects Technology. Currently his role is Corporate Vice President of Mobile Platforms Division.

22.2.4 Kim Rachmeler, Amazon.com

Kim Rachmeler is Vice President of Customer Service and Supply Chain for Amazon.com, an acknowledged leader in e-commerce. In her background as a software development engineer, Kim was a member of the Bay Area CHI and took extension courses at UC-Berkeley in usability, but today she is in charge of customer service and supply chain and manages the development of tools for customer service agents.

22.2.5 A Quick Thanks

I would like to thank these four software executives for their selfless, thoughtful, and timely responses to my questions. I believe strongly that the general arc of the usability of software and Web-based user interfaces (UIs) will be better—we will more quickly approximate a world of high and realized expectations for excellent user experiences—given that the readers of this chapter will do a better job than they otherwise would have of securing funding for their usability work and repaying that funding with superior, usability-engineered products.

None of these people or companies asked to be part of this chapter or book, and none asked for anything in return.

22.3 QUESTION #1: WHAT ARGUMENTS TO PAY FOR USABILITY SUPPORT HAVE YOU SAID "YES" TO?

Inovant's Sara Garrison offers that, in general, usability efforts target benefits such as increasing productivity, decreasing training, and leveraging brand assets. With regard to past projects she has been in charge of, the most effective arguments for usability have asserted that the usability improvements would:

- ✦ Increase throughput on heads-down, hands-on, full-time transactional applications, such as financial services back-office systems
- ✦ Integrate disparate legacy applications within a line of business through a simple and straightforward UI
- ✦ Perfect the look, tone, and feel of content-rich applications and Web sites in accordance with published brand standards
- ✦ Design multinational applications, especially when the development team doesn't represent the target cultures
- ✦ Comply with accessibility requirements

Joyce Durst, Infraworks CEO, answers Question 1 the following way:

> The arguments I have supported in the past to approve usability projects focus on two themes. The first is **sales**. A product that is highly usable will have a shorter testing and sales cycle and will result in increased sales. In addition, there may be competitive pressure to improve the usability of the product. Competitive losses quickly get the attention of the management team. The second theme is **costs**. It costs much less to code the interface in a customer acceptable way the first time than it does to introduce a poor UI to the field and then rework that UI in version 2. In addition a poor UI will increase support costs. The usability proposals I've said "yes" to have demonstrated clearly that the usability team's efforts will result in increased sales or in reduced costs in getting a usable product to market. Or both.

Microsoft's Bill Mitchell says:

> The best argument I've ever heard for usability support went something like this: "I've looked at the product support logs, and we're receiving an average of 2000 calls per month on our product. Let's assume that it is costing us $20 per call. Here's what the usability support will cost you—X people with the latitude to do testing. Plus maybe another month in your schedule."

Bill believes it is important to acknowledge that, although, in the long run the inclusion of usability engineering methods will shorten the amount of time it will take to get a usable product to market, the intelligent application of usability engineering *does* take time. "This is one of the most important things. Be up-front with that. I like to hear up front that it is likely to increase the length of the schedule for this cycle. This adds tremendous credibility to the usability manager's or engineer's argument."

As an Amazon.com executive, Kim Rachmeler inherited a set of tools for customer service agents. The tools were developed by engineers without the help of usability professionals, although they may have had some graphic design help.

Early on her goal was to acquire the right usability expertise. "We hired an interaction designer, then a good graphic designer," she says. They conducted focus groups with customer service agents, and field/nature studies of agents using the tools. They ended up reworking process paths. She believes

> you need a critical mass to have a usability team. Out of that investment you get three things:
>
> 1. Reduced training time to get people up on tool set. This is important when we need to add to the workforce during holidays or when we need to hire temporary employees.

2. Reduced mistakes. The tool set should *lead* people to the right answer. Customer service is a complex environment. We have "blurbs," or templates, that embody correct response in a customer service setting—1600 of them. This helps make the responses straightforward. The helps us eliminate "write back"—the "no, you moron" second e-mail from the same customer.
3. Increased productivity of customer service associates. Increased usability means straightforward navigation, less window shuffling.

22.4 QUESTION #2: WHICH ARGUMENTS HAVE YOU SAID "NO" TO?

In fact, there was a fifth respondent to my request for information from "the other side of the table." Steve Mills, Senior Vice President and Group Executive of IBM's Software Group, was also asked. His simple response was, "I have never said 'no' to a request to spend money on usability. Effectively, usability always makes products easier and faster to sell and deploy."

If all executives were like Steve, this book would be unnecessary. Those usability professionals not lucky enough to work in the IBM Software Group may be interested to hear why some other executives might reject their pleas for usability help, as presented by our other four respondents.

Let's start again with Inovant's Sara Garrison:

> Usability engineering is an established workflow within our systems development methodology, so all production applications always receive initial usability engineering of both screen interfaces and printed reports from on-staff UI specialists . . . and most receive additional usability evaluations, if only heuristic evaluations from the specialists' peers in our corporate UI Center of Excellence. The budgetary shortfall that is typical of all application development efforts may limit but doesn't eliminate the work, just as it might limit but wouldn't compromise QA testing. For example, only the most critical applications are rigorously tested in an outside laboratory because of the time and expense (and we can't justify the construction of UI testing facilities onsite). Applications that are limited proofs-of-concept for internal use and that aren't seen by internal or external "customers" aren't subjected to usability engineering.

At Joyce Durst's Infraworks,

> The driving reasons for rejecting a project are, not surprisingly, time and/or money. The costs associated with incorporating improved usability must pass the credibility test. What will I need to do in increased sales to justify this project or conversely, what current customers/revenue will I retain by improving the usability of our solutions? I have said "no" if the project presented did not include supporting customer/prospect data and, at the least, estimates on revenue impact. I have also said "no" if the project did not consider alternatives such as in-house versus a consultative approach, or a multiphased approach versus only a "do-it-all-at-once" improvement plan.

Microsoft's Bill Mitchell says he has

> . . . historically received mostly weak arguments for usability. It's like with user education—you need it, but how much? The temptation, when times are lean, is to say "how little usability can I get away with?" This is *not* the way to think about the problem. There is a tendency to staff usability last. I've learned the hard way that putting off usability involvement to the end is costly—it relegates their contribution to little or no benefit.

Kim Rachmeler of Amazon.com says she is predisposed towards usability, "however, some arguments made me fall off my chair laughing." She believes that some of the attention to a sci-fi sort of usability future has taken the discipline down unproductive paths.

22.5 QUESTION #3: WHAT WOULD YOU LIKE TO HEAR FROM YOUR WOULD-BE USABILITY SUPPORT (THE IN-HOUSE TEAM OR A CONSULTANT), TO HELP YOU MAKE YOUR DECISION TO "BUY" OR NOT?

Joyce Durst says:

> I would like to know that the usability support team understands the tradeoffs associated with this project. There are a number of ways to increase a company's revenues or decrease a company's costs. How does improving the usability stack up against the alternatives? What are the opportunity costs?

> If there is specific customer and revenue data presented with the project cost plan, the project has a much greater chance of being approved. If they are too unifocused, they will lose credibility.

> From Sara Garrison:

> The soft qualitative benefits of usability engineering are obvious, but the hard quantitative benefits need more exposition. Assertions of ROI need to be supported with a business analyst's numbers and not a marketer's platitudes.

Bill Mitchell and Joyce Durst both said usability should save you time, money, or both. Bill went on to point out that however you build your usability proposal, it is important to present it to the right audience. He echoes the sentiments of Joyce Durst and Sara Garrison, just quoted, when he says, "The message must be tied to financials and not just general feel-good aspects."

22.6 THREE FINAL CONSIDERATIONS

22.6.1 Starting Up

Bill Mitchell has been "sort of an internal entrepreneur" at Microsoft. In our interview he addressed the situation in which you start with no people. The manager of such a new group has to decide whom to hire first. Bill believes strongly that "a usability specialist needs to be hired early in the team development. I learned that the hard way." Relatedly, Sara Garrison has recently taken over a new set of tools, and asserts she's going to "steal time from the usability team" to help get the new group going in the right direction.

22.6.2 The Attribution Problem

Kim Rachmeler points out one of the biggest obstacles to a cost justified approach to usability engineering. She says that at Amazon.com they have a "very structured approach to development, including training, quality, productivity." They end up doing many things at once to improve the metrics connected to a usability approach. While this structured, integrated approach likely maximizes the chances of an excellent user experience, it also makes it hard or impossible to attribute particular improvements in certain metrics to particular programmatic changes. Thus, because she has been convinced of the general value of usability engineering, she tends not to demand to see clear increases as a result of each specific effort.

22.6.3 The "Compleat Angler" (for Usability Funding)

Bill Mitchell says that one key issue for him is the propensity for "usability people to be inexperienced on the business side." He says this problem "percolates up—it affects the individual usability specialist when the usability manager can't cast the team's benefits in terms of benefits for the business, and thus it affects performance reviews." It is the intention of this book to "tool up" the usability practitioner, and the usability manager, to rectify this problem.

22.7 AND SO . . .

The "users" of proposals for usability support are the executives who hold the purse strings and decide how to deploy resources. I have, in this chapter, pursued a user-centered design approach to help us maximize the chances that our proposals for funding will be seen as usable (and, more to the point, will be funded). In Chapter 1 of this book Clare-Marie Karat and I offered a lengthy list of methods and skills to which the usability professional should aspire. To that list these four software executives have added the following:

- An awareness of the need to balance a variety of business goals and accept tradeoffs
- The ability to tie usability gains to business goals—an ability to lay out a phased approach to usability
- An up-front admission of the possibility that although a course of usability engineering may shorten the time-to-market of a usable product, in long run it may extend the current release's schedule, and
- The ability to discern which products or Web sites may not require quite as much usability engineering as other current products or projects

The usability professional will be a productive change agent indeed, if he or she not only has the skill to employ usability engineering methods to gather user data, has the discernment to know which methods to employ when, has the communication skills and political wherewithal to know how to advocate for the gathered and analyzed user data, *but also* has the business acumen to understand how to place his or her usability efforts in context and to cost-justify the usability investment.

Index

A

I

J

K

L

M

S

T

V

About the Authors

Nigel Bevan is a research manager at Serco Usability Services. He earned a B.S. in physics and in psychology and a Ph.D. in man-machine interaction. Bevan provides consultancy and training in usability and user-centered design. He was manager of European projects that have demonstrated the value of incorporating user-centered design into the development processes of several large organizations, and he was manager of the UsabilityNet project that established a Web site of usability resources. He participates in several international standards groups, and he edited standards including ISO 9241-11 Guidance on Usability and ISO 20282-2 Ease of Operation of Everyday Products–Test Method. He was responsible for developing the new Common Industry Format standard for usability requirements.

Randolph G. Bias is an associate professor at the University of Texas at Austin School of Information. With a Ph.D. in cognitive psychology from the University of Texas at Austin, Bias spent two decades in industry as a human factors professional, addressing software usability for AT&T Bell Labs, IBM, and BMC Software, where he founded and managed the usability department. In 2000, Bias cofounded a usability consulting company. In January 2003, Bias left that company to join the School of Information faculty. While in industry, Bias published prolifically on topics such as usability engineering methods and worked to bridge the technology transfer gap between the academy and practice. He now resumes his work on that bridge from the other shore. Bias's interest in designing technology to fit the user fuels several research threads. His recent research interests include the effects of Microsoft's ClearType technology on on-screen reading performance, a correction to Fitts's Law, and variables influencing the perceived length of on-hold times in customer call centers. In the School of Information, Bias is teaching courses in usability methods and research design and statistics. He consulted with many companies, large and small, helping them design Web sites and other user interfaces by systematically taking into account the users and their tasks. Bias received his B.S. with honors in psychology from Florida State University. He is married to Cheryl Bias, and they dote on their two sons, Travis and Drew.

Tom Brinck is coauthor of the book *Usability for the Web: Designing Websites That Work* and founder of Diamond Bullet Design (diamondbullet.com), a Web design and usability firm that applies customer research and design processes to create simple, useful, and accessible interactive products. Brinck is an adjunct lecturer at the University of Michigan's School of Information, where he teaches user-interface design. He conducted research in educational software, speech interfaces, groupware, and user-interface toolkits at Apple Computer, Toshiba, and Bellcore.

David Crow is currently responsible for HR Systems Development at Ryerson University. He continues to work with early-stage software startups in Canada and in the United States. Before returning to the university sector, Crow worked with entrepreneurs (Zaplet, eLaw.com, Living.com, and MetalSite.net) at Reactivity, Inc. to turn new technologies into practical, market-altering innovations. He started his career as an interaction designer with Trilogy Software and as a research engineer with Rockwell Collins. Crow received an M.S. in human-computer interaction from Carnegie Mellon University and a B.S. in kinesiology from the University of Waterloo.

Susan M. Dray is the founder and coprincipal (with coauthor David A. Siegel) of Dray & Associates, Inc., a user-centered design (UCD) firm that consults with large and small companies around the world to make sure their product and software designs meet user needs. They are highly experienced practitioners of international UCD, carrying out both ethnography and usability projects all over the world. They have consulted in 16 countries and have taught UCD techniques in many more. Dray has published many articles on UCD and has given keynote addresses on three continents. Together with Siegel, she recently published chapters about planning international studies and about the relationship between international UCD and localization. They also coedit the business column of the Association of Computing Machines SIGCHI magazine, *interactions*. Dray received a Ph.D. in psychology from the University of California at Los Angeles.

Evan Feldman has been active in expanding user research activities since the early 1990s. He evaluated mice (computer pointing devices) in zero gravity, worked on detailed ethnographies of network administrators, modified existing techniques for studying how people manage their finances, and completed comprehensive field studies of how people manage their notes and use the Tablet PC. Currently employed at Microsoft, he is working toward building a better understanding of how people interact with multiple computers and devices within their ecosystems.

Carolyn Fuson leads the Design Anthropology group within the Customer Design Center at MSN/Microsoft Corporation. She taught computers and technology for 7 years in Bellevue, Washington, where she also wrote the curriculum and evaluated the classes for grades 1 through 6. In addition to teaching, she researched how different students with cognitive learning styles assimilated the courses and how they interacted with the technology. Before managing the Design Anthropology group, she was responsible for consumer research for MSN.

Douglas Gillan has focused on researching two major topics during the past 20 years—(1) the perceptual and cognitive processes that underlie reading of informational displays and (2) the application of psychological research to the design of usable technology. The applications from this research have included the design of computer and robotic interfaces for the NASA Space Station, assistive software for blind users, and interfaces for robotic control. Before his work in human factors, Gillan's research investigated animal and human taste perception, basic mechanisms of associative learning, and reasoning in chimpanzees. Gillan is a professor and department head in the Department of Psychology at New Mexico State University. He earned his B.A. in psychology from Macalester College in 1974 and his Ph.D. in psychology, with a specialization in biopsychology, from the University of Texas at Austin. He has authored more than 100 articles on various research topics.

Jonathan Grudin is a senior researcher in the Adaptive Systems and Interaction group at Microsoft Research. He was previously a professor of information and computer science at the University of California, Irvine, in the Computers, Organizations, Policy, and Society research group. He has worked in industry as a developer and researcher for more than 30 years, focused primarily on studies on the adoption and use of technologies to support communication, information sharing, and coordination. He recently completed a 6-year term as editor in chief of Association of Computing Machines (ACM) transactions on computer-human interaction. In April he was elected to the ACM SIGCHI Academy, which comprises 25 people recognized for scholarly achievement and influence in the field of human-computer interaction.

Richard L. Henneman is the chief information architect for Internet Security Systems (ISS), a computer security firm with headquarters in Atlanta, Georgia, where he is responsible for ensuring the usability of ISS's products. Before joining ISS, Henneman was the director of user experience at marchFIRST, a professional services firm focused on Internet applications. A major part of Henneman's career was spent at NCR Corporation, where he was the director of user-centered design services at that company's Human Interface Design Center in Atlanta. The focus of all these positions has been on the application of user-centered design methodologies to create easy-to-use user interfaces. Before joining NCR, Henneman held research and teaching positions at the Georgia Institute of Technology in the School of Industrial and Systems Engineering, the Center for Human-Machine Systems Research, and the Computer Integrated Manufacturing Systems Program. He received a Ph.D. in industrial and systems engineering from Georgia Tech and an M.S. and a B.S. in industrial engineering from the University of Illinois at Urbana-Champaign.

Clyde C. Heppner is senior manager of user-centered research and metrics with Sprint Corporation. He joined Sprint Corporation in 1996; there he is responsible for setting the strategic direction of product and usability research. He established research programs for usability measurement, technology adoption, portfolio analysis, return on investment analysis, and modeling to predict post-purchase behavior and attitude. Heppner also leads the user-centered design process, a precursor to Sprint Corporation's enterprise development process where high-priority and strategic product concepts are defined. He received his undergraduate degree from the

University of Minnesota, Morris, and his graduate degree from the University of Nebraska, Lincoln. His areas of specialization are in experimental design and applied statistics, with a minor emphasis in cognitive psychology. Before working for Sprint Corporation, he held research positions in various industries, including education, electric utilities, pharmaceuticals, and human nutrition. He has national and international research experience. Throughout his career, he worked to extend research methodologies and statistical procedures beyond commonly defined applications. He has presented at national conferences and authored book chapters and various journal articles.

Clare-Marie Karat is a research staff member at the IBM TJ Watson Research Center. She received a B.A. in psychology, with honors, from Stanford University and a Ph.D. in social psychology from the University of Colorado at Boulder. Karat conducts human-computer interaction (HCI) research in the areas of privacy, personalization, and conversational interface technologies. She is an international expert in cost-justifying human factors work, and she develops innovative user-interface designs and methodologies. Karat is a coeditor of the book *Designing Personalized User Experiences in eCommerce* and has published numerous articles and chapters in professional and technical journals, in conference proceedings, and in books. As an editorial board member of the Association of Computing Machines' *interactions,* the British Computer Society's *Interacting with Computers,* and Elsevier's *International Journal of Human Computer Studies* journals, and as a technical program committee member of the CHI, Human Factors and Ergonomics Society, and INTERACT conferences, she provides leadership and maintains an active network of communication with HCI professionals in the field.

Jesse Kates serves as a usability engineer in Sprint Corporation's User Experience Design team, based in Overland Park, Kansas. In 2001 Kates graduated from Carnegie Mellon University (CMU) with an M.S. in human-computer interaction and a B.S. in technical writing, with an additional major in creative writing. While at CMU, Kates also coauthored the Software Engineering Institute Technical Report "Achieving Usability through Software Architecture" with Len Bass and Bonnie E. John and served as research assistant to Dr. Robert Klatzky, head of the psychology department. In his spare time, Kates writes poetry and composes original music for guitar, banjo, and other stringed instruments.

Anne Kirah currently serves as a senior design anthropologist for the MSN Customer Design Center with the Microsoft Corporation. She is responsible for national and international field research and participatory design activities intended to influence current and future Microsoft product, software, and service designs to improve humans' interaction with technology. In addition she is tasked with aiding in strategy and vision decisions for MSN. Kirah recently won the award for MSN Contributor of the Year in July, 2004.

Jurek Kirakowski comes from a practical computer science and psychology background. His speciality is quantitative measurement in human-computer interaction, and he has contributed numerous books, articles, and workshops to this theme. His major research goal is to show and prove how the quality of use of information technology products can and should be quantitatively measured in an objective manner to make the technology more effective. He is the director of the Human Factors Research group at University College Cork in Ireland. This group, which he founded in 1984, developed three main objectives under his leadership: to expand and disseminate information about usability in the wider information technology community; to engage in projects with industry in a consultancy capacity; and to develop measurement tools. For the latter, Kirakowski and his group have contributed the Software Usability Measurement Inventory (SUMI), Measuring Usability of Multi-Media Software (MUMMS), and most recently Web site Analysis and Measurement Inventory (WAMMI) questionnaires. SUMI and WAMMI are by now *de facto* standards in their respective areas. His personal home page is found at www.ucc.ie/hfrg/jk.

Arnold Lund is the director of user experience for Microsoft's Mobile PC Division. Before joining Microsoft he was a director of user experience for Sapient and managed their global research and development (R&D). He spent 20 years managing design, user research, and engineering and software development for emerging technologies at various telecommunications companies (beginning with AT&T Bell Labs). He has taught and published widely on design for emerging technologies and on R&D management topics. His research interests currently include improved user research methods and mobile computing. He is a fellow of the Human Factors and Ergonomics Society, is on the Board of Directors for the Board of Certification in Professional Ergonomics, and is active with Association of Computing Machines SIGCHI. He received a Ph.D in experimental psychology from Northwestern University and a B.S. in chemistry from the University of Chicago.

Jeffrey S. Lynch is the assistant vice president of Sprint Corporation's User Experience Design organization. In this role, he leads Sprint Corporation's user-centered design program and oversees the company's product usability, product design research, and usability testing programs. He received a B.A. in political science from Boston University and an M.B.A. from Cornell University's Johnson Graduate School of Management. Before joining Sprint Corporation in 1999, he held several positions in capital markets and corporate finance, most recently at Intel Corporation.

Aaron Marcus is president of Aaron Marcus and Associates, Inc. (AM+A) in Berkeley, California. His research interests lie in user-interface design, semiotics, culture, mobile technology, and knowledge visualization. He graduated from Princeton University with a degree in physics and from Yale University with a degree in graphic design. In 1967 he became the world's first graphic designer to work full time with computer graphics. In the 1970s he programmed a prototype desktop-publishing page-layout application for the Picturephone at AT&T Bell Labs, programmed virtual reality spaces while a faculty member at Princeton University, and directed an international team of visual communicators as a research fellow at the East-West Center in Honolulu. In the early 1980s he was a staff scientist at Lawrence Berkeley Laboratory in Berkeley, California; he founded AM + A; and he began research as a Co-Principal Investigator of a project funded by the U.S. Department of Defense's Advanced Research Projects Agency. In 1992 he received the National Computer Graphics

Association's annual award for contributions to industry. His firm helped design the user interface for the first version of AOL and for Travelocity.com. Marcus wrote more than 200 articles and wrote/cowrote five books. Marcus has published, lectured, tutored, and consulted internationally for 36 years.

Charles L. Mauro is president and CEO of MauroNewMedia (MNM), a New York-based consulting firm offering services in usability science, formal user-centered design, and high performance user interface design. Mauro holds an M.S. in human factors engineering from New York University (NYU). At NYU he was a National Institute Occupational Safety and Health (NIOSH) research fellow at the Rusk Institute of Rehabilitation Medicine. He received grants and fellowships from the Ford Foundation, NIOSH, and the National Endowment for the Arts. Before forming MNM he worked directly with product design pioneers Henry Dreyfus and Raymond Loewy. He was directly responsible for research and development of numerous mission-critical user interfaces, including design of primary trading systems used on the floor of the New York Stock Exchange. His experience spans more than 25 years and includes consumer, commercial, military, and aerospace applications. He consults regularly with Fortune 500 clients and leading startups. Mauro has received numerous citations and awards, including the Alexander C. Williams Award from the Human Factors and Ergonomics Society and citations from the NASA and the Association of Computing Machines. He served on many national and international panels and has chaired two ANSI standards committees. Mauro also served on the Presidential Design Awards Program for the NEA and was a founding member of the Human Factors Society Special Interest Group on Consumer Products. He consults regularly with leading corporate CEOs on strategic issues of screen-based customer experience design, usability, interactive brand development, and fixed to virtual migration of products and services. He has been accepted in federal court as an expert witness for design patents, trade dress, and other intellectual property issues related to user-interface design. He holds numerous U.S. and international patents. Mauro is widely published in professional and popular literature and is quoted in *Fortune, Business Week*, the *Wall Street Journal, Science*, and other leading business publications. He routinely speaks at national and international conferences on design, usability, and user interface design and development. Mauro lectures at leading graduate programs, including Massachusetts Institute of Technology Sloan School, Stanford, and other engineering and M.B.A. programs. His first major book, *Usability: the Bottom Line,* will be published next year.

Deborah J. Mayhew, Ph.D., is an internationally recognized author, teacher, speaker, and consultant on software user interface design and usability engineering. Since 1986 she has been the owner and principal consultant of Deborah J. Mayhew & Associates, a consulting firm offering various services related to usability engineering, where she became one of the first independent consultants in her field. Clients have included IBM, AT&T, John Hancock Insurance Company, GE, Hewlett-Packard, Ford Motor Company, GTE, American Express, Apple, American Airlines, Texas Instruments, NASA, the National Cancer Institute, the New York City Police Department, Computer Sciences Corporation, Cisco Systems, the IRS, and many others. Mayhew is a frequent instructor in conference tutorial programs. She taught courses in many large organizations and authored and coauthored four books on topics in usability engineering, some of which have been adopted internationally as texts in university courses. Her most recent book is *The Usability Engineering Lifecycle.* She also coedited the first edition of *Cost-Justifying Usability* with Randolph G. Bias. Additionally she has contributed chapters to several other recent books. Mayhew lives with her 12-year-old daughter and works from Martha's Vineyard Island, where she was born and raised. Her Web site is at drdeb.vineyard.net.

Michael C. Medlock is a usability engineer at Microsoft in the human resources department. Before joining Microsoft, he spent 6 years as an engineer and manager helping to found the Microsoft Games Studios User-Testing group, one of the only usability groups in the world conducting practical research on video games. Some of the games that were researched include PC titles such as Age of Empires II, Dungeon Siege, and Flight Simulator 2000 and Xbox titles such as Project Gotham Racing, Top Spin, and Crimson Skies: High Road to Revenge. He authored articles on the Rapid Iterative Test and Evaluation method, research applied to game design and schedules of reinforcement that model self-control paradigms. He earned an M.A. in experimental psychology with a focus on animal behavior and learning theory.

Mick McGee is a senior usability engineer at Oracle Corporation. He has researched and worked in the usability and human factors profession since 1994 at IBM, Computing Devices Canada, and Oracle. He received a Ph.D. and an M.S. in industrial and systems engineering from Virginia Tech and a B.S. in psychology from the University of Illinois. His interests are in advanced information interfaces (voice, mobile, virtual, tele, neural, and desktop), improved research and evaluation methods, leadership research, and triathalons.

Robert R. Moritz is senior director of user experience design with Sprint Corporation, where, for the past 8 years, he has focused on the design of voice and mobile data products for the consumer mass market. Before joining Sprint Corporation, he worked in the petrochemical industry for Exxon Biomedical Sciences, Inc. as the lead human factors psychologist for Gulf Coast operations. He began his career in 1988 with Bell Communications Research (Bellcore). At Bellcore he concentrated on the user interface design of directory assistance systems, demonstrating the business value of voice processing and speech recognition technologies and facilitating the transfer of speech technology from the lab to the field. He earned an M.A. in applied experimental psychology from Miami University in Ohio and is a board certified professional ergonomist.

Janice Anne Rohn is vice president of user experience (UE) at World Savings Bank, where she leads the team that is responsible for the design and usability of the consumer online banking and other Web sites. In addition to developing her current department, Rohn has founded and built UE groups at Siebel Systems and Sun Microsystems. During her career, Rohn worked in various organizations, hired more than 60 UE professionals, and designed and built more than 15 usability labs. Rohn also worked at PeopleSoft, Apple, and Stanford University. Rohn is a leader in cost-benefit analysis and strategic UE, researching and using the most effective methods and organizational approaches to ensure optimal decision-making. Rohn was President and a founding board member of the Usability Professionals' Association (UPA). She also founded the Outreach effort, working with the U.S. Congress, the Consumer Protection Agency, and Vice President Gore's office on the benefits of user experience. Rohn has authored more than 40 publications and delivered many presentations at CHI, UPA, Interact, and other conferences. Rohn has also delivered several keynote speeches and taught courses at several universities.

Stephanie Rosenbaum is founder and president of Tec-Ed, Inc., a 15-person firm specializing in usability research and information design. Headquartered in Ann Arbor, Michigan, Tec-Ed maintains offices in Palo Alto, California; Milwaukee, Wisconsin; and Rochester, New York. Tec-Ed clients include eBay, Cisco Systems, AOL, the National Institutes of Health, Sun Microsystems, Yahoo!, and various smaller firms. A member of Association of Computing Machines (ACM) SIGCHI, the Human Factors and Ergonomics Society, and the Usability Professionals' Association (UPA), as well as a fellow and exemplar of the Society for Technical Communication, Rosenbaum was a vice-chair of ACM SIGDOC and headed the STC's Research Grants Committee for 5 years. Rosenbaum was awarded a Millennium Medal in 2000 by the Institute of Electrical and Electronics Engineers, Inc.; her many publications include a chapter in the Copenhagen Business School Press volume, *Software Design and Usability*. Her research background includes anthropology studies at Columbia University and experimental psychology research for the University of California at Berkeley. From 1997–2000, Stephanie copresented a series of ACM SIGCHI sessions on increasing the strategic position of usability within organizations. She moderated a panel on "Measuring Return on Investment for Usability" at the UPA 2002 Conference and facilitated a workshop on "Measuring Usability ROI to Justify Usability Investments" at the UPA 2003 Conference.

Merrill Sapp is a graduate student and instructor at New Mexico State University, where she is currently working on her Ph.D. She and Doug Gillan have worked together on basic and applied research concerning perceptual judgments of attributes such as area, depth, and speed.

David A. Siegel and his coauthor, Susan M. Dray, are coprincipals of Dray & Associates, Inc., a user-centered design (UCD) firm that consults with large and small companies around the world to make sure their product and software designs meet user needs. They are highly experienced practitioners of international UCD, carrying out both ethnography and usability projects internationally. They have consulted in 16 countries, and taught UCD techniques in many more. They also recently published chapters about planning international studies and about the relationship between international UCD and localization. Siegel has published many articles on UCD, and, together with Dray, he edits the business column of the Association of Computing Machines SIGCHI magazine, *interactions*. Siegel received a Ph.D. in psychology from University of California at Los Angeles.

Marilyn Tremaine is a chair of the Information Systems Department at the New Jersey Institute of Technology and director of the Human-Computer Interaction (HCI) degree program. She is also a research professor at the Center for Advanced Information Processing at Rutgers University–New Brunswick. Before moving to New Jersey, Tremaine was a professor of computer science at the University of Toronto where she conducted research on designing tactile and audio feedback input devices and collaborative systems design. Before moving to Toronto, she was an assistant professor of information systems at the Michigan Business School. Tremaine also worked as the vice president of research and development for three software startup companies and was senior research scientist at the EDS Center for Applied Research. She started their HCI program at the University of Michigan and was the first to work on HCI cost-benefit analysis. At EDS, she developed the Capture Lab, one of the first computer-supported meeting rooms. At the University of Toronto she headed the CAVECAT project, a video desktop conferencing research initiative, the Jabber project, an exploration into real time meeting indexing, and the Nonspeech audio project for understanding audio interface design. Her current research focuses on the design and development of multimodal interfaces to be used in collaborative settings, particularly in telemedicine.

Scott Weiss was an outspoken skeptic of online usability research methods until he started to conduct research for this chapter. He is the principal of Usable Products Company, which has designed information architectures and tested the usability of desktop and handheld products since 1996. Weiss's book, *Handheld Usability*, describes design, prototyping, and usability testing for mobile phones, personal digital assistants, and two-way e-mail pagers.

Dan Welsh is a usability engineer in the MSN Division. His work includes consulting on the design of consumer products using various methods, including rapid iterative testing and evaluation, qualitative lab and field research, and survey research and customer segmentation. Before joining Microsoft in 2001, Welsh worked as a usability consultant and a medical researcher studying the effects of behavior and lifestyle on chronic disease outcomes. Welsh is a member of the Association of Computing Machines SIGCHI.

Chauncey E. Wilson is founder and president of WilDesign Consulting, a firm specializing in user experience design and evaluation. WilDesign Consulting is headquartered in Wayland, Massachusetts, and clients include Microsoft, Cisco Systems, Pershing, IDX Corporation, Bentley College, Citizens Bank, Elsevier, and Michigan State University. Chauncey has been in the human factors and usability field for 30 years and has a background in physics, social psychology, and human factors engineering. In addition to providing consultancy with WilDesign Consulting, he is currently an adjunct professor in the Human Factors and Information Design program at Bentley College. Wilson was the first director of the Bentley College Design and Usability Testing Center. Before his work as director, Wilson was a product line development manager and usability engineer at BMC Software, Inc., where he was responsible for the development and usability of complex performance monitoring software. He was a human-computer interaction (HCI) architect and usability engineer for IDX Corporation, FTP Software, Dun & Bradstreet Software, Human Factors International, and Digital Equipment Corporation. Wilson coauthored chapters on usability engineering, return on investment, user-centered design, and criminal victimization. He has presented often at the Usability Professionals' Association (UPA), STC, and CHI conferences and is a member of UPA, STC, CHI, Human Factors Ergonomics Society, and Institute of Electrical and Electronics Engineers, Inc.

Dennis Wixon, Ph.D., is the usability manager for the Microsoft Games Studios (MGS) User-Testing group, a group of 20 usability engineers whose mission is to provide the user perspective throughout MGS. These efforts include consulting on game design, rapid iterative testing and evaluation, playtesting, and deep game instrumentation. The goal of this work is to make games that users experience as fun and worthwhile. He is also a member of Microsoft User Experience Leadership Team, a corporate-wide steering group for usability and design at Microsoft. Wixon, who joined Microsoft in 1998, previously worked as a usability manager at Digital Equipment Corporation, where he helped develop and define a number of important usability methods such as Usability Engineering and Contextual Inquiry. Wixon has been an active member of the human factors community for many years. He was one of the five founding members of the Greater Boston Chapter of the Association of Computing Machines SIGCHI and was general co-chair of the SIGCHI 2002 conference. Wixon has authored numerous articles on human-computer interaction methods and coedited a book *Field Methods Case Book for Software Design* with Professor Judy Ramey of the University of Washington.